Banach Algebras of
Ultrametric Functions

Banach Algebras of Ultrametric Functions

Alain Escassut

Université Clermont Auvergne, France

World Scientific

NEW JERSEY · LONDON · SINGAPORE · BEIJING · SHANGHAI · HONG KONG · TAIPEI · CHENNAI · TOKYO

Published by

World Scientific Publishing Co. Pte. Ltd.

5 Toh Tuck Link, Singapore 596224

USA office: 27 Warren Street, Suite 401-402, Hackensack, NJ 07601

UK office: 57 Shelton Street, Covent Garden, London WC2H 9HE

Library of Congress Cataloging-in-Publication Data

Names: Escassut, Alain, author.

Title: Banach algebras of ultrametric functions / Alain Escassut, Université Clermont Auvergne, France.

Description: New Jersey : World Scientific, [2022] | Includes bibliographical references and index.

Identifiers: LCCN 2021058964 | ISBN 9789811251658 (hardcover) |

    ISBN 9789811251665 (ebook for institutions) | ISBN 9789811251672 (ebook for individuals)

Subjects: LCSH: Banach algebras. | Functions.

Classification: LCC QA326 .E773 2022 | DDC 512/.554--dc23/eng20220208

LC record available at https://lccn.loc.gov/2021058964

British Library Cataloguing-in-Publication Data

A catalogue record for this book is available from the British Library.

For any available supplementary material, please visit
https://www.worldscientific.com/worldscibooks/10.1142/12707#t=suppl

Desk Editors: Jayanthi Muthuswamy/Rok Ting Tan

Typeset by Stallion Press
Email: enquiries@stallionpress.com

Printed in Singapore

About the Author

 **Alain Escassut**, after studies in Licence and Master, worked at University Bordeaux 1 from 1969 to 1987 (where he obtained a Ph.D. and a Doctorat d'Etat) as Maître-Assistant, Maître de conférence (and as Visiting Assistant Professor at Princeton University in Spring 1981). Next, he worked from 1987 to 2012 as Professor at Université Blaise Pascal. He is now Emeritus Professor in this university, become Université Clermont Auvergne.

Alain Escassut is a specialist of ultrametric analysis and first studied algebras of analytic elements long ago defined by Marc Krasner He examined many properties linked to them, such as Krasner–Tate algebras and (with Abdelbaki Boutabaa) properties linked to the p-adic Nevanlinna Theory. Their joint works and other works with C.C. Yang and Ta Thi Hoai An let them obtain results on problems of uniqueness. He also recently gave a new proof of the Hermite–Lindemann theorem in an ultrametric field.

A large part of his work has been devoted to ultrametric Banach algebras and to the ultrametric holomorphic functional calculus where he examined the major role of the multiplicative spectrum, particularly using circular filters and T-filters.

Contents

Introduction

Let $\mathbb{L}$ be a complete ultrametric field and let A be a Banach $\mathbb{L}$-algebra of functions defined in an ultrametric space E, with values in $\mathbb{L}$, satisfying certain properties (bounded continuous functions, Lipschitz functions, differentiable functions, analytic functions of several kinds). We examine algebraic properties and topologic properties they satisfy, gathering various results obtained during the last 50 years. Many properties are linked to the multiplicative spectrum of the algebras.

Admissible algebras (denoted by S) are Banach algebras of functions on an ultrametric space E whose spectral semi-norm is the norm of uniform convergence, where each closed open subset has a characteristic function and any function is invertible if and only if its minimum is strictly positive. Particularly, the algebra of bounded continuous function from E to $\mathbb{L}$ is admissible. We show the role of sticked ultrafilters and obtain an equivalence relation whose classes characterize the maximal ideals. A prime ideal is contained in a unique maximal ideal. A prime ideal of S is a maximal ideal if and only if it equals its closure with respect to the uniform convergence. That notion particularly applies to continuous functions. Two main topologies are defined on an algebra S: the classical topology and the spectral topology defined by the spectral semi-norm. Each ultrafilter on E defines a maximal ideal of the algebra S and two ultrafilters define the same maximal ideal if and only if they are sticked. On an admissible algebra S, a kind of Bezout–Corona theorem works and that shows the role of filters concerning all ideals of S. Given a prime ideal, its closure with respect to the classical topology of S is a maximal ideal. If the field $\mathbb{L}$ is perfect, then all maximal ideals of S are of codimension 1.

Every continuous multiplicative semi-norm of an algebra S has a kernel that is a maximal ideal and each maximal ideal of S is the kernel of a unique continuous multiplicative semi-norm (which is not true in certain ultrametric Banach algebras, details in what follows). The multiplicative spectrum appears as a compactification of the space E and the Shilov boundary of S is equal to the multiplicative spectrum of S. The stone space $St(E)$ is defined as usual and appears as a compactification of E that actually, is homeomorphic to the multiplicative spectrum.

If the field $\mathbb{L}$ is perfect, then all maximal ideals of S are of codimension 1. Every continuous multiplicative semi-norm of an algebra S has a kernel that is a maximal ideal and each maximal ideal of S is the kernel of a unique continuous multiplicative semi-norm (which is not true in certain ultrametric Banach algebras).

On an admissible algebra S, a kind of Bezout–Corona theorem works and that shows the role of filters concerning all ideals of S. The Shilov boundary of an algebra S is equal to its multiplicative spectrum. The Stone space of E is a compactification of E which is equivalent to the multiplicative spectrum of S.

Next, we define a class of algebras T, called compatible algebras, which is similar but a bit different from the admissible algebras: instead of assuming that characteristic functions of clopen sets belong to the algebra, here we assume that the characteristic functions of uniformly open sets belong to the algebra. Contiguous ultrafilters are defined (sticked ultrafilters are contiguous). Then in a compatible algebra, maximal ideals are characterized by classes of contiguous ultrafilters. A Bezout–Corona theorem applies as in admissible algebras. Here again, each continuous multiplicative semi-norms of an algebra T has a kernel that is a maximal ideal and each maximal ideal of T is the kernel of a unique continuous multiplicative semi-norm, the multiplicative spectrum of T appears also as a compactification of the space E and the Shilov boundary of T is equal to the multiplicative spectrum of T. A Stone space of T is defined by considering now the set of uniformly open subsets of E instead of the clopen subsets.

Let $\mathcal{B}$ be the set of bounded uniformly continuous functions from E to $\mathbb{L}$, let $\mathcal{L}$ be the set of bounded Lipschitz functions from E to $\mathbb{L}$, let $\mathcal{D}$ be the subset of $\mathcal{L}$ of derivable functions in E and let $\mathcal{E}$ be the subset of $\mathcal{L}$ of functions such that for every $a \in E$, $\frac{f(x)-f(y)}{x-y}$ has limit when x and y tend to a separately. All these sets are Banach algebras with respect to good norms, the norm of uniform convergence on E is the spectral norm of these algebras, all are compatible algebras and every maximal ideal of finite codimension in one of these algebras is of codimension 1. If E has no isolated point, in $\mathcal{L}$, the invertible elements are the ones that have no zero and are not topological divisors of zero.

We denote by $\mathbb{K}$ an algebraically closed complete ultrametric field (that might be an algebraic closure of $\mathbb{L}$). Many properties of ultrametric Banach algebras are proven with help of an ultrametric holomorphic functional calculus, which requires to use circular filters which make a tree for a certain order and is provided with the topology of pointwise convergence. Several topologies are examined on the set of multiplicative semi-norms on an algebra of analytic elements.

Next, we have to recall properties of T-filters and T-sequences, with many applications such as the problem of multbijectivity of $\mathbb{K}$-Banach algebras (which depends on the valuation of the field $\mathbb{K}$) and the non-injectivity of the p-adic Fourier transform.

Then various properties of spectral semi-norms are recalled. If a field extension E of the ground field $\mathbb{K}$ admits a semi-multiplicative norm and two distinct continuous absolute values, then the completion of E with respect to its norm is a ring with divisors of zero.

The Hensel Lemma is recalled with its long proof. The definition and the main algebraic properties of affinoid algebras are recalled. Given an affinoid algebra A, it is a Noetherian Jacobson ring, every maximal ideal is of finite codimension and when it is reduced, its spectral semi-norm is a norm equivalent to the Banach norm. Properties of restricted power series are recalled, such as factoriality (Salmon's theorem) and Krasner–Tate algebras are characterized.

The last section is dedicated to prove this famous property: if the multiplicative spectrum of a unital commutative $\mathbb{L}$-Banach algebra admits a partition in two open closed subsets, then each admits an associated idempotent which is proven to be unique. The property is first recalled in an affinoid algebra and then is generalized to all unital commutative $\mathbb{L}$-Banach algebras (this was stated by B. Guennebaud in his unpublished thesis but the proof was very long).

I am very grateful to Bertin Diarra for many pieces of advice in that work as in previous ones.

Chapter 1

Basic Properties in Algebra

1.1. Properties in commutative algebra

In this first chapter, we have to collect several classical results in commutative algebra that will be indispensable when studying affinoid algebras.

Definitions and notations: Throughout the chapter, B will denote a unital commutative ring with unity, $\mathbb{E}$ will denote a field and A will be a unital commutative $\mathbb{E}$-algebra.

Given an ideal I of B, we call *radical* of I the ideal of the $x \in B$ such that $x^n \in I$ for some $n \in \mathbb{N}^*$. We call *nilradical* of B the radical of $\{0\}$, i.e., the intersection of all prime ideals of B and *Jacobson radical* the intersection of all maximal ideals of B. Moreover, B is said to be *reduced* if its nilradical is equal to $\{0\}$, and *semi-simple* if its Jacobson radical is equal to $\{0\}$. Moreover, B is called a *Jacobson ring* if every prime ideal of B is equal to the intersection of all maximal ideals that contain it.

$\text{Max}(B)$ is the set of all maximal ideals of B. Henceforth, $\text{Max}_a(A)$ will denote the subset of $\text{Max}(A)$ which consists of all maximal ideals M such that $\frac{A}{M}$ is a finite extension of $\mathbb{E}$ and $\text{Max}_1(A)$ will denote the subset of $\text{Max}_a(A)$ which consists of all maximal ideals M such that $\frac{A}{M} = \mathbb{E}$.

We will denote by $\mathcal{X}(A, \mathbb{E})$ the set of $\mathbb{E}$-algebra homomorphisms from A onto $\mathbb{E}$, and by $\mathcal{X}(A)$ the set of all $\mathbb{E}$-algebra homomorphisms from A onto various fields $\frac{A}{M}$ when $M \in \text{Max}(A)$. Given $x \in A$, we denote by $sp_A(x)$ the set of $\lambda \in E$ such that $x - \lambda$ is not invertible in A, and by $sa_A(x)$ the set of $\lambda \in E$ which are images of x by $\mathbb{E}$-algebra homomorphisms from A onto algebraic extensions of $\mathbb{E}$. So, in particular, when $\mathbb{E}$ is algebraically closed, we have $sp_A(x) = sa_A(x) = \{\chi(x) \mid \chi \in \mathcal{X}(A, \mathbb{E})\}$.

When there is no risk of confusion on the $\mathbb{E}$-algebra A, we will only write $sp(x)$ instead of $sp_A(x)$ and $sa(x)$ instead of $sa_A(x)$.

Propositions 1.1.1–1.1.3 are classical in commutative algebra.

Proposition 1.1.1. *Let $I_1, \ldots, I_n$ be ideals of B such that $I_k + I_l = B \; \forall k \neq l$. Then $\frac{B}{I_1 \cdots I_n}$ is isomorphic to $\frac{B}{I_1} \times \cdots \times \frac{B}{I_n}$.*

Lemma 1.1.2. *Let B be a Noetherian integral domain. Then B is factorial if and only if for every irreducible element f of B, the ideal fB is prime.*

Lemma 1.1.3. *Let B be factorial and let $f \in B$. The ideal fB is equal to its radical if and only if any factorization of f into irreducible factors admits no factors with a power $q > 1$.*

Proposition 1.1.4. *Let $\mathbb{F}$ be a field which is an integral ring extension of B. Then B is a field.*

Proposition 1.1.5. *Suppose that for every $x \in B \setminus \{0\}$, the quotient ring $\frac{B}{xB}$ is Noetherian. Then B is Noetherian.*

Proof.   Let I be an ideal of B. Let $t \in I$, $t \neq 0$ and let $\tilde{B} = \frac{B}{tB}$. Then tB is also an ideal of I, and $\frac{I}{tB}$ is an ideal of $\frac{B}{tB}$. For every $x \in B$, we denote by $\tilde{x}$ its class in $\frac{B}{tB}$. Since $\tilde{B}$ is Noetherian, we find $x_1, \ldots, x_q \in I$ such that $\frac{I}{tB} = \tilde{B}\tilde{x}_1 + \cdots + \tilde{B}\tilde{x}_q$. Then $tB + \sum_{i=1}^{q} x_i B$ is an ideal of B included in I. Let $x \in I$. Since $\tilde{x}$ belongs to $\frac{I}{tB}$, we can find $a_1, \ldots, a_q \in B$ such that $\tilde{x} = \sum_{i=1}^{q} \tilde{a}_i \tilde{x}_i$. Therefore $x - \sum_{i=1}^{q} \tilde{a}_i \tilde{x}_i \in tB$, and then is equal to some ta, with $a \in B$. Thus, $x = tb + \sum_{i=1}^{q} a_i x_i$, and then $I \subset tB + \sum_{i=1}^{q} Ax_i$. Since by definition I contains $tB + \sum_{i=1}^{q} Ax_i$, this finishes showing that I is finitely generated, and thereby A is Noetherian.   □

Proposition 1.1.6 is a consequence of Kronnecker's Theorem.

Proposition 1.1.6. *Let B be an entire commutative ring with unity u, let B' be an integrally closed subring B containing u of B, and let $x \in B$, be integral over B'. Then, the minimal polynomial of x over the field of fractions of B' belongs to $B'[X]$.*

Notations: Let B' be an integrally closed subring of B containing the unity of B and let $x \in B$ be integral over B'. Then, the minimal polynomial of x over B' will be denoted by $irr(x, B')$. Hence by Proposition 1.1.6, $irr(x, B') \in B'[X]$.

Proposition 1.1.7. *Let B be an integral domain which is integrally closed and let $R \supset B$ be a unital commutative ring which is integral over B. Let $x \in B$ and let $irr(x, B) = x^n + \sum_{i=0}^{n-1} a_i x^i$. Then x is invertible in R if and only if a_0 is invertible in B.*

Proof. We have $x^n + \sum_{i=0}^{n-1} a_i x^i = 0$. If a_0 is invertible in B, then $x(\sum_{i=0}^{n-1} a_i(a_0)^{-1} x^{n-i-1}) = -1$, hence x is invertible in R. Now, suppose that x is invertible in R, and let $u = x^{-1}$. Then u satisfies $1 + \sum_{i=1}^{n} a_{n-i} u^i = 0$. Let $P(X) = irr(u, B)$. Since B is integrally closed, x is of degree n over the field of fractions of B, and so is u, hence P divides the polynomial $G(X) = 1 + \sum_{i=1}^{n} a_{n-i} X^i$, and is of same degree. Consequently $G = a_0 P$, and therefore $P(0) = \frac{1}{a_0}$, which shows that a_0 is invertible in B. $\qquad\square$

Notation and definition: We will call a *Luroth* $\mathbb{F}$-*algebra* a $\mathbb{E}$-algebra of the form $\mathbb{E}[h, x]$, with $h \in \mathbb{E}(x)$ and x transcendental over $\mathbb{F}$. Given $h(x) = \frac{P(x)}{Q(x)} \in \mathbb{E}(x)$, we call *degree of h* the number $\deg(P) - \deg(Q)$, and we will denote it by $\deg(h)$. Lemma 1.1.8 is immediate.

Lemma 1.1.8. *Let* $B = \mathbb{E}[h, x]$ *be a Luroth* $\mathbb{E}$-*algebra. For every pole* α *of* h, $x - \alpha$ *is invertible in* B.

Corollary 1.1.9. *Let* $h_j = \frac{P_j(x)}{Q_j(x)} \in F(x)$, $(1 \leq j \leq n)$, *with* P_j *and* Q_j *relatively prime. Let* $B = F[h_1, \ldots, h_n, x]$. *Then* $\frac{1}{Q_j(x)}$ *belongs to* B *for all* $j = 1, \ldots, n$.

Propositions 1.1.10 and 1.1.12 are classical in commutative algebra [7].

Theorem 1.1.10. *The set of prime ideals of* B *is inductive with respect to relation* $\supset$.

Corollary 1.1.11. *Every prime ideal of* B *contains a minimal prime ideal.*

Theorem 1.1.12. *If* B *is Noetherian it has finitely many minimal prime ideals.*

Lemma 1.1.13 is immediate.

Lemma 1.1.13. *Let* I *be an ideal of* A *and let* $A' = \frac{A}{I}$. *Let* θ *be the canonical surjection of* A *onto* A'. *The maximal ideals of* A' *are the* $\theta(\mathcal{M})$ *whenever* $\mathcal{M}$ *runs through maximal ideals of* A *containing* I. *Moreover, the maximal ideals of codimension 1 of* A' *are the* $\theta(\mathcal{M})$ *whenever* $\mathcal{M}$ *runs through maximal ideals of codimension 1 of* A *containing* I.

Lemma 1.1.14. *Let* S *be a subset of* $\mathcal{X}(A)$. *If there exists an idempotent* u *of* A *satisfying* $\chi(u) = 1 \ \forall \chi \in S$, *and* $\chi(u) = 0 \ \forall \chi \in \mathcal{X}(A) \setminus S$, *then it is unique.*

Proof. Suppose there exist two idempotents u and u' satisfying $\chi(u) = \chi(u') = 1 \ \forall \chi \in S$, and $\chi(u) = \chi(u') = 0 \ \forall \chi \in \mathcal{X}(A) \setminus S$. We notice that

$$(1) \qquad\qquad (u - u')(1 - u - u') = 0.$$

Let $\chi \in S$. Then $\chi(u') = 1$, hence $\chi(1 - u') = 0$, and therefore $\chi(1 - u - u') = \chi(-u) = -1$. Now, let $\chi \in \mathcal{X}(A) \setminus S$. Since $\chi(u') = 0$, we have $\chi(1 - u - u') = \chi(1) = 1$. Consequently, $\chi(1 - u - u') \neq 0$ whenever $\chi \in \mathcal{X}(A)$, and therefore $1 - u - u'$ is invertible in A. Hence by (1) we have $u - u' = 0$. $\qquad\square$

Lemma 1.1.15. *Let A be a $\mathbb{E}$-algebra having finitely many minimal prime ideals $\mathcal{P}_1, \ldots, \mathcal{P}_q$, and for each $j = 1, \ldots, q$, let θ_j be the canonical surjection from A onto $\frac{A}{\mathcal{P}_j}$. Then for every $x \in A$, $sp(x) = \bigcup_{j=1}^{q} sp(\theta_j(x))$.*

Proof. For each $j = 1, \ldots, q$, we put $A_j = \frac{A}{\mathcal{P}_j}$. Given $\chi \in \mathcal{X}(A_j)$ then $\chi \circ \theta_j$ belongs to $\mathcal{X}(A)$. In particular this is true when $\chi(\theta_j(x)) \in E$. Therefore, we have $\bigcup_{j=1}^{q} sp(\theta_j(x)) \subset sp(x)$. Conversely, let $\phi \in \mathcal{X}(A)$. By Corollary 1.1.11 $\text{Ker}(\phi)$ contains at least one minimal prime ideal $\mathcal{P}_h$. Consequently, ϕ factorizes in the form $\chi \circ \theta_h$, with $\chi \in \mathcal{X}(A_j)$. Particularly this is true when $\chi(x) \in \mathbb{E}$. Consequently, $sp(x) \subset \bigcup_{j=1}^{q} sp(\theta_j(x))$. $\qquad\square$

Normalization lemma is classical.

Theorem 1.1.16 (Normalization lemma). *Let A be a finite type $\mathbb{F}$-algebra. There exists a finite algebraically free set $\{y_1, \ldots, y_s\} \subset A$ such that A is finite over $\mathbb{F}[y_1, \ldots, y_s]$.*

When studying Krasner–Tate algebras, we will consider universal generators.

Definitions: Let A be a $\mathbb{E}$-algebra. An element $x \in A$ will be said to be *spectrally injective* if, for every $\lambda \in sp(x)$, $x - \lambda$ belongs to a unique maximal ideal. Moreover, x will be called *a universal generator* if, for every $\lambda \in sp(x)$, $(x - \lambda)A$ is a maximal ideal.

Lemma 1.1.17 is immediate by considering the fields of fractions of A and B.

Lemma 1.1.17. *Let A be a $\mathbb{E}$-algebra finite over a $\mathbb{E}$-algebra B. For all $x \in \mathcal{X}(B)$, there exists $\hat{\chi} \in \mathcal{X}(A)$ satisfying $\hat{\chi}(x) = \chi(x) \ \forall x \in B$.*

Definition: An element $a \neq 0$ of a ring R will be called *divisor of zero* if there exists $b \in R, b \neq 0$ such that $ab = 0$.

Theorem 1.1.18. *Let A be a $\mathbb{E}$-algebra without divisors of zero, admitting a spectrally injective element x. Let B be an integrally closed $\mathbb{E}$-subalgebra of A containing x and the unity of A such that is finite over B. Assume that all maximal ideals of A and B have codimension 1. Then for every $y \in B$, we have $sp_B(y) = sp_A(y)$ and x is a spectrally injective element of B. Moreover, for every $\mathcal{M} \in \mathrm{Max}(A)$, $\mathcal{M} \cap B$ belongs to $\mathrm{Max}(B)$ and the mapping ϕ from $\mathrm{Max}(A)$ into $\mathrm{Max}(B)$ defined as $\phi(\mathcal{M}) = \mathcal{M} \cap B$ is an injection from $\mathrm{Max}(A)$ into $\mathrm{Max}(B)$. Further, assume that B is semi-simple and let p be the characteristic of $\mathbb{E}$: either $p = 0$, then $A = B$, or $p \neq 0$, and then for every $y \in A$, $irr(y, B)$ is of the form $X^{p^t} - f$, $f \in B$.*

Proof. Since the unity of A lies in B it is easily seen that for each maximal ideal $\mathcal{M}$ of A, the $\mathbb{E}$-algebra homomorphism from A to F whose kernel is $\mathcal{M}$ has a restriction to B which is surjective on $\mathbb{E}$, so $\mathcal{M} \cap B$ is a maximal ideal of B. We will deduce that the mapping ϕ is injective. Indeed, let $\mathcal{M}_1$ and $\mathcal{M}_2$ be two different maximal ideals of A, and let λ_1, $\lambda_2 \in \mathbb{E}$ be such that $x - \lambda_1 \in \mathcal{M}_1$, $x - \lambda_2 \in \mathcal{M}_2$. Then $x - \lambda_1$ lies in $\mathcal{M}_1 \cap B$, but does not lie in $\mathcal{M}_2$, hence $\mathcal{M}_1 \cap B \neq \mathcal{M}_2 \cap B$, thereby ϕ is injective. The mapping is also surjective: let $\mathcal{J}$ be a maximal ideal of B and let $\chi \in \mathcal{X}(B)$ admit $\mathcal{J}$ for kernel, then by Lemma 1.1.17, there exists $\hat{\chi} \in \mathcal{X}(A)$ whose restriction to B is equal to χ and therefore $\mathcal{J} = \mathrm{Ker}(\hat{\chi}) \cap B = \phi(\mathrm{Ker}(\hat{\chi}))$. Consequently $sp_A(y) = sp_B(y) \; \forall y \in B$. Then it is clear that x is spectrally injective in B: given $\lambda \in sp_B(x) = sp_A(x)$, then there is a unique maximal ideal $\mathcal{M}$ of A which contains $x - \lambda$ and then $\phi(\mathcal{M})$ is the unique maximal ideal of B which contains $x - \lambda$. We now suppose that B is semi-simple. Let $y \in A \setminus B$, let $B' = B[y]$ and let $P(X) = \sum_{j=0}^n f_j X^j = irr(y, B)$. Then x is a spectrally injective element of B'. Let $\chi \in \mathcal{X}(B)$. Since ϕ is bijective there exists a unique $\hat{\chi} \in \mathcal{X}(A)$ whose restriction to B is equal to χ. Consequently, the polynomial $G(X) = \sum_{j=0}^n \chi(f_j) X^j$ admits a unique zero α of order n. Therefore we have $\chi(f_j) = (-\alpha)^j \binom{n}{j} \; \forall j = 0, \ldots, n-1$, and in particular $\chi(f_{n-1}) = -n\alpha$. Suppose p does not divide n. Since this is true for all $\chi \in \mathcal{X}(B)$ and since the Jacobson radical of B is null, this shows

$$f_j = \binom{n}{j} \left(\frac{-f_{n-1}}{n} \right)^j \quad \forall j = 1, \ldots, n-1.$$

Consequently,

$$G(X) = \left(X - \frac{f_{n-1}}{n} \right)^n,$$

therefore $n = 1$. Thus, if $p = 0$ then $B = A$. Similarly, if p divides n, we can see that P is of the form $X^{p^t} - f$, $f \in B$. $\qquad\qquad\square$

1.2.   Tree structure

Trees will be very helpful when studying multiplicative semi-norms on polynomials and rational functions in one variable. Most of results given here were published in [11, 26]. In this chapter, we define a basic notion of tree such that, given a strictly increasing function from the tree to $\mathbb{R}_+$, this function provides the tree with a distance.

Definition: Let E be a set equipped with an order relation $\leq$. Then E will be called *a tree* if:

(i) For every a, $b \in E$ there exists $\sup(a, b) \in E$.
(ii) For every $a \in E$ and b, $c \in E$ satisfying $a \leq b$, $a \leq c$, then b and c are comparable with respect to the order $\leq$.

Theorem 1.2.1. *Let $(E, \leq)$ be a tree and let f be a strictly increasing function from E to $\mathbb{R}$. For all a, $b \in E$ we put $\sigma_1(a, b) = (f(\sup(a, b)) - \min(f(a), f(b))$ and $\sigma_2(a, b) = 2f(\sup(a, b)) - f(a) - f(b)$. Then σ_1 and σ_2 are two equivalent distances such that $\sigma_1 \leq \sigma_2 \leq 2\sigma_1$. Moreover, if $b, c \in E$ satisfy $a \leq b$, $a \leq c$ and if $\sigma_j(a, b) = \sigma_j(a, c)$ for some j, then $b = c$.*

Proof.   It is easily seen that $\sigma_j(a, b) \geq 0$ $\forall a, b \in E$, $(j = 1, 2)$ and that $\sigma_j(a, b) = 0$ if and only if $a = b$ $(j = 1, 2)$ because f is strictly increasing. It is clear that $\sigma_1 \leq \sigma_2$ so the same properties hold for σ_2. Now, let $a, b, c \in E$. By definition of both σ_1 and σ_2 it is obviously seen that they satisfy $\sigma_j(a, b) = \sigma_j(b, a)$. Thus, it only remains us to check the triangular inequality.

Since E is a tree, $\sup(a, b)$ and $\sup(b, c)$ are comparable. For instance, suppose first $\sup(b, c) \geq \sup(a, b)$. Then it is easily seen that $\sup(b, c) = \sup(a, b, c) \geq \sup(a, c)$, and therefore we obtain

(2.1) $\quad \sigma_1(a, b) + \sigma_1(b, c) \; \geq \; f(\sup(a, b)) - f(a) + f(\sup(b, c)) - f(b) \; \geq$
$f(\sup(b, c)) - f(a) \geq f(\sup(a, c)) - f(a).$

(2.2) $\quad \sigma_1(a, b) + \sigma_1(b, c) \; \geq \; f(\sup(a, b)) - f(a) + f(\sup(b, c)) - f(c) \; \geq$
$f(\sup(b, c)) - f(c) \geq f(\sup(a, c)) - f(c).$

In the same way, suppose now $\sup(b, c) \leq \sup(a, b)$. Then we obtain

(2.3) $\quad \sigma_1(a, b) + \sigma_1(b, c) \; \geq \; f(\sup(a, b)) - f(a) + f(\sup(b, c)) - f(b) \; \geq$
$f(\sup(a, b)) - f(a) \geq f(\sup(a, c)) - f(a).$

(2.4) $\quad \sigma_1(a,b) + \sigma_1(b,c) \geq f(\sup(a,b)) - f(a) + f(\sup(b,c)) - f(c) \geq$
$f(\sup(a,b)) - f(c) \geq f(\sup(a,c)) - f(c)$.

So, by (2.1),(2.2) whenever $\sup(b,c) \geq \sup(a,b)$ and by (2.3), (2.4) whenever $\sup(b,c) \geq \sup(a,b)$, we obtain $\sigma_1(a,b) + \sigma_1(b,c) \geq \sigma_1(a,c)$. This finishes proving that σ_1 is a distance. $\qquad\square$

We now consider σ_2. We first suppose again $\sup(b,c) \geq \sup(a,b)$. Then $\sigma_2(a,b) + \sigma_2(b,c) = 2f(\sup(a,b)) - f(a) - f(b) + 2f(\sup(b,c)) - f(b) - f(c) = 2f(\sup(a,b,c)) - f(a) - f(c) + 2f(\sup(a,b)) - 2f(b) \geq 2f(\sup(a,b,c)) - f(a) - f(c) \geq 2f(\sup(a,c)) - f(a) - f(c) = \sigma_2(a,c)$. And in the same way, when $\sup(b,c) \geq \sup(a,b)$ then $\sigma_2(a,b) + \sigma_2(b,c) = 2f(\sup(a,b,c)) - f(a) - f(c) + 2f(\sup(b,c)) - 2f(b) \geq 2f(\sup(a,b,c)) - f(a) - f(c) \geq 2f(\sup(a,c)) - f(a) - f(c) = \sigma_2(a,c)$. This finishes proving that σ_2 is also a distance. Next, it is easily seen that the two distances are equivalent because they satisfy $\sigma_1(a,b) \leq \sigma_2(a,b) \leq 2\sigma_1(a,b)$. Now, suppose $a \leq b$ and $a \leq c$, and $\sigma_j(a,b) = \sigma_j(a,c)$ for some j. Since E is a tree, b and c are comparable. Without loss of generality we can assume $b \leq c$. Then we have $\sigma_j(a,b) = j(f(b) - f(a))$, $\sigma_j(a,c) = j(f(c) - f(a))$, $(j = 1,2)$, hence $f(b) = f(c)$. But since f is strictly increasing, that means $b = c$.

Definitions: Let E be a tree and let f be a strictly increasing mapping from E into $\mathbb{R}$. We will call *metric topology associated to f on E* the topology defined by the distances introduced in Theorem 1.2.1. The distance denoted by σ_1 in Theorem 1.2.1 will be called *supremum distance associated to f* and this denoted by σ_2 in Theorem 1.2.1 will be called *the whole distance associated to f*.

A topological space E is said to be *pathwise connected* or *arcwise connected* if for every $a, b \in E$ there exists a continuous mapping ϕ from an interval $[\alpha, \beta]$ into E such that $\phi(\alpha) = a$, $\phi(\beta) = b$. Theorem 1.2.2 is classical in basic topology.

Theorem 1.2.2. *An arcwise connected topological space is connected.*

Theorem 1.2.3. *Let $(E, \leq)$ be a tree and let f be a strictly increasing function from E to $\mathbb{R}$. Let a, $b \in E$ be such that $a \leq b$ and $\{f(x) \mid a \leq x \leq b\} = [f(a), f(b)]$. Then there exists a mapping from $[f(a), f(b)]$ into E bicontinuous with respect to the metric topology of E associated to f.*

Proof. For each $r \in [f(a), f(b)]$, we denote by $\psi(r)$ the unique $x \in E$ such that $a \leq x$ and $f(x) - f(a) = r$. Let σ_1 be the supremum distance associated to f. Then we have $\sigma_1(\psi(r), \psi(s)) = r - s \ \forall r, \ s \in [f(a), f(b)]$, which clearly shows that ψ is bicontinuous. $\qquad\qquad\qquad\qquad\qquad\qquad\qquad\qquad\square$

Theorem 1.2.4. *Let $(E, \leq)$ be a tree and let f be a strictly increasing function from E to $\mathbb{R}$. If for all $a, b \in E$ such that $a \leq b$, the equality $\{f(x) \mid a \leq x \leq b\} = [f(a), f(b)]$ holds, then E is arcwise connected with respect to its metric topology associated to f.*

Proof. Indeed, let $a, b \in E$ and let $s = \sup(a, b)$. By Theorem 1.2.3, there exists a bicontinuous mapping ϕ_1 from $[f(a), f(s)]$ into E such that $\phi_1(f(a)) = a$, $\phi_1(f(s)) = s$ and a bicontinuous mapping ϕ_2 from $[f(b), f(s)]$ into E such that $\phi_1(f(b)) = b$, $\phi_1(f(s)) = s$. Then we can obviously define a bicontinuous mapping ψ from an interval $[\alpha, \beta]$ into E such that $\psi(\alpha) = a$, $\psi(\beta) = b$. $\qquad\qquad\qquad\qquad\qquad\qquad\qquad\qquad\square$

1.3.   Ultrametric absolute values

In this chapter, we have to recall many basic definitions and results about valuations in ultrametric fields. Most of them were given in [26, 28] with full proofs. Here, when we state one with some additional considerations, it is followed by a new proof.

Notations: Throughout the book, $\mathbb{L}$ will denote a field equipped with a non-trivial ultrametric absolute value $| \, . \, |$ which is complete for this absolute value and $\mathbb{K}$ will be an algebraically closed complete ultrametric field with a non-trivial ultrametric absolute value $| \, . \, |$. It is convenient and useful to define the valuation v associated to the absolute value $| \, . \, |$. Let $\omega \in]1, +\infty[$ and let log be the real logarithm function of base ω. We put $v(x) = -\log|x|$, and v is named *the valuation associated to the absolute value $| \, . \, |$*.

The Archimedean absolute value defined on $\mathbb{R}$ will be denoted by $| \, . \, |_\infty$.

Lemmas 1.3.1 and 1.3.2 are classical, and proven in the same way regardless the absolute value.

Lemma 1.3.1. *Let $\mathbb{E}$ be a field equipped with two ultrametric absolute values whose associated valuations are v and w, respectively. They are equivalent if and only if there exists $r > 0$ such that $w(x) = rv(x)$ whenever $x \in \mathbb{E}$.*

Lemma 1.3.2. *Let $\mathbb{E}$ be a field equipped with an ultrametric absolute value. Let V be an $\mathbb{E}$-vector space of finite dimension equipped with two norms. These two norms are equivalent in each one of these two cases:*

    $\mathbb{E}$ is complete,
    The dimension of V is one.

Theorem 1.3.3 is an easy corollary.

Theorem 1.3.3. *Let $\mathbb{F}$ be an algebraic extension of $\mathbb{L}$, equipped with two absolute values extending this of $\mathbb{L}$. These absolute values are equal.*

We will sometimes use Hahn–Banach's Theorem for non-Archimedean vector spaces.

Theorem 1.3.4. *Let $\mathbb{F}$ be a $\mathbb{L}$-vector space which is complete for two norms $\| \cdot \|$ and $\| \cdot \|'$ such that $\|x\| \leq C\|x\|'$ $\forall x \in \mathbb{F}$, where C is a positive constant. Then the two norms are equivalent.*

Notation: The set of $x \in \mathbb{L}$ such that $|x| \leq 1$ will be denoted by $U_\mathbb{L}$, and the set of $x \in \mathbb{L}$ such that $|x| < 1$ will be denoted by $M_\mathbb{L}$.

Lemma 1.3.5. *$U_\mathbb{L}$ is a local subring of $\mathbb{L}$ whose maximal ideal is $M_\mathbb{L}$.*

Lemma 1.3.6. *Let B be a unital commutative ring equipped with an absolute value $| \cdot |$ such that $|x| \leq 1$ $\forall x \in B$, and such that every x satisfying $|x| = 1$ is invertible in B. The set of $x \in B$ such that $|x| < 1$ is the unique maximal ideal of B. Let $\mathbb{F}$ be the field of fractions of B. Then the absolute value of B has continuation to $\mathbb{F}$ and $B = U_\mathbb{F}$.*

Definitions and notations: We call *a valuation ring* a unital commutative ring equipped with an absolute value $| \cdot |$ such that $|x| \leq 1$ $\forall x \in B$, and such that every x satisfying $|x| = 1$ is invertible in B, and then the maximal ideal of B is called *the valuation ideal of B*. In the field $\mathbb{L}$, $U_\mathbb{L}$ is called *the valuation ring of $\mathbb{L}$*. The maximal ideal $M_\mathbb{L}$ of $U_\mathbb{L}$ is called *the valuation ideal* and the field $\mathcal{L} = \dfrac{U_\mathbb{L}}{M_\mathbb{L}}$ is called *the residue class field of $\mathbb{L}$*. For any $a \in \mathbb{L}$, the residue class of a will be denoted by $\bar{a}$. If D is a set in $\mathbb{L}$ we put $|D| = \{|x| \mid x \in D\}$, and $v(D) = \{v(x) \mid x \in D\}$.

The characteristic of $\mathcal{L}$ is named *the residue characteristic of $\mathbb{L}$* and will be denoted by p. We put $\mathbb{L}^* = \mathbb{L} \setminus \{0\}$. Then $|\mathbb{L}^*|$ is a subgroup of the multiplicative group $(\mathbb{R}_+^*, .)$. The image of L^* by the valuation v associated to $\mathbb{L}$ is then a subgroup of the additive group $(\mathbb{R}, +)$ called *the valuation group of $\mathbb{L}$*. The valuation of $\mathbb{L}$ is said to be *discrete* if its valuation group is a discrete subgroup of $\mathbb{R}$. Else, the valuation group is dense in $\mathbb{R}$, and then the valuation is said to be *dense*. The field $\mathbb{L}$ will be said to be *strongly valued* if at least one of the two sets $|\mathbb{L}|$, $\mathcal{L}$, is not countable. If $\mathbb{L}$ is not strongly valued it will be said to be *weakly valued*.

Let $a \in \mathbb{L}$ and let r be a positive number. We denote by $d(a, r)$ the disk $\{x \in \mathbb{L} \mid |x - a| \leq r\}$ and by $d(a, r^-)$ the disk $\{x \in \mathbb{L} \mid |x - a| < r\}$. Moreover we denote by $C(a, r)$ the circle $\{x \in \mathbb{L} \mid |x - a| = r\}$.

Lemma 1.3.7 is well known.

Lemma 1.3.7. *If $\mathbb{L}$ is algebraically closed, its valuation group is dense in $\mathbb{R}$.*

Remark 1. The most classical example of an ultrametric complete algebraically closed field is the field $\mathbb{C}_p$ (one can find a description in [28, Chapter 8, 41]). Then $\mathbb{C}_p$ is weakly valued. Now, consider a field $\mathbb{L}$ with a dense valuation, and let A be the L-algebra B of bounded analytic functions in $d(0, 1^-)$, equipped with the norm of uniform convergence in $d(0, 1^-)$. It is known and easily seen that $\{\|f\| \mid f \in B\} = [0, +\infty[$. Therefore, its field of fractions is strongly valued. When $p \neq 0$ it is useful to take $\omega = p$.

Theorems 1.3.8 and 1.3.9 are well known (proofs are given in [28, Theorems 5.4 and 5.14]).

Theorem 1.3.8. *Let $\mathbb{F}$ be an algebraic extension of $\mathbb{L}$. There exists a unique ultrametric absolute value on $\mathbb{F}$ extending this of $\mathbb{L}$. Moreover, given $x \in F$, $P(X) = irr(x, \mathbb{F})$ and $q = \deg(P)$, then $|x| = \sqrt[q]{|P(0)|}$.*

Theorem 1.3.9. *Let $\mathbb{F}$ be an algebraically closed field equipped with a nontrivial ultrametric absolute value. The completion of $\mathbb{F}$ is algebraically closed.*

Example. $\mathbb{C}_p$ is the completion of an algebraic closure of the field $\mathbb{Q}_p$, the completion of $\mathbb{Q}$ with respect to the p-adic absolute value. Then $\mathbb{C}_p$ is algebraically closed.

Chapter 2

Norms, Semi-norms and Multiplicative Spectrum

2.1. A metric on the set of the semi-norms bounded by the norm

Definitions and notations: Let $\mathbb{L}$ be a field equipped with an absolute value denoted $|\,.\,|$.

Let $(E, \|\,.\,\|)$ be a normed $\mathbb{L}$-vector space, let E_0 be the unit ball of E, let S be the set of semi-norms of E bounded by the norm, equipped with topology of pointwise convergence and let $\mathcal{T}$ be a closed subset of S with respect to the topology of pointwise convergence.

Given ϕ, $\psi \in \mathcal{T}$, we put $\Theta(\phi, \psi) = \sup\{|\phi(x) - \psi(x)|_\infty,\ x \in E_0\}$. Given $\phi \in \mathcal{T}$, $f_1, \ldots, f_q \subset E$ and $\epsilon \in \mathbb{R}_+^*$, we denote by $\mathcal{W}(\phi, f_1, \ldots, f_q, \epsilon)$ the neighborhoods of ϕ: $\{\zeta \in \mathcal{T} \mid |\phi(f_j) - \zeta(f_j)|_\infty \leq \epsilon,\ j = 1, \ldots, q\}$.

Moreover, we denote by $\mathbb{B}(\phi, \epsilon)$ the ball $\{\zeta \in \mathcal{T} \mid \Theta(\phi, \zeta) \leq \epsilon\}$.

Given $f_1, \ldots, f_q \in E_0$, we put $\omega(f_1, \ldots, f_q) = \max(1, \|f_1\|, \ldots, \|f_q\|)$.

By Tykhonov's Theorem [6, paragraph 9, n. 5, Theorem 3], we have Lemma 2.1.1.

Lemma 2.1.1. *Let g be a function from D to $\mathbb{R}_+$. Then the set of functions from D to $\mathbb{R}_+$ bounded by g is compact with respect to the topology of pointwise convergence.*

Corollary 2.1.2. *S and $\mathcal{T}$ are compact with respect to the topology of pointwise convergence.*

Lemma 2.1.3 is immediate.

Lemma 2.1.3. *Θ is a distance on $\mathcal{T}$.*

11

Now we given two subsets U, V of $\mathcal{T}$, we put $\Theta(U,V) = \inf\{\Theta(\phi,\psi)\ \phi \in U,\ \psi \in V\}$.

Lemma 2.1.4. *Let $\phi \in \mathcal{T}$, let $f_1, \ldots, f_q \in E$. There exists $t \in \mathbb{R}_+$ such that, for all $\epsilon > 0$ and $\psi \in \mathcal{W}(\phi, f_1, \ldots, f_q, \epsilon)$, $\mathcal{W}(\phi, f_1, \ldots, f_q, t\epsilon)$ contains the ball $B(\psi, \frac{\epsilon}{\omega(f_1,\ldots,f_q)})$.*

Proof. Let $b \in \mathbb{L}$ and $s \in \mathbb{R}_+$ satisfy $\omega(f_1, \ldots, f_q) \leq |b| \leq s\omega(f_1, \ldots, f_q)$ and for each $j = 1, \ldots, q$, set $g_j = \frac{f_j}{b}$. Then $\|g_j\| \leq 1\ \forall j = 1, \ldots, q$. Let $\zeta \in \mathbb{B}(\psi, \frac{\epsilon}{\omega(f_1,\ldots,f_q)})$. Then for every $g \in E_0$, we have $\left|\psi(g) - \zeta(g)\right|_\infty \leq \frac{\epsilon}{\omega(f_1,\ldots,f_q)}$ and particularly, $\left|\psi(g_j) - \zeta(g_j)\right|_\infty \leq \frac{\epsilon}{\omega(f_1,\ldots,f_q)}$ hence $\left|\psi(bg_j) - \zeta(bg_j)\right|_\infty = \left||b|(\psi(g_j) - \zeta(g_j))\right|_\infty \leq |b|(\frac{\epsilon}{\omega(f_1,\ldots,f_q)}) \leq s\epsilon$. Thus, $\left|\psi(f_j) - \zeta(f_j)\right|_\infty \leq s\epsilon\ \forall j = 1, \ldots, q$. But now, by hypothesis we have $\left|\psi(f_j) - \phi(f_j)\right|_\infty \leq \epsilon\ \forall j = 1, \ldots, q$, hence $\left|\zeta(f_j) - \phi(f_j)\right|_\infty \leq (s+1)\epsilon\ \forall j = 1, \ldots, q$. Then putting $t = s+1$, that ends the proof. $\square$

Theorem 2.1.5. *Let $(\phi_n)_{n\in\mathbb{N}}$ be a Cauchy sequence in $\mathcal{T}$, with respect to the metric Θ. Then the sequence has a unique cluster with respect to the topology of pointwise convergence and it converges in $\mathcal{T}$ to this cluster with respect to the topology of pointwise convergence.*

Proof. Since the sequence $(\phi_n)_{n\in\mathbb{N}}$ is a Cauchy sequence, there exists $\sigma(\epsilon) \in \mathbb{N}$, such that $\Theta(\phi_m, \phi_n) \leq \epsilon\ \forall m, n \in N$ such that $m \geq \sigma(\epsilon)$, $n \geq \sigma(\epsilon)$. Next, since $\mathcal{T}$ is compact, the sequence admits a cluster $\phi \in \mathcal{T}$. Let us fix $\epsilon > 0$ and $f_1, \ldots, f_q \in E$. There exists $T(\epsilon) \in \mathbb{N}$, with $T(\epsilon) \geq \sigma(\epsilon)$ such that $\phi_{T(\epsilon)} \in \mathcal{W}(\phi, f_1, \ldots, f_q, \epsilon)$ and hence, by Lemma 2.1.4, there exists $t > 0$ such that $\mathbb{B}(\phi_{T(\epsilon)}, \epsilon)$ is included in $\mathcal{W}(\phi, f_1, \ldots, f_q, t\epsilon)$. But then all terms ϕ_n with $n \geq T(\epsilon)$ satisfy $\Theta(\phi_{T(\epsilon)}, \phi_n) \leq \epsilon$, hence they all lie in $\mathcal{W}(\phi, f_1, \ldots, f_q, t\epsilon)$.

This holds for every finite set $f_1, \ldots, f_q$ of E and for every $\epsilon > 0$, hence the sequence $(\phi_n)_{n\in\mathbb{N}}$ converges to ϕ with respect to the topology of pointwise convergence. $\square$

Theorem 2.1.6. *$\mathcal{T}$ is complete with respect to the distance Θ.*

Proof. Let $(\phi_n)_{n\in\mathbb{N}}$ be a Cauchy sequence in $\mathcal{T}$, with respect to the metric Θ. By Theorem 2.1.5, the sequence converges to a limit ϕ with respect to the topology of pointwise convergence. Suppose that the sequence does not converge to ϕ with respect to the distance Θ. Then we can extract a subsequence $(\phi_{s(m)\in\mathbb{N}})$ of the sequence (ϕ_n) and find $\lambda > 0$ such that

$\Theta(\phi, \phi_{s(m)}) \geq \lambda \ \forall m \in \mathbb{N}$. Thus, without loss of generality, we can assume that the sequence $(\phi_n)_{n \in \mathbb{N}}$ is such that $\Theta(\phi, \phi_n) \geq \lambda \ \forall n \in \mathbb{N}$. Let $\sigma \in \mathbb{N}$ be such that $\Theta(\phi_n, \phi_m) \leq \frac{\lambda}{3} \ \forall m, \ n \in \mathbb{N}$ such that $m \geq S, n \geq \sigma$. Thus, we have $\Theta(\phi, \phi_\sigma) \geq \lambda$ but $\Theta(\phi_\sigma, \phi_n) \leq \frac{\lambda}{3} \ \forall n \geq \sigma$. On the other hand, there exists $f \in E_0$ such that $|\phi(f) - \phi_\sigma(f)|_\infty \geq \frac{5\lambda}{6}$. But since $\Theta(\phi_n, \phi_\sigma) \leq \frac{\lambda}{3} \ \forall n \geq \sigma$, this implies $|\phi(f) - \phi_n(f)|_\infty \geq \frac{\lambda}{2} \ \forall n \geq \sigma$. And since the sequence (ϕ_n) converges to ϕ with respect to the pointwise convergence, we must have $|\phi(f) - \phi_n(f)|_\infty \leq \frac{\lambda}{3}$ when n is big enough, which contradicts $|\phi(f) - \phi_n(f)|_\infty \geq \frac{\lambda}{2}$. That finishes proving that the sequence $(\phi_n)_{n \in \mathbb{N}}$ converges to ϕ with respect to the metric Θ and this proves that $\mathcal{T}$ is complete with respect to the distance Θ. $\qquad\square$

Theorem 2.1.7. *Let U, V be two compact subsets of $\mathcal{T}$ with respect to the pointwise topology, making a partition of $\mathcal{T}$. Then $\Theta(U, V) > 0$.*

Proof. Suppose $\Theta(U, V) = 0$. There exists a sequence $(\phi_n)_{n \in \mathbb{N}}$ of U and a sequence $(\psi_n)_{n \in \mathbb{N}}$ of V such that $\lim_{n \to \infty} \Theta(\phi_n, \psi_n) = 0$. Since U and V are compact with respect to the pointwise topology, the sequence $(\phi_n)_{n \in \mathbb{N}}$ has a cluster point ϕ in U with respect to the pointwise topology.

Let $\mathcal{W}(\phi, f_1, \ldots, f_q, \epsilon)$ be a neighborhood of ϕ in U with respect to the pointwise topology. There exists a rank $L \in \mathbb{N}$ such that $\phi_L \in \mathcal{W}(\phi, f_1, \ldots, f_q, \epsilon)$ and there are infinitely many $n \in \mathbb{N}, n \geq L$, such that $\phi_n \in \mathcal{W}(\phi, f_1, \ldots, f_q, \epsilon)$. Particularly, we can find $M > L$ such that $\phi_M \in \mathcal{W}(\phi, f_1, \ldots, f_q, \epsilon)$ and that $\Theta(\phi_M, \psi_M) \leq \frac{\epsilon}{\omega(f_1, \ldots, f_q)}$. But then, by Lemma 2.1.4, there exists $t > 0$ such that $\mathcal{V}\phi, f_1, \ldots, f_q, t\epsilon)$ contains the ball $\mathbb{B}(\phi_M, \frac{\epsilon}{\omega(f_1, \ldots, f_q)})$. That holds for every $\epsilon > 0$ and for every $f_1, \ldots, f_q \in E$, therefore we have infinitely many indices m such that $|\phi(f_j) - \psi_m(f_j)|_\infty \leq t\epsilon$. Consequently, ϕ is a cluster point of the sequence $(\psi_n)_{n \in \mathbb{N}}$. But since V is compact with respect to the pointwise topology, V is closed and hence ϕ lies in V, a contradiction since $U \cap V = \emptyset$. This ends the proof. $\qquad\square$

2.2. L-productal vector spaces

Definitions and notations: Recall that $\mathbb{L}$ denotes a complete ultrametric field. A normed $\mathbb{L}$-vector space E will be said to be $\mathbb{L}$-*productal* if every $\mathbb{L}$-subspace of $\mathbb{E}$, equipped with the topology defined by its norm, is homeomorphic to $\mathbb{L}^n$, equipped with the product topology.

Throughout this chapter, A will denote a unital commutative $\mathbb{L}$-algebra denoted by 1.

Let us recall that *a semi-norm* (resp., *a norm*) *of* $\mathbb{L}$*- algebra is a semi-norm* (resp., *a norm*) *of* $\mathbb{L}$-linear space φ that satisfies $\varphi(x.y) \leq \varphi(x)\,\varphi(y)$. Moreover, φ is said to be *semi-multiplicative* if $\varphi(x^n) = \varphi(x)^n$ whenever $x \in A$ and φ is said to be *multiplicative* if $\varphi(x.y) = \varphi(x)\,\varphi(y)$ whenever $x, y \in A$.

In particular, a multiplicative norm is just an absolute value extending that of $\mathbb{L}$ (when identifying the unity with 1 in $\mathbb{L}$).

Given a semi-norm of $\mathbb{L}$-algebra φ, we will denote by $\mathrm{Ker}(\varphi)$ the set of the $x \in A$ such that $\varphi(x) = 0$. A normed $\mathbb{L}$-vector space E will be said to be $\mathbb{L}$-*productal* if every $\mathbb{L}$-subspace of E, equipped with the topology defined by its norm, is homeomorphic to $\mathbb{L}^n$, equipped with the product topology.

Lemma 2.2.1. *Let E be a finite dimensional normed $\mathbb{L}$-vector space. Then E is homeomorphic to $\mathbb{L}^n$ if and only if for every $x \in E$, there exists $\phi \in E'$ satisfying $\phi(x) \neq 0$. Moreover, if E is homeomorphic to $\mathbb{L}^n$, then $E' = E^*$.*

Proof. If E is homeomorphic to $\mathbb{L}^n$ for every $x \in E$, there obviously exists $\phi \in E'$ satisfying $\phi(x) \neq 0$ because, given $u \in E$, $u \neq 0$, we can take a basis $(e_j)_{1 \leq j \leq n}$ of E such that $u = e_1$ and such that $\| \sum_{j=1}^{n} \lambda_j e_j \| \leq \max_{1 \leq j \leq n} |\lambda_j|$, and then, in the dual basis $(\phi_j)_{1 \leq j \leq n}$ of the basis (e_j), we check that ϕ_1 satisfies $|\phi_1(\sum_{j=1}^{n} \lambda_j e_j)| = |\lambda_1| \leq \| \sum_{j=1}^{n} \lambda_j e_j \|$.   □

Conversely, we assume that for every $x \in E$, there exists $\phi \in E'$ satisfying $\phi(x) \neq 0$. We will show the existence of a basis of E^* consisting of elements of E'. Let n be the dimension of E. By hypothesis, we can find $\phi_1 \in E'$, $\phi_1 \neq 0$. Suppose we have already constructed $\phi_1, \ldots, \phi_m \in E'$, linearly independent, with $m < n$. Then the intersection of the m hyperplans $\mathrm{Ker}(\phi_j)$ is not reduced to the null subspace. Hence, there exists $u \in \bigcap_{j=1}^{m} \mathrm{Ker}(\phi_j)$, with $u \neq 0$. By hypothesis, there exists $\phi_{m+1} \in E'$ such that $\phi_{m+1}(u) \neq 0$. Then, since $\phi_j(u) = 0 \; \forall j = 1, \ldots, m$, and $\phi_{m+1} \neq 0$, we easily check that $\phi_1, \ldots, \phi_{m+1}$ are linearly independent. Thus, by induction, we can obtain $\phi_1, \ldots, \phi_n \in E'$ which are linearly independent. Consequently, this is a basis of E^* and therefore $E^* = E'$. Now, consider the mapping ψ from E into $\mathbb{L}^n$ defined as $\psi(x) = (\phi_1(x), \ldots, \phi_n(x))$. We can easily check that ψ is an isomorphism from E onto $\mathbb{L}^n$, and is continuous by definition because so are the ϕ_j. We have to check that ψ^{-1} is also continuous. By considering the bidual of E we can obtain a basis $\{e_1, \ldots, e_n\}$ of E such that $\{\phi_1, \ldots, \phi_n\}$ is the dual basis of $\{e_1, \ldots, e_n\}$. Let $M = \max_{1 \leq j \leq n} \|e_j\|$. Let $(\lambda_1, \ldots, \lambda_n) \in \mathbb{L}^n$ and let $x = \sum_{j=1}^{n} \lambda_j e_j$. Then $\psi(x) = (\lambda_1, \ldots, \lambda_n)$, hence

$\psi^{-1}(\lambda_1, \ldots, \lambda_n) = x$, and therefore $\|\psi^{-1}(\lambda_1, \ldots, \lambda_n)\| \le M \max_{1 \le j \le n} |\lambda_j|$. Thus, ψ^{-1} is continuous.

Corollary 2.2.2. *A normed $\mathbb{L}$-vector space E is $\mathbb{L}$-productal if and only if for every $x \in E$, there exists $\phi \in E'$ satisfying $\phi(x) \ne 0$.*

Proof. If E is $\mathbb{L}$-productal, obviously for every $x \in E$, there exists $\phi \in E'$ satisfying $\phi(x) \ne 0$. Conversely, if for every $x \in E$, there exists $\phi \in E'$ satisfying $\phi(x) \ne 0$, this property is true in each finite-dimensional $\mathbb{L}$-subspace F of dimension n of E, and proves that F is homeomorphic to $\mathbb{L}^n$. □

Lemma 2.2.3. *Let E be a $\mathbb{L}$-productal normed $\mathbb{L}$-vector space and let $q \in \mathbb{N}^*$. Then E^q, equipped with the product topology, is $\mathbb{L}$-productal.*

Proof. Let W be a subspace of E^q of finite dimension n. For each $h = 1, \ldots, q$, we denote by E_h the subspace of E^q consisting of the $(x_1, \ldots x_q)$ such that $x_j = 0 \ \forall j \ne h$ and we put $W_j = W \cap E_j$. Then W is the direct sum of the W_j, and for each $j = 1, \ldots, q$, W_j has finite dimension n_j. Let $n = \sum_{j=1}^q n_j$. By hypothesis, W_j is homeomorphic to $\mathbb{L}^{n_j}$. Let $\| . \|$ be the norm of E, let $\| . \|$ be the product norm on E^q defined as $\|(x_1, \ldots x_q)\| = \max_{1 \le j \le q} \|x_j\|$. Now, in each subspace W_j, we denote by let $| . |_j$ the norm induced by $\mathbb{L}^{n_j}$ on W_j. Thus, there exist positive constants A and B such that $A|a_j|_j \le \|a_j\| \le B|a_j|_j \ \forall a_j \in W_j, \forall j = 1, \ldots, q$. But since E^q is equipped with the product topology, given $a \in W$, a is of the form $\sum_{j=1}^q a_j$, with $a_j \in W_j$, $(j = 1, \ldots, q)$, and therefore we have $A \max_{1 \le j \le q} \|a_j\| \le \|a\| \le B \max_{1 \le j \le q} \|x_j\|$. But it is seen that $\max_{1 \le j \le q} \|x_j\|$ is a norm equivalent to the product norm on $\mathbb{L}^n$, and this lets us conclude that $\mathbb{L}^q$, equipped with the norm $\| . \|$, is homeomorphic to $\mathbb{L}^q$. □

Lemma 2.2.4. *F be a subfield of $\mathbb{L}$ such that $\mathbb{L}$ is a F-productal F-vector space with respect to its absolute value. Then, every $\mathbb{L}$-productal $\mathbb{L}$-vector space is F-productal.*

Proof. Let E be a $\mathbb{L}$-productal $\mathbb{L}$-vector space. Then E is a F-vector space. Let W be a F-subspace of finite dimension over E and let $\{e_1, \ldots, e_q\}$ be a basis of W. Let V be the $\mathbb{L}$-subvector space generated by $\{e_1, \ldots, e_q\}$ in E. By hypothesis, W is homeomorphic to $\mathbb{L}^q$. But since $\mathbb{L}$ is F-productal, by Lemma 2.2.3 so is $\mathbb{L}^q$. Hence V is F-productal, and therefore, so is every F-subspace of V, hence so is W. □

Before ending this chapter, we can also recall this immediate and classical lemma:

Lemma 2.2.5. *Let $(E_n)_{n\in\mathbb{N}}$ be a sequence of Banach $\mathbb{L}$-vector spaces and let E be the $\mathbb{L}$-vector space of sequences $(a_n)_{n\in\mathbb{N}}$ such that $\lim_{n\to\infty}\|a_n\| = 0$. Then E is a Banach $\mathbb{L}$-vector space with respect to the norm $\|(a_n)_{n\in\mathbb{N}}\| = \sup_{n\in\mathbb{N}}\|a_n\|$.*

2.3. Multiplicative semi-norms

The idea of studying systematically the multiplicative semi-norms of an ultrametric normed algebra is due to Bernard Guennebaud [31, 32]. This theory found many applications in the study of analytic elements and actually gave the good explanation of several basic properties [21, 22]. More generally, this is also true in ultrametric Banach algebras, where multiplicative semi-norms play a role somewhat similar to that of maximal ideals in complex Banach algebras [8, 39]. Next, Vladimir Berkovich also used the set of multiplicative semi-norms in order to reconstruct Tate's Theory [3, 44]. In [10], Kamal Boussaf showed the existence of a Shilov boundary for the norm of an algebra of analytic elements and described it with help of circular filters that characterize multiplicative semi-norms. In [3], V. Berkovich showed the existence of a Shilov boundary for the spectral semi-norm of an affinoid algebra. This chapter is aimed at proving, in the general case, the existence of a Shilov boundary for a semi-multiplicative semi-norm of a L-algebra, i.e., the existence of a closed subset F of the set of continuous multiplicative semi-norms, minimal for inclusion, such that for every $x \in A$, there exists $\varphi \in F$ satisfying $\varphi(x) = \|x\|$. Such a process was outlined by B. Guennebaud in his unpublished Thèse d'Etat [32] and our way widely takes from it. However, certain intermediate results were missing, or suffered a lack of rigor, while hypotheses and definitions were often confusing. Besides, there was a confusion between two different problems: on one hand, finding a "Shilov boundary" for a semi-multiplicative function defined on a semi-group, inside a set of multiplicative functions, and on the other hand, finding a "Shilov boundary" for a semi-multiplicative semi-norm defined on an algebra, inside a set of multiplicative semi-norms. Here, we have reconstructed the framework of the proof and only considered the problem of a Shilov boundary for a semi-multiplicative semi-norm [25].

Definitions and notations: Throughout this chapter, $\mathbb{E}$ will denote a field equipped with a non-trivial absolute value $|\,.\,|$ which makes it complete, hence is either $\mathbb{R}$ or $\mathbb{C}$, or the field $\mathbb{L}$ equipped with a non-trivial ultrametric absolute value for which it is complete.

Given sets X, Y we will denote by $\mathcal{F}(X, Y)$ the set of mappings from X to Y. Let D a be subset of $\mathbb{E}$ and let F be a $\mathbb{E}$-vector space of bounded functions from D to $\mathbb{E}$. In the sequel, we will denote by $\| \cdot \|_D$ the norm of uniform convergence on D. The $\mathbb{R}$-vector space $\mathcal{F}(D, \mathbb{R})$ is equipped with the topology of pointwise convergence: the filter of neighborhoods of $\varphi \in \mathcal{F}(D, \mathbb{R})$ admits as a generating system the family of sets $\mathcal{W}(\varphi, a_1, \ldots, a_n, \epsilon) = \{\psi \in \mathcal{F}(D, \mathbb{R}) \mid |\psi(a_i) - \varphi(a_i)|_\infty \leq \epsilon\}$, with $a_1, \ldots, a_n \in D$ and $\epsilon > 0$. Given a topological space D we denote by $\mathcal{G}(D, \mathbb{R})$ the algebra of continuous functions from D into $\mathbb{R}$.

We call *semi-group* a set equipped with an associative internal law and of a neutral element for this law, and we will denote by B, B' semigroups.

In this chapter, we will denote by A (resp., A') a unital commutative $\mathbb{E}$-algebra.

Now, let θ be a function from D to $\mathbb{R}_+$ and let $\mathcal{B}(D, \theta)$ be the subset of $\mathcal{F}(D, \mathbb{R})$ consisting of the functions f from D to $\mathbb{R}_+$ such that $f(x) \leq \theta(x) \ \forall x \in D$.

Let $\mu \in \mathcal{F}(B, \mathbb{R}_+)$. The function μ will be said to be *submultiplicative* (resp., *multiplicative*) if $\mu(xy) \leq \mu(x)\mu(y) \ \forall x, y \in B$ (resp., $\mu(xy) = \mu(x)\mu(y) \ \forall x, y \in B$). And μ will be said to be *semi-multiplicative* if $\mu(x^n) = \mu(x)^n \ \forall x \in B, \forall n \in \mathbb{N}^*$.

Let S be a subsemi-group of B. A submultiplicative function θ on B will be said to be *S-multiplicative* if $\theta(xy) - \theta(x)\theta(y) \ \forall x, y \in S$.

Let us recall that *a semi-norm* (resp., *a norm*) of $\mathbb{E}$-algebra defined on A is a semi-norm (resp., a norm) of $\mathbb{E}$-linear space φ that satisfies $\varphi(x.y) \leq \varphi(x) \ \varphi(y) \ \forall x, \ y \in A$. Given a semi-norm of $\mathbb{E}$-algebra φ, we will denote by $\mathrm{Ker}(\varphi)$ the set of the $x \in A$ such that $\varphi(x) = 0$. Henceforth, we will denote by $\| \cdot \|$ (resp., $\| \cdot \|'$) a $\mathbb{E}$-algebra semi-norm on A (resp., A').

We will denote by $SM(A)$ the set of non-identically zero semi-multiplicative semi-norms of A, by $\mathrm{Mult}(A)$ the set of non-identically zero multiplicative semi-norms of A, by $\mathrm{Mult}_m(A)$ the set of the $\varphi \in \mathrm{Mult}(A)$ such that $\mathrm{Ker}(\varphi) \in \mathrm{Max}(A)$, by $\mathrm{Mult}_1(A)$ the set of the $\varphi \in \mathrm{Mult}(A)$ such that $\mathrm{Ker}(\varphi) \in \mathrm{Max}_1(A)$ and by $\mathrm{Mult}_a(A)$ the set of the $\varphi \in \mathrm{Mult}(A)$ such that $\mathrm{Ker}(\varphi)$ is a maximal ideal of algebraic codimension and by $\mathrm{Mult}'(A)$ the set of multiplicative norms of A [26, 28, 32]. Let A be equipped with a topology T of $\mathbb{E}$-vector space. We will denote by $SM(A, T)$ (resp., $\mathrm{Mult}(A, T)$, resp., $\mathrm{Mult}'(A, T)$, resp., $\mathrm{Mult}_m(A, T)$, resp. $\mathrm{Mult}_1(A, T)$,) the set of the $\varphi \in SM(A)$ (resp., $\varphi \in \mathrm{Mult}(A)$, resp. $\varphi \in \mathrm{Mult}'(A)$, resp., $\varphi \in \mathrm{Mult}_m(A)$, resp., $\varphi \in \mathrm{Mult}_1(A)$) that are continuous with respect to the topology T. Particularly, if the topology T on A is defined by a norm or a semi-norm $\| \cdot \|$, we will denote by $SM(A, \| \cdot \|)$ (resp., $\mathrm{Mult}(A, \| \cdot \|)$, resp. $\mathrm{Mult}'(A, \| \cdot \|)$, resp., $\mathrm{Mult}_m(A, \| \cdot \|)$, resp., $\mathrm{Mult}_1(A, \| \cdot \|)$) the set $SM(A, T)$ (resp., $\mathrm{Mult}(A, T)$, resp., $\mathrm{Mult}'(A, T)$, resp., $\mathrm{Mult}_m(A, T)$, resp., $\mathrm{Mult}_1(A, T)$).

Given $\varphi \in \mathrm{Mult}(A, \| \cdot \|)$ we denote by $\mathrm{Ker}(\varphi)$ the set of $x \in A$ such that $\varphi(x) = 0$.

By Corollary of Lemma 2.1.1, we have this obvious corollary.

Corollary 2.3.1. $\mathrm{Mult}(A, \| \cdot \|)$ *is compact with respect to the topology of pointwise convergence.*

Proof.   Indeed, $\mathrm{Mult}(A, \| \cdot \|)$ is a closed subset of the product of the sets $\{\phi(x) \| \phi \in \mathrm{Mult}(A, \| \cdot \|)$ whenever x runs in A. $\square$

Statements of Lemma 2.3.2 are classical, and come from properties in commutative algebra and in topological vector spaces theory.

Lemma 2.3.2. *Let φ be a semi-norm of $\mathbb{E}$-algebra of A. Then $\mathrm{Ker}(\varphi)$ is an ideal of A. If φ is multiplicative, $\mathrm{Ker}(\varphi)$ is a prime ideal of A.*

Let I be an ideal of A, let A' be the quotient $\mathbb{E}$-algebra $\frac{A}{I}$ and let Φ be the canonical homomorphism from A onto A'. Let φ' be the mapping from A' to $\mathbb{R}_+$ defined as $\varphi'(\Phi(x)) = \inf\{\varphi(t)| \ \Phi(t) = \Phi(x)\}$. Then φ' is a $\mathbb{E}$-algebra semi-norm on A'. If φ is ultrametric, so is φ'. Moreover, if φ is a norm, and if I is closed with respect to φ, then φ' also is a norm.

Lemma 2.3.3 is Lemma 1.7 in [28].

Lemma 2.3.3. *If A is a $\mathbb{L}$-algebra, every element of $SM(A)$ is ultrametric.*

Lemmas 2.3.4 is obvious.

Lemma 2.3.4. *Let S be a subset of $SM(A)$. If $\sup\{\varphi(x) \mid \varphi \in S\}$ is pointwise bounded, then the mapping ϕ defined on A by $\phi(x) = \sup\{\varphi(x) \mid \varphi \in S\}$, belongs to $SM(A)$.*

Lemma 2.3.5. *Both $\mathrm{Mult}(A)$ and $SM(A)$ are closed in $\mathcal{F}(A, \mathbb{R}_+)$ with respect to the topology of pointwise convergence.*

Proof.   Let ϕ belong to the closure of $\mathrm{Mult}(A)$ (resp., of $SM(A)$) in $\mathrm{Mult}(A)$ and let ϵ be > 0. Let $x, y \in A$ and let $\lambda \in \mathbb{L}$, let $n \in \mathbb{N}$ and let $m = \max(|x|, |y|, |\lambda x|)$. There exists $\psi \in \mathrm{Mult}(A)$ such that $\max(|\phi(x) - \psi(x)|_\infty, |\phi(x^n) - \psi(x^n)|_\infty, |\phi(y) - \psi(y)|_\infty, |\phi(\lambda x) - \psi(\lambda x)|_\infty, |\lambda|) < \epsilon$. Then $|\phi(\lambda x) - |\lambda|\phi(x)|_\infty \le |\phi(\lambda x) - \psi(\lambda x)|_\infty + |\psi(\lambda x) - |\lambda|\phi(x)|_\infty \le \epsilon(1 + |\lambda|)$. Since ϵ is arbitrary, we have $\phi(\lambda x) = |\lambda|\phi(x)$. In the same way, $\phi(x + y) \le$

$\psi(x + y) + \epsilon \leq \max(\psi(x), \psi(y)) + \epsilon \leq \max(\phi(x), \phi(y)) + 2\epsilon$. Consequently, $\phi(x+y) \leq \max(\phi(x), \phi(y))$. Thus, ϕ is an ultrametric semi-norm of $\mathbb{L}$-vector space. Now, suppose first that ϕ belongs to the closure of $\mathrm{Mult}(A)$. Then $|\phi(xy) - \phi(x)\phi(y)|_\infty \leq |\phi(xy) - \psi(xy)|_\infty + |\psi(x)\psi(y) - \phi(x)\psi(y)|_\infty + |\phi(x)\psi(y) - \phi(x)\phi(y)|_\infty \leq \epsilon(1 + 2m + \epsilon)$, therefore $|\phi(xy)|_\infty = |\phi(x)\phi(y)|_\infty$. This finishes showing that ϕ belongs to $\mathrm{Mult}(A)$. Consequently, $\mathrm{Mult}(A)$ is closed in $\mathcal{F}(A, \mathbb{R}_+)$. Now, assume that ϕ belongs to the closure of $SM(A)$. Similarly, $|\phi(x^n) - \phi(x)^n|_\infty \leq |\phi(x^n) - \psi(x^n)|_\infty + |\psi(x)^n - \phi(x)^n|_\infty \leq 2\epsilon$ therefore $|\phi(x^n)|_\infty = |\phi(x)^n|_\infty$. Consequently, $SM(A)$ is closed in $\mathcal{F}(A, \mathbb{R}_+)$. $\qquad \square$

Lemma 2.3.6. *Let A be equipped with a semi-norm $\| \cdot \|$ of $\mathbb{E}$-algebra and let $\varphi \in SM(A)$. If φ belongs to $SM(A, \| \cdot \|)$ then $\varphi(x) \leq \| x \|$ whenever $x \in A$.*

Moreover, if $\varphi \in \mathrm{Mult}(A, \| \cdot \|)$, then $\mathrm{Ker}(\varphi)$ is a prime closed ideal of A.

Proof. Suppose that for some $x \in A$ we have $\varphi(x) > \| x \|$. Since the absolute value $\mathbb{E}$ is not trivial, it contains a subgroup of the form $\{a^n \mid n \in \mathbb{Z}\}$, with $a \in R_+^*$, $a > 1$. We can find $q \in \mathbb{N}$ such that $q \log(\|x\|) < \log a \leq \log(\varphi(x))$, thereby there exists $\lambda \in \mathbb{E}$ such that $\|x^q\| < \lambda \leq \varphi(x^q)$. Putting $t = \frac{x^q}{\lambda}$, we have $\varphi(t) \geq 1 > \|t\|$, and therefore φ is not continuous. $\qquad \square$

The second assessment is immediate.

Corollary 2.3.7. *Let A be equipped with a semi-norm $\| \cdot \|$ of $\mathbb{E}$-algebra. Then $\mathrm{Mult}(A, \| \cdot \|)$ is closed in $\mathcal{F}(A, \mathbb{R}_+)$.*

Proof. Indeed, by Lemma 2.3.5, $\mathrm{Mult}(A)$ is closed in $\mathcal{F}(A, \mathbb{R})$. But consider ϕ in the closure of $\mathrm{Mult}(A, \| \cdot \|)$. Let $x \in A$ and let ϵ be > 0. There exists $\psi \in \mathrm{Mult}(A, \| \cdot \|)$ such that $\phi(x) \leq \psi(x) + \epsilon \leq \|x\| + \epsilon$, hence finally $\phi(x) \leq \|x\|$. $\qquad \square$

2.4. Shilov boundary

Definitions and notations: A set F equipped with an order relation $\leq$ is said to be *well ordered* with respect to the inverse order $\geq$ if every subset of F admits a maximum element, and then the order $\geq$ is called *a good order.*. Let $\geq$ be a good order on the set F and let $x \in F$. If there exists $y < x$, we call *follower of x* the element $\sup\{y \in F \mid y < x\}$. If there exists $z \in F$ such that x is the follower of z, then z is called *precedent of x*.

Let A be equipped with a $\mathbb{E}$-algebra semi-multiplicative semi-norm $\| \, . \, \|$. A subset F of $\mathrm{Mult}(A, \| \, . \, \|)$ will be called a *boundary for* $(A, \| \, . \, \|)$ if for every $x \in A$, there exists $\psi \in F$ such that $\psi(x) = \|x\|$. A closed boundary (with respect to the topology of pointwise convergence on A) is called *Shilov boundary for* $(A, \| \, . \, \|)$ if it is the smallest of all closed boundaries for $(A, \| \, . \, \|)$ with respect to inclusion.

Let S be a subsemi-group of B. A sub-multiplicative function θ on A will be said to be *S-multiplicative* if $\theta(xy) = \theta(x)\theta(y) \; \forall x, y \in S$. Let μ be a S-multiplicative function such that $\mu(s) \neq 0 \; \forall s \in S$. We will denote by μ^S the function defined in A by $\mu^S(x) = \inf\{\frac{\mu(sx)}{\mu(s)} | \; s \in S\}$.

Let N be a subset of $\mathcal{F}(B, \mathbb{R}_+)$. Then N will be said to be *constructible* if it satisfies the following three properties:

(i) N is well ordered with respect to the order $\geq$.
(ii) For all $\mu \in N$ whose follower is μ' (with respect to the good order $\geq$), there exists a subsemi-group S of $(A, .)$ such that μ is S-multiplicative and such that $\mu' = \mu^S$.
(iii) For every $\mu \in N$ having no precedent, μ is equal to $\inf\{\nu \in N \mid \mu < \nu\}$.

Let θ, ν be semi-multiplicative functions on B. Then ν will be said to be *constructible from* μ if there exists a constructible set N admitting μ as its first element and ν as its last element, with respect to the good order on N.

We will denote by $\mathrm{Min}(B, \theta)$ the set of multiplicative functions $\phi \in \mathcal{B}(B, \theta)$ which are constructible from θ and by $Z(B, \theta)$ the set of the $\phi \in \mathcal{B}(B, \theta)$ which are constructible from θ. Given a subsemigroup S of B we will denote by $Z(B, S, \theta)$ the set of $\phi \in Z(B, \theta)$ which are S-multiplicative.

Lemma 2.4.1 is obvious, and comes from the definition of constructible sets.

Lemma 2.4.1. *Let X be a constructible subset of $\mathcal{F}(B, \mathbb{R}_+)$ and let $\zeta = \inf\{\phi \in X\}$. Then $X \cup \{\zeta\}$ is a constructible subset of $\mathcal{F}(B, \mathbb{R}_+)$. Let $\psi \in X$ and let Y be a subset of X such that $\inf(Y) \geq \psi$. Then $\inf(Y)$ belongs to X.*

Lemma 2.4.2. *Let S be a subsemi-group of $(A, .)$ and let θ be a S-multiplicative $\mathbb{E}$-algebra semi-norm on A. Then θ^S also is a S-multiplicative $\mathbb{E}$-algebra semi-norm on A.*

Proof. We have to show that θ is an $\mathbb{E}$-algebra semi-norm. Let $x, y \in A$, and let ϵ be > 0. We can find $s, t \in S$ such that

$\frac{\theta(sx)}{\theta(s)} < \theta^S(x) + \epsilon$, $\frac{\theta(ty)}{\theta(t)} < \theta^S(y) + \epsilon$. On the other hand, since θ is a E-algebra semi-norm and is S-multiplicative, we check that

$$\theta^S(x+y) \leq \frac{\theta(st(x+y))}{\theta(st)} \leq \frac{\theta(stx) + \theta(sty)}{\theta(st)} \leq \frac{\theta(t)\theta(sx) + \theta(s)\theta(by)}{\theta(s)\theta(t)}$$

$$= \frac{\theta(sx)}{\theta(s)} + \frac{\theta(ty)}{\theta(t)}. \qquad (2.4.1)$$

Consequently, we have $\theta^S(x+y) \leq \theta^S(x) + \theta^S(y) + 2\epsilon$. Since ϵ is arbitrary, we have proven that $\theta^S(x+y) \leq \theta^S(x) + \theta^S(y)$. Next,

$$\theta^S(xy) \leq \frac{\theta(stxy)}{\theta(st)} \leq \left(\frac{\theta(sx)}{\theta(s)}\right)\left(\frac{\theta(ty)}{\theta(t)}\right) \leq (\theta^S(x) + \epsilon)(\theta^S(y) + \epsilon)$$

hence $\theta^S(xy) \leq \theta^S(x)\theta^S(y)$. Thus, we have proven that θ^S is an E-algebra semi-norm. Then, it is obviously seen that it is S-multiplicative in the same way as θ. □

Proposition 2.4.3. *Let $\mathcal{N}$ be a totally ordered family of E-algebra semi-norms of A and let ϕ be the function defined on A as $\phi(x) = \inf\{\varphi(x) \mid \varphi \in \mathcal{N}\}$. Then ϕ is a E-algebra semi-norm.*

Proposition 2.4.4. *Let $\| . \|$ be a semi-multiplicative E-algebra semi-norm of A. Then every element of $Z(A, \| . \|)$ is a E-algebra semi-norm.*

Proof. Suppose Proposition 2.4.4 is not true. Let $\phi \in Z(A, \| . \|)$ which is not a E-algebra semi-norm. Let $\mathcal{T}$ be a constructible set admitting $\| . \|$ as first element and ϕ as last element. Since $\mathcal{T}$ is well ordered, the subset $\mathcal{S}$ of the $\psi \in \mathcal{T}$ which are not E-algebra semi-norms admits a maximum element θ. If θ admits a precedent ξ, then there exists a subsemigroup S such that ξ is S-multiplicative and satisfies $\theta = \xi^S$. But by hypothesis ξ is a E-algebra semi-norm, and by Lemma 2.4.2 so is θ, a contradiction. Consequently, θ has no precedent, and then we have $\theta = \inf\{\psi \in \mathcal{T} \mid \theta < \psi\}$. But $\mathcal{T}$ is all ordered. Therefore, by Proposition 2.4.3, θ is a E-algebra semi-norm because so are all $\psi \in \mathcal{T}$ such that $\theta < \psi$. □

Corollary 2.4.5. *Let $\| . \|$ be a semi-multiplicative E-algebra semi-norm of A. Then $\mathrm{Min}(A, \| . \|)$ is included in $\mathrm{Mult}(A, \| . \|)$.*

Lemma 2.4.6. *Let S be a subsemi-group of B and let $\theta \in \mathcal{F}(B, \mathbb{R}_+)$ be S-multiplicative. There exists $f \in \mathrm{Min}(B, \theta)$ such that $f(x) = \theta(x) \, \forall x \in S$.*

Proof. Let $\mathcal{H}$ be the set of constructible subsets of $Z(B, S, \theta)$. Clearly, $Z(B, S, \theta)$ is not empty, hence $\mathcal{H}$ isn't either. On $\mathcal{H}$ we denote by $\preceq$ the order relation defined as $N \preceq N'$ if N is a beginning section of of N'. Then $\mathcal{H}$ is inductive for its order. Let N_0 be a maximal element of $\mathcal{H}$ and let $f_0 = \inf\{\sigma \mid \sigma \in N_0\}$. Clearly f_0 lies in $Z(B, S, \theta)$. The family G of subsemi-groups $J \supset S$ such that f_0 is J-multiplicative and satisfies $f(s) \neq 0 \ \forall s \in J$ is an inductive family with respect to inclusion. Let T_0 be a maximal element of G. Suppose $T_0 \neq B \setminus \mathrm{Ker}(f_0)$ and let $x \in B \setminus (T_0 \cup \mathrm{Ker}(f_0))$. Let T_1 be the subsemi-group generated by x and T_0. Then f_0 is not T_1-multiplicative. But since f_0 is sub-multiplicative and since T_0 is maximal, there exists $t \in T_0$ such that $f_0(tx) < f_0(t)f_0(x)$, hence $f_0^{T_0} \neq f_0$. On the other hand, since $f_0^{T_0}$ obviously is S-multiplicative, $f_0^{T_0}$ belongs to $Z(B, S, \theta)$. Hence, we can check that $N_0 \cup \{f_0^{T_0}\}$ is a constructible subset of $Z(B, S, \theta)$, a contradiction to the hypothesis "N_0 is maximal". Consequently, $T_0 = B \setminus \mathrm{Ker}(f_0)$. Thus, f_0 is T_0-multiplicative. On the other hand, f_0 trivially satisfies $f(xy) \leq f(x)f(y) = 0 \ \forall x \in \mathrm{Ker}(f_0), y \in B$, hence $f(xy) = f(x)f(y) \ \forall x \in \mathrm{Ker}(f_0), y \in B$. This finishes proving that f_0 is B-multiplicative, and therefore belongs to $\mathrm{Min}(B, \theta)$. Finally, by definition, all elements of $Z(B, S, \theta)$ satisfies $f(x) = \theta(x) \ \forall x \in S$, so f_0 is the f we looked for. $\square$

Theorem 2.4.7. *Let $\| \, . \, \|$ be a semi-multiplicative $\mathbb{E}$-algebra semi-norm of A. Then $\mathrm{Min}(A, \| \, . \, \|)$ is not empty and is a boundary for $(A, \| \, . \, \|)$.*

Proof.   Let $x \in A$ and let S_x be the subsemigroup generated by x in A. Since $\| \, . \, \|$ is semi-multiplicative, $\| \, . \, \|$ is obviously S_x-multiplicative. Hence by Lemma 2.4.6, there exists $f \in \mathrm{Min}(A, \| \, . \, \|)$ such that $f(x) = \|x\|$, and by Corollary 2.4.5, $\mathrm{Min}(A, \| \, . \, \|)$ is included in $\mathrm{Mult}(A, \| \, . \, \|)$, hence it is a boundary for $(A, \| \, . \, \|)$. $\square$

Notation: Let γ be a semi-group homomorphism from B into B'. We will denote by $\overline{\gamma}$ the mapping from $\mathcal{F}(B', \mathbb{R})$ into $\mathcal{F}(B, \mathbb{R})$ defined by $\overline{\gamma}(\phi) = \phi \circ \gamma, \ \forall \phi \in \mathcal{F}(B', \mathbb{R})$.

Lemma 2.4.8. *Let γ be a semi-group homomorphism from B into B' and let θ (resp., θ') be semi-multiplicative functions such that $\theta = \theta' \circ \gamma$. The restriction of $\overline{\gamma}$ to $Z(B', \theta')$ is a surjection onto $Z(B, \theta)$. Moreover, $\mathrm{Min}(B, \theta) \subset \overline{\gamma}(\mathrm{Min}(B', \theta'))$.*

Proof.   Suppose Proposition 2.4.8 is not true. Then, there exists $\nu \in Z(B, \theta)$ such that $\nu \neq \phi \circ \gamma \ \forall \phi \in Z(B', \theta')$. Let $\mathcal{T}$ be a constructible ordered set admitting θ as first element and ν as last element. Let $\mathcal{S}$ be the set of $\phi \in \mathcal{T}$ such that $\phi \neq \nu' \circ \gamma \ \forall \nu' \in Z(B', \theta)$, let ψ be the maximum element of $\mathcal{S}$, and let $\mathcal{N}$ be the set of $\phi \in \mathcal{T}$ which are of the form $\phi' \circ \gamma$, with $\phi' \in Z(B', \theta')$ and satisfy $\phi \geq \psi$. Let $\mathcal{Q}$ be the family of constructible subsets of $Z(B', \theta')$

admitting θ' as a first element, consisting of ϕ' such that $\phi' \circ \gamma \in \mathcal{N}$. Then $\mathcal{Q}$ is inductive. Let X be a maximal element of $\mathcal{Q}$ and let $\zeta = \inf\{\phi \in X\}$. By Lemma 2.4.1, $X \cup \{\zeta\}$ is a constructible subset of $Z(B', \theta')$. In $\mathcal{T}$, we notice that $\inf\{\xi \circ \gamma \in \mathcal{T} \mid \xi \in X\} = \zeta \circ \gamma$, hence of course $\psi \leq \zeta \circ \gamma$. Then, since $\psi \in \mathcal{T}$, by Lemma 2.4.1, $\zeta \circ \gamma$ belongs to $\mathcal{T}$, and therefore belongs to $\mathcal{N}$. But then, by Lemma 2.4.1, $X \cup \{\zeta\}$ belongs to $\mathcal{Q}$, hence $\zeta \in X$ because X is maximal. Consequently, $\zeta \circ \gamma > \psi$, and hence, $\zeta \circ \gamma$ has follower ν such that $\psi \leq \nu$. Let S be a sub-semigroup of B such that $\nu = (\zeta \circ \gamma)^S$ and let $\nu' = \zeta^{\gamma(S)}$. Then $\nu = \nu' \circ \gamma$. On the other hand, ν' is the follower of ζ in the set $X' = X \cup \{\nu'\}$, hence X' is a constructible set, so X' clearly belongs to $\mathcal{Q}$, a contradiction with the hypothesis X maximal. This proves that for every $\nu \in Z(B, \theta)$ there exists $\nu' \in Z(B', \theta')$ such that $\nu = \nu' \circ \gamma$.

Now, we can show that $\mathrm{Min}(B, \theta) \subset \overline{\gamma}(\mathrm{Min}(B', \theta'))$. Let $\zeta \in \mathrm{Min}(B, \theta)$. As we just showed, there exists $\xi \in Z(B', \theta')$ such that $\zeta = \xi \circ \gamma$. Let $Y = \gamma(B) \backslash \xi^{-1}(0)$. Then Y is a subsemi-group of $(B', .)$ and we can check that ξ is Y-multiplicative. By Lemma 2.4.6, there exists $\zeta' \in \mathrm{Min}(B', \xi)$ such that $\zeta'(x) = \zeta(x) \, \forall x \in Y$, hence $\zeta = \zeta' \circ \gamma$. But of course $\mathrm{Min}(B', \xi) \subset \mathrm{Min}(B', \theta')$, hence f belongs to $\overline{\gamma}(\mathrm{Min}(B', \theta'))$. $\qquad\square$

We must recall two classical results [6].

Theorem 2.4.9 (Urysohn). *Let D be a compact and let α, β be two different points of D. There exists a continuous mapping f from D into $\mathbb{R}_+$ such that $f(\alpha) = 0$ and $f(\beta) = 1$.*

Lemma 2.4.10. *Let D be a compact. The Gelfand mapping θ from D into $\mathcal{X}(\mathcal{G}(D, \mathbb{R}), \mathbb{R})$ defined as $\theta(a) = \phi_a$, with $\phi_a(f) = f(a)$, $f \in \mathcal{G}(D, \mathbb{R})$ actually is a bijection from D onto $\mathcal{X}(\mathcal{G}(D, \mathbb{R}), \mathbb{C})$ and the mapping $\overline{\theta}$ defined as $\overline{\theta}(a) = |\phi_a|$, with $\phi_a(f) = |f(a)|$, $f \in \mathcal{G}(D, \mathbb{R})$ is a bijection from D onto $\mathrm{Mult}(\mathcal{G}(D, \mathbb{R}), \| . \|_D)$.*

Lemma 2.4.11. *Let D be a compact, and let ϕ be the mapping from D into $\mathrm{Mult}(\mathcal{G}(D, \mathbb{R}), \| . \|_D)$ defined by $\phi(\alpha)(f) = |f(\alpha)|_\infty$. Then ϕ is a bijection from D onto $\mathrm{Mult}(\mathcal{G}(D, \mathbb{R}), \| . \|_D)$ which is equal to $\mathrm{Min}(\mathcal{G}(D, \mathbb{R}), \| . \|_D)$.*

Proof. By Lemma 2.4.10, the mapping ϕ from D into $\mathrm{Mult}(\mathcal{G}(D, \mathbb{R}), \| . \|_D)$ defined as $\phi(a) = \psi_a$, with $\psi_a(f) = |f(a)|$ $f \in \mathcal{G}(D, \mathbb{R})$, is a bijection from D onto $\mathrm{Mult}(\mathcal{G}(D, \mathbb{R}), \| . \|_D)$. By Corollary 2.4.5, $\mathrm{Min}(\mathcal{G}(D, \mathbb{R}), \| . \|_D)$ is included in $\mathrm{Mult}(\mathcal{G}(D, \mathbb{R}), \| . \|_D)$, so we just have to show that $\mathrm{Mult}(\mathcal{G}(D, \mathbb{R}), \| . \|_D) \subset \mathrm{Min}(\mathcal{G}(D, \mathbb{R}), \| . \|_D)$. Let $\alpha \in D$, and let us show that $\phi(\alpha)$ belongs to $\mathrm{Min}(\mathcal{G}(D, \mathbb{R}), \| . \|_D)$. Let $S = \{f \in \mathcal{G}(D, \mathbb{R}) \mid |f(\alpha)|_\infty = \|f\|_D = 1\}$. Then S is a subsemi-group of $\mathcal{G}(D, \mathbb{R})$ and the norm $\| . \|_D$ is

S-multiplicative, hence by Lemma 2.4.6, there exists $\zeta \in \text{Min}(\mathcal{G}(D, \mathbb{R}), \| . \|_D)$ such that $\zeta(f) = 1 \; \forall f \in S$. But there exists $\beta \in D$ such that $\phi(\beta) = \zeta$, hence $|f(\beta)|_\infty = 1$. Consequently, for every $f \in \mathcal{G}(D, \mathbb{R})$ such that $|f(\alpha)|_\infty = 1$, f must also satisfy $|f(\beta)|_\infty = 1$. Since D is compact, by Theorem 2.4.9, this implies $\alpha = \beta$ hence $\phi(\alpha)$ belongs to $\text{Min}(\mathcal{G}(D, \mathbb{R}), \| . \|_D)$. $\square$

Theorem 2.4.12. *Let $\| . \|$ be a semi-multiplicative $\mathbb{E}$-algebra semi-norm of A. Then there exists a Shilov boundary for $(A, \| . \|)$ which is the closure of $\text{Min}(A, \| . \|)$ in $\text{Mult}(A, \| . \|)$ [25].*

Proof. By Corollary 2.4.5 and Theorem 2.4.7, $\text{Min}(A, \| . \|)$ is not empty, is included in $\text{Mult}(A, \| . \|)$ and is a boundary for $(A, \| . \|)$. Now, let F be a closed boundary for $(A, \| . \|)$ and let H be the multiplicative semi-group $(\mathcal{G}(F, \mathbb{R}), .)$. Since F is a closed subset of $\text{Mult}(A, \| . \|)$, by Lemma 2.4.1, it is a compact. Let γ be the mapping defined on A, taking values in $\mathcal{G}(F, \mathbb{R})$, as $\gamma(x)(\alpha) = \alpha(x)$, $\alpha \in F$. Let $x \in A$.

On the one hand, by Lemma 2.3.6, every element ψ of $\text{Mult}(A, \| . \|)$ by definition satisfies $\|x\| \geq \sup\{\psi(x) \mid \psi \in F\}$. On the other hand, since F is a boundary for $(A, \| . \|)$ there exists $\varphi \in F$ such that $\|x\| = \sup\{\psi(x) \mid \psi \in F\}$, and consequently, $\|\gamma(x)\|_F = \|x\|$.

Now, since $\| . \|_F \circ \gamma = \| . \|$ and since γ is a semi-group homomorphism from $(A, .)$ into $(\mathcal{G}(F, \mathbb{R}), .)$, we can apply Lemma 2.4.8 and we have $\text{Min}(A, \| . \|) \subset \overline{\gamma}(\text{Min}(\mathcal{G}(F, \mathbb{R}), \| . \|_F))$. Let $\nu \in \text{Min}(A, \| . \|)$. So, there exists $\psi \in \text{Min}(\mathcal{G}(F, \mathbb{R}), \| . \|_F)$ such that $\nu = \psi \circ \gamma$. Now, let ϕ be the mapping from F into $\text{Mult}(\mathcal{G}(F, \mathbb{R}), \| . \|_F)$ defined as $\phi(\alpha)(f) = |f(\alpha)|_\infty$, $f \in \mathcal{G}(F, \mathbb{R})$. By Lemma 2.4.11, ϕ is a bijection from F onto $\text{Min}(\mathcal{G}(F, \mathbb{R}), \| . \|_F)$. Consequently, $\psi \circ \gamma$ is of the form $\phi(\alpha)$. Finally, for every $x \in A$ we have $\nu(x) = \overline{\gamma}(\phi(\alpha))(x) = (\phi(\alpha) \circ \gamma)(x) = |\gamma(x)(\alpha)|_\infty = |\alpha(x)|_\infty = \alpha(x)$, because by definition α lies in $\text{Mult}(A, \| . \|)$. Thus, ν belongs to F. Therefore $\text{Min}(A, \| . \|)$ is included in F and then the closure of $\text{Min}(A, \| . \|)$ is the smallest of all closed boundaries for $(A, \| . \|)$. $\square$

Corollary 2.4.13. *Let $\| . \|$ be a semi-multiplicative $\mathbb{L}$-algebra semi-norm of A. For every $x \in A$, there exists $\varphi \in \text{Mult}(A, \| . \|)$ such that $\varphi(x) = \|x\|$.*

2.5.   Spectral semi-norm

We shall recall the basic properties of continuous multiplicative semi-norms in an ultrametric normed algebra: the set of continuous multiplicative semi-norms is compact for the topology of pointwise convergence, and its superior envelope is a semi-multiplicative semi-norm called the spectral semi-norm [8, 31]. Applying Section 2.4, we will show that the set of continuous multiplicative semi-norms admits a Shilov boundary.

Notations: Throughout the chapter, $\mathbb{L}$ will denote a complete field with respect to a non-trivial ultrametric absolute value, and we will denote by $(A, \| \cdot \|)$, $(A', \| \cdot \|')$ a unital commutative normed $\mathbb{L}$-algebra. We will show the existence of a spectral semi-norm equal to the supremum of continuous multiplicative semi-norms [31].

ϕ will be a $\mathbb{L}$-algebra homomorphism from A into A' and we will denote by $^*\phi$ the mapping from $\mathrm{Mult}(A')$ into $\mathrm{Mult}(A)$ defined as $^*\phi(\varphi) = \varphi \circ \phi$.

Lemma 2.5.1. *For every subset W of $\mathrm{Mult}(A, \| \cdot \|)$ the mapping ϕ defined on A by $\phi(x) = \sup_{\varphi \in W} \varphi(x)$ belongs to $SM(A, \| \cdot \|)$. Moreover, if A has a unity u, and if φ is not identically 0, then $\varphi(\lambda u) = |\lambda|$ whenever $\lambda \in \mathbb{L}$.*

Proof. Given $W \subset SM(A, \| \cdot \|)$, clearly we have $\sup_{\varphi \in W} \varphi(x) \le \|x\|$, and therefore $\sup_{\varphi \in W} \varphi(x)$ belongs to $SM(A, \| \cdot \|)$. Finally, let u be the unity in A. Either $\varphi(u) = 0$ and then $\varphi(x) = 0$ whenever $x \in A$, or $\varphi(u) = 1$ and then we have $\varphi(\lambda u) = |\lambda|\varphi(u) = |\lambda|$ whenever $\lambda \in \mathbb{L}$. □

Lemma 2.5.2 is obvious.

Lemma 2.5.2. *Let ϕ be a $\mathbb{L}$-algebra homomorphism from A into another unital commutative $\mathbb{L}$-algebra A'. For each $\varphi \in \mathrm{Mult}(A')$, $\varphi \circ \phi$ belongs to $\mathrm{Mult}(A)$.*

Lemma 2.5.3. *If ϕ is surjective then $^*\phi$ is injective. If ϕ is an isomorphism, then $^*\phi$ is a homeomorphism from $\mathrm{Mult}(A')$ onto $\mathrm{Mult}(A)$ with respect to topologies of pointwise convergence on both sets.*

Proof. Suppose ϕ is surjective and let φ_1, $\varphi_2 \in \mathrm{Mult}(A')$ be such that $^*\phi(\varphi_1) = {}^*\phi(\varphi_2)$. Then $\varphi_1(\phi(f)) = \varphi_2(\phi(f))$ $\forall f \in A$. But since ϕ is surjective we have $\varphi_1(g) = \varphi_2(g)$ $\forall g \in A'$, hence $\varphi_1 = \varphi_2$.

Now suppose that ϕ is an isomorphism. It is obviously seen that $(^*\phi)^{-1} = {}^*(\phi^{-1})$ and that $^*\phi$ transforms the filter of neighborhoods of any $\varphi \in \mathrm{Mult}(A')$ into this of $\varphi \circ \phi$ in $\mathrm{Mult}(A)$. □

Lemma 2.5.4. *Let A be a direct product $A_1 \times \cdots \times A_q$ whose norm is the product norm of the norm $\| \cdot \|_j$ of algebras A_j, respectively. Then the mapping Ξ defined on $\bigcup_{j=1}^q \mathrm{Mult}(A_j)$ as $\Xi(\psi_j)(x_1, \dots x_q) = \psi_j(x_j)$ is a bijection from $\bigcup_{j=1}^q \mathrm{Mult}(A_j)$ onto $\mathrm{Mult}(A, \| \cdot \|)$.*

Proof. By Lemma 2.5.3 it is an injection. Putting $u_1 = (1, 0, \ldots, 0), \ldots,$ $u_q = (0, \ldots, 0, 1)$, for each $j = 1, \ldots, q$, we check that for every $\varphi \in \mathrm{Mult}(A, \| \cdot \|)$ we have $\varphi(u_k) = 1$ for a certain k and $\varphi(u_j) = 0 \ \forall j \neq k$. Consequently such a φ is of the form $\Xi(\psi_k)$, which shows that this mapping is surjective onto $\mathrm{Mult}(A, \| \cdot \|)$.

Lemma 2.5.5 is immediate. $\qquad \square$

Lemma 2.5.5. *Let ϕ be continuous. The restriction of $^*\phi^*$ to $\mathrm{Mult}(A', \| \cdot \|')$ takes values in $\mathrm{Mult}(A, \| \cdot \|)$.*

Notation: In the hypothesis of Lemma 2.5.5 we will denote by $^*\phi^*$ the restriction of $^*\phi$ to $\mathrm{Mult}(A', \| \cdot \|')$.

Recall that $\mathrm{Mult}(A, \| \cdot \|)$ is equipped with the topology of pointwise convergence [6, 28, 32], which means that a basic neighborhood of some $\psi \in \mathrm{Mult}(A, \| \cdot \|)$ is a set of the form $\mathcal{W}(\psi, f_1, \ldots, f_q, \epsilon)$, with $f_j \in A$ and $\epsilon > 0$ and this is the set of $\phi \in \mathrm{Mult}(A, \| \cdot \|)$ such that $|\psi(f_j) - \phi(f_j)|_\infty \leq \epsilon \ \forall j = 1, \ldots, q$.

Theorem 2.5.6. *Suppose that A' is a dense $\mathbb{L}$-subalgebra of A and assume that ϕ is continuous. Then $^*\phi$ is injective. Moreover if ϕ is injective and bicontinuous, then $^*\phi^*$ is a homeomorphism from $\mathrm{Mult}(A', \| \cdot \|')$ onto $\mathrm{Mult}(A, \| \cdot \|)$.*

Proof. Let $A'' = \phi(A)$. Let $\varphi_1, \varphi_2 \in \mathrm{Mult}(A, \| \cdot \|)$ satisfy $\varphi_1 \circ \phi = \varphi_2 \circ \phi$. Then $\varphi_1(\phi(f)) = \varphi_2(\phi(f)) \ \forall f \in A$, hence $\varphi_1(g) = \varphi_2(g) \ \forall g \in A''$. But since A'' is dense in A, and since φ_1, φ_2 are continuous with respect to the norm of A', we conclude that $\varphi_1(g) = \varphi_2(g) \ \forall g \in A'$, hence $\varphi_1 = \varphi_2$.

Now, suppose ϕ is a bicontinuous injection from A into A'. Let $\psi \in \mathrm{Mult}(A, \| \cdot \|)$. We can define $\varphi \in \mathrm{Mult}(A'', \| \cdot \|')$ as $\varphi(\phi(f)) = \psi(f)$ and extend by continuity the definition of φ to A'. Since ϕ is bicontinuous from $(A, \| \cdot \|)$ onto $(A'', \| \cdot \|')$, φ belongs to $\mathrm{Mult}(A'', \| \cdot \|')$ and therefore its expansion to A' belongs to $\mathrm{Mult}(A', \| \cdot \|')$. Thus, $^*\phi^*$ is a bijection from $\mathrm{Mult}(A', \| \cdot \|')$ onto $\mathrm{Mult}(A, \| \cdot \|)$. We will check that $^*\phi^*$ is bicontinuous. Let $\varphi \in \mathrm{Mult}(A', \| \cdot \|')$ and let $\psi = \varphi \circ \phi$. Let $\mathcal{V}(\varphi \circ \phi, f_1, \ldots, f_n, \epsilon)$ be a neighborhood of $\varphi \circ \phi$. In $\mathrm{Mult}(A', \| \cdot \|')$, $\mathcal{W}(\psi, \phi \circ f_1, \ldots, \phi \circ f_n, \epsilon)$ is a neighborhood of ψ whose image by $^*\phi^*$ is just $\mathcal{W}(\varphi \circ \phi, f_1, \ldots, f_n, \epsilon)$. Consequently $^*\phi^*$ is continuous.

Conversely, let $\mathcal{W}(\varphi, g_1, \ldots, g_n, \epsilon)$ be a neighborhood of φ. Since the multiplicative semi-norms are ultrametric and since A'' is dense in A' we can find $g'_1, \ldots, g'_n \in A''$ such that $\mathcal{W}(\varphi, g_1, \ldots, g_n, \epsilon) = \mathcal{W}(\varphi, g'_1, \ldots, g'_n, \epsilon)$.

Each g_i' is of the form $\phi(f_i)$ $(1 \leq i \leq n)$, so $\mathcal{W}(\varphi \circ \phi, f_1, \ldots, f_n, \epsilon) = (^*\phi^*)^{-1}(\mathcal{W}(\varphi, g_1', \ldots, g_n', \epsilon))$, which finishes proving that $(^*\phi^*)^{-1})$ also is bicontinuous. $\qquad\square$

Notation: When A' is a dense $\mathbb{L}$-subalgebra of A henceforth, we will consider that $\mathrm{Mult}(A, \| \cdot \|)$ and $\mathrm{Mult}(A', \| \cdot \|)$ are equal.

Theorem 2.5.7. *For every $x \in A$ the sequence $\left(\| \, x^n \, \|^{\frac{1}{n}} \right)_{n \in \mathbb{N}}$ has a limit denoted by $\| \, x \, \|_{\mathrm{sp}}$, satisfying $\|x\|_{\mathrm{sp}} \leq \|x\|$ $\forall x \in A$ and the mapping f defined in A as $f(x) = \| \, x \, \|_{\mathrm{sp}}$ belongs to $SM(A, \| \cdot \|)$ and is ultrametric. Moreover $\| \, x \, \|_{\mathrm{sp}} < 1$ if and only if $\lim_{n \to \infty} \|x^n\| = 0$.*

Proof. The statement is well known and proven, for instance, in [28] as Theorem 1.13 when the semi-norm $\| \cdot \|$ is a norm. Actually the proof holds all the same when it is just a semi-norm. $\qquad\square$

By Theorem 2.4.12, we have Corollary 2.5.8.

Corollary 2.5.8. *There exists a Shilov boundary for $(A, \| \cdot \|_{\mathrm{sp}})$ which is included in $\mathrm{Mult}(A, \| \cdot \|)$ and is equal to the closure of $\mathrm{Min}(A, \| \cdot \|_{\mathrm{sp}})$ in $\mathrm{Mult}(A, \| \cdot \|)$.*

Theorem 2.5.9. *Let $f \in A$. Then $\|f\|_{\mathrm{sp}} = 0$ if and only if $\lim_{n \to \infty} \|(\lambda f)^n\| = 0$ $\forall \lambda \in \mathbb{L}$.*

Proof. If $\|f\|_{\mathrm{sp}} = 0$, then of course $\|\lambda f\|_{\mathrm{sp}} = 0$, hence $\lim_{n \to \infty} \|(\lambda f)^n\| = 0$ $\forall \lambda \in \mathbb{L}$. Now, suppose $\lim_{n \to \infty} \|(\lambda f)^n\| = 0$ $\forall \lambda \in \mathbb{L}$. Suppose $\|f\|_{\mathrm{sp}} \neq 0$, and let $\lambda \in E$ be such that $|\lambda| > \frac{1}{\|f\|_{\mathrm{sp}}}$. Then $\lim_{n \to \infty} \|(\lambda f)^n\|_{\mathrm{sp}} = +\infty$, which contradicts the hypothesis $\lim_{n \to \infty} \|(\lambda f)^n\| = 0$. $\qquad\square$

Definition: $(A, \| \cdot \|)$ will be said to be *uniform* if the norm $\| \cdot \|$ is equivalent to $\| \cdot \|_{\mathrm{sp}}$.

In the same way as in Archimedean analysis, we must also mention this proposition:

Proposition 2.5.10. *$\| \cdot \|$ and $\| \cdot \|_{\mathrm{sp}}$ are two equivalent norms on A if and only if there exists $C \in \mathbb{R}_+$ such that $\|x^2\| \leq C\|x\|^2$ $\forall x \in A$. And they are equal if and only if $\|x^n\| \leq \|x\|^n$ $\forall n \in \mathbb{N}^*$, $\forall x \in A$.*

Theorem 2.5.11. *Let $\| \cdot \|$ be a norm of $\mathbb{L}$-algebra on A. Then $\| \cdot \|$ is semi-multiplicative if and only if the semi-norm $\| \cdot \|_{\mathrm{sp}}$ associated to $\| \cdot \|$ is equal to $\| \cdot \|$.)*

On the other hand, if A is complete for its norm, then for every $x \in A$ satisfying $\|x\|_{\mathrm{sp}} < 1$, $1 - x$ is invertible in A, the set of invertible elements

*in A is open, each maximal ideal $\mathcal{M}$ of A is closed. Finally, for every $\phi \in$
$\mathcal{X}(A, \mathbb{L})$, the mapping $|\psi|$ from A into $\mathbb{R}_+$ defined as $|\phi|(x) = |\phi(x)|$ belongs
to $\mathrm{Mult}(A, \| \cdot \|_{\mathrm{sp}})$.*

Proof. By Theorem 2.5.7, if $\| \cdot \|_{\mathrm{sp}}$ is the $\mathbb{L}$-algebra norm, then it is semi-multiplicative. Conversely, if the $\mathbb{L}$-algebra norm is semi-multiplicative, it is equal to $\| \cdot \|_{\mathrm{sp}}$ because $\| x^n \|^{\frac{1}{n}} = \|x\|$ for every $x \in A$. $\qquad \square$

Now assume that A is complete. Let $x \in A$ satisfy $\|x\|_{\mathrm{sp}} < 1$. By Theorem 2.5.7, we have $\lim_{n \to \infty} x^n = 0$ and then the series $\sum_{n=0}^{\infty} x^n$ converges in A to a limit S that obviously satisfies $S(1 - x) = 1$. As a consequence, the set of invertible elements in A is open. Since a maximal ideal is either closed or dense, here each maximal ideal of A is closed. Moreover, for every $\lambda \in \mathbb{L}$ satisfying $\|x\|_{\mathrm{sp}} < |\lambda|$, $1 - \frac{x}{\lambda}$ is invertible in A.

Now, $|\psi|$ obviously belongs to $\mathrm{Mult}(A)$. Assume that $|\psi(x)| > \| x \|_{\mathrm{sp}}$ for some $x \in A$. Let $\lambda = \psi(x) \in \mathbb{L}$ and let $y = 1 - \dfrac{x}{\lambda}$. Then $\psi(y) = 0$, hence $1 - \dfrac{x}{\lambda} \in \mathrm{Ker}(\psi)$. But $|\lambda| > \| x \|_{\mathrm{sp}}$ hence $\|\dfrac{x}{\lambda}\|_{\mathrm{sp}} < 1$, so $1 - \dfrac{x}{\lambda}$ is invertible. This just contradicts $1 - \dfrac{x}{\lambda} \in \mathrm{Ker}(\psi)$. Finally let $\phi \in \mathcal{X}(A, \mathbb{L})$. Then $|\phi(x)| \leq \| x \|_{\mathrm{sp}}$ for every $x \in A$, hence $|\phi| \in \mathrm{Mult}(A, \| \cdot \|_{\mathrm{sp}})$.

Theorem 2.5.12 was given in several works [26, 28].

Theorem 2.5.12. *Let F be a field extension of $\mathbb{L}$ equipped with a $\mathbb{L}$-algebra semi-norm $\| \cdot \|$. Then $\| \cdot \|$ is a $\mathbb{L}$-algebra norm and there exists a unique ultrametric absolute value φ on F extending that of $\mathbb{L}$, such that $\varphi(x) \leq \|x\|$ whenever $x \in F$.*

Theorem 2.5.13 is an immediate consequence ([26, Theorem 6.5]).

Theorem 2.5.13. *Let $(A, \| \cdot \|)$ be complete. For every maximal ideal $\mathcal{M}$ of A, there exists $\varphi \in \mathrm{Mult}(A, \| \cdot \|)$ such that $\mathrm{Ker}(\varphi) = \mathcal{M}$, φ is of the form $|\chi|$, with $\chi \in A$ and $| \cdot |$ an absolute value on the field $\chi(A)$ expanding that of $\mathbb{L}$.*

Remark 1. As shown in [9, 23, Theorem 40.5] there exist uniform ultrametric Banach unital commutative algebras without divisors of zero, admitting no continuous absolute values and therefore, there exist Banach ultrametric unital commutative algebras admitting closed prime ideal which are not the kernel of any continuous multiplicative semi-norm. Thus, as far as ultrametric Banach algebras are concerned, Theorem 2.5.13 is specific to maximal ideals, in the general case.

Corollary 2.5.14. *Given $x \in A$, then $|sp(x)| \subset \{\phi(x) \mid \phi \in \mathrm{Mult}_m(A, \|\cdot\|)\}$.*

Theorem 2.5.15. *Let F, G be a partition of $\mathrm{Mult}_m(A, \|\cdot\|)$ and suppose that there exists an idempotent u such that $\phi(u) = 1 \; \forall \phi \in F$ and $\phi(u) = 0 \; \forall \phi \in G$. Then u is unique satisfying these relations.*

Proof. Let $\phi \in F$. By Theorem 2.5.13, ϕ is of the form $|\chi|$ with $\chi \in \mathcal{X}(A)$, the field image of A by χ being equipped with an absolute value extending that of $\mathbb{L}$. Since $\chi(u)$ must be equal to 0 or 1, we have $\chi(u) = 1$. And similarly, if $\phi \in G$, it is of the form $|\chi|$ with $\chi \in \mathcal{X}(A)$, and of course $\chi(u) = 0$. Consequently, by Lemma 1.1.14, u is unique to satisfy $\chi(u) = 1 \; \forall \chi \in F$, $\chi(u) = 0 \; \forall \chi \in G$. $\qquad\square$

Proposition 2.5.16 is classical (in particular, we can obtain it by Proposition 2.2.1).

Proposition 2.5.16. *Let A be a normed $\mathbb{L}$-algebra whose norm $\|\cdot\|$ is ultrametric, let $r \in]0, +\infty[$ and let B be the set of the series $\sum_{n=0}^{\infty} a_n x^n$ such that $\lim_{n\to\infty} \|a_n\| r^n = 0$. Then B, equipped with the multiplication of series, is a unital algebra which contains A and admits 1 for unity. Let $\|\cdot\|_r$ be defined on B by $\|\sum_{n=0}^{\infty} a_n x^n\|_r = \sup_{n\in\mathbb{N}} \|a_n\| r^n$. Then $\|\cdot\|_r$ is a norm of $\mathbb{L}$-algebra on B. Moreover, if the norm of $\mathbb{L}$ is multiplicative, then so is $\|\cdot\|_r$. Further, if A is complete, then so is B.*

Theorem 2.5.17. *Given $\varphi \in \mathrm{Mult}(A)$, φ belongs to $\mathrm{Mult}(A, \|\cdot\|)$ if and only if it satisfies $\varphi(t) \leq \|t\|_{\mathrm{sp}}$ whenever $t \in A$, and we have $\mathrm{Mult}(A, \|\cdot\|) = \mathrm{Mult}(A, \|\cdot\|_{\mathrm{sp}})$. Further, for every $t \in A$ we have*

$$\|t\|_{\mathrm{sp}} = \sup\{\varphi(t) \mid \varphi \in \mathrm{Mult}(A, \|\cdot\|)\}$$

and there exists $\varphi \in \mathrm{Mult}(A, \|\cdot\|)$ such that $\varphi(t) = \|t\|_{\mathrm{sp}}$ [30].

Proof. Since $\|t\|_{\mathrm{sp}} \leq \|t\|$ for all $t \in A$, each $\varphi \in \mathrm{Mult}(A)$ satisfying $\varphi(t) \leq \|t\|_{\mathrm{sp}}$ whenever $t \in A$, obviously belongs to $\mathrm{Mult}(A, \|\cdot\|)$. Conversely, suppose that $\varphi \in \mathrm{Mult}(A)$ satisfies $\varphi(t) > \|t\|_{\mathrm{sp}}$ for certain $t \in A$. As in the proof of Lemma 2.5.1, there exists $q \in \mathbb{N}^*$ and $\lambda \in E$ such that $\varphi(t)^q > |\lambda| > \|t\|_{\mathrm{sp}}^q$. Let $u = \frac{t}{\lambda}$. Then we have $\|u\|_{\mathrm{sp}} < 1 < \varphi(u)$, hence by Theorem 2.5.7, $\lim_{n\to\infty} u^n = 0$, whereas $\lim_{n\to\infty} \varphi(u^n) = +\infty$. This shows $\varphi \notin \mathrm{Mult}(A, \|\cdot\|)$. As a consequence, we have $\mathrm{Mult}(A, \|\cdot\|) = \mathrm{Mult}(A, \|\cdot\|_{\mathrm{sp}})$. The last statements come from Corollary 2.5.8. $\qquad\square$

Corollary 2.5.18. *Let* $\mathbb{F}$ *be a complete field extension of* $\mathbb{L}$ *equipped with a* $\mathbb{L}$-*algebra norm* $\| \cdot \|$. *Then there exists an ultrametric absolute value* φ *on* $\mathbb{F}$ *extending that of* $\mathbb{L}$, *such that* $\varphi(x) \leq \|x\|$ *whenever* $x \in A$.

Corollary 2.5.19. *Let* $(A, \| \cdot \|)$ *be a Banach* $\mathbb{L}$ *algebra. For every maximal ideal* $\mathcal{M}$, *there exists* $\varphi \in \mathrm{Mult}(A, \| \cdot \|)$ *such that* $\mathrm{Ker}(\varphi) = \mathcal{M}$ *and* φ *is of the form* $|\chi|$ *where* χ *is a homomorphism from* A *to the field* $\frac{A}{\mathcal{M}}$, *with an absolute value* $| \cdot |$ *defined by* $|\chi|$, *extending the one of* $\mathbb{L}$. *Moreover, if* $\mathcal{M}$ *is of codimension 1, then this absolute value is the one defined on* $\mathbb{L}$ *and coincides with the quotient norm of the norm* $\| \cdot \|$ *of* T.

2.6. Topological divisors of zero

Now, we shall state an easy characterization of the Shilov boundary, taking from what was made on $\mathbb{C}$. Throughout the chapter, $\mathbb{L}$ is a complete ultrametric field and A is a commutative unital Banach $\mathbb{L}$-algebra.

Theorem 2.6.1. *The Shilov boundary* $\mathcal{S}$ *for* A *is equal to the set of the* $\psi \in \mathrm{Mult}(A, \| \cdot \|)$ *such that for every neighborhood* $\mathcal{U}$ *of* ψ *in* $\mathrm{Mult}(A, \| \cdot \|)$, *there exists* $\theta \in \mathcal{U}$ *and* $g \in A$ *satisfying* $\|g\|_{\mathrm{sp}} = \theta(g)$ *and* $\gamma(g) < \|g\|_{\mathrm{sp}}$ $\forall \gamma \in \mathcal{S} \setminus \mathcal{U}$.

Proof.   Let ψ belong to the Shilov boundary for A and suppose that there exists a neighborhood $\mathcal{U}$ of ψ in $\mathrm{Mult}(A, \| \cdot \|)$ such that for every $g \in A$, we have $\|g\|_{\mathrm{sp}} = \sup\{\gamma(g) \mid \gamma \in \mathcal{S} \setminus \mathcal{U}\}$. Without loss of generality, we may assume that $\mathcal{U}$ is open in $\mathrm{Mult}(A, \| \cdot \|)$, hence $\mathcal{S} \setminus \mathcal{U}$ is compact and is a closed subset of $\mathrm{Mult}(A, \| \cdot \|)$. Thus, $\mathcal{S} \setminus \mathcal{U}$ is a closed boundary which does not contain ψ, and therefore is strictly included in $\mathcal{S}$, a contradiction. This shows that for all neighborhood $\mathcal{U}$ of ψ in $\mathrm{Mult}(A, \| \cdot \|)$, there exists $\theta \in \mathcal{U} \cap \mathcal{S}$ and $g \in A$ satisfying $\|g\|_{\mathrm{sp}} = \theta(g)$ and $\gamma(g) < \|g\|_{\mathrm{sp}}$ $\forall \gamma \in \mathcal{S} \setminus U$.

Now, suppose that $\psi \in \mathrm{Mult}(A, \| \cdot \|)$ is such that for all neighborhood $\mathcal{U}$ of ψ in $\mathrm{Mult}(A, \| \cdot \|)$, there exists $\theta \in \mathcal{U} \cap \mathcal{S}$ and $g \in A$ satisfying $\|g\|_{\mathrm{sp}} = \theta(g)$ and $\gamma(g) < \|g\|_{\mathrm{sp}}$ $\forall \gamma \in \mathcal{S} \setminus \mathcal{U}$. Thus, given any neighborhood $\mathcal{U}$ of ψ, $\mathcal{S}$ admits at least one point in $\mathcal{U}$ and since $\mathcal{S}$ is closed, ψ belongs to $\mathcal{S}$.   □

The following basic lemma will help us prove Theorem 2.6.3.

Lemma 2.6.2. *Let* $a, b, \omega, \sigma, u, w \in \mathbb{R}_+^*$ *satisfy* $u < w$ *and* $\sigma < \omega$. *There exist* $t \in \mathbb{N}$ *and* $d \in \mathbb{Z}$ *such that* $a - tw < d\sigma < -b - tw + \omega$ *and* $d\sigma < -b - tu$.

Theorem 2.6.3 is an application of Theorem 2.6.1.

Theorem 2.6.3. *Let* $\beta > 1$ *satisfy* $\beta > \inf\{|x| \mid |x| > 1\}$. *Let* $\mathcal{S}$ *be the Shilov boundary for* A *and let* $\psi \in \mathrm{Mult}(A, \| \cdot \|)$. *Let* $\mathcal{U}$ *be a neighborhood of* ψ *and* let $M \in \mathbb{R}_+^*$. There exists $h \in A$ and $\gamma \in \mathcal{U}$ such that $M < \gamma(h) = \|h\| < \beta M$ and $\gamma(h) < \|h\|_{sp} \ \forall \gamma \in \mathcal{S} \setminus \mathcal{U}$.

Proof. Without loss of generality we may assume that $\mathcal{U}$ is open. Hence $\mathcal{S} \setminus \mathcal{U}$ is closed. By Theorem 2.6.1, there exists $\gamma \in \mathcal{U}$ and $g \in A$ satisfying $\|g\|_{sp} = \gamma(g)$ and $\theta(g) < \|g\|_{sp} \ \forall \theta \in \mathcal{S} \setminus U$. Since $\mathcal{S} \setminus \mathcal{U}$ is closed, it is compact, hence $\sup\{\theta(g) \mid \theta \in \mathcal{S} \setminus \mathcal{U}\} < \|g\|_{sp}$. Let $m = \sup\{\theta(g) \mid \theta \in \mathcal{S} \setminus \mathcal{U}\}$. By hypothesis, there exists $\alpha \in E$ satisfying $1 < |\alpha| < \beta$. Putting $a = \log M$, $b = -\log \epsilon$, $u = \log m$, $w = \log(\|g\|)$, $\omega = \log \beta, \sigma = \log |\alpha|$, then thanks to Lemma 2.6.2 we can find $t \in \mathbb{N}$ and $d \in \mathbb{Z}$ such that $M < \|\alpha^d g^t\| < \beta M$ and $|\alpha|^d m^t < \epsilon$. Now we can put $h = \alpha^d g^t$, and we check that $M < \|h\| = \gamma(h) < \beta M$ and $\theta(h) < \epsilon \ \forall \theta \in \mathcal{S} \setminus \mathcal{U}$. $\square$

Definitions and notations: An element $x \in A$ different from 0 will be called *a topological divisor of zero* if $\inf_{y \in A, 1 \leq \|y\|} \|xy\| = 0$.

Given $f \in A$, we will denote by $\mathcal{Z}_A(f)$ the set of $\phi \in \text{Mult}(A, \| \cdot \|)$ such that $\phi(f) = 0$. However, when there is no risk of confusion on the algebra, we shall only write $\mathcal{Z}(f)$ instead of $\mathcal{Z}_A(f)$.

On a complex Banach algebra, it is known that the topological divisors of zero are the elements $f \in A$ such that there exists an element ϕ of the Shilov boundary satisfying $\phi(f) = 0$ [15]. Here we mean to extend this property to ultrametric algebras

Theorem 2.6.4. *Let A be uniform and let $\mathcal{S}$ be the Shilov boundary for $(A, \| \cdot \|)$. An element $f \in A$ is a topological divisor of zero if and only if there exists $\psi \in \mathcal{S}$ such that $\psi(f) = 0$.*

Proof. Let $\mathcal{S}$ be the Shilov boundary for $(A, \| \cdot \|)$. Let $f \in A$ be a topological divisor of zero and let $(U_n)_{n \in \mathbb{N}}$ be a sequence in A such that $\|U_n\| > m > 0 \ \forall n \in \mathbb{N}$ and $\lim_{n \to \infty} \|fU_n\| = 0$. Since A is uniform, its norm is equal to $\| \cdot \|_{sp}$, so for every $n \in \mathbb{N}$, we can find $\phi_n \in \mathcal{S}$ such that $\phi_n(U_n) = \|U_n\|$. Let ϕ be a point of adherence of the sequence $(\phi_n)_{n \in \mathbb{N}}$. Thus, ϕ belongs to $\mathcal{S}$. On the other hand, we have $\phi_n(fU_n) \geq m\phi_n(f) \ \forall n \in \mathbb{N}$. But by hypothesis $\lim_{n \to \infty} \|fU_n\| = 0$, hence $\lim_{n \to \infty} \phi_n(fU_n) = 0$, hence $\lim_{n \to \infty} \phi_n(f) = 0$. Suppose $\phi(f) \neq 0$ and let $l = \phi(f)$. Therefore, we can find $t \in \mathbb{N}$ such that

(1) $\quad \phi_n(f) < \frac{l}{2} \ \forall n \geq t$.

But since ϕ is a point of adherence of the sequence (ϕ_n) with respect to the topology of pointwise convergence, we can find $k \in \mathbb{N}$, $k > t$, such that $|\phi(f) - \phi_k(f)| < \frac{l}{2}$, hence by (1) we check that $\phi(f) < l$, a contradiction.

Conversely, let $f \in A$ and assume that there exists $\psi \in \mathcal{S}$ such that $\psi(f) = 0$. Let us fix $n \in \mathbb{N}$ and let $\mathcal{U}$ be a neighborhood of ψ such that $\phi(f) \leq \frac{1}{n} \ \forall \phi \in \mathcal{U}$. Let $\beta > 1$ satisfy $\beta > \inf\{|x| \mid |x| > 1\}$. By Theorem 2.6.3, there exists $U_n \in A$ and $\gamma \in \mathcal{U}$ such that $1 < \gamma(U_n) = \|U_n\| < \beta$ and $\gamma(U_n) < \frac{1}{n} \ \forall \gamma \in \mathcal{S} \setminus \mathcal{U}$. We check that if $\phi \in \mathcal{U}$ then $\phi(fU_n) \leq \frac{\beta}{n}$ and if $\phi \in \mathcal{S} \setminus \mathcal{U}$ then $\phi(fU_n) \leq \frac{\|f\|}{n}$. Consequently, by putting $M = \max(\|f\|, \beta)$, we have $\phi(fU_n) \leq \frac{M}{n} \ \forall \phi \in \mathcal{S}$, hence $\|fU_n\| \leq \frac{M}{n}$, which finishes showing that f is a topological divisor of zero. $\qquad\square$

Corollary 2.6.5. *Let A be uniform and let $\mathcal{S}$ be the Shilov boundary for $(A, \| \cdot \|)$. There exists no topological divisor of zero if and only if all elements of $\mathcal{S}$ are absolute values.*

Corollary 2.6.6. *Let A be uniform. If A has no continuous absolute values, then there exist topological divisors of zero.*

Remark 2. In any ring having divisors of zero, there exist no absolute values. In [9], it was shown that certain Krasner algebras have no divisor of zero but admit no continuous absolute value. However, as we will see, among affinoid algebras, every affinoid algebra having no divisor of zero admits at least one absolute value.

Theorem 2.6.7 is easy.

Theorem 2.6.7. *Let $f \in A$. If f is not a divisor of zero, then f is a topological divisor of zero if and only if fA is not closed.*

Proof. Let ϕ be the linear mapping from A to fA defined as $\phi(x) = xf$. Since f is not a divisor of zero, ϕ is a continuous $\mathbb{L}$-vectorial space isomorphism from A to fA. First, suppose that f is not a topological divisor of zero. Then there exists $m > 0$ such that $m\|x\| \leq \|xf\| \leq \|x\|\|f\|$. Therefore, fA is provided with two equivalent norms: on one hand, the norm induced by A, and on the other hand the norm $\|| \cdot \||$ defined on fA as $\||xf\|| = \|f\|$. Since fA is obviously complete for the last one, it is also complete for the first, hence fA is closed in A.

   Conversely suppose that fA is closed in A. Then ϕ is a continuous $\mathbb{L}$-Banach space isomorphism from A onto fA. Hence, by Hahn–Banach's Theorem, ϕ is bicontinuous, therefore $\{x \in A \mid \|sf\| \leq 1\}$ is bounded, which proves that f is not a topological divisor of zero. $\qquad\square$

Corollary 2.6.8. *Let $f \in A$. If fA is not closed, then f is a topological divisor of zero.*

Theorem 2.6.9. *Let A be uniform and let S be the Shilov boundary for $(A, \| \, . \, \|)$. Let $f \in A$. Let $\mathcal{T}(f)$ be the closure of the interior of $\mathcal{Z}(f)$ in $\mathrm{Mult}(A, \| \, . \, \|)$. If $(\mathcal{Z}(f) \cap S) \setminus \mathcal{T}(f) \neq \emptyset$, then the ideal fA is not closed.*

Proof. Let $\beta > 1$ satisfy $\beta > \inf\{|x| \mid x \in \mathbb{L}, |x| > 1\}$ and let $\phi \in (\mathcal{Z}(f) \cap S) \setminus \mathcal{T}(f)$. We shall construct a sequence of pairs (U_n, ψ_n) of $A \times \mathrm{Mult}(A, \| \, . \, \|)$ satisfying $\beta^{(n-1)} < \|U_n\| < \beta^n$, $\beta^{(n-1)} < \psi_n(U_n) < \beta^n$ and $\psi_j \notin \mathcal{U}_m$ for every $j < m \leq n$, whenever $n \in \mathbb{N}$.

Suppose for each $j = 1, \dots, n$ we have already constructed pairs (U_j, ψ_j) of $A \times \mathrm{Mult}(A, \| \, . \, \|)$ and neighborhoods $\mathcal{U}_j$ of ϕ satisfying $U_j \in A$ such that $\beta^{j-1} < \|U_j\| < \beta^j$, $\beta^{j-1} < \psi_j(U_j) < \beta^j$ and $\psi_j \notin \mathcal{U}_m$ for all $j < m \leq n$.

Since $\phi \notin \mathcal{T}$, we can take a neighborhood $\mathcal{U}_{n+1}$ of ϕ included in $\mathcal{U}_n \setminus \{\psi_1, \dots, \psi_n\}$ such that $\mathcal{U}_{n+1} \cap \mathcal{T}(f) = \emptyset$ and such that $\theta(f) \leq \frac{1}{(n+1)\beta^{n+1}}$ $\forall \theta \in \mathcal{U}_{n+1}$. By Theorem 2.6.3, there exists $\gamma \in \mathcal{U}_{n+1}$ and $U_{n+1} \in A$ such that $\beta^n < \|U_{n+1}\| = \gamma(U_{n+1}) < \beta^{(n+1)}$ and $\theta(U_{n+1}) \leq \frac{1}{n+1}$ $\forall \theta \in \mathrm{Mult}(A, \| \, . \, \|) \setminus \mathcal{U}_{n+1}$. If $\gamma(f) \neq 0$, we just put $\psi_{n+1} = \gamma$.

Suppose now that $\gamma(f) = 0$. We can obviously find a neighborhood $\mathcal{W}$ of γ included in $\mathcal{U}_{n+1}$ such that $\beta^n < \|U_{n+1}\| = \theta_{n+1}(U_{n+1}) < \beta^{(n+1)}$ $\forall \theta \in \mathcal{W}$. Since $\mathcal{U}_{n+1} \cap \mathcal{T}(f) = \emptyset$ we can obviously find $\psi_{n+1} \in \mathcal{W} \setminus \mathcal{Z}(f)$ such that $\psi_{n+1}(f) \neq 0$. Thus, we have defined $U_{n+1}, \psi_{n+1}, \mathcal{U}_{n+1}$ such that the finite sequence $(U_j, \psi_j, \mathcal{U}_j)$ satisfies $U_j \in A$ $\beta^{j-1} < \|U_j\| < \beta^j$, $\beta^{j-1} < \psi_j(U_j) < \beta^j$ and $\psi_j \notin \mathcal{U}_m$ for all $j < m \leq n+1$.

We can check that $\theta(fU_n) \leq \frac{1}{n}$ $\forall \theta \in S$, hence $\|fU_n\| \leq \frac{1}{n}$. This is true for each $n \in \mathbb{N}$, hence the series $\sum_{n=0}^{\infty} fU_n$ converges in A to an element h which obviously belongs to the closure of fA in A. Now, suppose that fA is closed. There exists $g \in A$ such that $h = fg$. For every $n \in \mathbb{N}$ we put $a_n = \psi_n(f)$. Hence $a_n > 0$ $\forall n \in \mathbb{N}$. Next, we notice that $\psi_n(fU_n) > a_n \beta^{(n-1)}$, and that $\psi_n(fU_j) \leq a_n \|U_j\| < \beta^{(n-1)}$ $\forall j \leq n$. Moreover, since $\psi_n \notin \mathcal{U}_s$ $\forall s > n$, we have $\psi_n(fU_s) \leq a_n \|U_s\| \leq \frac{a_n}{s}$ $\forall s > n$. Thus, we have $\psi_n(fU_n) > \psi_n(fU_j)$ $\forall j \neq n$, hence $\psi_n(\sum_{j=1}^{\infty} fU_j) = \psi_n(fU_n) > a_n \beta^{(n-1)}$. But since $\psi(fg) = a_n \psi_n(g)$, we obtain $\psi_n(g) > \beta^{(n-1)}$. This is true for every $n \in \mathbb{N}$, and is absurd because $\psi_n(g) \leq \|g\|$. This shows that fA is not closed and finishes the proof of Theorem 2.6.9. $\qquad \square$

Proposition 2.6.10. *Let E be a normed $\mathbb{L}$-vector space which also is a A-module. Let $t \in M_{\mathbb{L}}$. Let E_1 be a A_0-submodule defining the topology of E. Assume that there exists $e_1, \dots, e_q \in E_1$ such that $E_1 = \sum_{i=1}^{q} A_0 e_i + t E_1$. Let ϕ be the mapping from A^q into E defined as $\phi(t_1, \dots, t_q) = \sum_{i=1}^{q} t_i e_i$. Then ϕ is surjective and E is finite over A.*

Proof.   Let $y \in E_1$. We can write y in the form $\sum_{i=1}^{q} f_{1,i} e_i + t y_1$, with $f_{1,i} \in A_0$ and $y_1 \in E_1$, and in the same way, we can write $y_1 = \sum_{i=1}^{q} f_{2,i} e_i + t y_2$, with $f_{2,i} \in A_0$, and $y_2 \in E_1$. By induction, it clearly appears that for every $m \in \mathbb{N}^*$, we obtain $y = \sum_{i=1}^{q} (\sum_{j=1}^{m} f_{m,i} t^{m-1}) e_i + t^m y_m$. And since all terms $f_{j,i}$ lie in A_0, and all y_j lie in E_1, each series $\sum_{j=1}^{m} f_{m,i} t^{m-1}$ converges in the Banach algebra A to an element f_i. Then, since each set $t^m E_1$ is closed, $y - \sum_{i=1}^{q} f_i e_i$ lies in $t^m E_1$ for all $m \in \mathbb{N}^*$. Therefore, $y = \sum_{i=1}^{q} f_i e_i$. Consequently $\phi(A^q)$ contains E_1 and therefore ϕ is surjective on E. Moreover, E is finite over A. $\qquad \square$

Proposition 2.6.11. *Let F be a normed A-algebra which is a A-module whose completion is finite over A. Then F is complete.*

Proof.   Let E be the completion of F, and let $e_1, \ldots, e_q \in F$ be such that $E = \sum_{i=1}^{q} A e_i$. Let ϕ be the canonical surjection from A^q onto E : $\phi(a_1, \ldots, a_q) = \sum_{i=1}^{q} a_i e_i$. On A^q we can define the product norm of A^q: $\||(a_1, \ldots, a_q)\|| = \max_{1 \leq i \leq q} \|a_i\|$. Putting then $c = \max_{1 \leq i \leq q} \|e_i\|$, we have $\||(a_1, \ldots, a_q)\|| \geq c \|\phi(a_1, \ldots, a_q)\|$. Thus, ϕ is continuous, hence $\mathrm{Ker}(\phi)$ is closed and therefore E is equipped with the quotient norm $\| . \|'$ of $\|| . \||$. Then, E is complete for the norm $\| . \|'$. On the other hand, given $f \in A^q$, we have $\|\phi(f)\|' = \inf_{x \in \mathrm{Ker}(\phi)} \||f + x\|| \geq c \inf_{x \in \mathrm{Ker}(\phi)} \|\phi(f + x)\|$. But since $\phi(f + x) = \phi(f)$, we check that $\|\phi(f)\|' \geq c \|\phi(f)\|$. Thus, E is complete for both norms $\| . \|$ and $\| . \|'$ and therefore by Hahn–Banach's theorem (Theorem 1.3.4), the two norms are equivalent. By definition, ϕ is open with respect to the norm $\| . \|'$ on E. Consequently $\sum_{i=1}^{q} A_0 e_i$ is a A_0-submodule E_1 that defines the topology of E. Since F is dense in E, each e_i is of the form $f_i + t h_i$, with $h_i \in E_1$. Therefore E_1 is included in $\sum_{i=1}^{q} A_0 f_i + t E_1$, and then by Proposition 2.6.10, we have $E = \sum_{i=1}^{q} A_0 f_i \subset F$. $\qquad \square$

Corollary 2.6.12. *Let A be Noetherian and let F be a complete normed A-algebra which is a finite A-module. Then every A-submodule of F is closed.*

Proof.   Indeed, since A is Noetherian and since F is finite, F is Noetherian, hence any A-submodule E has a finite closure in F, hence by Proposition 2.6.11 is closed. $\qquad \square$

Corollary 2.6.13. *Let A be Noetherian. Every ideal of A is closed.*

Proof.   Indeed, the closure of an ideal I is of finite type, and therefore, I is closed. $\qquad \square$

Corollary 2.6.14. *Let A be Noetherian and let $f \in A$. If f is not a divisor of zero, then it is not a topological divisor of zero.*

Proof. Indeed, if f were a topological divisor of zero, fA wouldn't be closed, what is impossible since A is Noetherian. $\square$

Corollary 2.6.15. *Let A be Noetherian and let $f \in A$. Any topological divisor of zero of A actually is a divisor of zero.*

Notation: Given a normed $\mathbb{L}$-algebra E and some $\chi \in \mathcal{X}(E)$, we denote by $\mathbb{B}_E(\chi, r)$ the set of the $\zeta \in \mathcal{X}(E)$ such that $|\chi(x) - \zeta(x)| \leq r \ \forall x \in E$, $\|x\| \leq 1$.

Lemma 2.6.16. *Let A be a normed $\mathbb{L}$-algebra, let $\chi \in \mathcal{X}(A)$ and let $f_1, \dots f_q \in A$ and let $\lambda = \inf_{1 \leq j \leq q} \frac{1}{\|f_j\|}$. Then $\{\gamma \in \mathcal{X}(A) \mid |\gamma| \in \mathcal{W}_A(|\chi|, f_1, \dots, f_q, r)\}$ contains $\mathbb{B}_A(\chi, r\lambda)$.*

Proof. Let us take $\gamma \in \mathbb{B}_A(\chi, \lambda r)$. For each $j = 1, \dots, q$, we have $\big| |\gamma(f_j)| - |\chi(f_j)| \big|_\infty \leq |\gamma(f_j) - \chi(f_j)|_\infty \leq \|f_j\| \|\gamma - \chi\| \leq r$. Thus, $|\gamma|$ belongs to $\mathcal{W}_A(\chi, f_1, \dots, f_q, r)$. $\square$

Theorem 2.6.17. *Let A be an entire Banach $\mathbb{K}$-algebra and let B be a finite integral extension of A. For each $\chi \in \mathcal{X}(B)$, we denote by $\hat{\chi}$ its restriction to A. Then, by providing $\mathcal{X}(A)$ (resp., $\mathcal{X}(B)$) with the uniform convergence norm on the unit ball of A (resp., B), the mapping ϕ from $\mathcal{X}(B)$ to $\mathcal{X}(A)$ defined as $\phi(\chi) = \hat{\chi}$ is open.*

Proof. By induction, we can easily reduce ourselves to the case when B is of the form $A[x]$. Let r be > 0 and let $P(X) = X^d + a_{d-1}X^{d-1} + \cdots + a_0$ be the minimal polynomial of x over A. Let us fix $\chi \in \mathcal{X}(B)$ and let $\alpha = \chi(x)$. Then we have $\alpha^d + \chi(a_{d-1})\alpha^{d-1} + \cdots + \chi(a_0) = 0$. Let $\alpha_1, \dots, \alpha_d$ be the zeros of the polynomial $X^d + \chi(a_{d-1})X^{d-1} + \cdots + \chi(a_0)$, with $\alpha = \alpha_1$. Let $\mathbb{K}$ be an ultrametric algebraically closed extension of $\mathbb{L}$ By results on the continuity of the zeros, [1, 28, 34], there exists $s > 0$ such that, if a polynomial $Q(X) = X^d + b_{d-1}X^{d-1} + \cdots + b_0 \in \mathbb{K}[X]$ satisfies $|b_j - \chi(a_j)| \leq s$, the zeros β_j $(1 \leq j \leq d)$, once correctly ordered, satisfy $|\alpha_j - \beta_j| \leq r$. Now, let $\gamma \in \mathcal{X}(A)$ satisfy $\|\gamma - \hat{\chi}\|_{A_0} \leq s$. So, we have $|\gamma(a_j) - \hat{\chi}(a_j)| \leq s$, and therefore, the polynomial $X^d + \gamma(a_{d-1})X^{d-1} + \cdots + \gamma(a_0)$ admits a zero $\beta \in \mathbb{K}$ such that $|\beta - \alpha| \leq r$. Then γ admits a continuation $\overline{\gamma}$ to B such that $\overline{\gamma}(x) = \beta$. Thus, we have $\gamma = \phi(\hat{\overline{\gamma}})$, which proves that $\phi(\mathbb{B}_A(\chi, r))$ contains $\mathbb{B}_B(\phi(\chi), s))$. Consequently, ϕ is open. $\square$

Chapter 3

Admissible $\mathbb{L}$-Algebras

3.1. Definition of admissible $\mathbb{L}$-algebras

Definitions and notations: Recall that throughout the book, $\mathbb{L}$ denotes a complete valued field with respect to an ultrametric absolute value $| . |$ and henceforth, E denotes a metric space whose distance δ is ultrametric. Given $a \in E$ and $r > 0$, we denote by $B(a, r)$ the closed ball $\{x \in E | \delta(a, x) \leq r\}$.

Henceforth, we will denote by $\mathcal{C}$ the $\mathbb{L}$-Banach algebra of bounded continuous functions from E to $\mathbb{L}$ equipped with the norm of uniform convergence $\| . \|_0$ on E.

We denote by $| . |_\infty$ the Archimedean absolute value of $\mathbb{R}$.

Given a subset H of E, we denote by $\overline{H}$ the closure of H in E and the function u defined on E by $u(x) = 1$ if $x \in H$ and $u(x) = 0$ otherwise, is called the *characteristic function of H.*

A subset of a topological space is said to be *clopen* if it is open and closed.

We will denote by $\mathbb{B}(E)$ the Boolean ring of clopen subsets of E with respect to the two classical laws Δ and $\cap$.

Let us recall this obvious lemma.

Lemma 3.1.1. *Let F be a subset of E and let u be its characteristic function. Then u is continuous if and only if F is clopen.*

The following lemma is also clear since each ball of $\mathbb{L}$ is clopen.

Lemma 3.1.2. *Let f be a continuous function from E to $\mathbb{L}$ and let $M > 0$. Given $M > 0$, the sets $E_1 = \{x \in E \, |f(x)| \geq M\}$ and $E_2 = \{x \in E \, |f(x)| \leq M\}$ are clopen.*

Proof. Let $a \in E_1$. Since f is continuous and E is ultrametric, there exists $r > 0$ such that $|f(x)| = |f(a)| \; \forall x \in B(a, r)$, hence E_1 is open. And given a converging sequence $(a_n)_{n\in\mathbb{N}}$ of limit $a \in E$ such that $|f(a_n)| \geq M$, obviously $|f(a)| \geq M$. □

Corollary 3.1.3. *Let f be a continuous function from E to $\mathbb{L}$, let $M > 0$ and let $h > 0$. Then $\{x \in E \mid \left|\,|f(x)| - M\,\right|_\infty \leq h\}$ is clopen.*

Definition and notations: Throughout the book, we will denote by $\mathcal{C}$ the Banach $\mathbb{L}$-algebra of all bounded continuous functions from E to $\mathbb{L}$ and by S a $\mathbb{L}$-subalgebra of $\mathcal{C}$.

The Banach $\mathbb{L}$-algebra $\mathcal{C}$ of all bounded continuous functions from E to $\mathbb{L}$ is equipped with the norm $\|\,.\,\|_0$.

Recall the following:

Proposition 3.1.4. *Let T be a unital commutative Banach $\mathbb{L}$-algebra of bounded continuous functions from E to $\mathbb{L}$. Then $\|f\|_0 \leq \|f\|_{\mathrm{sp}} \leq \|f\|$ $\forall f \in T$. Moreover, given $f \in T$ satisfying $\|f\|_{\mathrm{sp}} < 1$, then $\lim_{n\to+\infty} \|f^n\| = 0$.*

Proof.   The norm $\|\,.\,\|_0$ is power multiplicative and classically it is bounded by the norm $\|\,.\,\|$ of T, it is then bounded by $\|\,.\,\|_{\mathrm{sp}}$. The last claim is immediate. □

Definition and notations: Let $(S, \|\,.\,\|)$ be a $\mathbb{L}$-subalgebra of $\mathcal{C}$. We say that $(S, \|\,.\,\|)$ is *semi-admissible* if it is a Banach algebra satisfying the following two properties:

(1) For every $f \in S$ such that $\inf\{|f(x)| \mid x \in E\} > 0$, f is invertible in S.
(2) For every $O \in \mathbb{B}(E)$, the characteristic function of O belongs to S.

Moreover, the semi-admissible algebra S will be called *admissible* if it satisfies:

(3) $\|f\|_0 = \|f\|_{\mathrm{sp}}$ $\forall f \in S$ [15].

Given a subset X of S, we call *spectral closure of X* denoted by $\mathrm{spcl}(X)$ the closure of X with respect to the semi-norm $\|\,.\,\|_{\mathrm{sp}}$ and X will be said to be *spectrally closed* if $X = \mathrm{spcl}(X)$. Moreover, X will be said to be *uniformly closed* if it is closed with respect to the uniform norm and we call *uniform closure of X* the closure of X with respect to the semi-norm $\|\,.\,\|_0$.

Remark 1. Given an admissible $\mathbb{L}$-algebra $(S, \|\,.\,\|)$, the norm $\|\,.\,\|$ is not supposed to be the uniform convergence norm on E.

Lemma 3.1.5. *Given two subsets A and B of E such that $\delta(A, B) > 0$, there exist a clopen subset F such that $A \subset F$ and $B \subset E \setminus F$.*

Proof. Let $\delta(A, B) = r > 0$ then take $F = \cup_{a \in A} B(a, r)$. The set F is clopen since F contains the ball $< b(x, r)$ whenever $x \in F$. And clearly $A \subset F$ and $B \subset E \setminus F$. $\qquad \square$

Let $f \in \mathcal{C}$ be such that $\inf\{|f(x)| \mid x \in E\} > 0$, it is clear that $\frac{1}{f}$ belongs to $\mathcal{C}$. On the other hand, $\mathcal{C}$ is complete with respect to the uniform norm, then we have the following statement:

Proposition 3.1.6. *The Banach L-algebra $\mathcal{C}$ is admissible.*

The following Theorem 3.1.7 shows another example of admissible algebra which is a bit less immediate. In Theorem 3.1.11, we can see that in various cases, this algebra is strictly included in $\mathcal{C}$.

Lemma 3.1.7. *Let $(O_i)_{i=1,\ldots,n}$ be a finite cover of E with clopen sets. Then there exists a finite cover $(B_j)_{j=1,\ldots,p}$ of E where the sets B_j are not empty, clopen, pairwise disjoint and such that every B_j is contained in some O_i.*

Proof. To the system $(O_i)_{i=1,\ldots,n}$, let us associate the system $(O'_i)_{i=1,\ldots,2n}$ where $O'_i = O_i$ if $1 \leq i \leq n$ and $O'_i = X \setminus O_{i-n}$ otherwise. For every $x \in E$ define $I_x = \{i \in \{1, \ldots, n\} : x \in O'_i\}$ and consider the following equivalence relation on E: $x(R)y$ if and only if $I_x = I_y$. For any $x \in E$ the equivalence class of x is equal to $\cap_{i \in I_x} O'_i$ and it is clopen since so are the O'_i. Then the equivalence classes $(B_j)_{j=1,\ldots,p}$ satisfy the expected properties. $\qquad \square$

Theorem 3.1.8. *Let $\mathcal{T}$ be the L-subalgebra of $\mathcal{C}$ generated by the characteristic functions of all clopen sets of E and let $\overline{\mathcal{T}}$ be its closure in $\mathcal{C}$ (for the uniform convergence $\| \cdot \|_0$ on E). Then $\overline{\mathcal{T}}$ is admissible.*

Proof. One just has to prove Property (2) in the definition of a semi-admissible algebra.

First, we check that if $g \in \mathcal{T}$ is such that $\inf\{|g(x)| : x \in E\} = m > 0$ then $\frac{1}{g} \in \mathcal{T}$. Since $g \in \mathcal{T}$ there exists a finite cover $(O_i)_{i=1,\ldots,n}$ of E with clopen sets and scalars $(\lambda_i)_{1 \leq i \leq n}$ in $\mathbb{L}$ such that $g = \sum_{i=1}^{n} \lambda_i u_i$ where u_i is the characteristic function of the clopen O_i. Using the preceding lemma we get a finite cover $(B_j)_{j=1,\ldots,p}$ of E where the sets B_j are not empty, clopen, pairwise disjoint and such that every B_j is contained in some O_i. Then there exist scalars $(\beta_j)_{1 \leq j \leq p}$ in $\mathbb{L}$ such that $g = \sum_{j=1}^{p} \beta_j e_j$ where e_j is the characteristic function of the clopen B_j. For every j we get $|\beta_j| \geq m > 0$, then it is clear that

$$\frac{1}{g} = \sum_{j=1}^{p} \frac{1}{\beta_j} e_j \quad \text{and} \quad \frac{1}{g} \in \mathcal{T}.$$

Now consider any $f \in \mathcal{T}$ such that $\inf\{|f(x)| : x \in E\} = m > 0$. For every $\epsilon > 0$ such that $\epsilon < \frac{1}{m}$ we have $\epsilon m^2 < m$ and we can consider some $g \in \mathcal{T}$ such that $\|f - g\|_0 \leq \epsilon m^2$. Since $\epsilon m^2 < m$, we get $|f(x)| = |g(x)|$ for every $x \in E$ and then $\inf\{|g(x)| : x \in E\} = m$.

Next $\frac{1}{g} \in \mathcal{T}$ and we have for every

$$x \in E: \left| \frac{1}{f(x)} - \frac{1}{g(x)} \right| = \frac{|f(x) - g(x)|}{|f(x)g(x)|} \leq \frac{\|f - g\|_0}{m^2} \leq \epsilon.$$

This proves that $\frac{1}{f} \in \overline{\mathcal{T}}$, which ends the proof.   □

Corollary 3.1.9. *The algebra $\overline{\mathcal{T}}$ defined in Theorem 3.1.8 is the $\mathbb{L}$-algebra of the continuous functions f from E to $\mathbb{L}$ such that the closure of $f(E)$ in $\mathbb{L}$ is compact. In particular when $\mathbb{L}$ is locally compact or E is compact then $\overline{\mathcal{T}} = \mathcal{C}$.*

In order to prove Theorem 3.1.11, we must recall the following classical proposition.

Proposition 3.1.10. *The field $\mathbb{L}$ is locally compact if and only if its valuation group is discrete and its residue class field is finite* [1 , *Proposition* 2.3.3].

Theorem 3.1.11. *Suppose that E contains a sequence $(a_n)_{n \in \mathbb{N}}$ such that $\inf_{n \neq m}(\delta(a_n, a_m)) > 0$ and that $\mathbb{L}$ is not locally compact. Then $\mathcal{G}$ is strictly included in $\mathcal{C}$.*

Proof.   We put $s = \inf_{n \neq m}(\delta(a_n, a_m))$. Suppose first that the valuation group of $\mathbb{L}$ is dense. We can consider a partition of E by an infinite family of balls $d_E(b_i, s^-)$.

Suppose first that the valuation group of $\mathbb{L}$ is dense. Then we can define a bounded mapping ψ from E into $\mathbb{L}$ such that $\psi(x)$ is constant in each ball $d_E(b_i, s^-)$, such that $|\psi(a_n) - \psi(a_m)| \geq 1$ and such that $|\psi(x)| \geq 1 \, \forall x \in E$. Particularly, $|\psi(x)|$ takes infinitely many values. Suppose that $\mathcal{T}$ is dense in $\mathcal{C}$. Following the same process as in the proof of Theorem 1.1.6, we can construct a function $g \in \mathcal{T}$ such that $|\psi(x)| = |g(x)| = \lambda_j \, \forall x \in E$. But $|\psi(x)|$ then only takes finitely many values, a contradiction.

Similarly, suppose now that the residue class field of $\mathbb{L}$ is infinite. Let us consider a sequence of distinct disks $(d(\mu_n, 1^-))_{n \in \mathbb{N}}$ in the unit circle and

and put $B_n = d(\mu_n, 1^-)$. Now, consider a sequence of balls $d_E(a_n, s^-)$ in E and an element f of $\mathcal{C}$ constant in $d_E(a_n, s)$ and such that $f(a_n)$ belongs to B_n. Suppose that f is in the closure of $\mathcal{T}$. Then there exists $g \in \mathcal{T}$ such that $|f(x) - g(x)| < 1$ for any $x \in E$. In particular, we get that $g(a_n) \in B_n$ for every n. Thus, g should take infinitely many values, a contradiction. $\square$

Theorem 3.1.12. *Suppose that E has no isolated points. Let S be an admissible Banach L-algebra complete with respect to the norm $\| \cdot \|_0$. An element of S having no zero in E is a topological divisor of zero if and only if it is not invertible.*

Proof. It is obvious that an invertible element of S is not a topological divisor of zero. Now, consider an element $f \in \mathcal{T}$ that is not invertible. Then $\inf_{x \in E} |f(x)| = 0$. Therefore, there exists a sequence of disks $(d_E(a_n, r_n))_{n \in \mathbb{N}}$ with $\lim_{n \to \infty} r_n = 0$, such that $|f(x)| \leq \frac{1}{n}, \forall x \in d_E(a_n, r_n), \forall n \in \mathbb{N}^*$. For each $n \in \mathbb{N}$, let h_n be the characteristic function of $d_E(a_n, r_n)$. Then h_n belongs to $\mathcal{T}$ and satisfies $\|h_n\|_0 = 1 \ \forall n \in \mathbb{N}^*$. On the other hand, we have $\|f h_n\|_0 \leq \frac{1}{n}$, hence $\lim_{n \to +\infty} f h_n = 0$. $\square$

3.2. Sticked ultrafilters

Definitions and notations: Let $\mathcal{F}$ be a filter on E. Given a function f from E to $\mathbb{L}$ admitting a limit along $\mathcal{F}$, we will denote by $\lim_{\mathcal{F}} f(x)$ this limit.

Given a filter $\mathcal{F}$ on E, we will denote by $\mathcal{I}(\mathcal{F}, S)$ the ideal of the $f \in S$ such that $\lim_{\mathcal{F}} f(x) = 0$. Notice that the unity does not belong to $\mathcal{I}(\mathcal{F}, S)$, so $\mathcal{I}(\mathcal{F}, S) \neq S$.

Given $a \in E$, we will denote by $\mathcal{I}(a, S)$ the ideal of the $f \in S$ such that $f(a) = 0$ and by $\mathcal{I}'(a, S)$ the ideal of the $f \in S$ such that there exists an open neighborhood G of a such that $f(x) = 0 \ \forall x \in G$.

We will denote by $\mathrm{Max}(S)$ the set of maximal ideals of S and by $\mathrm{Max}_E(S)$ the set of maximal ideals of S of the form $\mathcal{I}(a, S), \ a \in E$.

Given a set F, we will denote by $U(F)$ the set of ultrafilters on F.

Two ultrafilters $\mathcal{U}, \mathcal{V}$ on E will be said to be *sticked* if for every closed subsets $H \in \mathcal{U}, \ G \in \mathcal{V}$, we have $H \cap G \neq \emptyset$.

We will denote by $(\mathcal{R})$ the relation defined on $U(E)$ as $\mathcal{U}(\mathcal{R})\mathcal{V}$ if $\mathcal{U}$ and $\mathcal{V}$ are sticked [14, 27].

Throughout this chapter, we will denote by S a semi-admissible L-algebra.

Remark 2. Relation $(\mathcal{R})$ is not the equality between ultrafilters, even when the ultrafilters are not convergent. In [33], Labib Haddad introduced the following equivalence relation $(\mathcal{H})$ on ultrafilters. Given two ultrafilters $\mathcal{U}$, $\mathcal{V}$ we write $\mathcal{U}(\mathcal{H})\mathcal{V}$ if there exists an ultrafilter $\mathcal{W}$ such that every closed set G lying in $\mathcal{W}$ also lies in $\mathcal{U}$ and similarly, every closed set G lying in $\mathcal{W}$ also lies in $\mathcal{V}$. So, Relation $(\mathcal{H})$ is clearly thinner than Relation $(\mathcal{R})$. However, it is shown that two ultrafilters $\mathcal{U}$, $\mathcal{V}$ satisfying $\mathcal{U}(\mathcal{H})\mathcal{V}$ may be distinct without converging.

The following lemma is classical [6].

Lemma 3.2.1. *Given* $\mathcal{U} \in U(E)$ *and a subset* X *of* E, *then either* $X \in \mathcal{U}$ *or* $(E \setminus X) \in \mathcal{U}$.

Theorem 3.2.2.

(1) *if* F *and* G *are disjoint closed subsets of* E *then there exists a clopen* O *such that* $F \subset O$ *and* $G \subset (E \setminus O)$.
    This is the case when $\delta(F, G) > 0$.
(2) *If* $\mathcal{U}$ *and* $\mathcal{V}$ *are ultrafilters on* E *then they are sticked if and only if they contain the same clopen sets.*
    In particular if $\mathcal{U}$, $\mathcal{V}$ *are not sticked, there exist disjoint clopen subsets* H *and* J *of* E *such that* $H \in \mathcal{U}$, $H \notin \mathcal{V}$ *and* $J \in \mathcal{V}$, $G \notin \mathcal{U}$.

Proof. (1) For each $x \in F$ take $r_x > 0$ such that $d(x, r_x^-) \cap G = \emptyset$ and define the open set $O = \bigcup_{x \in F} d(x, r_x^-)$. We clearly have $F \subset O$ and $G \subset E \backslash O$. Let us prove that O is closed. Let $y \in \overline{O}$. For every $n \in \mathbb{N}^*$, there exists $x_n \in F$ such that $d(y, \frac{1}{n}^-) \cap d(x_n, r_{x_n}^-) \neq \emptyset$, then let $y_n \in d(y, \frac{1}{n}^-) \cap d(x_n, r_{x_n}^-)$.

First, assume that $\inf\{r_{x_n} : n \in \mathbb{N}^*\} = m > 0$. Take $n \in \mathbb{N}^*$ such that $\frac{1}{n} < m$. Since the distance is ultrametric we then have: $d(y, \frac{1}{n}^-) = d(y_n, \frac{1}{n}^-) \subset d(y_n, r_{x_n}^-) = d(x_n, r_{x_n}^-)$. Finally $y \in O$.

Assume now that $\inf\{r_{x_n} : n \in \mathbb{N}^*\} = 0$. There exists a subsequence $(x_{n_k})_k$ such that $(r_{x_{n_k}})_k$ tends to 0. Then we immediately get that $(x_{n_k})_k$ tends to y since $(y_{n_k})_k$ tends to y. So $y \in \overline{F} = F$ and again $y \in O$.

(2) If $\mathcal{U}$ and $\mathcal{V}$ are sticked then for every clopen $O \in \mathcal{U}$ we necessarily have $O \in \mathcal{V}$. Otherwise, using the preceding lemma the clopen $E \setminus O$ is in $\mathcal{V}$ so $\mathcal{U}$ and $\mathcal{V}$ cannot be sticked. Conversely, if $\mathcal{U}$ and $\mathcal{V}$ contain the same clopen sets then using the preceding property (1), for every closed sets $F \in \mathcal{U}$ and $G \in \mathcal{V}$ we necessarily get $F \cap G \neq \emptyset$, otherwise taking a clopen O such in (1) we have $O \in \mathcal{U}$ and $O \notin \mathcal{V}$ since $E \setminus O \in \mathcal{V}$.

In particular, if $\mathcal{U}$ and $\mathcal{V}$ are not sticked then taking some clopen H in $\mathcal{U}$ which is not in $\mathcal{V}$, we have $(E \setminus H) \in \mathcal{V}$ and putting $J = E \setminus H$, H and J are clopen sets satisfying the expected property. $\qquad\square$

Corollary 3.2.3. *Let* $\mathcal{U}$, $\mathcal{V}$ *be two ultrafilters on* E *that are not sticked. There exist clopen subsets* $H \in \mathcal{U}$, $J \in \mathcal{V}$ *and* $f \in S$ *such that* $f(x) = 1 \; \forall x \in H$, $f(x) = 0 \; \forall x \in J$.

Lemma 3.2.4 is classical.

Lemma 3.2.4. *Let* $\mathcal{U}$ *be an ultrafilter on* E. *Let* f *be a bounded function from* E *to* $\mathbb{L}$. *The function* $|f|$ *from* E *to* $\mathbb{R}_+$ *defined as* $|f|(x) = |f(x)|$ *admits a limit along* $\mathcal{U}$. *Moreover, if* $\mathbb{L}$ *is locally compact, then* $f(x)$ *admits a limit along* $\mathcal{U}$.

Recall that for any normed $\mathbb{L}$-algebra $(B, \|\,.\,\|)$, the closure of an ideal of B is an ideal of B. Lemmas 3.2.5 and 3.2.6 are immediate.

Lemma 3.2.5. *The spectral closure of an ideal of* S *is an ideal of* S.

Lemma 3.2.6. *Let* $X \subset S$ *be spectrally closed. Then* X *is closed with respect to the norm of* S. *Let* $Y \subset S$ *be uniformly closed. Then it is spectrally closed.*

Now we can recall a classical result known in ultrametric analysis as in Archimedean analysis.

Proposition 3.2.7. *Every maximal ideal* $\mathcal{M}$ *of* S *is spectrally closed.*

Proof. By Lemma 3.2.5, the spectral closure $\mathrm{spcl}(\mathcal{M})$ of $\mathcal{M}$ is an ideal. If $\mathcal{M}$ is not spectrally closed, then $\mathrm{spcl}(\mathcal{M}) = S$, hence there exists $t \in S$ such that $1 - t \in \mathcal{M}$ and $\|t\|_{\mathrm{sp}} < 1$. Consequently, by Proposition 3.1.4, $\lim_{n \to +\infty} \|t^n\| = 0$, therefore the series $\left(\sum_{n=0}^{\infty} t^n\right)$ converges and $\left(\sum_{n=0}^{\infty} t^n\right)(1 - t) = 1$ and hence the unity belongs to $\mathcal{M}$, a contradiction. $\square$

Proposition 3.2.8 now is easy.

Proposition 3.2.8. *Given an ultrafilter* $\mathcal{U}$ *on* E, $\mathcal{I}(\mathcal{U}, S)$ *is a prime ideal. Moreover,* $\mathcal{I}(\mathcal{U}, S)$ *is uniformly closed and hence is spectrally closed and closed for the topology of* S.

Proof. Since $\mathcal{U}$ is an ultrafilter, it is obvious that $\mathcal{I}(\mathcal{U}, S)$ is prime. Indeed, given $f \in S$, by Lemma 3.2.4, $|f(x)|$ admits a limit along $\mathcal{U}$ and hence, if f, $g \in S$ are such that $\lim_{\mathcal{U}} f(x) g(x) = 0$, then either $\lim_{\mathcal{U}} f(x) = 0$ or $\lim_{\mathcal{U}} g(x) = 0$, hence either f or g belongs to $\mathcal{I}(\mathcal{U}, S)$.

Let us now check that $\mathcal{I}(\mathcal{U}, S)$ is uniformly closed. Indeed let g in the closure of $\mathcal{I}(\mathcal{U}, S)$ with respect to $\| \cdot \|_0$, let $b = \lim_{\mathcal{U}} |g(x)|$ and suppose $b > 0$. There exists $f \in \mathcal{I}(\mathcal{U}, S)$ such that $\|f - g\|_0 < b$ and then

$$b = \left| \lim_{\mathcal{U}} |f(x)| - \lim_{\mathcal{U}} |g(x)| \right|_{\infty} \leq \lim_{\mathcal{U}} |f(x) - g(x)| \leq \|f - g\|_0 < b,$$

a contradiction showing that $\mathcal{I}(\mathcal{U}, S)$ is uniformly closed. Therefore, it is spectrally closed and closed for the topology of S. $\square$

The following Theorem 3.2.9 is proved in [14].

Theorem 3.2.9. *Let $\mathcal{U}$, $\mathcal{V}$ be two ultrafilters on E. Then $\mathcal{I}(\mathcal{U}, S) = \mathcal{I}(\mathcal{V}, S)$ if and only if $\mathcal{U}$ and $\mathcal{V}$ are sticked.*

Proof. First, if $\mathcal{U}$ and $\mathcal{V}$ are not sticked, by Corollary 3.2.3, we have $\mathcal{I}(\mathcal{U}, S) \neq \mathcal{I}(\mathcal{V}, S)$. Now, suppose that $\mathcal{U}$, $\mathcal{V}$ are sticked. By Theorem 3.2.2, then they contain the same clopen sets. But for every $f \in S$ and $\epsilon > 0$ the set $L_\epsilon = \{x \in E : |f(x)| \leq \epsilon\}$ is clopen and we have: $f \in \mathcal{I}(\mathcal{U}, S) \iff \forall \epsilon > 0$, $L_\epsilon \in \mathcal{U}$ and hence $L_\epsilon \in \mathcal{V}$. Consequently, $\forall \epsilon > 0$, $L_\epsilon \in \mathcal{V}$ and hence f belongs to $\mathcal{I}(\mathcal{V}, S)$. Thus, $\mathcal{I}(\mathcal{U}, S) \subset \mathcal{I}(\mathcal{V}, S)$ and similarly, $\mathcal{I}(\mathcal{V}, S) \subset \mathcal{I}(\mathcal{U}, S)$, therefore $\mathcal{I}(\mathcal{V}, S) = \mathcal{I}(\mathcal{U}, S)$ $\square$

Corollary 3.2.10. *Relation $(\mathcal{R})$ is an equivalence relation on $U(E)$.*

Theorem 3.2.11 looks like certain Bezout–Corona statements [28].

Theorem 3.2.11. *Let $f_1, \ldots, f_q \in S$ satisfy $\inf_{x \in E}(\max_{1 \leq j \leq q} |f_j(x)|) > 0$. Then there exist $g_1, \ldots, g_q \in S$ such that $\sum_{j=1}^{q} f_j(x)g_j(x) = 1 \ \forall x \in E$.*

Proof. Let $M = \inf_{x \in E}(\max_{1 \leq j \leq q} |f_j(x)|)$. Let $E_j = \{x \in E \mid |f_j(x)| \geq M\}$, $j = 1, \ldots, q$ and let $F_j = \bigcup_{m=1}^{j} E_m$, $j = 1, \ldots, q$. Let $g_1(x) = \frac{1}{f_1(x)} \forall x \in E_1$ and $g_1(x) = 0 \ \forall x \in E \setminus E_1$. Since $|f_1(x)| \geq M \ \forall x \in E_1$, $|g_1(x)|$ is clearly bounded. By Lemma 3.1.2, each E_j is obviously clopen and so is each F_j. And since S is semi-admissible, g_1 belongs to S.

Suppose now we have constructed $g_1, \ldots, g_k \in S$ satisfying $\sum_{j=1}^{k} f_j g_j(x) = 1 \ \forall x \in F_k$ and $\sum_{j=1}^{k} f_j g_j(x) = 0 \ \forall x \in E \setminus F_k$. Let g_{k+1} be defined on E by $g_{k+1}(x) = \frac{1}{f_{k+1}(x)} \ \forall x \in F_{k+1} \setminus F_k$ and $g_{k+1}(x) = 0 \ \forall x \in E \setminus (F_{k+1} \setminus F_k)$. Since F_k and F_{k+1} are clopen, so is $E \setminus (F_{k+1} \setminus F_k)$ and consequently, g_{k+1} is continuous. Similarly as for g_1, since $|f_{k+1}(x)| \geq M \ \forall x \in E_{k+1}$, $|g_{k+1}(x)|$ is clearly bounded, hence g_{k+1} belongs to S. Now we can check that $\sum_{j=1}^{k+1} f_j g_j(x) = 1 \ \forall x \in F_{k+1}$ and $\sum_{j=1}^{k} f_j g_j(x) = 0 \ \forall x \in E \setminus F_{k+1}$. So, by a finite induction, we get functions $g_1, \ldots, g_q \in S$ such that $\sum_{j=1}^{q} f_j g_j(x) = 1 \ \forall x \in E$, which ends the proof. $\square$

Notation: Let $f \in S$ and let $\epsilon > 0$. We set $D(f, \epsilon) = \{x \in E \mid |f(x)| \leq \epsilon\}$.

Corollary 3.2.12. *Let I be an ideal of S different from S. The family of sets $\{D(f, \epsilon), \ f \in I, \ \epsilon > 0\}$ generates a filter $\mathcal{F}_{I,S}$ on E such that $I \subset \mathcal{I}(\mathcal{F}_{I,S}, S)$.*

3.3. Maximal and prime ideals of S

Except Theorem 3.3.7 and its corollaries, most of the results of this paragraph were given in [13] for the algebra $\mathcal{C}$.

Theorem 3.3.1. *Let $\mathcal{M}$ be a maximal ideal of S. There exists an ultrafilter $\mathcal{U}$ on E such that $\mathcal{M} = \mathcal{I}(\mathcal{U}, S)$. Moreover, $\mathcal{M}$ is of codimension 1 if and only if every element of S converges along $\mathcal{U}$. In particular if $\mathcal{U}$ is convergent, then $\mathcal{M}$ is of codimension 1.*

Proof. Indeed, by Corollary 3.2.12, we can consider the filter $\mathcal{F}_{\mathcal{M},S}$ and we have $\mathcal{M} \subset \mathcal{I}(\mathcal{F}_{\mathcal{M},S}, S)$. Let $\mathcal{U}$ be an ultrafilter thinner than $\mathcal{F}_{\mathcal{M},S}$. So, we have $\mathcal{M} \subset \mathcal{I}(\mathcal{F}_{\mathcal{M},S}, S) \subset \mathcal{I}(\mathcal{U}, S)$. But since $\mathcal{M}$ is a maximal ideal, either $\mathcal{M} = \mathcal{I}(\mathcal{U}, S)$, or $\mathcal{I}(\mathcal{U}, S) = S$. But obviously, $\mathcal{I}(\mathcal{U}, S) \neq S$, hence $\mathcal{M} = \mathcal{I}(\mathcal{U}, S)$.

Now assume that $\mathcal{M}$ is of codimension 1 and let χ be the $\mathbb{L}$-algebra homomorphism from S to $\mathbb{L}$ admitting $\mathcal{M}$ for kernel. Let $f \in S$ and let $b = \chi(f)$. Then $f - b$ belongs to the kernel of $\mathcal{M}$, hence $\lim_{\mathcal{U}} f(x) - b = 0$ that is $\lim_{\mathcal{U}} f(x) = b$ therefore every element of S converges along $\mathcal{U}$.

Conversely if every element of S admits a limit along $\mathcal{U}$ then the mapping χ which associates to each $f \in S$ its limit along $\mathcal{U}$ is a $\mathbb{L}$-algebra homomorphism from S to $\mathbb{L}$ admitting $\mathcal{M}$ for kernel, therefore $\mathcal{M}$ is of codimension 1.

In particular, if $\mathcal{U}$ converges to a point a, then each f in S converges to $f(a)$ along $\mathcal{U}$. $\qquad\square$

By Lemma 3.2.4 and Theorem 3.3.1, the following corollary is immediate.

Corollary 3.3.2. *Let $\mathbb{L}$ be a locally compact field. Then every maximal ideal of S is of codimension 1.*

Remark 3. If $\mathbb{L}$ is locally compact, a maximal ideal of codimension 1 of S is not necessarily of the form $\mathcal{I}(\mathcal{U}, S)$ where $\mathcal{U}$ is a Cauchy ultrafilter. Suppose that E admits a sequence $(a_n)_{n \in \mathbb{N}}$ such that either it satisfies $|a_n - a_m| = r \ \forall n \neq m$, or the sequence $|a_{n+1} - a_n|$ is strictly increasing. Let $\mathcal{U}$ be an ultrafilter thinner than the sequence $(a_n)_{n \in \mathbb{N}}$.

Consider now a function $f \in S$. Since $\mathbb{L}$ is locally compact, $f(x)$ does converge along $\mathcal{U}$ to a point $b \in \mathbb{L}$. In that way, we can define a homomorphism χ from S onto $\mathbb{L}$ as $\chi(g) = \lim_{\mathcal{U}} g(x)$ and therefore $\mathbb{L}$ is the quotient $\frac{S}{\mathrm{Ker}(\chi)}$. So, $\mathcal{I}(\mathcal{U}, S)$ is a maximal ideal of codimension 1.

Notation: Following notations of [14], we will denote by $Y_{(S)}(E)$ the set of equivalence classes on $U(E)$ with respect to Relation $(\mathcal{R})$.

By Theorem 3.2.9, we can get Corollary 3.3.3.

Corollary 3.3.3. *Let $\mathcal{M}$ be a maximal ideal of S. There exists a unique $\mathcal{H} \in Y_{(S)}(E)$ such that $\mathcal{M} = \mathcal{I}(\mathcal{U}, S)$ for every $\mathcal{U} \in \mathcal{H}$.*

Conversely, Theorem 3.3.4 now characterizes all maximal ideals of S.

Theorem 3.3.4. *Let $\mathcal{U}$ be an ultrafilter on E. Then $\mathcal{I}(\mathcal{U}, S)$ is a maximal ideal of S.*

Proof. Let $I = \mathcal{I}(\mathcal{U}, S)$ and let $\mathcal{M}$ be a maximal ideal of S containing I. Then, by Theorem 3.3.1, there exists an ultrafilter $\mathcal{V}$ such that $\mathcal{M} = \mathcal{I}(\mathcal{V}, S)$. Suppose now $\mathcal{I}(\mathcal{U}, S) \neq \mathcal{I}(\mathcal{V}, S)$. Then, $\mathcal{U}$ and $\mathcal{V}$ are not sticked. Consequently, by Theorem 3.2.2, there exists a clopen subset $F \in \mathcal{V}$ that does not belong to $\mathcal{U}$ and hence its characteristic function $u \in S$ belongs to $\mathcal{I}(\mathcal{U}, S)$ but does not belong to $\mathcal{I}(\mathcal{V}, S)$. Thus, u belongs to I but does not belong to $\mathcal{M}$, a contradiction to the hypothesis. $\qquad\square$

Definition: A field $\mathbb{E}$ is said to be *perfect* if every algebraic extension of $\mathbb{E}$ is algebraically separable.

By Corollary 3.3.3 and Theorem 3.3.4, we can derive the following Corollary 3.3.5.

Corollary 3.3.5. *The mapping that associates to each maximal ideal $\mathcal{M}$ of S the class with respect to $(\mathcal{R})$ of ultrafilters $\mathcal{U}$, such that $\mathcal{M} = \mathcal{I}(\mathcal{U}, S)$, is a bijection from $\mathrm{Max}(S)$ onto $Y_{(S)}(E)$.*

Remark 4. Let $\mathcal{F}$ be a Cauchy filter on E admitting a limit limit $a \in E$ and let $\mathcal{M} = \mathcal{I}(\mathcal{F}, S)$. Then every function $f \in S$ converges to a limit $\theta(f)$ along $\mathcal{F}$ and $\mathcal{M}$ is a maximal ideal of codimension 1. Indeed, let $f \in S$. Since f is continuous, then $f(x)$ converges to a point $\theta(f) = f(a)$ in $\mathbb{L}$. Consider now the mapping θ from S into $\mathbb{L}$: it is an algebra morphism whose kernel is $\mathcal{M}$ and whose image is $\mathbb{L}$. Consequently, $\frac{S}{\mathcal{M}} = \mathbb{L}$, therefore $\mathcal{M}$ is a maximal ideal of codimension 1.

Notations: For any subset F of E, we denote by u_F its characteristic function. Let $\mathcal{M}$ be a maximal ideal of S and let $\mathcal{U} \in U(E)$ be such that $\mathcal{M} = \mathcal{I}(\mathcal{U}, S)$. By Theorems 3.2.9, we can define the set $\mathcal{O}_{\mathcal{M}}$ of all clopen subsets of E which belong to $\mathcal{U}$. We then denote by $\mathcal{C}_{\mathcal{M}}$ the set $\{u_{E\setminus L} \mid L \in \mathcal{O}_{\mathcal{M}}\}$ and by $\mathcal{J}_{\mathcal{M}}$ the set of all functions $f \in S$ which are equal to 0 on some $L \in \mathcal{O}_{\mathcal{M}}$.

Given $a \in E$, we will denote by $\mathcal{I}'(a, S)$ the ideal of the functions $f \in S$ equal to 0 on an open subset of E containing a.

Theorem 3.3.6. *Let* $\mathcal{M}$ *be a maximal ideal of* S.

(1) $\mathcal{J}_{\mathcal{M}}$ *is an ideal of* S *containing* $\mathcal{C}_{\mathcal{M}}$,
(2) $\mathcal{J}_{\mathcal{M}}$ *is the ideal of* S *generated by* $\mathcal{C}_{\mathcal{M}}$ *and* $\mathcal{J}_{\mathcal{M}} = \{fu \mid f \in S, u \in \mathcal{C}_{\mathcal{M}}\}$,
(3) *If* $\mathcal{P}$ *is a prime ideal of* S *contained in* $\mathcal{M}$, *then* $\mathcal{J}_{\mathcal{M}} \subset \mathcal{P}$.
(4) *If* $\mathcal{M} = \mathcal{I}(a, S)$, *then* $\mathcal{I}'(a, S) = \mathcal{J}_{\mathcal{M}}$.

Proof. (1) Let us check that $\mathcal{J}_{\mathcal{M}}$ is an ideal of S. Let $f, g \in \mathcal{J}_{\mathcal{M}}$. So, there exist $F, G \in \mathcal{O}_{\mathcal{M}}$ such that $f(x) = 0 \ \forall x \in F$, $g(x) = 0 \ \forall x \in G$, hence $f(x) - g(x) = 0 \ \forall x \in F \cap G$. Since $F \cap G$ belongs to $\mathcal{O}_{\mathcal{M}}$, $f - g$ lies in $\mathcal{J}_{\mathcal{M}}$. And obviously, for every $h \in S$, we have $h(x)f(x) = 0 \ \forall x \in F$, hence fh lies in $\mathcal{J}_{\mathcal{M}}$.

Next, $\mathcal{J}_{\mathcal{M}}$ contains $\mathcal{C}_{\mathcal{M}}$ because given $F \in \mathcal{O}_{\mathcal{M}}$, the set $E \setminus F$ is clopen, then $u_{E\setminus F}$ belongs to S and is equal to 0 on F.

(2) Notice that if $f \in S$ and $u \in \mathcal{C}_{\mathcal{M}}$, then by 1) fu belongs to $\mathcal{J}_{\mathcal{M}}$. Conversely, if $f \in \mathcal{J}_{\mathcal{M}}$ and $F \in \mathcal{O}_{\mathcal{M}}$ are such that $f(x)$ is equal to 0 on F, then $u_{E\setminus F}$ belongs to $\mathcal{C}_{\mathcal{M}}$ and $f = fu_{E\setminus F}$. This proves that $\mathcal{J}_{\mathcal{M}} = \{fu \mid f \in S, u \in \mathcal{C}_{\mathcal{M}}\}$ and that $\mathcal{J}_{\mathcal{M}}$ is the ideal generated by $\mathcal{C}_{\mathcal{M}}$.

(3) It is sufficient to prove that $\mathcal{C}_{\mathcal{M}}$ is included in $\mathcal{P}$. Indeed, let $\mathcal{U} \in U(E)$ be such that $\mathcal{M} = \mathcal{I}(\mathcal{U}, S)$ and let $F \in \mathcal{O}_{\mathcal{M}}$. Then, $F \in \mathcal{U}$ and $u_F \notin \mathcal{M}$. So, $u_F \notin \mathcal{P}$. But $u_F.u_{E\setminus F} = 0$. Thus, $u_{E\setminus F}$ belongs to $\mathcal{P}$ since $\mathcal{P}$ is prime.

(4) Just notice that $\mathcal{J}_{\mathcal{M}}$ is the set of all functions in S which are equal to 0 on some clopen containing a and that each open neighborhood of a contains a disk $B(a, r)$, which is clopen. $\square$

Corollary 3.3.7. *Let* $\mathcal{U}$ *be an ultrafilter on* E *and let* $\mathcal{P}$ *be a prime ideal included in* $\mathcal{I}(\mathcal{U}, S)$. *Let* $F \in \mathcal{U}$ *be clopen and let* $H = E \setminus F$. *Then the characteristic function* u *of* H *belongs to* $\mathcal{P}$.

Theorem 3.3.8. *Let* $\mathcal{M}$ *be a maximal ideal of* S. *The uniform closure of* $\mathcal{J}_{\mathcal{M}}$ *is equal to* $\mathcal{M}$.

Proof. Let $f \in \mathcal{M} = \mathcal{I}(\mathcal{U}, S)$. Then for every $\epsilon > 0$ the set $F = D(f, \epsilon)$ belongs to $\mathcal{U}$ and F is clopen. Therefore, F belongs to $\mathcal{O}_{\mathcal{M}}$ and the characteristic function u of $E \setminus F$ lies in $\mathcal{C}_{\mathcal{M}}$, so that $fu \in \mathcal{J}_{\mathcal{M}}$. But $f(x) - uf(x) = 0 \ \forall x \notin F$ and $|f(x) - uf(x)| = |f(x)| \le \epsilon \ \forall x \in F$, so

$\|f - uf\|_0 \leq \epsilon$. Hence $\mathcal{M}$ is the uniform closure of $\mathcal{J}_{\mathcal{M}}$ since, by Proposition 3.2.7, $\mathcal{M}$ is spectrally closed, hence uniformly closed. $\square$

Corollary 3.3.9. *Let $\mathcal{P}$ be a prime ideal contained in a maximal ideal $\mathcal{M}$. Then $\mathcal{M}$ is the uniform closure of $\mathcal{P}$.*

Corollary 3.3.10. *The uniform closure of a prime ideal of S is a maximal ideal of S and a prime ideal of S is contained in a unique maximal ideal of S.*

Corollary 3.3.11. *A prime ideal of S is a maximal ideal if and only if it is uniformly closed.*

Using Property 4 of Theorem 3.3.6 we get Corollary 3.3.12.

Corollary 3.3.12. *The uniform closure of $\mathcal{I}'(a, S)$ is $\mathcal{I}(a, S)$.*

Corollary 3.3.13. *Let $\mathcal{M}$ be a maximal ideal of S. Then:*

(1) *$\mathcal{M}$ is the spectral closure of $\mathcal{J}_{\mathcal{M}}$ and the spectral closure of any prime ideal contained in $\mathcal{M}$;*
(2) *a prime ideal is maximal if and only if it is spectrally closed.*

3.4. Maximal ideals of finite codimension for admissible algebras

The main results of this paragraph were already obtained in [14]. We recall them with all proofs in order to make easy the conclusions of this paragraph.

Notations: Let $\mathbb{H}$ be a finite algebraic extension of $\mathbb{L}$ equipped with the absolute value which extends that of $\mathbb{L}$ and let $t = [\mathbb{H} : \mathbb{L}]$. Let $\mathcal{C}^e$ be equal to the $\mathbb{H}$-algebra of the bounded continuous functions from E into $\mathbb{H}$ and $\widehat{\mathcal{C}} = \mathbb{H} \otimes_{\mathbb{L}} \mathcal{C}$. Since $\mathbb{H}$ is of finite dimension over $\mathbb{L}$, one obtains an immediate identification of $\mathcal{C}^e$ with $\widehat{\mathcal{C}}$.

The following Theorem 3.4.2 holds on all complete valued fields and is proven in [14, Lemma 7.2]. First we must state Lemma 3.4.1.

Lemma 3.4.1. *Let $\mathbb{H}$ be of the form $\mathbb{H} = \mathbb{L}[a]$. Let $f \in \mathcal{C}^e$. Then f is of the form $\sum_{j=0}^{t-1} a^j f_j$, $j = 0, \ldots, t-1$, with $f_j \in \mathcal{C}$.*

Theorem 3.4.2. *Let $T = \mathcal{C}$. Suppose there exists a morphism of $\mathbb{L}$-algebra, χ from T onto $\mathbb{H}$. Then χ has continuation to a surjective morphism $\widehat{\chi}$ of $\mathbb{H}$-algebra from $\widehat{T}$ to $\mathbb{H}$.*

Proof. Suppose first that $\mathbb{H}$ is of the form $\mathbb{L}[a]$. Let $f, g \in \widehat{T}$. Then by Lemma 3.4.1, f is of the form $\sum_{j=0}^{t-1} a^j f_j$, $j = 0, \ldots, t-1$ and g is of the form $\sum_{j=0}^{t-1} a^j g_j$, $j = 0, \ldots, t-1$, where the f_j and the g_j are functions from E to $\mathbb{L}$.

We can now define $\widehat{\chi}$ on $\widehat{T}$ as $\widehat{\chi}(f) = \sum_{j=0}^{t-1} a^j \chi(f_j)$. Then obviously, $\widehat{\chi}$ is $\mathbb{L}$-linear. On the other hand, for each $q \in \mathbb{N}$, a^q is of the form $P_q(a)$ where $P_q \in \mathbb{K}[x]$, $\deg(P_q) \leq t-1$. Then $\widehat{\chi}(a^q) = \widehat{\chi}(P_q(a)) = P_q(\widehat{\chi}(a)) = P_q(a) = a^q$. Next,

$$\widehat{\chi}(fg) = \widehat{\chi}\left(\left(\sum_{j=0}^{t-1} a^j f_j\right)\left(\sum_{j=0}^{t-1} a^j g_j\right)\right) = \widehat{\chi}\left(\sum_{\substack{0 \leq m \leq t-1 \\ 0 \leq n \leq t-1}} a^{m+n} f_m g_n\right)$$

$$= \sum_{\substack{0 \leq m \leq t-1 \\ 0 \leq n \leq t-1}} a^{m+n} \chi(f_m) \chi(g_n) = \left(\sum_{j=0}^{t-1} a^j \chi(f_j)\right)\left(\sum_{j=0}^{t-1} a^j \chi(g_j)\right)$$

$$= \chi(f)\chi(g).$$

Thus, the extension of χ is proved whenever $\mathbb{H}$ is of the form $\mathbb{L}[a]$. It is then immediate to check that $\widehat{\chi}$ is surjective: since $\widehat{T}$ is a $\mathbb{H}$-algebra, it contains the field $\mathbb{H}$ and every morphism $\widehat{\chi}$ from $\widehat{T}$ obviously satisfies $\widehat{\chi}(c) = c \ \forall c \in \mathbb{H}$.

Consider now the general case. We can obviously write $\mathbb{H}$ in the form $\mathbb{L}[b_1, \ldots, b_q]$. Writing $\mathbb{H}_j$ for the extension $\mathbb{L}[b_1, \ldots, b_j]$ we have $\mathbb{H}_j = \mathbb{H}_{j-1}[b_j]$.

By induction on j, using the preceding just proved result, we get that for each $j = 1, \ldots, q$, χ has continuation to a surjective morphism of $\mathbb{H}_j$-algebra, $\widehat{\chi}_j$, from $\mathbb{H}_j \otimes T$ onto $\mathbb{H}_j$. Taking $j = q$ ends the proof. $\square$

We are now able to prove that maximal ideals of finite codimension of S are of codimension 1 in two cases: when $S = \mathcal{A}$ and when the field $\mathbb{K}$ is perfect.

Theorem 3.4.3. *Every maximal ideal* $\mathcal{M}$ *of finite codimension of* $\mathcal{C}$ *is of codimension 1.*

Proof. Let $\mathbb{H}$ be the field $\frac{\mathcal{C}}{\mathcal{M}}$ and let $\mathcal{C}'$ be the $\mathbb{H}$-algebra of bounded continuous functions from E to $\mathbb{H}$. Then $\mathcal{C}'$ is semi-admissible. Now, let χ be the $\mathbb{L}$-algebra morphism from $\mathcal{C}$ onto $\mathbb{H}$ whose kernel is $\mathcal{M}$. Let $g \in \mathcal{C}$ and let $b = \chi(g) \in \mathbb{H}$. By Theorem 3.4.2, χ admits an extension to a morphism $\widehat{\chi}$ from $\mathcal{C}'$ to $\mathbb{H}$. Now, since S' is semi-admissible and since the kernel of $\widehat{\chi}$ is a maximal ideal $\widehat{\mathcal{M}}$ of $\mathcal{C}'$, there exists an ultrafilter $\mathcal{U}$ on E such that

$\widehat{\mathcal{M}} = \mathcal{I}(\mathcal{U}, \mathcal{C}')$. Then we have $\widehat{\chi}(g - b) = 0$, hence $g - b$ belongs to $\widehat{\mathcal{M}}$, therefore $\lim_{\mathcal{U}} g(x) - b = 0$ i.e., $\lim_{\mathcal{U}} g(x) = b$. But since $g \in \mathcal{C}$, $g(x)$ belongs to $\mathbb{L}$ for all $x \in E$. Therefore, since $\mathbb{L}$ is complete, b belongs to $\mathbb{L}$. But by definition χ is a surjection from $\mathcal{C}$ onto $\mathbb{H}$, hence every value b of $\mathbb{H}$ is the image of some $f \in \mathcal{C}$ and hence it lies in $\mathbb{L}$, therefore $\mathbb{L} = \mathbb{H}$.   □

Theorem 3.4.3 can be generalized to all semi-admissible algebras equipped $\mathbb{L}$ is a perfect field [13].

Definition and notation: Given a field $\mathbb{E}$ of characteristic $p \neq 0$, we denote by $\mathbb{E}^{\frac{1}{p}}$ the extension of $\mathbb{E}$ containing all pth-roots of elements of $\mathbb{E}$. Let $\mathbb{L}$ be a subfield of $\mathbb{K}$.

Henceforth, for the rest of this chapter, the field $\mathbb{L}$ is supposed to be perfect.

Proposition 3.4.4. *Let $\mathbb{H} = \mathbb{L}[a]$ be a finite extension of $\mathbb{L}$ of degree t, equipped with the unique absolute value extending that of $\mathbb{L}$ and let $a_2, \ldots, a_t$ be the conjugates of a over $\mathbb{L}$, with $a_1 = a$. Let $\widehat{S} = \mathbb{H} \otimes_{\mathbb{L}} S$ and let $g = \sum_{j=0}^{t-1} a^j f_j$, $f_j \in S$ be such that $\inf_E |g(x)| > 0$. For every $k = 1, \ldots, t$, let $g_k = \sum_{j=0}^{t-1} a_k^j f_j$, $f_j \in S$. Then $\prod_{k=1}^{t} g_k$ belongs to S and $\prod_{k=2}^{t} g_k$ belongs to $\widehat{S}$.*

Proof.   Since $\mathbb{H} = \mathbb{L}[a_1]$ and since $\mathbb{L}$ is perfect, the extension $N = \mathbb{L}[a_1, \ldots, a_t]$ is of the form

$$N = \mathbb{L}[a_1][a_2, \ldots, a_t] = \mathbb{H}[a_2, \ldots, a_t]$$

and N is a normal extension of $\mathbb{L}$ and of $\mathbb{H}$, respectively.

Thus, assuming that $a_1, \ldots, a_s$ belong to $\mathbb{H}$, we have $\mathbb{H} = \mathbb{L}[a_1, \ldots, a_s]$ and $N = \mathbb{L}[a_1 \ldots, a_s, a_{s+1}, \ldots, a_t] = \mathbb{L}[a_1, \ldots, a_s][a_{s+1}, \ldots, a_t] = \mathbb{H}[a_{s+1}, \ldots, a_t]$.

Let G be the Galois group of N over $\mathbb{L}$ and put $G' = \{\sigma \in G : \sigma(x) = x, \forall x \in \mathbb{H}\}$ where the extension N over $\mathbb{H}$ is Galoisian, whose Galois group $G(N|\mathbb{H}) =$ is G'. The subfield $\mathbb{H}$ of N then corresponds to the subgroup G' of G through the Galois correspondence.

Now, given $\sigma \in G$, set $\sigma(g) = \sum_{j=0}^{t-1} (\sigma(a))^j f_j$. Let $F = \prod_{k=1}^{t} g_k = \prod_{\sigma \in G} \sigma(g)$. Then F belongs to S if and only if for every $\tau \in G$, $\tau(F) = F$. Now, we have

$$\tau(F) = \prod_{\sigma \in G} \tau \circ \sigma(g) = \prod_{\zeta \in G} \zeta(g) = F,$$

therefore F belongs to S.

On the other hand, the roots a_i, for $s + 1 \le i \le t$, are conjugate over $\mathbb{H}$. Therefore, if $s + 1 \le i \le t$, there exists $\theta \in G'$ such that $a_i = \theta(a_{s+1})$. It follows that $g_i = \sum_{j=0}^{t-1} a_i^j f_j = \sum_{j=0}^{t-1} \theta(a_{s+1})^j f_j = \theta(\sum_{j=0}^{t-1} a_{s+1}^j f_j) = \theta(g_{s+1})$.

Let $P = \prod_{i=s+1}^{t} g_i = \prod_{\theta \in G'} \theta(g_{s+1})$. Then P belongs to $\widehat{S}$ if and only if $\tau(P) = P \ \forall \tau \in G'$. Now, we have $\tau(P) = \prod_{\theta \in G'} \tau \circ \theta(g_{s+1}) = \prod_{\zeta \in G'} \zeta(g_{s+1}) = P$, therefore P belongs to $\widehat{S}$. Consequently, since $\prod_{i=2}^{s} g_i$ belongs to $\widehat{S}$, one gets that $(\prod_{i=2}^{s} g_i) \cdot P = \prod_{i=2}^{t} g_i$ is an element of $\widehat{S}$, which ends the proof. $\qquad \square$

We can now establish the following Proposition 3.4.5.

Proposition 3.4.5. *Let* $\mathbb{H} = \mathbb{L}[a]$ *be a finite extension of* $\mathbb{L}$ *of degree* t, *equipped with the unique absolute value extending that of* $\mathbb{L}$ *and let* $a_2, \ldots, a_t$ *be the conjugates of* a *over* $\mathbb{L}$, *with* $a_1 = a$. *Let* $\widehat{S} = \mathbb{H} \otimes_{\mathbb{L}} S$ *and let* $g \in \widehat{S}$ *be such that* $\inf_{sE} |g(x)| > 0$. *Then* g *is invertible in* $\widehat{S}$.

Proof. Let $g = \sum_{j=0}^{t} a^j f_j$, $f_j \in S$ and for every $k = 1, \ldots, t$, let $g_k = \sum_{j=0}^{t} a_k^j f_j$, $f_j \in S$. Then, by Proposition 3.4.4, $\prod_{k=1}^{t} g_k$ belongs to S and in the same way, $\prod_{k=2}^{t} g_k$ belongs to $\widehat{S}$. Now, since $\inf_E |g(x)|$ is a number $m > 0$, we have $|\prod_{k=1}^{t} g_k| \ge m^t$ because in $\mathbb{H}$, we have $|g_k(x)| = |g_1(x)| \ \forall k = 1, 2, \ldots, t$, $\forall x \in E$. Consequently, $\prod_{k=1}^{t} g_k$ is invertible in S. Thus, there exists $f \in S$ such that $\prod_{k=1}^{t} g_k . f = 1$. But since, by Proposition 3.4.4, $\prod_{k=2}^{t} g_k$ belongs to $\widehat{S}$, one sees that $\prod_{k=2}^{t} g_k . f$ is an element of $\widehat{S}$. Hence $g = g_1$ is invertible in $\widehat{S}$ with inverse $g^{-1} = \prod_{k=2}^{t} g_k . f$ $\qquad \square$

Definition and notation: In the following Proposition 3.4.7 and in the theorems we will have to consider the *tensor product norm* [16]. We remind here some general facts. Let $\mathbb{H}$ be a complete valued field extension of $\mathbb{L}$ and A be a unital, ultrametric $\mathbb{L}$-Banach algebra. Given $z \in \mathbb{H} \otimes_{\mathbb{L}} A$, we put

$$\|z\|_\otimes = \inf \left\{ \max_{i \in I} |b_i| . \|x_i\| \ \middle| \ \sum_{i \in I} b_i \otimes_{\mathbb{K}} x_i = z, \text{Infinite} \right\}.$$

This norm $\| \, . \, \|_\otimes$ will be called *the (projective) tensor product norm*. It is an ultrametric norm.

In any unital $\mathbb{L}$-algebra B, let 1_B be the unity of B. Then for $b \in \mathbb{H}$ and $x \in B$, one has $\|b \otimes x\|_\otimes = |b| . \|x\|$. In particular for any $b \in \mathbb{H}$, (resp., $x \in B$), one has $\|b \otimes 1_B\|_\otimes = |b|$ (resp., $\|1_{\mathbb{L}} \otimes x\|_\otimes = \|x\|$.) Hence one has an isometric identification of $\mathbb{H}$ (resp., B) with $\mathbb{H} \otimes_{\mathbb{L}} 1_B$ (resp., $1_{\mathbb{H}} \otimes_{\mathbb{L}} B$).

Furthermore, one verifies that with the tensor norm, the tensor product $\mathbb{H} \otimes_{\mathbb{L}} B$, of the two unital $\mathbb{L}$-algebras $\mathbb{K}$ and B is a normed unital $\mathbb{L}$-algebra.

It is also a unital $\mathbb{H}$-algebra (obtained by extension of scalars). The completion $\mathbb{H}\widehat{\otimes}_{\mathbb{L}}B$ of $\mathbb{H}\otimes_{\mathbb{L}}B$ with respect to the tensor product norm $\|\,.\,\|_{\otimes}$ (called the topological tensor product) is a unital $\mathbb{L}$-Banach algebra as well as a $\mathbb{H}$-Banach algebra.

Now assume that $\mathbb{H}$ is of finite dimension d over $\mathbb{L}$. Fix a $\mathbb{L}$-basis $(e_j)_{1\leq j\leq d}$ of $\mathbb{H}$. It is readily seen that any $z\in\mathbb{H}\otimes_{\mathbb{L}}B$ can be written in the unique form $z=\sum_{j=1}^{d}e_j\otimes y_j$ and $\|z\|_{\otimes}=\|\sum_{j=1}^{d}e_j\otimes y_j\|_{\otimes}\leq\max_{1\leq j\leq d}|e_j|.\|y_j\|$.

On the other hand, given $b=\sum_{j=1}^{d}\beta_j e_j\in\mathbb{H}$, let us consider the norm $\|b\|_1=\max_{1\leq j\leq d}|\beta_j|.|e_j|$. One has $|b|\leq\|b\|_1$ and since $\mathbb{H}$ is finite dimensional, there exists $\alpha>0$ such that $\alpha\max_{1\leq j\leq d}|\beta_j|.|e_j|\leq|b|\leq\max_{1\leq j\leq d}|\beta_j|.|e_j|$. Considering the dual basis $(e_j')_{1\leq j\leq d}$ of $(e_j)_{1\leq j\leq d}$ and the continuous linear operators $e_j'\otimes id_B$ of $\mathbb{H}\otimes_{\mathbb{L}}B$ into $\mathbb{L}\otimes_{\mathbb{L}}B=B$, one proves that $\alpha\max_{1\leq j\leq d}|e_j|.\|y_j\|\leq\|z\|_{\otimes}\leq\max_{1\leq j\leq d}|e_j|.\|y_j\|=\|z\|_1$.

This means that the norms $\|\,.\,\|_{\otimes}$ and $\|\,.\,\|_1$ of $\mathbb{H}\otimes_{\mathbb{L}}B$ are equivalent. One immediately sees that $\mathbb{H}\otimes_{\mathbb{L}}B$ equipped with the norm $\|z\|_1=\max_{1\leq j\leq d}|e_j|\|y_j\|$ is isomorphic to the product $\mathbb{L}$-Banach space B^d and then it is complete. It follows that $(\mathbb{H}\otimes_{\mathbb{L}}B,\|\,.\,\|_{\otimes})$ is complete and $\mathbb{H}\otimes_{\mathbb{L}}B=\mathbb{H}\widehat{\otimes}_{\mathbb{L}}B$.

One then has the following Theorem 3.4.6 contained in [13, Theorem 3.6].

Theorem 3.4.6. *If $\mathbb{H}$ is a finite extension of $\mathbb{L}$ and B is a unital commutative $\mathbb{L}$-Banach algebra, then with the tensor product norm $\|\,.\,\|_{\otimes}$, the tensor product $\mathbb{H}\otimes_{\mathbb{L}}B$ of the $\mathbb{L}$-algebras $\mathbb{H}$ and B is a $\mathbb{L}$-Banach algebra as well as a Banach algebra over $\mathbb{H}$.*

Taking $B=S$, we can now conclude.

Proposition 3.4.7. *Let $\mathbb{H}=\mathbb{L}[a]$ be a finite extension of $\mathbb{L}$ of degree t, equipped with the unique absolute value extending that of $\mathbb{L}$. Then the algebra $\widehat{S}=\mathbb{H}\otimes_{\mathbb{L}}S$ equipped with the tensor product norm $\|\,\|_{\otimes}$, is complete. Moreover, $\widehat{S}$ can be identified with the Banach $\mathbb{H}$-algebra of functions f from E to $\mathbb{H}$ of the form $f=\sum_{j=0}^{t-1}a^j f_j$ with $f_j\in S$ and $\widehat{S}$ is a semi-admissible $\mathbb{H}$-algebra.*

Proof. By construction, $\widehat{S}$ is the set of functions $f=\sum_{j=0}^{t-1}a^j f_j$ with $f_j\in S$. Since each f_j is continuous, so is f. By Theorem 3.4.6, $\widehat{S}$ is a Banach $\mathbb{H}$-algebra. Next, given a clopen subset D of E, the characteristic function u of D exists in S and hence it belongs to $\widehat{S}$. Finally, given an element $g\in\widehat{S}$ such that $\inf_{x\in E}|g(x)|>0$, by Proposition 3.4.5, g is invertible in $\widehat{S}$. Therefore, $\widehat{S}$ is semi-admissible. $\qquad\square$

Theorem 3.4.8. *Let* $\mathbb{H}$ *be a finite extension of* $\mathbb{L}$ *of degree* t, *equipped with the unique absolute value extending that of* $\mathbb{L}$ *and let* $\widehat{S} = \mathbb{H} \otimes_{\mathbb{L}} S$ *be equipped with the tensor product norm. Then* $\widehat{S}$ *is a semi-admissible Banach* $\mathbb{H}$-*algebra.*

Proof. By definition, $\mathbb{H}$ is of the form $\mathbb{L}[b_1, \ldots, b_q]$ with $\mathbb{L}[b_1, \ldots, b_j]$ strictly included in $\mathbb{L}[b_1, \ldots, b_{j+1}]$, $j = 1, \ldots, q - 1$. Put $\mathbb{H}_j = \mathbb{L}[b_1, \ldots, b_j]$, $j = 1, \ldots, q$ and $\widehat{S}_j = \mathbb{H}_j \otimes_{\mathbb{L}} S$. Suppose we have proved that $\widehat{S}_j$ is semi-admissible for some $j < q$. Next, since $\mathbb{H}_{j+1} = \mathbb{H}_j[b_{j+1}]$, by Proposition 3.4.7, $\widehat{S}_{j+1}$ is semi-admissible. Therefore, by induction, $\widehat{S}_q = \widehat{S}$ is a semi-admissible Banach $\mathbb{H}$-algebra. $\qquad \square$

Theorem 3.4.9. *Let* $\mathcal{M}$ *be a maximal ideal of finite codimension of* S. *Then* $\mathcal{M}$ *is of codimension* 1.

Proof. Let $\mathbb{H}$ be the field $\frac{S}{\mathcal{M}}$ and let $\widehat{S} = \mathbb{H} \otimes_{\mathbb{L}} S$ be equipped with the tensor product norm. By Theorem 3.4.8, $\widehat{S}$ is semi-admissible. Now, let χ be the morphism from S onto $\mathbb{H}$ whose kernel is $\mathcal{M}$. Let $g \in S$ and let $b = \chi(g) \in \mathbb{H}$. By Theorem 3.4.2, χ admits an extension to a morphism $\widehat{\chi}$ from $\widehat{S}$ to $\mathbb{H}$. But since $\widehat{S}$ is semi-admissible and since the kernel of $\widehat{\chi}$ is a maximal ideal $\widehat{\mathcal{M}}$ of $\widehat{S}$, by Theorem 3.3.1, there exists an ultrafilter $\mathcal{U}$ on E such that $\widehat{\mathcal{M}} = \mathcal{I}(\mathcal{U}, S)$. Take $g \in S$ and let $b = \chi(g)$. Then we have $\widehat{\chi}(g - b) = 0$, hence $g - b$ belongs to $\widehat{\mathcal{M}}$, therefore $\lim_{\mathcal{U}} g(x) - b = 0$, i.e., $\lim_{\mathcal{U}} g(x) = b$. But since $g \in S$, $g(x)$ belongs to $\mathbb{L}$ for all $x \in E$. Therefore, since $\mathbb{L}$ is complete, b belongs to $\mathbb{L}$. But by definition χ is a surjection from S onto $\mathbb{H}$, hence every value b of $\mathbb{H}$ actually lies in $\mathbb{L}$ and hence $\mathbb{L} = \mathbb{H}$. $\qquad \square$

3.5. Multiplicative spectrum of admissible algebras

Notation: Let A be a normed $\mathbb{L}$-algebra. We denote by $\Upsilon(A)$ the set of $\mathbb{L}$-algebra homomorphisms from A to $\mathbb{L}$.

Particularly, considering the algebra S, we denote by $\mathrm{Mult}_E(S, \| \cdot \|)$ the set of continuous multiplicative semi-norms of T whose kernel is a maximal ideal of the form $\mathcal{I}(a, S)$, $a \in E$.

Let us recall that in S, we have $\| \cdot \|_0 \leq \| \cdot \|_{\mathrm{sp}}$ and that if S is admissible, then $\| \cdot \|_0 = \| \cdot \|_{\mathrm{sp}}$. Theorem 3.5.1 is classical [26, 28].

Theorem 3.5.1. *For each* $f \in S$, $\|f\|_{\mathrm{sp}} = \sup\{\phi(f) \mid \phi \in \mathrm{Mult}(S, \| \cdot \|)\}$. *For every* $\chi \in \Upsilon(S)$, *we have* $|\chi(f)| \leq \|f\|_{\mathrm{sp}} \ \forall f \in S$.

More notations: For any ultrafilter $\mathcal{U} \in U(E)$ and any $f \in S$, $|f(x)|$ has a limit along $\mathcal{U}$ since f is bounded. Given $a \in E$ we denote by φ_a the mapping from S to $\mathbb{R}$ defined by $\varphi_a(f) = |f(a)|$ and for any ultrafilter $\mathcal{U} \in U(E)$, we denote by $\varphi_{\mathcal{U}}$ the mapping from S to $\mathbb{R}$ defined by $\varphi_{\mathcal{U}}(f) = \lim_{\mathcal{U}} |f(x)|$ (see Lemma 3.2.4). These maps belong to $\text{Mult}(S, \| \cdot \|)$ since $\| \cdot \|_0 \leq \| \cdot \|_{\text{sp}} \leq \| \cdot \|$. Particularly, the elements of $\text{Mult}_E(S, \| \cdot \|)$ are the φ_a, $a \in E$.

Proposition 3.5.2. *Let* $a \in E$. *Then* $\mathcal{I}(a, S)$ *is a maximal ideal of* S *of codimension 1 and* φ_a *belongs to* $\text{Mult}_1(S, \| \cdot \|)$. *Conversely, for every algebra homomorphism* χ *from* S *to* $\mathbb{L}$, *its kernel is a maximal ideal of the form* $\mathcal{I}(a, S)$ *with* $a \in E$ *and* χ *is defined as* $\chi(f) = f(a)$, *while* $\varphi_a(f) = |\chi(f)|$.

Theorem 3.5.3. *Let* $\mathcal{U}$ *be an ultrafilter on* E. *Then* $\varphi_{\mathcal{U}}$ *belongs to the closure of* $\text{Mult}_E(S, \| \cdot \|)$, *with respect to the topology of* $\text{Mult}(S, \| \cdot \|)$.

Proof. Let $\psi = \varphi_{\mathcal{U}}$, take $\epsilon > 0$ and let $f_1, \ldots, f_q \in S$. There exists $H \in \mathcal{U}$ such that $|\psi(f_j) - |f_j(x)| \, |_\infty \leq \epsilon \, \forall x \in H$, $\forall j = 1, \ldots, q$. Therefore, taking $a \in H$, we have $|\varphi_a(f_j) - \psi(f_j)|_\infty \leq \epsilon \, \forall j = 1, \ldots, q$ which shows that φ_a belongs to the neighborhood W of ψ of the form $\{\phi \mid |\psi(f_j) - \phi(f_j)|_\infty \leq \epsilon \, \forall j = 1, \ldots, q\}$ and this proves the claim. $\square$

Remark 5. In E, we call *monotonous distances sequence* a sequence $(a_n)_{n \in \mathbb{N}}$ of E such that the sequence $\delta(a_n, a_{n+1})_{n \in \mathbb{N}}$ is strictly monotonous. We call *constant distances sequence* a sequence $(a_n)_{n \in \mathbb{N}}$ of E such that $\delta(a_n, a_m)$ is constant when n, m are big enough and $n \neq m$. According to Remark 3, if $\mathbb{L}$ is locally compact and E admits monotonous distances sequences or constant distances sequences, we can define $\varphi_{\mathcal{U}} \in \text{Mult}_1(S, \| \cdot \|)$ which does not belong to $\text{Mult}_E(S, \| \cdot \|)$.

Theorem 3.5.4. *For each* $\phi \in \text{Mult}(S, \| \cdot \|)$, $\text{Ker}(\phi)$ *is a prime spectrally closed ideal.*

Proof. Let $\phi \in \text{Mult}(S, \| \cdot \|)$ and let f belong to the spectral closure of $Ker(\phi)$. There exists a sequence $(f_n)_{n \in \mathbb{N}}$ of $\text{Ker}(\phi)$ such that $\lim_{n \to \infty} \| f_n - f \|_{\text{sp}} = 0$. By Theorem 3.5.1, since $\phi(g) \leq \|g\|_{\text{sp}} \, \forall g \in S$, we have $\lim_{n \to \infty} \phi(f_n - f) = 0$. But $\phi(f_n) = 0 \, \forall n \in \mathbb{N}$, hence it follows that $\phi(f) = \lim_{n \to +\infty} \phi(f_n - f) = 0$. Therefore, f belongs to $Ker(\phi)$, which means that $\text{spcl}(\text{Ker}(\phi)) = \text{Ker}(\phi)$. $\square$

By Theorem 3.5.4 and Corollary 3.3.14, we have the following Corollary 3.5.5.

Corollary 3.5.5. *Then* $\text{Mult}(S, \| \cdot \|) = \text{Mult}_m(S, \| \cdot \|)$.

Theorem 3.5.6 is classical [26, Theorem 6.15].

Theorem 3.5.6. *Let A be a commutative unital ultrametric Banach* 𝕃*-algebra. For every maximal ideal $\mathcal{M}$ of A, there exists $\phi \in \mathrm{Mult}_m(S, \| \cdot \|)$ such that $\mathcal{M} = \mathrm{Ker}(\phi)$.*

Recall that a unital commutative Banach 𝕃-algebra is said to be *multbijective* if every maximal ideal is the kernel of only one continuous multiplicative semi-norm.

Remark 6. There exist some rare cases of ultrametric Banach algebras that are not multbijective [21, 23].

Theorem 3.5.7. *Suppose that S is admissible. Then S is multbijective. Precisely if $\psi \in \mathrm{Mult}(S, \| \cdot \|)$ and $\mathrm{Ker}(\psi) = \mathcal{M}$ then $\psi = \varphi_{\mathcal{U}}$ for every ultrafilter $\mathcal{U}$ such that $\mathcal{M} = \mathcal{I}(\mathcal{U}, S)$.*

Proof. Let $\psi \in \mathrm{Mult}_m(S, \| \cdot \|)$, let $\mathcal{M} = \mathrm{Ker}(\psi)$ and $\mathcal{U}$ be an ultrafilter such that $\mathcal{M} = \mathcal{I}(\mathcal{U}, S)$.

Let $f \in S$. Notice that if $f \in \mathcal{M}$ then $\psi(f) = \varphi_{\mathcal{U}}(f) = 0$. Now we assume that $f \notin \mathcal{M}$. So $\psi(f)$ and $\varphi_{\mathcal{U}}(f)$ are both strictly positive. We prove that they are equal.

First let $\epsilon > 0$ and consider the set $L = \{x \in E : |f(x)| \leq \varphi_{\mathcal{U}}(f) + \epsilon\}$. This set belongs to $\mathcal{U}$ and by Lemma 3.1.2, it is clopen. Therefore, its characteristic function u lies in S. We have $\varphi_{\mathcal{U}}(u) = 1$. Consequently, we can derive that $\psi(u) = 1$ because u is idempotent and does not belong to $\mathcal{M}$. Therefore, $\psi(uf) = \psi(f)$ and $\varphi_{\mathcal{U}}(uf) = \varphi_{\mathcal{U}}(f)$. By Theorem 3.5.1, we have $\psi(f) = \psi(uf) \leq \|uf\|_{\mathrm{sp}} = \|uf\|_0$ because S is admissible. But by definition of L we have: $\|uf\|_0 \leq \varphi_{\mathcal{U}}(f) + \epsilon$. Therefore, $\psi(f) \leq \varphi_{\mathcal{U}}(f) + \epsilon$. This holds for every $\epsilon > 0$. Consequently we may conclude that $\psi(f) \leq \varphi_{\mathcal{U}}(f)$ for every $f \in S$.

We prove now the inverse inequality. We have $\varphi_{\mathcal{U}}(f) > 0$, so consider the set $W = \{x \in E : |f(x)| \geq \frac{\varphi_{\mathcal{U}}(f)}{2}\}$. This is a clopen set which belongs to $\mathcal{U}$. Let w be the characteristic function of W and put $g = wf + (1 - w)$. We have $\varphi_{\mathcal{U}}(w) = 1$ and $\varphi_{\mathcal{U}}(1 - w) = 0$ so $w \notin \mathcal{M}$ and $1 - w \in \mathcal{M}$. Since $\mathcal{M} = \mathrm{Ker}(\psi)$ we then have $\psi(1 - w) = 0$ and $\psi(w) = 1$ because w is idempotent. Finally $\psi(g) = \psi(f)$ and $\varphi_{\mathcal{U}}(g) = \varphi_{\mathcal{U}}(f)$.

On the other hand, we can check that $|g(x)| \geq \min\left(1, \frac{\varphi_{\mathcal{U}}(f)}{2}\right)$ for all $x \in E$, hence g is invertible in S. Putting $h = \frac{1}{g}$, using the first inequality yet proved, we have

$$\psi(f) = \psi(g) = \frac{1}{\psi(h)} \geq \frac{1}{\varphi_{\mathcal{U}}(h)} = \varphi_{\mathcal{U}}(g) = \varphi_{\mathcal{U}}(f).$$

That concludes the proof. □

Remark 7. Thus, if S is admissible, $\mathrm{Mult}(S, \| \cdot \|)$ can be identified to $\mathrm{Mult}(S, \| \cdot \|_0)$.

Corollary 3.5.8. *For every* $\phi \in \mathrm{Mult}(S, \| \cdot \|)$ *there exists a unique* $\mathcal{H} \in Y_{(S)}(E)$ *such that* $\phi(f) = \lim_{\mathcal{U}} |f(x)| \ \forall f \in S, \ \forall \mathcal{U} \in \mathcal{H}$.

Moreover, the mapping $\mathbb{J}$ *that associates to each* $\phi \in \mathrm{Mult}(S, \| \cdot \|)$ *the unique* $\mathcal{H} \in Y_{\mathcal{R}}(E)$ *such that* $\phi(f) = \lim_{\mathcal{U}} |f(x)| \ \forall f \in S, \ \forall \mathcal{U} \in \mathcal{H}$, *is a bijection from* $\mathrm{Mult}(S, \| \cdot \|)$ *onto* $Y_{(S)}(E)$.

By Theorem 3.5.7, since each element $\phi \in \mathrm{Mult}(S, \| \cdot \|)$ is of the form $\varphi_{\mathcal{U}}$, Corollary 3.5.9 is immediate from Theorem 3.5.7.

Corollary 3.5.9. $\mathrm{Mult}_E(S, \| \cdot \|)$ *is dense in* $\mathrm{Mult}(S, \| \cdot \|)$.

Proof. Indeed, $\mathrm{Mult}(S, \| \cdot \|) = \mathrm{Mult}_m(S, \| \cdot \|)$ and, since S is multbijective, every element element of $\mathrm{Mult}(S, \| \cdot \|)$ is of the form $\varphi_{\mathcal{U}}$ with $\mathcal{U}$ an ultrafilter on E. $\qquad\square$

Theorem 3.5.10. *The topological space* E, *equipped with its distance* δ, *is homeomorphic to* $\mathrm{Mult}_E(S, \| \cdot \|)$ *equipped with the restricted topology from that of* $\mathrm{Mult}(S, \| \cdot \|)$.

Proof. For every $a \in E$, put $Z(a) = \varphi_a$, take $f_1, \ldots, f_q \in S$ and $\epsilon > 0$. We set $\mathcal{W}'(\varphi_a, f_1, \ldots, f_q, \epsilon) = \mathcal{W}(\varphi_a, f_1, \ldots, f_q, \epsilon) \cap \mathrm{Mult}_E(S, \| \cdot \|)$. Considering the natural topology on E, the filter of neighborhoods of a admits for basis the family of disks $B(a, r)$, $r > 0$. We will show that it is induced through the mapping Z by the filter admitting for basis the family of neighborhoods of φ_a in $\mathrm{Mult}(S, \| \cdot \|)$. Indeed, take $r \in]0, 1[$ and let u be the characteristic function of $B(a, r)$. Then $\mathcal{W}'(\varphi_a, u, r)$ is the set of φ_b such that $|\varphi_a(u) - \varphi_b(u)|_\infty < r$. But since $r < 1$, we can see that this is satisfied if and only if $b \in B(a, r)$. Therefore the topology induced on E by $\mathrm{Mult}(S, \| \cdot \|)$ is thinner than its metric topology.

Conversely, take some neighborhood $\mathcal{W}'(\varphi_a, f_1, \ldots, f_q, \epsilon)$ of φ_a in $\mathrm{Mult}_E(S, \| \cdot \|)$. For each $j = 1, \ldots, q$, the set of the $x \in E$ such that $|\varphi_x(f_j) - \varphi_a(f_j)|_\infty \leq \epsilon$ is the set of the x such that $\left| |f_j(x)| - |f_j(a)| \right|_\infty \leq \epsilon$. But now, since each f_j is continuous, the set of the x such that $\left| |f_j(x)| - |f_j(a)| \right|_\infty \leq \epsilon \ \forall j = 1, \ldots, q$ is a neighborhood of a in E. Consequently, the metric topology of E is thinner than the topology induced by $\mathrm{Mult}_E(S, \| \cdot \|)$ and that finishes proving that the two topological spaces are homeomorphic. $\qquad\square$

Corollary 3.5.11. $\mathrm{Mult}(S, \| \cdot \|)$ *is a compactification of the topological space* E.

Notation: Let $X \subset \text{Mult}(S, \| \cdot \|)$. We denote by $\widetilde{X}$ the subset of the $a \in E$ such that $\varphi_a \in X$.

By Theorem 3.5.10, we can state this corollary.

Corollary 3.5.12. *If X, Y are two clopen subsets of $\text{Mult}(S, \| \cdot \|)$ making a partition, then $\widetilde{X}$, $\widetilde{Y}$ are two clopens of E making a partition of E.*

Corollary 3.5.13. *If X, Y are two open closed subsets of $\text{Mult}(S, \| \cdot \|)$ making a partition, there exist idempotentsu, $v \in S$ such that $\phi(u) = 1$, $\phi(v) = 0, \forall \phi \in X$, $\phi(u) = 0$, $\phi(v) = 1, \forall \phi \in Y$.*

Theorem 3.5.14. *Let $\phi = \varphi_\mathcal{U} \in \text{Mult}_m(S, \| \cdot \|)$, with $\mathcal{U}$ an ultrafilter on E, let $\mathcal{K}$ be the field $\frac{S}{\text{Ker}(\phi)}$ and let θ be the canonical surjection from S onto $\mathcal{K}$. Then, the mapping $\| \cdot \|'$ defined on $\mathcal{K}$ by $\|\theta(f)\|' = \phi(f) \; \forall f \in S$, is the quotient norm $\| \cdot \|'$ of $\| \cdot \|_0$ defined on $\mathcal{K}$ and is an absolute value on $\mathcal{K}$. Moreover, if $\text{Ker}(\phi)$ is of codimension 1, then this absolute value is the one defined on $\mathbb{L}$ and coincides with the quotient norm of the norm $\| \cdot \|$ of S.*

Proof. Let $\mathcal{M} = \text{Ker}(\varphi_\mathcal{U})$. Let $t \in \mathcal{K}$ and let $f \in S$ be such that $\theta(f) = t$. So, $\|t\|' \geq \lim_\mathcal{U} |f(s)|$. Conversely, take $\epsilon > 0$ and let $V = \{x \in E : |f(x)| \leq \lim_\mathcal{U} |f(s)| + \epsilon\}$. By Lemma 3.1.2, the set V is clopen and belongs to $\mathcal{U}$. The characteristic function u of $E \setminus V$ belongs to $\mathcal{M}$ and so does uf. But by construction, $(f - uf)(x) = 0 \; \forall x \in E \setminus V$ and $(f - uf)(x) = f(x) \; \forall x \in V$. Consequently, $\|f - uf\|_0 \leq \lim_\mathcal{U} |f(s)| + \epsilon$ and therefore $\|t\|' \leq \|f - uf\|_0 \leq \lim_\mathcal{U} |f(s)| + \epsilon$. This finishes proving the equality $\|\theta(f)\|' = \lim_\mathcal{U} |f(o)|$ and hence the mapping defined by $|\theta(f)| = \phi(f)$, $f \in S$ is the quotient norm $\| \cdot \|'$ of $\| \cdot \|_0$. Then it is multiplicative, hence it is an absolute value on $\mathcal{K}$.

Now, suppose that $\mathcal{M}$ is of codimension 1. Then $\mathcal{K}$ is isomorphic to $\mathbb{L}$ and its absolute value $\| \cdot \|'$ is continuous with respect to the topology of $\mathbb{L}$, hence it is equal to the absolute value of $\mathbb{L}$. Finally consider the quotient norm $\| \cdot \|_q$ of the norm $\| \cdot \|$ of S: that quotient norm of course bounds the quotient norm $\| \cdot \|'$ which is the absolute value of $\mathbb{L}$. If $f \in S$ and $b = \theta(f)$, we have $f - b \in \mathcal{M}$ and $\|\theta(f)\|_q \leq \|b\| = |b| = |\theta(f)| = \|\theta(f)\|'$, which ends the proof. $\square$

Corollary 3.5.15. *Let $\phi \in \text{Mult}(S, \| \cdot \|)$, let $\mathcal{K}$ be the field $\frac{S}{\text{Ker}(\phi)}$ and let θ be the canonical surjection from S onto $\mathcal{K}$. Then, the mapping defined on $\mathcal{K}$ by $|\theta(f)| = \phi(f), \forall f \in S$ is the quotient norm $\| \cdot \|'$ of $\| \cdot \|_0$ on $\mathcal{K}$ and is an absolute value on $\mathcal{K}$. Moreover, if $\text{Ker}(\phi)$ is of codimension 1, then this absolute value is the one defined on $\mathbb{L}$ and coincides with the quotient norm of the norm $\| \cdot \|$ of S.*

Remark 8. It is not clear whether an algebra S can admit a prime closed ideal $\mathcal{P}$ (with respect to the norm $\| \cdot \|$) which is not a maximal ideal. If it admits such a prime closed ideal, then it is not the kernel of a continuous multiplicative semi-norm since $Mult(S, \| \cdot \|) = Mult_m(S, \| \cdot \|)$. In such a case, the quotient algebra by $\mathcal{P}$ has no continuous absolute value extending that of $\mathbb{L}$, although it has no divisors of zero. Such a situation can happen in certain ultrametric Banach algebras [9].

By Theorem 2.4.12, we have Theorem 3.5.16 [25].

Theorem 3.5.16. *Every normed $\mathbb{L}$-algebra admits a Shilov boundary.*

Notation: Given a normed $\mathbb{L}$-algebra G, we denote by $Shil(G)$ the Shilov boundary of G.

Lemma 3.5.17. *Let us fix $a \in E$. For every $r > 0$, let $Z(a, r)$ be the set of multiplicative semi-norms $\varphi_{\mathcal{U}}$, with $\mathcal{U} \in U(E)$, such that $d_E(a, r)$ belongs to $\mathcal{U}$. The family $\{Z(a, r) \mid r \in]0, 1[\}$ forms a basis of the filter of neighborhoods of φ_a.*

Proof. Let $\mathcal{W}(\varphi_a, f_1, \ldots, f_q, \epsilon)$ be a neighborhood of φ_a in $Mult(S, \| \cdot \|)$. There exists $r > 0$ such that, whenever $\delta(a, x) \leq r$ we have $|f_j(x) - f_j(a)| \leq \epsilon \; \forall j = 1, \ldots, q$ and therefore, clearly, $|\varphi_{\mathcal{U}}(f_j) - \varphi_a(f_j)|_\infty \leq \epsilon \; \forall j = 1, \ldots, q$ for every $\mathcal{U}$ containing $d_E(a, r)$. Thus, $Z(a, r)$ is included in $\mathcal{W}(\varphi_a, f_1, \ldots, f_q, \epsilon)$.

Conversely, consider a set $Z(a, r)$ with $r \in]0, 1[$, let u be the characteristic function of $d_E(a, r)$ and consider $\mathcal{W}(\varphi_a, u, r)$. Given $\psi = \varphi_{\mathcal{U}} \in \mathcal{W}(\varphi_a, u, r)$, we have $|\psi(u) - \varphi_a(u)|_\infty \leq r$. But $|\psi(u) - \varphi_a(u)|_\infty = |\psi(u) - 1|_\infty = |\lim_{\mathcal{U}} |u(x)| - 1|_\infty$. If $d_E(a, r)$ belongs to $\mathcal{U}$, then $\lim_{\mathcal{U}} |u(x)| = 1$ and therefore $|\lim_{\mathcal{U}} |u(x)| - 1|_\infty = 0$. But if $d_E(a, r)$ does not belong to $\mathcal{U}$, then $\lim_{\mathcal{U}} |u(x)| = 0$ and therefore $|\lim_{\mathcal{U}} |u(x)| - 1|_\infty = 1$. Consequently, since $r < 1$, $\mathcal{W}(\varphi_a, u, r)$ is included in $Z(a, r)$, which finishes proving that the family of $Z(a, r)$, $r \in]0, 1[$ is a basis of the filter of neighborhoods of φ_a. $\square$

Theorem 3.5.18. *The Shilov boundary of S is equal to $Mult(S, \| \cdot \|)$* [12, 14, 27].

Proof. We will show that for every $a \in E$, φ_a belongs to $Shil(S)$. So, let us fix $a \in E$ and suppose that φ_a does not belong to $Shil(S)$. Since $Shil(S)$ is a closed subset of $Mult(S, \| \cdot \|)$, there exists a neighborhood of φ_a that contains no element of $Shil(S)$. Therefore, by Lemma 3.5.11, there exists $s > 0$ such that $Z(a, s)$ contains no element of $Shil(S)$. Now, let $D = d_E(a, s)$ and let u be the characteristic function of D. Since any $\phi \in Mult(S, \| \cdot \|)$ satisfies either $\phi(u) = 1$ or $\phi(u) = 0$, there exists $\theta \in Shil(S)$ such that

$\theta(u) = \|u\|_{\mathrm{sp}} = 1$. Then, θ is of the form $\varphi_{\mathcal{U}}$, with $\mathcal{U} \in U(E)$ and $\mathcal{U}$ does not contain D. But since $u(x) = 0 \; \forall x \in E \backslash D$, we have $\theta(u) = 0$, a contradiction. Consequently, for every $a \in E$, φ_a belongs to Shil(S) which is a closed subset of Mult$(S, \| \cdot \|)$ and since, by Corollary 3.5.9, Mult$_E(S, \| \cdot \|)$ is dense in Mult$(S, \| \cdot \|)$, then Shil(S) is equal to Mult$(S, \| \cdot \|)$. $\qquad \square$

3.6.   The Stone space $\mathbb{B}(E)$ for admissible algebras

Here, we can show that for the algebra $\mathcal{C}$ of continuous bounded functions from E to $\mathbb{L}$, the Banaschewski compactification of E is homeomorphic to Mult$(\mathcal{C}, \| \cdot \|_0)$.

Recall that we denote by $\mathbb{B}(E)$ the Boolean ring of clopen subsets of E equipped with the laws Δ for the addition and $\cap$ for the multiplication. As usually called the *Stone space of the Boolean ring* $\mathbb{B}(E)$ is the space $St(E)$ of non-zero ring homomorphisms from $\mathbb{B}(E)$ onto $\mathbb{F}_2$, equipped with the topology of pointwise convergence. This space is a compactification of E and is called the *Banaschewski compactification* of E (see, for example, [26] for further details).

For every $\mathcal{U} \in U(E)$, we denote by $\zeta_{\mathcal{U}}$ the ring homomorphism from $\mathbb{B}(E)$ onto the field of 2 elements $\mathbb{F}_2$ defined by $\zeta_{\mathcal{U}}(O) = 1$ for every $O \in \mathbb{B}(E)$ that belongs to $\mathcal{U}$ and $\zeta_{\mathcal{U}}(O) = 0$ for every $O \in \mathbb{B}(E)$ that does not belong to $\mathcal{U}$.

Particularly, given $a \in E$, we denote by ζ_a the ring homomorphism from $\mathbb{B}(E)$ onto $\mathbb{F}_2$ defined by $\zeta_a(O) = 1$ for every $O \in \mathbb{B}(E)$ that contains a and $\zeta_a(O) = 0$ for every $O \in \mathbb{B}(E)$ that does not contain a.

Remark 9. Let $St'(E)$ be the set of ζ_a, $a \in E$. The mapping that associates $\zeta_a \in E$ to $a \in E$ defines a surjective mapping from E onto $St'(E)$. That mapping is also injective because given $a, \; b \in E$, there exists a clopen subset F such that $a \in F$ and $b \notin F$.

By Corollary 3.5.8, we have a bijection Ψ from Mult$(S, \| \cdot \|)$ onto $Y_{(S)}(E)$ associating to each $\phi \in$ Mult$(S, \| \cdot \|)$ the unique $\mathcal{H} \in Y_{(S)}(E)$ such that $\phi(f) = \lim_{\mathcal{U}} |f(x)|$, $\mathcal{U} \in \mathcal{H}$, $f \in S$, i.e., $\phi = \varphi_{\mathcal{U}}$ for every $\mathcal{U} \in \mathcal{H}$.

On the other hand, let us take some $\mathcal{H} \in Y_{(S)}(E)$ and ultrafilters $\mathcal{U}, \; \mathcal{V}$ in $\mathcal{H}$. Since $\mathcal{U}, \mathcal{V}$ own the same clopen subsets of E, we have $\zeta_{\mathcal{U}} = \zeta_{\mathcal{V}}$ and hence we can define a mapping Ξ from $Y_{\mathcal{R}}(E)$ into $St(E)$ which associates to each $\mathcal{H} \in Y_{\mathcal{R}}(E)$ the Boolean homomorphism $\zeta_{\mathcal{U}}$ independent from $\mathcal{U} \in \mathcal{H}$.

Lemma 3.6.1. Ξ *is a bijection from* $Y_{\mathcal{R}}(E)$ *onto* $St(E)$.

Proof. Indeed, let $\mathcal{H}, \mathcal{K} \in Y_{(S)}(E)$ and suppose that $\mathcal{H} \neq \mathcal{K}$. Take ultrafilters $\mathcal{U} \in \mathcal{H}$ and $\mathcal{V} \in \mathcal{K}$. They are not sticked, therefore by Theorem 3.2.2, there exists clopens $L \in \mathcal{H}$, $M \in \mathcal{K}$ such that $H \cap K = \emptyset$. Then, $\Xi(\mathcal{H}) \neq \Xi(\mathcal{K})$, which proves the injectivity.

Now, let us check that Ξ is surjective. Let $\theta \in St(E)$. Since θ is a ring homomorphism for the Boolean laws, the family of clopen sets X satisfying $\theta(X) = 1$ generates a filter $\mathcal{F}$. Let $\mathcal{U} \in U(E)$ be thinner than $\mathcal{F}$ and let $\mathcal{H}$ be the class of $\mathcal{U}$ with respect to $(\mathcal{R})$. We will check that $\theta = \Xi(\mathcal{H}) = \zeta_{\mathcal{U}}$. Let O be a clopen subset that belongs to $\mathcal{U}$. Then $E \setminus O$ does not belong to $\mathcal{U}$ and therefore it does not belong to $\mathcal{F}$, so $\theta(E \setminus O) = 0$, consequently $\theta(O) = 1$. And now, let O be a clopen subset that does not belong to $\mathcal{U}$. Then O does not belong to $\mathcal{F}$, hence $\theta(O) = 0$, which ends the proof. $\quad\square$

Notations: We put $\Pi = \Xi \circ \mathbb{J}$ and hence Π is a bijection from $\text{Mult}(S, \| \cdot \|)$ onto $St(E)$. Notice that for every ultrafilter $\mathcal{U}$, $(\varphi_{\mathcal{U}})$ is the class $\mathcal{H}$ of $\mathcal{U}$ with respect to $(\mathcal{R})$ and $\Xi(\mathcal{H}) = \zeta_{\mathcal{U}}$ so $\Pi(\varphi_{\mathcal{U}}) = \zeta_{\mathcal{U}}$.

Theorem 3.6.2. Π *is a homeomorphism once* $St(E)$ *and* $\text{Mult}(S, \| \cdot \|)$ *are equipped with topologies of pointwise convergence.*

Proof. Recall that for any $\mathcal{U} \in U(E)$, a neighborhoods basis of $\varphi_{\mathcal{U}}$ in $\text{Mult}(S, \| \cdot \|)$ is given by the family of sets of the form $\mathcal{W}(\varphi_{\mathcal{U}}, f_1, \ldots, f_q, \epsilon)$ with $f_1, \ldots, f_q \in S$, $\epsilon > 0$ and

$$\mathcal{W}(\varphi_{\mathcal{U}}, f_1, \ldots, f_q, \epsilon) = \left\{ \varphi_{\mathcal{V}} \mid \left| \lim_{\mathcal{U}} |f_j(x)| - \lim_{\mathcal{V}} |f_j(x)| \right|_\infty \leq \epsilon, \ j = 1, \ldots, q \right\}.$$

On the other hand, for any $\mathcal{U} \in U(E)$, a neighborhood basis for $\zeta_{\mathcal{U}}$ in $St(E)$ is given by the family of sets $V(\zeta_{\mathcal{U}}, O_1, \ldots, O_q)$ where $O_1, \ldots, O_q$ belong to $\mathbb{B}(E)$ and

$$V(\zeta_{\mathcal{U}}, O_1, \ldots, O_q) = \{ \zeta_{\mathcal{V}} \mid \zeta_{\mathcal{U}}(O_j) = \zeta_{\mathcal{V}}(O_j), j = 1, \ldots, q \}.$$

Notice also that if F belongs to $\mathbb{B}(E)$ and if u is its characteristic function, then for any $\mathcal{U} \in U(E)$, we have $\zeta_{\mathcal{U}}(F) = 1$ if and only if $F \in \mathcal{U}$, i.e., if and only if $\lim_{\mathcal{U}} |u(x)| = 1$. Otherwise, both $\zeta_{\mathcal{U}}(F)$ and $\lim_{\mathcal{U}} |u(x)|$ are equal to 0. Therefore, the relation

$$\left| \lim_{\mathcal{U}} |u(x)| - \lim_{\mathcal{V}} |u(x)| \right|_\infty \leq \frac{1}{2}$$

holds if and only if $\zeta_{\mathcal{U}}(F) = \zeta_{\mathcal{V}}(F)$. Recall that for every $\mathcal{U} \in U(E)$ we have $\Pi(\varphi_{\mathcal{U}}) = \zeta_{\mathcal{U}}$.

We will show that Π is continuous. Consider $O_1, \ldots, O_q \in \mathbb{B}(E)$, $\mathcal{U} \in U(E)$ and the neighborhood $V(\zeta_{\mathcal{U}}, O_1, \ldots, O_q)$ of $\zeta_{\mathcal{U}}$. From the preceding remark, $\zeta_\mathcal{V}$ belongs to $V(\zeta_\mathcal{U}, O_1, \ldots, O_q)$ if and only if for every $j = 1, \ldots, q$, $\zeta_\mathcal{U}(O_j) = \zeta_\mathcal{V}(O_j)$, i.e., if for every $j = 1, \ldots, q$,

$$\left| \lim_\mathcal{U} |u_j(x)| - \lim_\mathcal{V} |u_j(x)| \right|_\infty \leq \frac{1}{2},$$

i.e., if $\varphi_\mathcal{V}$ belongs to $\mathcal{W}(\varphi_\mathcal{U}, u_1, \ldots, u_q, \frac{1}{2})$. Consequently, this proves that Π is continuous. We can now deduce that Π is a homeomorphism because it is a continuous bijection between compact spaces [6]. □

Corollary 3.6.3. *The space* $St(E)$ *is a compactification of* E *which is equivalent to the compactification* $\mathrm{Mult}(S, \| . \|)$.

Remark 10. For an admissible algebra S, the Banaschewski compactification $St(E)$ coincides with the Guennebaud–Berkovich multiplicative spectrum.

Chapter 4

Compatible Algebras

4.1. Definition of compatible algebras

In this chapter and in the next ones, we will consider compatible algebras, as we did with admissible algebras, and we will make a study comparable to this in Sections 4.1–4.5.

Given a subset F of $\mathbb{L}$ such that $F \neq \emptyset$ and $F \neq \mathbb{L}$, we call *codiameter of* F the number $\delta(F, \mathbb{L} \setminus F)$. If $F = \emptyset$ or $F = \mathbb{L}$, we say that its codiameter is infinite. The set F will be said to be *uniformly open* if its codiameter is strictly positive.

Recall that E is a metric space. We will denote by $\mathbb{G}(E)$ the family of uniformly open subsets of E. In Section 3.6 of Chapter 3, dealing with the Banaschewski compactification of E we considered the Boolean ring of clopen sets of E (with the usual addition Δ and multiplication $\cap$). Here we will consider the Boolean ring of uniformly open sets. Actually, we have the following lemmas that are easily checked:

Lemma 4.1.1. *Given two uniformly open subsets F, G, then $F \cup G$, $F \cap G$, $E \setminus G$, $F \Delta G$ are uniformly open.*

Corollary 4.1.2. $\mathbb{G}(E)$ *is a Boolean ring with respect to the addition Δ and the multiplication $\cap$.*

Lemma 4.1.3. *Given two subsets A and B of E, there exists a uniformly open subset F such that $A \subset F$ and $B \subset E \setminus F$ if and only if $\delta(A, B) > 0$.*

Lemma 4.1.4. *Let f be a uniformly continuous function from E to $\mathbb{K}$ and let $M > 0$.*

(1) *If D is a uniformly open subset of $\mathbb{K}$, then so is the set $F = \{x \in E \mid f(x) \in D\}$.*

(2) *Given $M > 0$, the sets $E_1 = \{x \in E \mid |f(x)| \geq M\}$ and $E_2 = \{x \in E \mid |f(x)| \leq M\}$ are uniformly open.*

Corollary 4.1.5. *Let f be a uniformly continuous function from E to $\mathbb{K}$, let $M > 0$ and let $h > 0$. Then $\{x \in E \mid \left||f(x)| - M\right|_{\infty} \leq h\}$ is uniformly open.*

We can easily derive the following the Lemma.

Lemma 4.1.6. *Let F be a subset of E and let u be its characteristic function. The three following statements are equivalent:*

(1) *F is uniformly open,*
(2) *u is Lipschitz,*
(3) *u is uniformly continuous.*

Definitions and notations: We denote by $\mathcal{B}$ the algebra of bounded uniformly continuous functions from E to $\mathbb{L}$.

We will call *semi-compatible algebra* a unital Banach $\mathbb{L}$-algebra T of uniformly continuous bounded functions f from E to $\mathbb{L}$ satisfying the two following properties:

(1) every function $f \in T$ such that $\inf_{x \in E} |f(x)| > 0$ is invertible in S,
(2) for every subset $F \subset E$, the characteristic function of F belongs to T if and only if F is uniformly open,

   Moreover, a semi-compatible algebra T will be said to be *compatible* if it satisfies
(3) the spectral semi-norm of T is equal to the norm $\| \cdot \|_0$.

Remark 1. $\mathcal{B}$ equipped with the norm $\| \cdot \|_0$ is easily seen to be compatible.

Throughout the following sections, we will denote by T a semi-compatible $\mathbb{L}$-algebra.

More definitions and notations: Two ultrafilters $\mathcal{U}$, $\mathcal{V}$ on E are said to be *contiguous* if for every $H \in \mathcal{U}$, $J \in \mathcal{V}$, we have $\delta(H, J) = 0$. We shall denote by $(\mathcal{S})$ the relation defined on $U(E)$ as $\mathcal{U}(\mathcal{S})\mathcal{V}$ if $\mathcal{U}$ and $\mathcal{V}$ are contiguous.

The following lemma is immediate.

Lemma 4.1.7. *Let $\mathcal{U}$, $\mathcal{V}$ be two sticked ultrafilters. Then $\mathcal{U}$ and $\mathcal{V}$ are contiguous.*

Remark 2. The contiguity relation on ultrafilters on E is a particular case of the relation on ultrafilters defined by Labib Haddad and in other terms by

Pierre Samuel in a uniform space. This relation on a uniform space actually is an equivalence relation.

Remark 3. Two contiguous ultrafilters are not necessarily sticked. Suppose $\mathbb{L}$ has dense valuation, let $(a_n)_{n \in \mathbb{N}}, (b_n)_{n \in \mathbb{N}}$ be two sequences of the disk $d(0,1)$ of $\mathbb{L}$ such that $|a_n| < |a_{n+1}|$, $|a_n| = |b_n|$, $a_n \neq b_n$, $\lim_{n \to +\infty} |a_n| = 1$, $\lim_{n \to +\infty} a_n - b_n = 0$. For every $n \in \mathbb{N}$, set $A_n = \{a_n, a_{n+1}, \ldots\}$, $B_n = \{b_n, b_{n+1}, \ldots\}$. Then each A_n, each B_n is a closed set of $\mathbb{L}$ and $\delta(A_n, B_n) \leq |a_q - b_q| \; \forall q \geq n$, hence $\delta(A_n, B_n) = 0$. Consider now an ultrafilter $\mathcal{F}$ on $\mathbb{N}$. It defines an ultrafilter $\mathcal{U}$ thinner than the sequence (a_n) and an ultrafilter $\mathcal{V}$ thinner than the sequence (b_n) which are contiguous. But clearly $a_n \neq b_m \forall n$, $m \in \mathbb{N}$ and hence $A_n \cap B_m = \emptyset \; \forall n, m \in \mathbb{N}$, which proves that the ultrafilters $\mathcal{U}$ and $\mathcal{V}$ are not sticked.

Lemma 4.1.8 is classical.

Lemma 4.1.8. *Let $\mathcal{U}$ be an ultrafilter on E. Let f be a bounded function from E to $\mathbb{L}$. The function $|f|$ from E to $\mathbb{R}_+$ defined as $|f|(x) = |f(x)|$ admits a limit along $\mathcal{U}$. Moreover, if $\mathbb{L}$ is locally compact, then $f(x)$ admits a limit along $\mathcal{U}$.*

The following lemmas are immediate:

Lemma 4.1.9. *The spectral closure of an ideal of T is an ideal of T.*

Lemma 4.1.10. *If a subset Y of T is spectrally closed, it is closed with respect to the norm $\| \, . \, \|$ of T.*

Lemma 4.1.11. *Every maximal ideal $\mathcal{M}$ of T is spectrally closed.*

Proof. By Lemma 4.1.9, the spectral closure $\mathrm{spcl}(\mathcal{M})$ of $\mathcal{M}$ is an ideal. If $\mathcal{M}$ is not spectrally closed, then $\mathrm{spcl}(\mathcal{M}) = T$, hence there exists $t \in T$ such that $1 - t \in \mathcal{M}$ and $\|t\|_{\mathrm{sp}} < 1$. Consequently, $\lim_{n \to +\infty} \|t^n\| = 0$, therefore the series $(\sum_{n=0}^{\infty} t^n)$ converges and $(\sum_{n=0}^{\infty} t^n)(1 - t) = 1$ and hence the unity belongs to $\mathcal{M}$, a contradiction. $\qquad \square$

Lemma 4.1.12 is now easy.

Lemma 4.1.12. *Given an ultrafilter $\mathcal{U}$ on E, $\mathcal{I}(\mathcal{U}, T)$ is a prime ideal. Moreover, $\mathcal{I}(\mathcal{U}, T)$ is closed with respect to the norm $\| \cdot \|_0$ and then is spectrally closed.*

Proof. Since $\mathcal{U}$ is an ultrafilter, it is straightforward that $\mathcal{I}(\mathcal{U}, T)$ is prime. Let us now check that $\mathcal{I}(\mathcal{U}, T)$ is closed with respect to the norm $\| \, . \, \|_0$. Indeed let g belong to the closure of $\mathcal{I}(\mathcal{U}, S)$ with respect to $\| \, . \, \|_0$, let $b = \lim_{\mathcal{U}} |g(x)|$

and suppose $b > 0$. There exists $f \in \mathcal{I}(\mathcal{U}, T)$ such that $\|f - g\|_0 < b$ and then

$$b = \left| \lim_{\mathcal{U}} |f(x)| - \lim_{\mathcal{U}} |g(x)| \right|_\infty \leq \lim_{\mathcal{U}} |f(x) - g(x)| \leq \|f - g\|_0 < b,$$

a contradiction showing that $\mathcal{I}(\mathcal{U}, T)$ is closed with respect to the norm $\| \cdot \|_0$. Therefore, since $\| \cdot \|_0 \leq \| \cdot \|_{\mathrm{sp}}$, it is closed with respect to the norm $\| \cdot \|_{\mathrm{sp}}$. $\qquad\square$

By Lemma 4.1.3, we have the following lemma.

Lemma 4.1.13. *Let $\mathcal{U}$, $\mathcal{V}$ be ultrafilters on E. Then $\mathcal{U}$ and $\mathcal{V}$ are not contiguous if and only if there exists a uniformly open set $H \in \mathcal{U}$ such that $E \setminus H \subset \mathcal{V}$.*

Corollary 4.1.14. *Let $\mathcal{U}, \mathcal{V}$ be ultrafilters on E. Then $\mathcal{U}$ and $\mathcal{V}$ are contiguous if and only if they contain the same uniformly open sets.*

Corollary 4.1.15. *Relation $(\mathcal{S})$ is an equivalence relation on $U(E)$.*

Theorem 4.1.16. *Let $\mathcal{U}$, $\mathcal{V}$ be ultrafilters on E. Then $\mathcal{I}(\mathcal{U}, T) = \mathcal{I}(\mathcal{V}, T)$ if and only if $\mathcal{U}$ and $\mathcal{V}$ are contiguous.*

Proof. Suppose that $\mathcal{U}$, $\mathcal{V}$ are not contiguous. By Lemma 4.1.13, there exists a uniformly open set $H \in \mathcal{U}$ such that $E \setminus H \in \mathcal{V}$. Then the characteristic function u of H belongs to $\mathcal{I}(\mathcal{V}, T)$ and does not belong to $\mathcal{I}(\mathcal{U}, T)$.

Conversely, suppose that $\mathcal{I}(\mathcal{U}, T) \neq \mathcal{I}(\mathcal{V}, T)$. Without loss of generality, we can assume that there exists $f \in \mathcal{I}(\mathcal{U}, T) \setminus \mathcal{I}(\mathcal{V}, T)$. Then $\lim_{\mathcal{V}} |f(x)|$ is a number $l > 0$. There exists $L \in \mathcal{V}$ such that $\left| |f(x)| - l \right|_\infty < \frac{l}{3}$ $\forall x \in L$ and then $|f(x)| \geq \frac{2l}{3}$ $\forall x \in L$. Therefore, by Lemma 4.1.4, the set $L' = \{x \in E \mid |f(x)| \geq \frac{2l}{3}\}$ is a uniformly open set that belongs to $\mathcal{V}$. But the set $H = \{x \in E \mid |f(x)| \leq \frac{l}{3}\}$ is a uniformly open set of $\mathcal{U}$ since $\lim_{\mathcal{U}} |f(x)| = 0$ and clearly $H \cap L' = \emptyset$. Consequently, $\mathcal{U}$ and $\mathcal{V}$ are not contiguous. $\qquad\square$

Now we can obtain a Bezout–Corona statement for semi-compatible algebras similar to Theorem 3.2.11 of Chapter 3:

Theorem 4.1.17. *Let $f_1, \ldots, f_q \in T$ satisfy $\inf_{x \in E}(\max_{1 \leq j \leq q} |f_j(x)|) > 0$. Then there exist $g_1, \ldots, g_q \in T$ such that $\sum_{j=1}^{q} f_j(x)g_j(x) = 1$ $\forall x \in E$.*

Proof. Let $M = \inf_{x \in E}(\max_{1 \leq j \leq q} |f_j(x)|)$. Let $E_j = \{x \in E \mid |f_j(x)| \geq M\}$, $j = 1, \ldots, q$ and let $F_j = \bigcup_{m=1}^{j} E_m$, $j = 1, \ldots, q$. Let $g_1(x) = \frac{1}{f_1(x)} \forall x \in E_1$ and $g_1(x) = 0$ $\forall x \in E \setminus E_1$. Since $|f_1(x)| \geq M$ $\forall x \in E_1$, $|g_1(x)|$ is clearly bounded. By Lemma 4.1.4 each E_j is uniformly open and so is each F_j. And since T is semi-compatible, g_1 belongs to T.

Suppose now we have constructed $g_1, \ldots, g_k \in T$ satisfying $\sum_{j=1}^{k} f_j g_j(x) = 1 \ \forall x \in F_k$ and $\sum_{j=1}^{k} f_j g_j(x) = 0 \ \forall x \in E \setminus F_k$. Let g_{k+1} be defined on E by $g_{k+1}(x) = \frac{1}{f_{k+1}(x)} \ \forall x \in F_{k+1} \setminus F_k$ and $g_{k+1}(x) = 0 \ \forall x \in E \setminus (F_{k+1} \setminus F_k)$. Since F_k and F_{k+1} are uniformly open, so is $E \setminus (F_{k+1} \setminus F_k)$ and consequently, g_{k+1} is uniformly continuous. Similarly as for g_1, since $|f_{k+1}(x)| \geq M \ \forall x \in E_{k+1}$, $|g_{k+1}(x)|$ is clearly bounded, hence g_{k+1} belongs to T. Now we can check that $\sum_{j=1}^{k+1} f_j g_j(x) = 1 \ \forall x \in F_{k+1}$ and $\sum_{j=1}^{k} f_j g_j(x) = 0 \ \forall x \in E \setminus F_{k+1}$. So, by a finite induction, we get functions $g_1, \ldots, g_q \in T$ such that $\sum_{j=1}^{q} f_j g_j(x) = 1 \ \forall x \in E$, which ends the proof. $\square$

Notation: Let $f \in T$ and let $\epsilon > 0$. We set $D(f, \epsilon) = \{x \in E \mid |f(x)| \leq \epsilon\}$.

Corollary 4.1.18. *Let I be an ideal of T different from S. The family of sets $\{D(f, \epsilon), \ f \in I, \ \epsilon > 0\}$ generates a filter $\mathcal{F}_{I,T}$ on E such that $I \subset \mathcal{I}(\mathcal{F}_{I,T}, T)$.*

4.2. Maximal and prime ideals of T

Theorem 4.2.1. *Let $\mathcal{M}$ be a maximal ideal of T. There exists an ultrafilter $\mathcal{U}$ on E such that $\mathcal{M} = \mathcal{I}(\mathcal{U}, T)$. Moreover, $\mathcal{M}$ is of codimension 1 if and only if every element of T converges along $\mathcal{U}$. In particular if $\mathcal{U}$ is convergent, then $\mathcal{M}$ is of codimension 1.*

Proof. Indeed, by Corollary 4.1.18, we can consider the filter $\mathcal{F}_{\mathcal{M},T}$ and we have $\mathcal{M} \subset \mathcal{I}(\mathcal{F}_{\mathcal{M},T}, T)$. Let $\mathcal{U}$ be an ultrafilter thinner than $\mathcal{F}_{\mathcal{M},T}$. So, we have

$$\mathcal{M} \subset \mathcal{I}(\mathcal{F}_{\mathcal{M},T}, T) \subset \mathcal{I}(\mathcal{U}, T).$$

But since $\mathcal{M}$ is a maximal ideal, either $\mathcal{M} = \mathcal{I}(\mathcal{U}, T)$, or $\mathcal{I}(\mathcal{U}, T) = T$. And obviously, $\mathcal{I}(\mathcal{U}, T) \neq T$, hence $\mathcal{M} = \mathcal{I}(\mathcal{U}, T)$. $\square$

Now assume that $\mathcal{M}$ is of codimension 1 and let χ be the $\mathbb{L}$-algebra homomorphism from T to $\mathbb{L}$ admitting $\mathcal{M}$ for kernel. Let $f \in T$ and let $b = \chi(f)$. Then $f - b$ belongs to the kernel of $\mathcal{M}$, hence $\lim_{\mathcal{U}} f(x) - b = 0$, hence $\lim_{\mathcal{U}} f(x) = b$ and every element of T converges along $\mathcal{U}$.

Conversely if every element of T admits a limit along $\mathcal{U}$ then the mapping χ which associates to each $f \in T$ its limit along $\mathcal{U}$ is a $\mathbb{L}$-algebra homomorphism from T to $\mathbb{L}$ admitting $\mathcal{M}$ for kernel.

In particular if $\mathcal{U}$ converges to a point a then each f in T converges to $f(a)$ along $\mathcal{U}$.

By Lemma 4.1.8 and Theorem 4.2.1, the following corollary is immediate.

Corollary 4.2.2. *Let* $\mathbb{L}$ *be locally compact. Then every maximal ideal of* T *is of codimension 1.*

Remark 4. If $\mathbb{L}$ is locally compact, a maximal ideal of codimension 1 of T is not necessarily of the form $\mathcal{I}(\mathcal{U}, S)$ for some Cauchy ultrafilter $\mathcal{U}$, as shown in [12].

Notation: We will denote by $Y_{(T)}(E)$ the set of equivalence classes on $U(E)$ with respect to the relation $(\mathcal{S})$.

By Theorem 4.1.16, we can get Corollary 4.2.3.

Corollary 4.2.3. *Let* $\mathcal{M}$ *be a maximal ideal of* T*. There exists a unique* $\mathcal{H} \in Y_{(T)}(E)$ *such that* $\mathcal{M} = \mathcal{I}(\mathcal{U}, T)$ *for every* $\mathcal{U} \in \mathcal{H}$*.*

Now, the following theorem together with Theorem 4.2.1 characterizes all maximal ideals of T.

Theorem 4.2.4. *Let* $\mathcal{U}$ *be an ultrafilter on* E*. Then* $\mathcal{I}(\mathcal{U}, T)$ *is a maximal ideal of* T*.*

Proof. Let $I = \mathcal{I}(\mathcal{U}, T)$ and let $\mathcal{M}$ be a maximal ideal of T containing I. Then by Theorem 4.2.1, there exists an ultrafilter $\mathcal{V}$ such that $\mathcal{M} = \mathcal{I}(\mathcal{V}, T)$. Suppose now $\mathcal{I}(\mathcal{U}, T) \neq \mathcal{I}(\mathcal{V}, T)$. Then, $\mathcal{U}$ and $\mathcal{V}$ are not contiguous. Consequently, by Corollary 4.1.14, there exists a uniformly open subset $F \in \mathcal{V}$ that does not belong to $\mathcal{U}$ and hence its characteristic function $u \in T$ belongs to $\mathcal{I}(\mathcal{U}, T)$ but does not belong to $\mathcal{I}(\mathcal{V}, T)$. Thus, u belongs to I but does not belong to $\mathcal{M}$, a contradiction to the hypothesis. $\square$

By Corollary 4.2.3 and Theorem 4.2.4, we can derive Corollary 4.2.5.

Corollary 4.2.5. *The mapping that associates to each maximal ideal* $\mathcal{M}$ *of* T *the class with respect to* $(\mathcal{S})$ *of ultrafilters* $\mathcal{U}$*, such that* $\mathcal{M} = \mathcal{I}(\mathcal{U}, T)$*, is a bijection from* $\mathrm{Max}(T)$ *onto* $Y_{(T)}(E)$*.*

The following theorem is quite easy.

Theorem 4.2.6. *Let* $\mathcal{F}$ *be a Cauchy filter on* E *and let* $\mathcal{M} = \mathcal{I}(\mathcal{F}, T)$*. Then every function* $f \in T$ *converges to a limit* $\theta(f)$ *along* $\mathcal{F}$ *and* $\mathcal{M}$ *is a maximal ideal of codimension 1.*

Notations: For any subset F of E, we denote by u_F its characteristic function. Let $\mathcal{M}$ be a maximal ideal of T and let $\mathcal{U} \in U(E)$ be such that $\mathcal{M} = \mathcal{I}(\mathcal{U}, T)$. By Theorem 4.1.16 and Corollary 4.1.14, we can define the set $\mathcal{O}_{\mathcal{M}}$ of all uniformly open subsets of E which belong to $\mathcal{U}$. We denote by $\mathcal{C}_{\mathcal{M}}$ the set $\{u_{E \setminus H} \mid H \in \mathcal{O}_{\mathcal{M}}\}$ and by $\mathcal{J}_{\mathcal{M}}$ the set of all functions $f \in T$ which are equal to 0 on some $F \in \mathcal{O}_{\mathcal{M}}$.

Given $a \in E$, we denote by $\mathcal{I}(a, T)$ the ideal of the $f \in T$ such that $f(a) = 0$ and by $\mathcal{I}'(a, T)$ the ideal of the $f \in$ such that there exists an open neighborhood G of a such that $f(x) = 0 \ \forall x \in G$.

Theorem 4.2.7. *Let* $\mathcal{M}$ *be a maximal ideal of* T.

(1) $\mathcal{J}_{\mathcal{M}}$ *is an ideal of* T *containing* $\mathcal{C}_{\mathcal{M}}$,
(2) $\mathcal{J}_{\mathcal{M}}$ *is the ideal of* T *generated by* $\mathcal{C}_{\mathcal{M}}$ *and* $\mathcal{J}_{\mathcal{M}} = \{fu \mid f \in T, \ u \in \mathcal{C}_{\mathcal{M}}\}$,
(3) *If* $\mathcal{P}$ *is a prime ideal of* T *contained in* $\mathcal{M}$, *then* $\mathcal{J}_{\mathcal{M}} \subset \mathcal{P}$,
(4) *if* $\mathcal{M} = \mathcal{I}(a, T)$, *then* $\mathcal{I}'(a, T) = \mathcal{J}_{\mathcal{M}}$.

Proof.

(1) Let us check that $\mathcal{J}_{\mathcal{M}}$ is an ideal of T. Let $f, \ g \in \mathcal{J}_{\mathcal{M}}$. So, there exist $F, \ G \in \mathcal{O}_{\mathcal{M}}$ such that $f(x) = 0 \ \forall x \in F$, $g(x) = 0 \ \forall x \in G$, hence $f(x) - g(x) = 0 \ \forall x \in F \cap G$. Since $F \cap G$ belongs to $\mathcal{O}_{\mathcal{M}}$, $f - g$ lies in $\mathcal{J}_{\mathcal{M}}$. And obviously, for every $h \in T$, we have $h(x) f(x) = 0 \ \forall x \in F$, hence fh lies in $\mathcal{J}_{\mathcal{M}}$.
Next, $\mathcal{J}_{\mathcal{M}}$ contains $\mathcal{C}_{\mathcal{M}}$ because given $H \in \mathcal{O}_{\mathcal{M}}$, the set $E \setminus H$ is uniformly open then $u_{E \setminus H}$ belongs to S and is equal to 0 on L.
(2) Notice that if $f \in T$ and $u \in \mathcal{C}_{\mathcal{M}}$, then by 1) fu belongs to $\mathcal{J}_{\mathcal{M}}$. Conversely, if $f \in \mathcal{J}_{\mathcal{M}}$ and $H \in \mathcal{O}_{\mathcal{M}}$ are such that $f(x)$ is equal to 0 on L, then $u_{E \setminus H}$ belongs to $\mathcal{C}_{\mathcal{M}}$ and $f = fu_{E \setminus H}$. This proves that $\mathcal{J}_{\mathcal{M}} = \{fu \mid f \in S, \ u \in \mathcal{C}_{\mathcal{M}}\}$ and that $\mathcal{J}_{\mathcal{M}}$ is the ideal generated by $\mathcal{C}_{\mathcal{M}}$.
(3) It is sufficient to prove that $\mathcal{C}_{\mathcal{M}}$ is included in $\mathcal{P}$. Indeed, let $\mathcal{U} \in U(E)$ be such that $\mathcal{M} = \mathcal{I}(\mathcal{U}, T)$ and let $H \in \mathcal{O}_{\mathcal{M}}$. Then $H \in \mathcal{U}$ and $u_H \notin \mathcal{M}$. So, $u_H \notin \mathcal{P}$. But $u_H . u_{E \setminus H} = 0$. Thus, $u_{E \setminus H}$ belongs to $\mathcal{P}$ since $\mathcal{P}$ is prime, which concludes the proof.
(4) Each open neighborhood of a contains a disk that also is a uniformly open neighborhood of a, which ends the proof. $\qquad\square$

Corollary 4.2.8. *Let* $\mathcal{U}$ *be an ultrafilter on* E *and let* $\mathcal{P}$ *be a prime ideal included in* $\mathcal{I}(\mathcal{U}, S)$. *Let* $H \in \mathcal{U}$ *be uniformly open and let* $B = E \setminus H$. *Then the characteristic function* u *of* B *belongs to* $\mathcal{P}$.

Recall that for any normed $\mathbb{K}$-algebra $(G, \| \cdot \|)$, the closure of an ideal of G is an ideal of G.

Theorem 4.2.9. *The closure of $\mathcal{J}_\mathcal{M}$ with respect to the norm $\| \cdot \|_0$ is equal to $\mathcal{M}$.*

Proof. Let $f \in \mathcal{M} = \mathcal{I}(\mathcal{U}, T)$. Then for every $\epsilon > 0$ the set $F = D(f, \epsilon)$ belongs to $\mathcal{U}$ and F is uniformly open. Therefore, F belongs to $\mathcal{O}_\mathcal{M}$ and the characteristic function u of $E \setminus F$ lies in $\mathcal{C}_\mathcal{M}$, so that $fu \in \mathcal{J}_\mathcal{M}$. But $f(x) - uf(x) = 0 \; \forall x \notin F$ and $|f(x) - uf(x)| = |f(x)| \le \epsilon \; \forall x \in F$, so $\|f - uf\|_0 \le \epsilon$ and hence $\mathcal{M}$ is the closure of $\mathcal{J}_\mathcal{M}$ with respect to the norm $\| \cdot \|_0$ since, by Proposition 4.1.12, $\mathcal{M}$ is closed with respect to the norm $\| \cdot \|_0$. $\qquad \square$

Corollary 4.2.10. *Let $\mathcal{P}$ be a prime ideal contained in $\mathcal{M}$. Then $\mathcal{M}$ is the closure of $\mathcal{P}$ with respect to $\| \cdot \|_0$.*

Corollary 4.2.11. *The closure with respect to $\| \cdot \|_0$ of a prime ideal of T is a maximal ideal of T and a prime ideal of T is contained in a unique maximal ideal of T.*

Corollary 4.2.12. *A prime ideal of T is a maximal ideal if and only if it is closed with respect to $\| \cdot \|_0$.*
If $\mathcal{M} = \mathcal{I}(a, T)$, then $\mathcal{J}_\mathcal{M} = \mathcal{I}'(a, T)$. Therefore we have the following Corollary:

Corollary 4.2.13. *The closure of $\mathcal{I}'(a, T)$ with respect to $\| \cdot \|_0$ is $\mathcal{I}(a, T)$.*

Remark 5. $\mathcal{I}(a, T)$ is not necessarily the closure of $\mathcal{I}'(a, T)$ in T with respect to the norm $\| \cdot \|$ of T.

Corollary 4.2.14. (1) *$\mathcal{M}$ is the spectral closure of $\mathcal{J}_\mathcal{M}$ and the spectral closure of any prime ideal is contained in $\mathcal{M}$;*
(2) *A prime ideal is maximal if and only if it is spectrally closed.*

4.3. Multiplicative spectrum of T

Definitions and notations: Let A be a normed $\mathbb{L}$-algebra. Let us recall that in A, we have $\| \cdot \|_0 \le \| \cdot \|_{\mathrm{sp}}$ and that if A is compatible, then $\| \cdot \|_0 = \| \cdot \|_{\mathrm{sp}}$.

For any ultrafilter $\mathcal{U} \in U(E)$ and any $f \in S$, $|f(x)|$ has a limit along $\mathcal{U}$ since f is bounded. Then we denote by $\varphi_\mathcal{U}$ the mapping from T to $\mathbb{R}$ defined by $\varphi_\mathcal{U}(f) = \lim_\mathcal{U} |f(x)|$. Given $a \in E$ we denote by φ_a the mapping from T to $\mathbb{R}$ defined by $\varphi_a(f) = |f(a)|$. These maps belong to $\mathrm{Mult}(T, \| \cdot \|)$ since

$\| \cdot \|_0 \leq \| \cdot \|_{\mathrm{sp}} \leq \| \cdot \|$. Particularly, the elements of $\mathrm{Mult}_E(T, \| \cdot \|)$ are the φ_a, $a \in E$.

Proposition 4.3.1. *Let $a \in E$. Then $\mathcal{I}(a, T)$ is a maximal ideal of T of codimension 1 and φ_a belongs to $\mathrm{Mult}_1(T, \| \cdot \|)$.*

Moreover, for every algebra homomorphism χ from T to $\mathbb{L}$, its kernel is a maximal ideal of codimension 1 of the form $\mathcal{I}(\mathcal{U}, T)$ with $\mathcal{U} \in U(E)$ and χ is defined as $\chi(f) = \lim_{\mathcal{U}} f(x)$, while $\varphi_{\mathcal{U}}(f) = |\chi(f)|$.

Proof. The first part is clear. The second comes from the proof of Theorem 4.2.1: actually we proved that if $\mathcal{M}$ is of codimension 1 then every $f \in T$ has a limit along $\mathcal{U}$ and $\lim_{\mathcal{U}} f(x) = \chi(f)$. $\qquad \square$

The following Proposition 4.3.2 is immediate.

Proposition 4.3.2. *Let $\mathcal{U}$ be an ultrafilter on E. Then $\varphi_{\mathcal{U}}$ belongs to the closure of $\mathrm{Mult}_E(T, \| \cdot \|)$.*

Remark 6. According to Remark 4, if $\mathbb{K}$ is locally compact we have $\mathrm{Max}_1(T) \neq \mathrm{Max}_E(T)$ and hence $\mathrm{Mult}_1(T, \| \cdot \|) \neq \mathrm{Mult}_E(T, \| \cdot \|)$ because given a maximal ideal $\mathcal{I}(\mathcal{U}, T)$ which does not belong to Max_E, we can define $\varphi_{\mathcal{U}} \in \mathrm{Mult}(T, \| \cdot \|)$ which does not belong to $\mathrm{Mult}_E(T, \| \cdot \|)$.

Given a normed $\mathbb{L}$-algebra A and $\phi \in \mathrm{Mult}(A, \| \cdot \|)$, it is well known that $Ker(\phi)$ is a prime closed ideal, with respect to the norm $\| \cdot \|$ of A. Actually we have the following proposition:

Proposition 4.3.3. *For each $\phi \in \mathrm{Mult}(T, \| . \|)$, $\mathrm{Ker}(\phi)$ is a maximal ideal.*

Proof. Let $\phi \in \mathrm{Mult}(T, \| . \|)$ and let f belong to the spectral closure of $\mathrm{Ker}(\phi)$. There exists a sequence $(f_n)_{n \in \mathbb{N}}$ of $\mathrm{Ker}(\phi)$ such that $\lim_{n \to \infty} \| f_n - f \|_{\mathrm{sp}} = 0$. By Theorem 2.5.17 of Chapter 2, since $\phi(g) \leq \| g \|_{sp} \, \forall g \in T$, we have $\lim_{n \to \infty} \phi(f_n - f) = 0$. But $\phi(f_n) = 0 \, \forall n \in \mathbb{N}$, hence $\phi(f_n - f) = \phi(f)$ and hence $\phi(f) = 0$. Therefore, f belongs to $\mathrm{Ker}(\phi)$, which means that $\widetilde{\mathrm{Ker}(\phi)} = \mathrm{Ker}(\phi)$. Therefore, by Theorem 4.2.4, $\mathrm{Ker}(\phi)$ is a maximal ideal of T. $\qquad \square$

Corollary 4.3.4. $\mathrm{Mult}(T, \| . \|) = \mathrm{Mult}_m(T, \| . \|)$.

Theorem 4.3.5. *T is multbijective. Precisely if $\psi \in \mathrm{Mult}(T, \| . \|)$ and $\mathrm{Ker}(\psi) = \mathcal{M}$ then $\psi = \varphi_{\mathcal{U}}$ for every ultrafilter $\mathcal{U}$ such that $\mathcal{M} = \mathcal{I}(\mathcal{U}, T)$.*

Proof. Let $\psi \in \mathrm{Mult}_m(T, \| \cdot \|)$, let $\mathcal{M} = \mathrm{Ker}(\psi)$ and $\mathcal{U}$ be an ultrafilter such that $\mathcal{M} = \mathcal{I}(\mathcal{U}, T)$. We shall prove that $\psi = \varphi_{\mathcal{U}}$.

Let $f \in T$. Notice that if $f \in \mathcal{M}$ then $\psi(f) = \varphi_{\mathcal{U}}(f) = 0$. Now we assume that $f \notin \mathcal{M}$. So $\psi(f)$ and $\varphi_{\mathcal{U}}(f)$ are both strictly positive.

First, take $\epsilon > 0$ and consider the set $F = \{x \in E : |f(x)| \leq \varphi_{\mathcal{U}}(f) + \epsilon\}$. This set belongs to $\mathcal{U}$ and by Lemma 4.1.4, it is uniformly open. Therefore, its characteristic function u lies in T, so we have $\varphi_{\mathcal{U}}(u) = 1$. Consequently, we can derive that $\psi(u) = 1$ because u is idempotent and does not belong to $\mathcal{M}$. Therefore $\psi(uf) = \psi(f)$ and $\varphi_{\mathcal{U}}(uf) = \varphi_{\mathcal{U}}(f)$. By Theorem 2.5.11 of Chapter 2, we have $\psi(f) = \psi(uf) \leq \|uf\|_{sp} = \|uf\|_0$ because T is compatible. But by definition of F we have: $\|uf\|_0 \leq \varphi_{\mathcal{U}}(f) + \epsilon$. Therefore, $\psi(f) \leq \varphi_{\mathcal{U}}(f) + \epsilon$. This holds for every $\epsilon > 0$. Consequently we may conclude that $\psi(f) \leq \varphi_{\mathcal{U}}(f)$ for every $f \in F$.

We now prove the inverse inequality. We have $\varphi_{\mathcal{U}}(f) > 0$, so consider the set $W = \{x \in E : |f(x)| \geq \frac{\varphi_{\mathcal{U}}(f)}{2}\}$. Again this is a uniformly open set which belongs to $\mathcal{U}$. Let w be the characteristic function of W and put $g = wf + (1 - w)$. We have $\varphi_{\mathcal{U}}(w) = 1$ and $\varphi_{\mathcal{U}}(1 - w) = 0$ so $w \notin \mathcal{M}$ and $1 - w \in \mathcal{M}$. Since $\mathcal{M} = \mathrm{Ker}(\psi)$ we then have $\psi(1 - w) = 0$ and $\psi(w) = 1$ because w is idempotent. Finally $\psi(g) = \psi(f)$ and $\varphi_{\mathcal{U}}(g) = \varphi_{\mathcal{U}}(f)$.

On the other hand, we can check that $|g(x)| \geq \min\left(1, \frac{\varphi_{\mathcal{U}}(f)}{2}\right)$ for all $x \in E$, hence g is invertible in T. Putting $h = \frac{1}{g}$, using the first inequality yet proved, we have

$$\psi(f) = \psi(g) = \frac{1}{\psi(h)} \geq \frac{1}{\varphi_{\mathcal{U}}(h)} = \varphi_{\mathcal{U}}(g) = \varphi_{\mathcal{U}}(f).$$

That concludes the proof. $\square$

Remark 7. It follows from the previous theorem that for a compatible algebra T, two ultrafilters on E that are not contiguous define two distinct continuous multiplicative semi-norms on T, this particularly applies to the algebra $\mathcal{B}$ of all uniformly continuous functions.

Corollary 4.3.6. *For every $\phi \in \mathrm{Mult}(T, \| \cdot \|)$ there exists a unique $\mathcal{H} \in Y_{(T)}(E)$ such that $\phi(f) = \lim_{\mathcal{U}} |f(x)| \ \forall f \in T, \ \forall \mathcal{U} \in \mathcal{H}$.*

Moreover, the mapping Ψ that associates to each $\phi \in \mathrm{Mult}(T, \| \cdot \|)$ the unique $\mathcal{H} \in Y_{\mathcal{R}}(E)$ such that $\phi(f) = \lim_{\mathcal{U}} |f(x)| \ \forall f \in T, \ \forall \mathcal{U} \in \mathcal{H}$, is a bijection from $\mathrm{Mult}(T, \| \cdot \|)$ onto $Y_{(T)}(E)$.

Assuming that T is compatible, since by Theorem 4.3.5 and Corollary 4.3.4, each element $\phi \in \mathrm{Mult}(T, \| \cdot \|)$ is of the form $\varphi_{\mathcal{U}}$, Corollary 4.3.7 is obvious.

Corollary 4.3.7. $\text{Mult}_E(T, \| \cdot \|)$ *is dense in* $\text{Mult}(T, \| \cdot \|)$.

Theorem 4.3.8. *The topological space* E, *equipped with its distance* δ, *is homeomorphic to* $\text{Mult}_E(T, \| \cdot \|)$ *equipped with the restricted topology from that of* $Mult(T, \| \cdot \|)$.

Proof. The proof is similar to the one of Theorem 3.5.10 of Chapter 3 but with however some difference.

For every $a \in E$, put $Z(a) = \varphi_a$, take $f_1, \ldots, f_q \in T$ and $\epsilon > 0$. We set $\mathcal{W}'(\varphi_a, f_1, \ldots, f_q, \epsilon) = \mathcal{W}(\varphi_a, f_1, \ldots, f_q, \epsilon) \cap \text{Mult}_E(T, \| \cdot \|)$. Considering the natural topology on E, the filter of neighborhoods of a admits for basis the family of disks $B(a, r)$, $r > 0$. We will show that it is induced through the mapping Z by the filter admitting for basis the family of neighborhoods of φ_a in $\text{Mult}(T, \| \cdot \|)$. Indeed, take $r \in]0, 1[$. Since $B(a, r)$ is obviously uniformly open, its characteristic function u belongs to T.

Then $\mathcal{W}'(\varphi_a, u, r)$ is the set of φ_b such that $|\varphi_a(u) - \varphi_b(u)|_\infty < r$. But since $r < 1$, we can see that this is satisfied if and only if $b \in B(a, r)$. Therefore the topology induced on E by $\text{Mult}(T, \| \cdot \|)$ is thinner than its metric topology.

Conversely, take some neighborhood $\mathcal{W}'(\varphi_a, f_1, \ldots, f_q, \epsilon)$ of φ_a in $\text{Mult}_E(T, \| \cdot \|)$. For each $j = 1, \ldots, q$, the set of the $x \in E$ such that $|\varphi_x(f_j) - \varphi_a(f_j)|_\infty \leq \epsilon$ is the set of the x such that $\left| |f_j(x)| - |f_j(a)| \right|_\infty \leq \epsilon$. But now, since each f_j is continuous, the set of the x such that $\left| |f_j(x)| - |f_j(a)| \right|_\infty \leq \epsilon \; \forall j = 1, \ldots, q$ is a neighborhood of a in E. Consequently, the metric topology of E is thinner than the topology induced by $\text{Mult}_E(T, \| \cdot \|)$ and that finishes proving that the two topological spaces are homeomorphic. $\square$

Corollary 4.3.9. $\text{Mult}(T, \| \cdot \|)$ *is a compactification of the topological space* E.

Notation: Let $X \subset \text{Mult}(S, \| \cdot \|)$. As in Section 3.5 of Chapter 3, we denote by $\text{spcl}(X)$ the subset of the $a \in E$ such that $\varphi_a \in X$.

By Theorem 4.3.8, we can state this corollary.

Theorem 4.3.10. *If* X, Y *are two open closed subsets of* $\text{Mult}(T, \| \cdot \|)$ *making a partition, then* $\text{spcl}(X)$, $\text{spcl}(Y)$ *are two uniformly open subsets of* E *making a partition of* E.

Proof. As in Corollary 3.5.12, $\text{spcl}(X)$, $\text{spcl}(Y)$ are two clopens of E making a partition of E. Suppose that $\text{spcl}(X)$ is not uniformly open. Then there exists a sequence $(a_n)_{n \in \mathbb{N}}$ of $\text{spcl}(X)$ and a sequence $(b_n)_{n \in \mathbb{N}}$ of $\text{spcl}(Y)$

such that $\lim_{n\to+\infty} a_n - b_n = 0$. Since both X and Y are closed, hence compact, we can extract from the sequence $(\varphi_{(a_n)})_{n\in\mathbb{N}}$ a converging subsequence, hence up to an extraction, we can assume that $\lim_{n\to+\infty} \varphi_{(a_n)} = \psi \in X$. $\square$

Now, consider a neighborhood of ψ in $\mathrm{Mult}(T, \| \cdot \|)$: $\mathcal{W}(\psi, f_1, \ldots, f_q, \epsilon)$ with $f_1, \ldots, f_q \in T$ and $\epsilon > 0$. Since the elements of T are uniformly continuous, there is $\eta > 0$ such that $|f_j(x) - f_j(y)| \leq \epsilon \ \forall j = 1, \ldots, q, \ \forall x, y \in E$ such that $|x - y| \leq \eta$. Then, there exists a rank N such that $|a_n - b_n| \leq \eta \ \forall n \geq N$, therefore $|f_j(a_n) - f_j(b_n)| \leq \epsilon \ \forall j = 1, \ldots, q$, hence $\left| |f_j(a_n)| - |f_j(b_n)| \right|_\infty \leq \epsilon \ \forall j = 1, \ldots, q$ and hence $|\varphi_{a_n}(f_j) - \varphi_{b_n}(f_j)|_\infty \leq \epsilon \ \forall j = 1, \ldots, q, \ \forall n \geq N$.

Next, there exists a rank M such that $|\varphi_{b_n}(f_j) - \psi(f_j)|_\infty \leq \epsilon \ \forall j = 1, \ldots, q$. Putting $P = \max(N, M)$ we see that for every $n \geq P$ we have $|\varphi_{b_n}(f_j) - \psi(f_j)|_\infty \leq 2\epsilon \ \forall j = 1, \ldots, q$ and therefore $\varphi_{b_n} \in \mathcal{W}(\psi, f_1, \ldots, f_q, 2\epsilon)$, which proves that $\lim_{n\to+\infty} \varphi_{b_n} = \psi$. Consequently, ψ belongs to Y, a contradiction.

Corollary 4.3.11. *If X, Y are two open closed subsets of $\mathrm{Mult}(T, \| \cdot \|)$ making a partition, then there exist idempotents u, $v \in T$ such that $\phi(u) = 1$, $\psi(u) = 0$, $\phi(v) = 0$, $\psi(v) = 1 \ \forall \phi \in X$, $\psi \in Y$.*

Theorem 4.3.12. *Let $\phi = \varphi_{\mathcal{U}} \in \mathrm{Mult}_m(T, \| \cdot \|)$, with $\mathcal{U}$ an ultrafilter on E, let $\mathcal{K}$ be the field $\frac{T}{\mathrm{Ker}(\phi)}$ and let θ be the canonical surjection from T onto the field $\mathcal{K}$. Then, the mapping defined on $\mathcal{K}$ by $|\theta(f)| = \phi(f) \ \forall f \in T$, is the quotient norm $\| \cdot \|'$ of $\| \cdot \|_0$ defined on $\mathcal{K}$ and is an absolute value on $\mathcal{K}$. Moreover, if $\mathrm{Ker}(\phi)$ is of codimension 1, then this absolute value is the one defined on $\mathbb{L}$ and coincides with the quotient norm of the norm $\| \cdot \|$ of T.*

Proof. Let $\mathcal{M} = \mathrm{Ker}(\varphi_{\mathcal{U}})$. Let $t \in \mathcal{K}$ and let $f \in T$ be such that $\theta(f) = t$. So, $\|t\|' \geq \lim_{\mathcal{U}} |f(s)|$. Conversely, take $\epsilon > 0$ and let $V = \{x \in E : |f(x)| \leq \lim_{\mathcal{U}} |f(s)| + \epsilon\}$. By Lemma 4.1.4, the set V is uniformly open and belongs to $\mathcal{U}$. The characteristic function u of $E \setminus V$ belongs to $\mathcal{M}$ and so does uf. But by construction, $(f - uf)(x) = 0 \ \forall x \in E \setminus V$ and $(f - uf)(x) = f(x) \ \forall x \in V$. Consequently, $\|f - uf\|_0 \leq \lim_{\mathcal{U}} |f(s)| + \epsilon$ and therefore $\|t\|' \leq \|f - uf\|_0 \leq \lim_{\mathcal{U}} |f(s)| + \epsilon$. This finishes proving the equality $\|\theta(f)\|' = \lim_{\mathcal{U}} |f(s)|$ and hence the mapping defined by $|\theta(f)| = \phi(f)$, $f \in T$ is the quotient norm $\| \cdot \|'$ of $\| \cdot \|_0$. Then it is multiplicative, hence it is an absolute value on $\mathcal{K}$.

Now, suppose that $\mathcal{M}$ is of codimension 1. Then $\mathcal{K}$ is isomorphic to $\mathbb{L}$ and its absolute value $\| \cdot \|'$ is continuous with respect to the topology of $\mathbb{L}$, hence it is equal to the absolute value of $\mathbb{L}$. Finally, consider the quotient norm $\| \cdot \|_q$ of the norm $\| \cdot \|$ of T: that quotient norm of course bounds the quotient norm $\| \cdot \|'$ which is the absolute value of $\mathbb{L}$. If $f \in T$ and $b = \theta(f)$, we have $f - b \in \mathcal{M}$ and $\|\theta(f)\|_q \leq \|b\| = |b| = |\theta(f)| = \|\theta(f)\|'$, which ends the proof. $\qquad\qquad\square$

Remark 8. Here we can make the same remark as in Chapter 3: It is not clear whether an algebra T can admit a prime closed ideal $\mathcal{P}$ (with respect to the norm $\| \cdot \|$) which is not a maximal ideal. If it admits such a prime closed ideal $\mathcal{P}$, then it is not the kernel of a continuous multiplicative semi-norm because $Mult(T, \| \cdot \|) = Mult_m(T, \| \cdot \|)$. Such a situation may happen as noticed in Remark 8 of Chapter 3.

Lemma 4.3.13. *Let us fix $a \in E$. For every $r > 0$, let $Z(a, r)$ be the set of $\varphi_{\mathcal{U}}$, $\mathcal{U} \in U(E)$, such that $B(a, r)$ belongs to $\mathcal{U}$. The family $\{Z(a, r) \, | r \in]0, 1[\}$ makes a basis of the filter of neighborhoods of φ_a.*

Proof. Let $\mathcal{W}(\varphi_a, f_1, \ldots, f_q, \epsilon)$ be a neighborhood of φ_a in $Mult(T, \| \cdot \|)$. There exists $r > 0$ such that, whenever $\delta(a, x) \leq r$ we have $|f_j(x) - f_j(a)| \leq \epsilon \, \forall j = 1, \ldots, q$ and therefore, clearly, $|\varphi_{\mathcal{U}}(f_j) - \varphi_a(f_j)|_\infty \leq \epsilon \, \forall j = 1, \ldots, q$, for every $\mathcal{U}$ containing $B(a, r)$. Thus, $Z(a, r)$ is included in $\mathcal{W}(\varphi_a, f_1, \ldots, f_q, \epsilon)$.

Conversely, consider a set $Z(a, r)$ with $r \in]0, 1[$, let u be the characteristic function of $B(a, r)$ and consider $\mathcal{W}(\varphi_a, u, r)$. Given $\psi = \varphi_{\mathcal{U}} \in \mathcal{W}(\varphi_a, u, r)$, we have $|\psi(u) - \varphi_a(u)|_\infty \leq r$. But $|\psi(u) - \varphi_a(u)|_\infty = |\psi(u) - 1|_\infty = |\lim_{\mathcal{U}} |u(x)| - 1|_\infty$. If $B(a, r)$ belongs to $\mathcal{U}$, then $\lim_{\mathcal{U}} |u(x)| = 1$ and therefore $|\lim_{\mathcal{U}} |u(x)| - 1|_\infty = 0$. But if $B(a, r)$ does not belong to $\mathcal{U}$, then $\lim_{\mathcal{U}} |u(x)| = 0$ and therefore $|\lim_{\mathcal{U}} |u(x)| - 1|_\infty = 1$. Consequently, since $r < 1$, $\mathcal{W}(\varphi_a, u, r)$ is included in $Z(a, r)$, which finishes proving that the family of $Z(a, r)$, $r \in]0, 1[$ is a basis of the filter of neighborhoods of φ_a. $\quad\square$

We now have a result for compatible algebras that is similar to this for admissible algebras.

Theorem 4.3.14. *The Shilov boundary of T is equal to $Mult(T, \| \cdot \|)$ [14].*

Proof. We will show that for every $a \in E$, φ_a belongs to $Shil(T)$. So, let us fix $a \in E$ and suppose that φ_a does not belong to $Shil(T)$. Since $Shil(T)$ is a closed subset of $Mult(T, \| \cdot \|)$, there exists a neighborhood of φ_a that contains no element of $Shil(T)$. Therefore, by the Lemma 4.3.13, there exists $s > 0$ such that $Z(a, s)$ contains no element of $Shil(T)$. Now, let $D = B(a, s)$, let u be the characteristic function of D. Since any $\phi \in Mult(T, \| \cdot \|)$ satisfies either $\phi(u) = 1$ or $\phi(u) = 0$, there exists $\theta \in Shil(T)$ be such that $\theta(u) = \|u\|_{sp} = 1$. Then, θ is of the form $\varphi_{\mathcal{U}}$, with $\mathcal{U} \in U(E)$ and $\mathcal{U}$ does not contain D. But since $u(x) = 0 \, \forall x \in E \backslash D$, we have $\theta(u) = 0$, a contradiction.

Consequently, for every $a \in E$, φ_a belongs to $\mathrm{Shil}(T)$ which is a closed subset of $\mathrm{Mult}(T, \| \cdot \|)$ and since, by Corollary 4.3.7, $\mathrm{Mult}_E(S, \| \cdot \|)$ is dense in $\mathrm{Mult}(T, \| \cdot \|)$, $\mathrm{Shil}(T)$ is equal to $\mathrm{Mult}(T, \| \cdot \|)$. $\qquad\square$

4.4.   The Stone space of $\mathbb{G}(E)$

It was proved in Section 3.6 of Chapter 3 and in [12] that for the algebra $\mathcal{C}$ of continuous bounded functions from E to $\mathbb{L}$, the Banaschewski compactification of E is homeomorphic to $\mathrm{Mult}(\mathcal{C}, \| \cdot \|_0)$. Here we get some similar version for compatible algebras.

Let $\mathcal{P}(E)$ be the set of non-zero ring homomorphisms from $\mathbb{G}(E)$ onto the field $\mathbb{F}_2$ equipped with the topology of pointwise convergence. This is the Stone space of the Boolean ring $\mathbb{G}(E)$ and hence it is a compact space. Thus, we have similar properties as those shown by $\mathbb{B}(E)$: thus, $\mathbb{B}(E)$ is equipped with a structure of ring whose addition is Δ and the multiplication is $\cap$.

For every $\mathcal{U} \in U(E)$, we denote by $\zeta_\mathcal{U}$ the ring homomorphism from $\mathbb{G}(E)$ onto $\mathbb{F}_2$ defined by $\zeta_\mathcal{U}(O) = 1$ for every $O \in \mathbb{G}(E)$ that belongs to $\mathcal{U}$ and $\zeta_\mathcal{U}(O) = 0$ for every $O \in \mathbb{G}(E)$ that does not belong to $\mathcal{U}$.

Particularly, given $a \in E$, we denote by ζ_a the ring homomorphism from $\mathbb{G}(E)$ onto $\mathbb{F}_2$ defined by $\zeta_a(O) = 1$ for every $O \in \mathbb{G}(E)$ that contains a and $\zeta_a(O) = 0$ for every $O \in \mathbb{G}(E)$ that does not contain a.

Remark 9. Let $\mathcal{P}'(E)$ be the set of ζ_a, $a \in E$. The mapping that associates ζ_a to any $a \in E$ defines a surjective mapping from E onto $\mathcal{P}'(E)$. That mapping is also injective because given a, $b \in E$, there exists a uniformly open subset F such that $a \in F$ and $b \notin F$.

By Corollary 4.3.6, we have a bijection $\mathcal{Y}$ from $\mathrm{Mult}(T, \| \cdot \|)$ onto $Y_{(T)}(E)$ associating to each $\phi \in \mathrm{Mult}(T, \| \cdot \|)$ the unique $\mathcal{H} \in Y_{(T)}(E)$ such that $\phi(f) = \lim_\mathcal{U} |f(x)|$, $\mathcal{U} \in \mathcal{H}$, $f \in T$, i.e., $\phi = \varphi_\mathcal{U}$ for every $\mathcal{U} \in \mathcal{H}$.

On the other hand, let us take some $\mathcal{H} \in Y_{(T)}(E)$ and ultrafilters $\mathcal{U}$, $\mathcal{V}$ in $\mathcal{H}$. Since $\mathcal{U}, \mathcal{V}$ own the same uniformly open subsets of E, we have $\zeta_\mathcal{U} = \zeta_\mathcal{V}$ and hence we can define a mapping Ξ from $Y_T(E)$ into $\mathcal{P}(E)$ which associates to each $\mathcal{H} \in Y_T(E)$ the $\zeta_\mathcal{U}$ such that $\mathcal{U} \in \mathcal{H}$.

Lemma 4.4.1. Ξ *is a bijection from* $Y_{(T)}(E)$ *onto* $\mathcal{P}(E)$.

Proof.   Indeed, Ξ is clearly injective by Corollary 4.1.14. Now, let us check that Ξ is surjective. Let $\theta \in \mathcal{P}(E)$. Since θ is a ring homomorphism for the Boolean laws, the family of uniformly open sets X satisfying $\theta(X) = 1$ generates a filter $\mathcal{F}$. Let $\mathcal{U} \in U(E)$ be thinner than $\mathcal{F}$ and let $\mathcal{H}$ be the class of $\mathcal{U}$ with respect to $\mathcal{T}$. We will check that $\theta = \Xi(\mathcal{H}) = \zeta_{\mathcal{U}}$. Let O be a uniformly open subset that belongs to $\mathcal{U}$. Then $E \setminus O$ does not belong to $\mathcal{U}$ and therefore it does not belong to $\mathcal{F}$, so $\theta(E \setminus O) = 0$, consequently $\theta(O) = 1$. And now, let O be a uniformly open subset that does not belong to $\mathcal{U}$. Then O does not belong to $\mathcal{F}$, hence $\theta(O) = 0$, which ends the proof. $\qquad\square$

Notations: We put $\mathcal{Z} = \Xi \circ \mathcal{Y}$ and hence $\mathcal{Z}$ is a bijection from $\mathrm{Mult}(T, \| \cdot \|)$ onto $\mathcal{P}(E)$. Notice that for every ultrafilter $\mathcal{U}$, $\mathbb{U}(\varphi_{\mathcal{U}})$ is the class $\mathcal{H}$ of $\mathcal{U}$ with respect to $(\mathcal{T})$ and $\Xi(\mathcal{H}) = \zeta_{\mathcal{U}}$ so $\Phi(\varphi_{\mathcal{U}}) = \zeta_{\mathcal{U}}$.

Theorem 4.4.2. *$\mathcal{Z}$ is a homeomorphism once $\mathcal{P}(E)$ and $\mathrm{Mult}(T, \| \cdot \|)$ are equipped with topologies of pointwise convergence.*

Proof. Recall that for any $\mathcal{U} \in U(E)$, a neighborhoods basis of $\varphi_{\mathcal{U}}$ in $\mathrm{Mult}(T, \| \cdot \|)$ is given by the family of sets of the form $\mathcal{W}(\varphi_{\mathcal{U}}, f_1, \ldots, f_q, \epsilon)$ with $f_1, \ldots, f_q \in T$, $\epsilon > 0$ and

$$\mathcal{W}(\varphi_{\mathcal{U}}, f_1, \ldots, f_q, \epsilon) = \left\{ \varphi_{\mathcal{V}} \mid \left| \lim_{\mathcal{U}} |f_j(x)| - \lim_{\mathcal{V}} |f_j(x)| \right|_{\infty} \le \epsilon, \ j = 1, \ldots, q \right\}.$$

On the other hand, for any $\mathcal{U} \in U(E)$, a neighborhood basis for $\zeta_{\mathcal{U}}$ in $\mathcal{P}(E)$ is given by the family of sets $V(\zeta_{\mathcal{U}}, O_1, \ldots, O_q)$ where $O_1, \ldots, O_q$ belong to $\mathbb{G}(E)$ and

$$V(\zeta_{\mathcal{U}}, O_1, \ldots, O_q) = \{ \zeta_{\mathcal{V}} \mid \zeta_{\mathcal{U}}(O_j) = \zeta_{\mathcal{V}}(O_j), j = 1, \ldots, q \}. \qquad\square$$

Notice also that if F belongs to $\mathbb{G}(E)$ and if u is its characteristic function, then for any $\mathcal{U} \in U(E)$, we have $\zeta_{\mathcal{U}}(F) = 1$ if and only if $F \in \mathcal{U}$, i.e., if and only if $\lim_{\mathcal{U}} |u(x)| = 1$. Otherwise, both $\zeta_{\mathcal{U}}(F)$ and $\lim_{\mathcal{U}} |u(x)|$ are equal to 0. Therefore, the relation

$$\left| \lim_{\mathcal{U}} |u(x)| - \lim_{\mathcal{V}} |u(x)| \right|_{\infty} \le \frac{1}{2}$$

holds if and only if $\zeta_{\mathcal{U}}(F) = \zeta_{\mathcal{V}}(F)$. Recall that for every $\mathcal{U} \in U(E)$ we have $\mathcal{Z}(\varphi_{\mathcal{U}}) = \zeta_{\mathcal{U}}$.

We will show that $\mathcal{Z}$ is continuous. Consider $O_1, \ldots, O_q \in \mathbb{G}(E)$, $\mathcal{U} \in U(E)$ and the neighborhood $V(\zeta_{\mathcal{U}}, O_1, \ldots, O_q)$ of $\zeta_{\mathcal{U}}$. From the preceding remark, $\zeta_{\mathcal{V}}$ belongs to $V(\zeta_{\mathcal{U}}, O_1, \ldots, O_q)$ if and only if for every $j = 1, \ldots, q$, $\zeta_{\mathcal{U}}(O_j) = \zeta_{\mathcal{V}}(O_j)$, i.e., if for every $j = 1, \ldots, q$,

$$\left| \lim_{\mathcal{U}} |u_j(s)| - \lim_{\mathcal{V}} |u_j(x)| \right|_{\infty} \le \frac{1}{2},$$

i.e., if $\varphi_\mathcal{V}$ belongs to $\mathcal{W}(\varphi_\mathcal{U}, u_1, \ldots, u_q, \frac{1}{2})$. Consequently, this proves that $\mathcal{Z}$ is continuous.

We will now prove that $\mathcal{Z}^{-1}$ is also continuous. Consider $f_1, \ldots, f_q \in S$, $\epsilon > 0$, $\mathcal{U} \in U(E)$ and the neighborhood $\mathcal{W}(\varphi_\mathcal{U}, f_1, \ldots, f_q, \epsilon)$ which is obviously $\bigcap_{j=1}^{q} \mathcal{W}(\varphi_\mathcal{U}, f_j, \epsilon)$. Let us fix $i \in \{1, \ldots, q\}$. Put $a_i = \lim_\mathcal{U} |f_i(x)|$ and $O_i = \{x \in E \mid \left| |f_i(x)| - a_i \right|_\infty \leq \frac{\epsilon}{2}\}$. By Lemma 4.1.4, O_i is uniformly open and of course it belongs to $\mathcal{U}$. Thus, we have $V(\zeta_\mathcal{U}, O_i) = \{\zeta_\mathcal{V} \mid O_i \in \mathcal{V}\}$. Now let $\mathcal{V} \in U(E)$ be such that $O_i \in \mathcal{V}$ and put $O_i' = \{x \in E \mid \left| |f_i(x)| - \lim_\mathcal{V} |f_i(x)| \right|_\infty \leq \frac{\epsilon}{2}\}$. Then O_i' is uniformly open also and it belongs to $\mathcal{V}$. Therefore, $O_i \cap O_i'$ belongs to $\mathcal{V}$. Take $x \in O_i \cap O_i'$. We have $\left| |f_i(x)| - a_i \right|_\infty \leq \frac{\epsilon}{2}$ and $\left| |f_i(x)| - \lim_\mathcal{V} |f_i(x)| \right|_\infty \leq \frac{\epsilon}{2}$, so $\left| \lim_\mathcal{V} |f_i(x)| - a_i \right|_\infty \leq \epsilon$ and hence $\varphi_\mathcal{V}$ belongs to $\mathcal{W}(\varphi_\mathcal{U}, f_i, \epsilon)$. This holds for every $i = 1, \ldots, q$. Therefore, we can conclude that if $\zeta_\mathcal{V}$ belongs to $V(\zeta_\mathcal{U}, O_1, \ldots, O_q)$, which is $\bigcap_{i=1}^{q} V(\zeta_\mathcal{U}, O_i)$, then $\varphi_\mathcal{V}$ belongs to $\bigcap_{i=1}^{q} \mathcal{W}(\varphi_\mathcal{U}, f_i, \epsilon)$ which is $\mathcal{W}(\varphi_\mathcal{U}, f_1, \ldots, f_q, \epsilon)$. This finishes proving that $\mathcal{Z}^{-1}$ is continuous too, and hence is a homeomorphism.

Corollary 4.4.3. *The space $\mathcal{P}(E)$ is a compactification of E which is equivalent to the compactification $\mathrm{Mult}(T, \| \cdot \|)$.*

Remark 10. For a compatible algebra T, the compactification $\mathcal{P}(E)$ coincides with the Guennebaud-Berkovich multiplicative spectrum of T. This is not the Banaschewski compactification, which corresponds to the Stone space associated to the Boolean ring of clopen sets of E.

4.5. About the completion of E

Notations: We denote by $\overline{\overline{E}}$ the completion of E and by $\overline{\overline{\delta}}$ the continuation of δ to $\overline{\overline{E}}$. We then identify E with a dense subset of $\overline{\overline{E}}$.

The following theorem is well known.

Theorem 4.5.1. *Every uniformly continuous function f from E to $\mathbb{L}$ has a unique extension to a uniformly continuous function $\overline{\overline{f}}$ from $\overline{\overline{E}}$ to $\mathbb{L}$ and we have $\|f\|_0 = \|\overline{\overline{f}}\|_0$.*

Notations: We denote by $\overline{\overline{T}}$ the set of functions $\overline{\overline{f}}$, $f \in T$. Given $f \in T$, we put $\|\overline{\overline{f}}\| = \|f\|$.

We have the following proposition.

Proposition 4.5.2. *The normed* $\mathbb{L}$-*algebra* $(\overline{\overline{T}}, \| \cdot \|)$ *is isomorphic to* $(T, \| \cdot \|)$ *and it is semi-compatible with respect to* $\overline{\overline{E}}$. *Moreover, if* T *is compatible, so is* $\overline{\overline{T}}$.

Proof. Obviously $(\overline{\overline{S}}, \| \cdot \|)$ is isomorphic to $(S, \| \cdot \|)$ and therefore it is a Banach $\mathbb{L}$-algebra. Now we prove that it is semi-compatible.

(1) Take $f \in T$ such that $\inf_{\overline{\overline{E}}}\{|\overline{\overline{f}}(x)| \mid x \in \overline{\overline{E}}\} > 0$. Then $\inf_E\{|f(x)| \mid x \in E\} > 0$ and hence f is invertible in T. Now, if $g \in T$ and $fg = 1$, then $\overline{\overline{fg}} = 1$ hence $\overline{\overline{f}}$ is invertible in $\overline{\overline{T}}$.

(2) Let $\overline{\overline{F}}$ be a subset of $\overline{\overline{E}}$ and let u be its characteristic function. Obviously, if $u \in \overline{\overline{T}}$, then, using Lemma 4.1.6, $\overline{\overline{F}}$ is uniformly open in $\overline{\overline{E}}$ since u is uniformly continuous.

Assume now that $\overline{\overline{F}}$ is uniformly open in $\overline{\overline{E}}$. Put $F = \overline{\overline{F}} \cap E$. If $\overline{\overline{F}} = \emptyset$, then $F = \emptyset$ and $u = 0$. Next, if $\overline{\overline{F}} \neq \emptyset$, then $F \neq \emptyset$ because $\overline{\overline{F}}$ is open and E is dense in $\overline{\overline{E}}$. So, we have $E \setminus F = (\overline{\overline{E}} \setminus \overline{\overline{F}}) \cap E$ and $\delta(F, E \setminus F) \geq \overline{\overline{\delta}}(\overline{\overline{F}}, \overline{\overline{E}} \setminus \overline{\overline{F}}) > 0$. Therefore, F is uniformly open in E, hence its characteristic function u' lies in S. But u' is the restriction of u to E and u is uniformly continuous. Consequently, by Theorem 2.1.1 of Chapter 2, $u = \overline{\overline{u'}}$ and hence u belongs to $\overline{\overline{T}}$.

(3) Suppose now that T is compatible. Clearly, the spectral norm on $\overline{\overline{T}}$ is induced by this of T and hence is $\| \cdot \|_0$. Therefore $\overline{\overline{T}}$ is compatible. $\square$

Remark 11. Thus, every result obtained in Chapter 3 also holds for the algebra $(\overline{\overline{T}}, \| \cdot \|)$. Particularly, we have Corollary 4.5.3.

Corollary 4.5.3. *The mapping which associates to every* $\overline{\overline{M}} \in \mathrm{Max}(\overline{\overline{S}})$ *the equivalence class* $\mathcal{H} \in Y_{(\mathcal{R})}(\overline{\overline{E}})$ *such that* $\mathcal{M} = \mathcal{I}(\mathcal{U}, \overline{\overline{S}})$ *for all* $\mathcal{U} \in \mathcal{H}$, *is a bijection from* $\mathrm{Max}(\overline{\overline{S}})$ *onto* $Y_{(\mathcal{R})}(\overline{\overline{E}})$.

On another hand, the algebras $(T, \| \cdot \|)$ and $(\overline{\overline{T}}, \| \cdot \|)$ are isometric and so are $(T, \| \cdot \|_0)$ and $(\overline{\overline{T}}, \| \cdot \|_0)$. Therefore, the mapping from $\mathrm{Max}(T)$ to $\mathrm{Max}(\overline{\overline{T}})$ which associates to each maximal ideal $\mathcal{M}$ of T the ideal $\overline{\overline{\mathcal{M}}} = \{\overline{\overline{f}} \mid f \in \mathcal{M}\}$ is bijective and the ideals $\overline{\overline{\mathcal{M}}}$ and $\mathcal{M}$ are isometric. Furthermore, since $(T, \| \cdot \|)$ is multbijective, so is $(\overline{\overline{T}}, \| \cdot \|)$ and $\mathrm{Mult}(\overline{\overline{T}}, \| \cdot \|)$ can be identified with $\mathrm{Mult}(T, \| \cdot \|)$.

Notice that any ultrafilter $\mathcal{U}$ on E generates an ultrafilter $\overline{\overline{\mathcal{U}}}$ on $\overline{\overline{E}}$ and for any $f \in T$, we have $\lim_{\mathcal{U}} |f(x)| = \lim_{\overline{\overline{\mathcal{U}}}} |\overline{\overline{f}}(x)|$. Thus, for each ultrafilter $\mathcal{U}$

on E, the ideal $\mathcal{I}(\mathcal{U}, T)$ corresponds to the ideal $\mathcal{I}(\overline{\overline{\mathcal{U}}}, \overline{\overline{T}})$ of $\overline{\overline{T}}$. Then, using results of Chapter 3 concerning T, we have the following theorem.

Theorem 4.5.4. *For each maximal ideal $\overline{\overline{\mathcal{M}}}$ of $\overline{\overline{T}}$.*

(1) *There exists an ultrafilter $\mathcal{U}$ on E such that $\overline{\overline{\mathcal{M}}} = \mathcal{I}(\overline{\overline{\mathcal{U}}}, \overline{\overline{T}})$,*
(2) *There exists a unique equivalence class $\mathcal{H}$ of $Y_{(T)}(E)$ such that $\mathcal{U}$ belongs to $\mathcal{H}$ if and only if $\overline{\overline{\mathcal{M}}} = \mathcal{I}(\overline{\overline{\mathcal{U}}}, \overline{\overline{T}})$.*

By Corollary 4.5.3 and Theorem 4.5.4, we get Corollary 4.5.5.

Corollary 4.5.5. *The mapping from $Y_{(T)}(E)$ to $Y_{(T)}(\overline{\overline{E}})$ that associates to the equivalence class of any ultrafilter $\mathcal{U}$ in $Y_{(\mathcal{R})}(E)$ the equivalence class of $\overline{\overline{\mathcal{U}}}$ in $Y_{(\mathcal{R})}(\overline{\overline{E}})$ is bijective. In particular every ultrafilter of $\overline{\overline{E}}$ is contiguous in $\overline{\overline{E}}$ to a certain ultrafilter $\overline{\overline{\mathcal{U}}}$ of E.*

4.6.   Algebras $\mathcal{B}$, $\mathcal{L}$, $\mathcal{D}$, $\mathcal{E}$

As recalled in Section 4.3, we denote by $\mathcal{B}$ the Banach $\mathbb{K}$-algebra of bounded uniformly continuous functions from E to $\mathbb{L}$. Next, we denote by $\mathcal{L}$ the set of bounded Lipschitz functions from E to $\mathbb{L}$. Whenever E is a subset of $\mathbb{L}$, we denote by $\mathcal{D}$ the subset of $\mathcal{L}$ of derivable functions in E and by $\mathcal{E}$ the subset of $\mathcal{L}$ of functions such that for every $a \in E$, $\frac{f(x)-f(y)}{x-y}$ has limit when x and y tend to a separately. Following [14] the functions of $\mathcal{E}$ are called *strictly differentiable* [29].

Given $f \in \mathcal{L}$, we put $\|f\|_1 = \sup_{\substack{x,y \in E \\ x \neq y}} \frac{|f(x)-f(y)|}{\delta(x,y)}$ and $\|f\| = \max(\|f\|_0, \|f\|_1)$. In particular, if $f \in \mathcal{D}$, then $\|f\|_1 = \sup_{\substack{x,y \in E \\ x \neq y}} \frac{|f(x)-f(y)|}{|x-y|}$.

Remark 12. If $E \subset \mathbb{L}$, then $\mathcal{E} \subset \mathcal{D} \subset \mathcal{L} \subset \mathcal{B}$.

As noticed in Chapter 3, $(\mathcal{B}, \| . \|_0)$ is a semi-compatible algebra. Here we will carefully study algebras $\mathcal{L}$, $\mathcal{D}$ and $\mathcal{E}$.

Nest, we can derive that $\mathcal{L}$, $\mathcal{D}$ and $\mathcal{E}$ also are Banach algebras. First we will prove that so is $\mathcal{L}$ and then we will show that $\mathcal{D}$ and $\mathcal{E}$ are closed in $\mathcal{L}$ when $E \subset \mathbb{L}$.

Theorem 4.6.1. *$\mathcal{B}$ is a Banach $\mathbb{L}$-algebra with respect to the norm $\| . \|_0$. $\mathcal{L}$, $\mathcal{D}$ and $\mathcal{E}$ are Banach $\mathbb{L}$-algebras with respect to the norm $\| . \|$.*

Proof.   The statement concerning $\mathcal{B}$ is immediate. Next, it is easily checked that $| . \|$ is a $\mathbb{L}$-algebra norm on $\mathcal{L}, \mathcal{D}, \mathcal{E}$. Let us check that these algebras are $\mathbb{L}$-Banach algebras.

Let $(f_n)_{n \in \mathbb{N}}$ be a Cauchy sequence of $\mathcal{L}$. Take $\epsilon > 0$ and let $N(\epsilon) \in \mathbb{N}$ be such that $\|f_n - f_m\| \leq \epsilon \; \forall m, \, n \geq N(\epsilon)$. Since $\|f_n - f_m\|_0 \leq \epsilon \; \forall m, \, n \geq N(\epsilon)$, the sequence $(f_n)_{n \in \mathbb{N}}$ converges with respect to the norm $\| \, . \, \|_0$ to a function g such that $\|f_n - g\|_0 \leq \epsilon \; \forall n \geq N(\epsilon)$. On the other hand, since the sequence $(f_n)_{n \in \mathbb{N}}$ is a Cauchy sequence for the norm $\| \, . \, \|_1$, then for all $x, \, y \in E$, such that $x \neq y$, we have

$$\frac{|f_n(x) - f_m(x) - (f_n(y) - f_m(y))|}{\delta(x, y)} \leq \epsilon \quad \forall m, n \geq N(\epsilon)$$

and therefore, fixing n and passing to the limit on m, for all $x, \, y \in E$, such that $x \neq y$ we get

$$\frac{|f_n(x) - g(x) - (f_n(y) - g(y))|}{\delta(x, y)} \leq \epsilon \quad \forall \, n \geq N(\epsilon).$$

This is true for all $x, \, y \in E$, $x \neq y$ and shows that $f_n - g$ belongs to $\mathcal{L}$. Consequently, g also belongs to $\mathcal{L}$. Particularly we notice that $\|g - f_n\|_1 \leq \epsilon$, hence $\|g - f_n\| \leq \epsilon$. Thus, the sequence $(f_n)_{n \in \mathbb{N}}$ does converge to g in $\mathcal{L}$. $\square$

Suppose now that $E \subset \mathbb{L}$ and let us show that $\mathcal{D}$ is closed in $\mathcal{L}$. Take a sequence $(f_n)_{n \in \mathbb{N}}$ converging to a limit $f \in \mathcal{L}$ and let us show that f belongs to $\mathcal{D}$. As noticed above, since the sequence $(f_n)_{n \in \mathbb{N}}$ is a Cauchy sequence with respect to the norm $\| \, . \, \|_1$, the sequence $(f_n')_{n \in \mathbb{N}}$ is a Cauchy sequence with respect to the norm $\| \, . \, \|_0$. Let h be its limit for this norm. This limit is then bounded in E. We will show that f is derivable and that $f' = h$. Fix $a \in E$ and $\epsilon > 0$. For all $x \in E$ and for every $n \in \mathbb{N}$, we have

$$\left| \frac{f(x) - f(a)}{x - a} - h(a) \right|$$

$$= \left| \frac{f(x) - f(a) - (f_n(x) - f_n(a))}{x - a} + \frac{f_n(x) - f_n(a)}{x - a} \right.$$

$$\left. - f_n'(a) + f_n'(a) - h(a) \right|$$

$$\leq \max \left(\left| \frac{f(x) - f(a) - (f_n(x) - f_n(a))}{x - a} \right|, \left| \frac{f_n(x) - f_n(a)}{x - a} \right. \right.$$

$$\left. \left. - f_n'(a) \right|, |f_n'(a) - h(a)| \right)$$

$$\leq \max \left(\|f - f_n\|, \left| \frac{f_n(x) - f_n(a)}{x - a} - f_n'(a) \right|, \|f_n' - h\|_0 \right).$$

We can fix $N \in \mathbb{N}$ such that $\|f - f_N\| \leq \epsilon$ and $\|f_N' - h\|_0 \leq \epsilon$. Then there exists $r > 0$ such that, for all $x \in B(a, r)$, we have

$$\left| \frac{f_N(x) - f_N(a)}{x - a} - f_N'(a) \right| \leq \epsilon$$

and thus, we have

$$\left| \frac{f(x) - f(a)}{x - a} - h(a) \right| \leq \epsilon.$$

This proves that $f'(a) = h(a)$. Therefore f is derivable and $f' = h$.

A similar proof shows that $\mathcal{E}$ also is closed in $\mathcal{D}$.

Remark 13. Concerning $\mathcal{D}$, one can ask why we do not consider a norm of the form $\|f\|_d = \max(\|f\|_0, \|f'\|_0)$ instead of the above norm $\| . \|$ where $\| . \|_1$ is defined with the help of the Lipschitz inequality. Indeed, $\| . \|_d$ is a norm of $\mathbb{K}$-algebra. But the problem is that the algebra $\mathcal{D}$ is not complete with respect to that norm, in the general case. The example given in [29] (Remark 2) shows that we can't obtain a Banach algebra in that way because a sequence that converges with respect to that norm may have a limit which is not derivable at certain points.

Remark 14. Now, suppose that every non-empty circle $C(0, r)$ has at least two classes and consider a function f derivable in E and a a point of $\overline{\overline{E}} \setminus E$. Then in general, $\overline{\overline{f}}$ is not derivable at a, as the following example shows. Let E be the set $\{x \in \mathbb{K} \mid 0 < |x| \leq 1\}$ and let $(a_n)_{n \in \mathbb{N}}$ be a sequence in E such that

$$|a_n| < |a_{n-1}|, \quad \lim_{n \to +\infty} |a_n| = 0.$$

For each $n \in \mathbb{N}$, put $r_n = |a_n|$. Let g be the function defined on E by $g(x) = a_n \; \forall x \in d(a_n, r_n^-)$ and

$$g(x) = 0 \; \forall x \in E \setminus \left(\bigcup_{n=1}^{\infty} d(a_n, r_n^-) \right).$$

We can check that g is derivable and Lipschitz in E. But $\overline{\overline{g}}(0) = 0$ and $\overline{\overline{g}}$ is not derivable at 0. Indeed, let $(b_n)_{n \in \mathbb{N}}$ be a sequence of E such that

$|b_n| = |b_n - a_n| = r_n \; \forall n \in \mathbb{N}$. Now,

$$\frac{g(a_n) - \overline{\overline{g}}(0)}{a_n - 0} = 1 \quad \forall n \in \mathbb{N},$$

whereas

$$\frac{g(b_n) - \overline{\overline{g}}(0)}{b_n - 0} = 0 \quad \forall n \in \mathbb{N},$$

which shows that $\frac{\overline{\overline{g}}(x) - \overline{\overline{g}}(0)}{x}$ has no limit at 0. Therefore, g is not derivable in $\overline{\overline{E}}$. In the same way, we can show that $\overline{\overline{g}}$ is strictly differentiable in E but not in $\overline{\overline{E}}$.

Theorem 4.6.2. *An element of $\mathcal{B}, \mathcal{L}, \mathcal{D}, \mathcal{E}$ is invertible if and only if*

$$\inf\{|f(x)| \mid x \in E\} > 0.$$

Proof. Suppose that $\inf\{|f(x)| \mid x \in E\} > 0$ and put $g(x) = \frac{1}{f(x)}$. Let us first show that f belongs to $\mathcal{L}$ (resp., to $\mathcal{B}$). Indeed, let $m = \inf\{|f(x)| \mid x \in E\}$. Then

$$\frac{|g(x) - g(y)|}{\delta(x,y)} = \frac{|f(y) - f(x)|}{|f(x)f(y)|\delta(x,y)} \leq \frac{|f(y) - f(x)|}{m^2 \delta(x,y)},$$

which proves that g belongs to $\mathcal{L}$ (resp., to $\mathcal{B}$). Similarly, if $f \in \mathcal{D}$ (resp., $f \in \mathcal{E}$), then g belongs to $\mathcal{D}$ (resp., to $\mathcal{E}$). $\square$

Theorem 4.6.3. *In each algebra $\mathcal{B}, \mathcal{L}, \mathcal{D}, \mathcal{E}$, the spectral norm $\| \cdot \|_{sp}$ is $\| \cdot \|_0$.*

Proof. Concerning $\mathcal{B}$, by definition its norm is $\| \cdot \|_0$. Now, take $f \subset \mathcal{L}$ and $n \subset \mathbb{N}$. Without loss of generality, we can suppose that $\|f\|_0 \geq 1$. We have

$$\|f^n\| = \max \left(\|f^n\|_0, \sup_{x,y \in E, x \neq y} \frac{|(f(x))^n - (f(y))^n|}{\delta(x;y)} \right).$$

We notice that $|(f(x))^n - (f(y))^n| \leq |f(x) - f(y)|(\|f\|_0)^{n-1}$ and hence

$$\sup_{x,y \in E, x \neq y} \frac{|(f(x))^n - (f(y))^n|}{\delta(x;y)} \leq (\|f\|_0)^{n-1} \sup_{x,y \in E, x \neq y} \frac{|(f(x)) - (f(y))|}{\delta(x;y)}$$

$$= (\|f\|_0)^{n-1}\|f\|_1.$$

Consequently, we have

$$\|f\|_0 \leq \|f\|_{\text{sp}} \leq \sqrt[n]{\|f\|_1}(\|f\|_0)^{\frac{n-1}{n}}.$$

Then when n tends to $+\infty$, we get $\|f\|_{\text{sp}} = \|f\|_0 \ \forall f \in \mathcal{L}$. This is then true in $\mathcal{D}$ and $\mathcal{E}$ too. $\qquad\square$

Theorem 4.6.4. *The $\mathbb{L}$-algebras $\mathcal{B}, \mathcal{L}, \mathcal{D}, \mathcal{E}$ are compatible algebras.*

Proof. Let T be one of the algebras $\mathcal{B}, \mathcal{L}, \mathcal{D}, \mathcal{E}$. By Lemma 4.1.6, the characteristic function of an open set belongs to T if and only if it is uniformly open. Next, by Theorem 4.6.2, an element of $\mathcal{B}, \mathcal{L}, \mathcal{D}, \mathcal{E}$ admitting a strictly positive low bound admits an inverse. And finally, by Theorem 4.6.3, its spectral norm is $\| \cdot \|_0$. $\qquad\square$

4.7. Particular properties of the algebras $\mathcal{B}, \mathcal{L}, \mathcal{D}, \mathcal{E}$

A first specific property of the algebras $\mathcal{B}, \mathcal{L}, \mathcal{D}, \mathcal{E}$ concerns maximal ideals of finite codimension: they are of codimension 1.

Notations: Throughout this section, $\mathbb{F}$ will denote a finite algebraic extension of $\mathbb{L}$, equipped with the absolute value extending that of $\mathbb{L}$. For convenience, we will denote here by T one of the algebras $\mathcal{B}, \mathcal{L}, \mathcal{D}, \mathcal{E}$ and by T^* the $\mathbb{F}$-algebra of bounded uniformly continuous functions (resp., Lipschitz functions, resp. derivable functions, resp. strictly differentiable functions) from E to $\mathbb{F}$.

Remark 15. By definition, T^* is compatible.

Lemmas 4.7.1 and 4.7.2 are basic results in algebra.

Lemma 4.7.1. *Let $\mathbb{F}$ be of the form $\mathbb{L}[a]$ of degree d equipped with the absolute value which extends that of $\mathbb{L}$. Let $f \in T^*$. Then f is of the form $\sum_{j=0}^{d-1} a^j f_j, \ j = 0, \ldots, d-1$, with $f_j \in T$. So, T^* is isomorphic to $T \otimes \mathbb{F}$.*

Lemma 4.7.2. *Let $\mathbb{F}$ be equipped with the absolute value which extends that of $\mathbb{L}$. Suppose there exists a morphism of $\mathbb{L}$-algebra, χ, from T to $\mathbb{F}$. Then χ has continuation to a surjective morphism of $\mathbb{F}$-algebra χ^* from T^* onto $\mathbb{F}$.*

Proof. Let $d = [\mathbb{F} : \mathbb{L}]$. Suppose first that $\mathbb{F}$ is of the form $\mathbb{L}[a]$. By Lemma 4.7.1, any f in T^* is of the form $\sum_{j=0}^{d-1} a^j f_j, \ j = 0, \ldots, d-1$ where the f_j are functions from E to $\mathbb{L}$. We then set $\chi^*(f) = \sum_{n=0}^{d-1} a_j \chi(f_j)$. Then χ^* is obviously surjective onto $\mathbb{F}$.

Consider now the general case. We can obviously write $\mathbb{F}$ in the form $\mathbb{L}[b_1, \ldots, b_q]$. Then we have $\mathbb{L}[b_1, \ldots, b_j] = \mathbb{L}[b_1, \ldots, b_{j-1}][b_j]$. By induction on j, using the just proved preceding result we get that for each $j = 1, \ldots, q$, χ has continuation to a surjective morphism of $\mathbb{L}[b_1, \ldots, b_j]$-algebra χ_j^*, from $(T \otimes \mathbb{L}[b_1, \ldots, b_j])$ onto $\mathbb{L}[b_1, \ldots, b_j]$. Taking $j = q$ ends the proof. $\qquad\square$

We can now state the following theorem whose proof is similar to that of Theorem 3.4.9 of Chapter 3 but here concerns algebras $\mathcal{B}, \mathcal{L}, \mathcal{D}, \mathcal{E}$. Actually, that result may be generalized to all semi-compatible algebras, equipped that $\mathbb{K}$ is a perfect field, a hypothesis that we can avoid here.

Theorem 4.7.3. *Every maximal ideal of T of finite codimension is of codimension 1.*

Proof. Let $\mathcal{M}$ be a maximal ideal of finite codimension of T and let $\mathbb{F}$ be the field $\frac{T}{\mathcal{M}}$. Now, let χ be the quotient morphism from T over $\mathbb{F}$ whose kernel is $\mathcal{M}$ and let T^* be defined as above, relatively to the field $\mathbb{F}$. By Lemma 4.7.2, χ admits an extension to a surjective morphism χ^* from T^* to $\mathbb{F}$. Since the kernel of χ^* is a maximal ideal $\mathcal{M}^*$ of T^*, there exists an ultrafilter $\mathcal{U}$ on E such that $\mathcal{M}^* = \mathcal{I}(\mathcal{U}, T^*)$.

Let $g \in T$ and let $b = \chi(g) \in \mathbb{L}$. Then we have $\chi^*(g - b) = 0$, hence $g - b$ belongs to $\mathcal{M}^*$, therefore $\lim_{\mathcal{U}} g(x) - b = 0$, i.e., $\lim_{\mathcal{U}} g(x) = b$. But since $g \in T$, $g(x)$ belongs to $\mathbb{L}$ for all $x \in E$. Therefore, since $\mathbb{L}$ is complete, b belongs to $\mathbb{L}$. But by definition χ is a surjection from T onto $\mathbb{F}$, hence every value b of $\mathbb{F}$ is the image of some $g \in T$ and hence it lies in $\mathbb{L}$, therefore $\mathbb{F} = \mathbb{L}$. $\qquad\square$

Remark 16. In [6], it is shown that in the algebra of bounded analytic functions in the open unit disk of a complete ultrametric algebraically closed field, any maximal ideal which is not defined by a point of the open unit disk is of infinite codimension. Here, we may ask whether the same holds. In the general case, no answer is obvious. We can only answer a particular case.

Theorem 4.7.4. *Let $\mathcal{M} = \mathcal{I}(\mathcal{U}, T)$ be a maximal ideal of T where $\mathcal{U}$ is an ultrafilter on E. If $\mathcal{U}$ is a Cauchy filter, then $\mathcal{M}$ is of codimension 1. Else, $\mathcal{M}$ is of infinite codimension.*

Proof. Suppose first that $\mathcal{U}$ is a Cauchy ultrafilter. Let $f \in T$. Then f is uniformly continuous, hence $f(x)$ converges along $\mathcal{U}$ because $\mathbb{L}$ is complete. Consequently, by Theorem 4.2.1, $\mathcal{M}$ is of codimension 1.

Now, suppose that $\mathcal{U}$ is not a Cauchy filter and consider the identical function g defined on E. Then g has no limit on $\mathcal{U}$, therefore by

Theorem 4.2.1, $\mathcal{M}$ is not of codimension 1. But then by Theorem 4.7.3, $\mathcal{M}$ is not of finite codimension. $\qquad\square$

Corollary 4.7.5. *Suppose E is complete. Then* $\mathrm{Max}_1(T) = \mathrm{Max}_E(T)$.

Remark 17. Consider $\phi \in \mathrm{Mult}(\mathcal{L}, \| \, . \, \|)$, let $\mathcal{M} = \mathrm{Ker}(\phi)$. By theorem 4.3.4, all compatible algebras satisfy $\mathrm{Mult}(T, \| \, . \, \|) = \mathrm{Mult}_m(T, \| \, . \, \|)$. Now, let θ be the canonical surjection from $\mathcal{L}$ onto $\frac{\mathcal{L}}{\mathcal{M}}$. By Theorem 4.3.12, the quotient norm of the quotient field $\frac{\mathcal{L}}{\mathcal{M}}$ is just the quotient norm of uniform convergence norm $\| \, . \, \|_0$ and is equal to the absolute value of $\mathbb{L}$. In the case of a maximal ideal of infinite codimension, we can't apply Hahn–Banach's Theorem and there is no reason to think that the quotient norm is equivalent to the absolute value defined as $|\theta(f)| = \lim_\mathcal{U} |f(x)|$.

Theorem 4.7.6. *Suppose that E has no isolated points. Then an element of an algebra $\mathcal{L}$ is a topological divisor of zero if and only if it is not invertible.*

Proof. It is obvious that an invertible element of $\mathcal{L}$ is not a topological divisor of zero.

Now, consider an element $f \in \mathcal{B}$ or $f \in \mathcal{L}$ that is not invertible. By Theorem 4.6.2, we have $\inf_{x \in E} |f(x)| = 0$. Therefore, there exists a sequence of disks $(B(a_n, r_n))_{n \in \mathbb{N}}$ with $\lim_{n \to \infty} r_n = 0$, such that $|f(x)| \leq \frac{1}{n}, \forall x \in B(a_n, r_n), \forall n \in \mathbb{N}^*$.

Since the points a_n are not isolated, for every $n \in \mathbb{N}$ we can fix $b_n \in B(a_n, r_n)$ such that $b_n \neq a_n$.

For each $n \in \mathbb{N}^*$, let $t_n = \delta(a_n, b_n)$ and h_n be the characteristic function of $B(a_n, t_n)$. Notice that $0 < t_n \leq r_n$ so $\lim_{n \to \infty} t_n = 0$. Now h_n belongs to $\mathcal{L}$, hence to $\mathcal{B}$ and clearly satisfies

$$(1) \qquad \frac{|h_n(x) - h_n(y)|}{\delta(x, y)} \leq \frac{1}{t_n} \, \forall x, \quad y \in E, \ x \neq y.$$

Then concerning $\mathcal{B}$, we just have $\|fh_n\|_0 \leq \frac{1}{n}$ and hence f is a divisor of zero.

Suppose now $f \in \mathcal{L}$. We notice that $\frac{|h_n(a_n) - h_n(b_n)|}{\delta(a_n, b_n)} = \frac{1}{t_n}$ hence

$$(2) \qquad \|h_n\|_1 = \frac{1}{t_n} \, \forall n \in \mathbb{N}^*.$$

Let $l \in \mathbb{L}$ be such that $|l| \in \,]0, 1[$. Since the valuation on $\mathbb{L}$ is not trivial, for each $n \in \mathbb{N}$, we can find an element $\tau_n \in \mathbb{L}$ such that $|l|t_n \leq |\tau_n| \leq t_n$. We put $w_n = \tau_n h_n$ for all $n \in \mathbb{N}$. Then clearly we have

$$(3) \qquad \|w_n\|_0 = |\tau_n| \|h_n\|_0 = |\tau_n| \leq t_n,$$

and by (2), we have

$$(4) \qquad |l| \leq \|w_n\|_1 = \frac{|\tau_n|}{t_n} \leq 1.$$

Hence

$$(5) \qquad |l| \leq \|w_n\| \leq \max(1, |\tau_n|) \quad \forall n \in \mathbb{N}^*.$$

Consider now the sequence $(fw_n)_{n\in\mathbb{N}^*}$. By (3), we have $\|fw_n\|_0 \leq t_n\|f\|_0$, hence

$$(6) \qquad \lim_{n\to\infty} \|fw_n\|_0 = 0.$$

Furthermore for all $x, y \in E$, we have

$$\frac{|f(x)w_n(x) - f(y)w_n(y)|}{\delta(x,y)}$$

$$\leq \max\left(|f(x)| . \frac{|w_n(x) - w_n(y)|}{\delta(x,y)}, |w_n(y)| . \frac{|f(x) - f(y)|}{\delta(x,y)}\right)$$

and by (3) it is easily seen that

$$(7) \qquad |w_n(y)| . \frac{|f(x) - f(y)|}{\delta(x,y)} \leq t_n\|f\|_1.$$

On the other hand, if $x \in B(a_n, r_n)$, we have : $|f(x)| \leq \frac{1}{n}$, hence by (4),

$$(8) \qquad |f(x)| . \frac{|w_n(x) - w_n(y)|}{\delta(x,y)} \leq \frac{1}{n},$$

and by (7) and (8), we obtain

$$(9) \qquad \frac{|f(x)w_n(x) - f(y)w_n(y)|}{\delta(x,y)} \leq \max\left(\frac{1}{n}, t_n\|f\|_1\right).$$

Similarly, since x and y play the same role, if y belongs to $B(a_n, r_n)$, we obtain the same inequality.

Suppose now that neither x nor y belongs to $B(a_n, r_n)$. Then $w_n(x) = w_n(y) = 0$, therefore

(10)
$$\frac{|f(x)w_n(x) - f(y)w_n(y)|}{\delta(x,y)} \leq t_n \|f\|_1.$$

Consequently, by (9) and (10) we have proved that $\|fw_n\|_1 \leq \max(\frac{1}{n}, t_n\|f\|_1)$. Hence $\lim_{n\to\infty} \|fw_n\|_1 = 0$ and by (6) $\lim_{n\to\infty} \|fw_n\| = 0$ which, together with (5), finishes proving that f is a divisor of zero in $\mathcal{L}$ and this ends the proof of Theorem 4.7.6.

Finally, we can prove this last result: given $a \in E$, $\mathcal{I}(a, \mathcal{L})$ is not necessarily the closure of $\mathcal{I}'(a, \mathcal{L})$, while by Corollary 4.2.13, it is its spectral closure. □

Proposition 4.7.7. *Suppose that the set E is included in an ultrametric field $\mathbb{F}$ and contains a disk $d(0, R)$ of the field $\mathbb{F}$. There exists $f \in \mathcal{I}(0, \mathcal{L})$ that does not belong to the closure of $\mathcal{I}'(0, \mathcal{L})$ with respect to the norm $\| . \|$ of $\mathcal{L}$.*

Proof. Let $\omega \in \mathbb{F}$ be such that $0 < |\omega| < 1$. For every $n \in \mathbb{N}$, set $r_n = |\omega|^{-n}$, let $a_n \in C(0, r_n)$, let $F_n = d(a_n, r_n^-)$ and let $H = \bigcup_{n=1}^{\infty} F_n$. Let f be the function defined in E as $f(x) = 0 \ \forall x \in E \setminus H$ and $f(x) = a_n \ \forall x \in F_n$, $n \in \mathbb{N}$.

We notice that f belongs to $\mathcal{L}$. Indeed, let $x, y \in E$ with $x \neq y$. If $f(x) \neq f(y)$, then at least one of the points x and y belongs to H. Suppose that $y \in H$.

Suppose first that $x \notin H$. Then $f(x) = 0$ and y belongs to some disk $d(a_n, r_n^-)$ and hence $|f(y)| = |a_n| = r_n$, whereas $|x - y| \geq r_n$, therefore $\left|\frac{f(x)-f(y)}{x-y}\right| \leq 1$.

Suppose now that x and y belong to H. Say, x belongs to $d(a_m, r_m^-)$ and y belongs to $d(a_n, r_n^-)$ with $m < n$ since $f(x) \neq f(y)$. Then $|f(x)| = r_m < r_n = |f(y)|$, hence $|f(x) - f(y)| = |f(y)| = r_n$ and $|x - y| = |y| = r_n$ therefore $\left|\frac{f(x)-f(y)}{x-y}\right| \leq 1$. Thus, we have checked that $\left|\frac{f(x)-f(y)}{x-y}\right| \leq 1 \ \forall x, y \in E$, $x \neq y$. That finishes proving that f belongs to $\mathcal{L}$.

Now, by construction, we can see that f belongs to $\mathcal{I}(0, \mathcal{L})$. However, we will check that f does not belong to the closure of $\mathcal{I}'(0, \mathcal{L})$. Let $h \in \mathcal{I}'(0, \mathcal{L})$. There exists a disk $d(0, r_q)$ such that $h(x) = 0 \ \forall x \in d(0, r_q)$. Consequently, $f(x) - h(x) = f(x) \ \forall x \in d(0, r_q)$. But we notice that $f(x) = 0 \ \forall x \in C(0, r_q) \setminus F_q$. So, when x belongs to F_q and y belongs to $C(0, r_q) \setminus F_q$, we have $\left|\frac{f(x)-f(y)}{x-y}\right| = 1$, therefore $\|f - h\| \geq \|f - h\|_1 \geq 1$. This proves that $\mathcal{I}(0, \mathcal{L})$ is not the closure of $\mathcal{I}'(0, \mathcal{L})$ with respect to the norm $\| \cdot \|$. $\qquad \square$

4.8. A kind of Gelfand transform

A Gelfand transform is not easy on ultrametric Banach algebras, due to maximal ideals of infinite codimension. However, here we can obtain a kind of Gelfand transform under certain hypotheses on the multiplicative spectrum in order to find again an algebra of bounded Lipschitz functions on some ultrametric space.

Notations: Let $(A, \| \cdot \|)$ be a commutative unital Banach $\mathbb{L}$-algebra which is not a field. Let $\mathcal{X}(A, \mathbb{L})$ be the set of algebra homomorphisms from A onto $\mathbb{L}$ and let λ_A be the mapping from $A \times A$ to $\mathbb{R}_+$ defined by $\lambda_A(\chi, \zeta) = \sup\{|\chi(f) - \zeta(f)| \mid \|f\| \leq 1\}$.

Given $\chi \in \mathcal{X}(A, \mathbb{L})$, we denote by $|\chi|$ the element of $\mathrm{Mult}(A, \| \cdot \|)$ defined as $|\chi|(f) = |\chi(f)|$, $f \in A$. Given $D \subset \mathcal{X}(A, \mathbb{L})$, we put $|D| = \{|\chi|, \chi \in D\}$.

The following Lemma is easily checked.

Lemma 4.8.1. λ_A *is an ultrametric distance on* $\mathcal{X}(A, \mathbb{L})$ *such that* $\lambda_A(\chi, \xi) \leq 1 \ \forall \chi, \xi \in A$.

Definitions: The algebra $(A, \| \cdot \|)$ will be said to be $\mathbb{L}$-*based* if it satisfies the following:

(a) $\mathrm{Mult}_1(A, \| \cdot \|)$ is dense in $\mathrm{Mult}(A, \| \cdot \|)$,
(b) the spectral semi-norm $\| \cdot \|_{\mathrm{sp}}$ is a norm,
(c) for every uniformly open subset D of $\mathcal{X}(A, \mathbb{L})$ with respect to λ_A, the closures of $|D|$ and $|\mathcal{X}(A, \mathbb{L}) \setminus D|$ are disjoint open subsets of $\mathrm{Mult}(A, \| \cdot \|)$.

Proposition 4.8.2. *Let* $(A, \| \cdot \|)$ *be a unital commutative algebra satisfying properties* (a) *and* (b) *above. Then the algebra* A *is algebraically isomorphic to an algebra* $\mathcal{G}$ *of bounded Lipschitz functions from the ultrametric space* $F = (\mathcal{X}(A, \mathbb{L}), \lambda_A)$ *to* $\mathbb{L}$. *Identifying* A *with* $\mathcal{G}$, *the following are true:*

(i) *the spectral norm* $\| \cdot \|_{\mathrm{sp}}$ *of* A *is equal to the uniform convergence norm* $\| \cdot \|_0$ *on* F.
(ii) *every* $f \in A$ *such that* $\inf\{|\chi(f)| : \chi \in \mathcal{X}(A, \mathbb{L})\} > 0$ *is invertible in* A.

(iii) *there exists a constant $c \geq 1$ such that the Lipschitz semi-norm defined as*

$$\|f\|_1 = \sup\left(\left\{\frac{|f(x) - f(y)|}{\lambda_A(x,y)} \;\middle|\; x,\, y \in F,\, x \neq y\right\}\right)$$

satisfies $\|f\|_1 \leq c\|f\|$ for all $f \in A$ and the topology defined by $\|\cdot\|$ on A is at least as strong as the topology induced by the norm $\|\cdot\|_{\mathcal{L}}$ of the Banach $\mathbb{L}$-algebra $\mathcal{L}$ of all bounded Lipschitz functions from F to $\mathbb{L}$, where $\|f\|_{\mathcal{L}} = \max(\|f\|_0, \|f\|_1),\; f \in \mathcal{L}$.

Proof. We first show that A is isomorphic to a sub-$\mathbb{L}$-algebra of the algebra of bounded functions from F to $\mathbb{L}$. For each $f \in A$ and $\chi \in F$, we put $f^\circ(\chi) = \chi(f)$ and then we define a bounded function f° from F to $\mathbb{L}$. Let us check that this mapping G associating to each $f \in A$ the function f° is injective. Indeed, G is obviously an algebra homomorphism whose kernel is the intersection $\mathcal{J}$ of all maximal ideals of codimension 1. But thanks to Properties (a), (b) and to Theorem 2.5.17 of Chapter 2, we can check that $\mathcal{J} = (0)$. Consequently G is injective and hence A is isomorphic to a subalgebra of the algebra of bounded functions from F to $\mathbb{L}$. Hencefore we will identify an element f of A with the function it defines on F.

Let us now show that every $g \in A$ is Lipschitz, with respect to the distance λ_A. Let $\chi,\, \zeta \in F$ and let $g \in A$ be such that $\|g\| \leq 1$. Then we have

$$|\chi(g) - \zeta(g)| \leq \sup\{|\chi(f) - \zeta(f)| \mid \|f\| \leq 1\} = \lambda_A(\chi, \zeta).$$

Now in general, take $g \in A$ and $\nu \in \mathbb{L}$ such that $|\nu| \geq \|g\|$. Let $h = \frac{g}{\nu}$. Then $|\chi(h) - \zeta(h)| \leq \lambda_A(\chi, \zeta)$ hence $|\chi(g) - \zeta(g)| \leq |\nu|\lambda_A(\chi, \zeta)$, therefore g is Lipschitz.

Consequently, A can be identified with a $\mathbb{L}$-subalgebra of bounded Lipschitz functions from the ultrametric space F to $\mathbb{L}$.

We will now show that the statements (i)–(iii) are true. Thanks to Property (a) and Theorem 2.5.17 of Chapter 2, it is immediately seen that the spectral norm $\|\cdot\|_{sp}$ of A is the uniform convergence norm $\|\cdot\|_0$ on F hence (i) is true.

Let us now show that whenever $|\chi(f)| \geq m > 0$ for all $\chi \in \mathcal{X}(A, \mathbb{L})$, then f is invertible in A. Indeed, suppose that f is not invertible. Then there exists a maximal ideal $\mathcal{M}$ that contains f. By Theorem 2.5.17 of Chapter 2, there exists $\phi \in \text{Mult}(A, \|\cdot\|)$ such that $\mathcal{M} = \text{Ker}(\phi)$ and then $\phi(f) = 0$. Given $r > 0$, let us denote again by $\mathcal{W}(\phi, f, r)$ the neighborhood of ϕ: $\{\psi \in \text{Mult}(A, \|\cdot\|) \mid |\psi(f) - \phi(f)|_\infty \leq r\}$. Now $\text{Mult}_1(A, \|\cdot\|)$ is just the set of

$|\chi|$, $\chi \in \mathcal{X}(A, \mathbb{L})$. Thus, since $\text{Mult}_1(A, \| . \|)$ is dense in $\text{Mult}(A, \| . \|)$, there exists a sequence χ_n of $\mathcal{X}(A, \mathbb{L})$ such that for every $n \in \mathbb{N}$, $|\chi_n|$ belongs to the neighborhood $\mathcal{W}(\phi, f, \frac{1}{n})$ and hence, $\phi(f) = \lim_{n \to +\infty} |\chi_n(f)|$, i.e., $\phi(f) = \lim_{n \to +\infty} \chi_n(f) = 0$, a contradiction because $|\chi(f)| \geq m \ \forall \chi \in \mathcal{X}(A, \mathbb{L})$. Consequently, f is invertible in A, i.e., (ii) holds.

Finally let us prove (iii). Let $f \in A$. Notice that if $\|f\| \leq 1$ then for every $x, y \in F$ with $x \neq y$ we have, by definition of λ_A, $\frac{|f(x) - f(y)|}{\lambda_A(x,y)} \leq 1$ and hence $\|f\|_1 \leq 1$.

Suppose first that the valuation of $\mathbb{L}$ is dense. Take $\epsilon > 0$ and $\nu \in \mathbb{L}$ such that $\frac{\|f\|}{|\nu|} \leq 1$ and $\|f\| \leq |\nu| \leq \|f\| + \epsilon$. Then $\left\|\frac{f}{\nu}\right\| \leq 1$ and hence $\left\|\frac{f}{\nu}\right\|_1 \leq 1$, i.e., $\|f\|_1 \leq |\nu|$ and $\|f\|_1 \leq \|f\| + \epsilon$. This holds for all $\epsilon > 0$ and hence we have $\|f\|_1 \leq \|f\|$.

Suppose now that $\mathbb{L}$ has a discrete valuation. Let $\mu = \sup\{|x| \mid x \in \mathbb{L}, |x| < 1\}$ and take $\nu \in \mathbb{L}$ such that $|\nu| = \mu$.

If $\|f\| = 1$ then $\|f\|_1 \leq \|f\|$. If $\|f\| < 1$ then we can find $n \in \mathbb{N}$ such that $\mu^{n+1} \leq \|f\| \leq \mu^n$. Hence putting $g = \frac{f}{\nu^n}$, we have $\mu \leq \|g\| \leq 1$ and hence $\|g\|_1 \leq 1$. Therefore, $\frac{\|g\|_1}{\|g\|} \leq \frac{\|g\|_1}{\mu} \leq \frac{1}{\mu}$. But $\frac{\|g\|_1}{\|g\|} = \frac{\|f\|_1}{\|f\|}$ hence $\frac{\|f\|_1}{\|f\|} \leq \frac{1}{\mu}$ which finishes proving $\frac{\|f\|_1}{\|f\|} \leq c$, with $c = \frac{1}{\mu} \geq 1$. Consequently, we have $\|f\|_1 \leq c\|f\|$ for all $f \in A$ such that $\|f\| \leq 1$.

If $\|f\| > 1$ then there exists $n \in \mathbb{N}$ such that $\|f\| \leq \frac{1}{\mu^{n+1}}$. Putting $h = \nu^{n+1} f$ we have $\|h\| \leq 1$ hence $\|h\|_1 \leq c\|h\|$ which gives again $\|f\|_1 \leq c\|f\|$. Finally $\|f\|_1 \leq c\|f\|$ for every $f \in A$.

On the other hand, by Theorem 2.5.7 of Chapter 2, we have $\|f\|_{\text{sp}} \leq \|f\| \ \forall f \in A$, hence $\|f\| \geq \max(\|f\|_{\text{sp}}, \frac{1}{c}\|f\|_1) \geq \frac{1}{c} \max(\|f\|_0, \|f\|_1)$ for all $f \in A$, which proves that the norm $\| . \|$ of A is at least as strong as the norm of the Banach $\mathbb{L}$-algebra of all bounded Lipschitz functions on F. That finishes proving (iii). $\qquad \square$

Theorem 4.8.3. *Let E be a complete ultrametric space equipped with its distance δ and let T be a Banach algebra of bounded Lipschitz functions form E to $\mathbb{L}$. Let Z be the mapping from E into $\mathcal{X}(T, \mathbb{L})$ that associates to each point $a \in E$ the element of $\mathcal{X}(T, \mathbb{L})(T)$ whose kernel is $\mathcal{I}(a, T)$. Then Z is a bijection from E onto $\mathcal{X}(T, \mathbb{L})(T)$. Moreover, we have $\delta(a, b) \geq \lambda_T(a, b) \ \forall a, b \in E$.*

Proof. Z obviously is an injection from E into $\mathcal{X}(T, \mathbb{L})(T)$. Now, let $\chi \in \mathcal{X}(T, \mathbb{L})(T)$ and let $\mathcal{M} = \text{Ker}(\chi)$. By Theorem 4.2.1, there exists an ultrafilter $\mathcal{U}$ on E such that $\text{Ker}(\chi) = \mathcal{I}(\mathcal{U}, T)$. Since $\mathcal{M}$ is of codimension 1 and since E is complete, by Theorem 4.2.1 again, $\mathcal{U}$ converges in E to a point $c \in E$. Consequently, Z is surjective. $\qquad\square$

Now we will show that $\delta(a,b) \geq \lambda_T(a,b) \ \forall a,b \in E$. Let us take a, $b \in E$, with $a \neq b$ and consider $\lambda_T(a,b) = \sup\{|f(a) - f(b)|,$ $\|f\| \leq 1\}$. Recall that we have defined the Lipschitz semi-norm $\|f\|_1$ as $\|f\|_1 = \sup\{\frac{|f(x)-f(y)|}{\delta(x,y)}, \ x \neq y\}$. For every $f \in T$ such that $\|f\| \leq 1$, by definition of the norm $\| \ . \ \|$, we have $\|f\|_1 \leq \|f\|$ and hence $\|f\|_1 \leq 1$, therefore $|f(a) - f(b)| \leq \delta(a,b)$. Consequently, $\delta(a,b) \geq \lambda_T(a,b)$.

Corollary 4.8.4. *Let E be a complete ultrametric space equipped with its distance δ and let T be a Banach algebra of bounded Lipschitz functions form E to $\mathbb{L}$. Every uniformly open subset of $\mathcal{X}(A,\mathbb{L})$ with respect to λ_T is also a uniformly open subset of E with respect to the distance δ of E.*

Theorem 4.8.5. *Let E be a complete ultrametric space equipped with its distance δ and let T be the Banach algebra of all bounded Lipschitz functions form E to $\mathbb{L}$. Then T is a $\mathbb{L}$-based algebra.*

Proof. By Corollary 4.7.5, $\text{Mult}_1(T, \| \ . \ \|) = \text{Mult}_E(T, \| \ . \ \|)$. By Theorems 4.6.4, T is compatible, hence by Corollary 4.3.7, $\text{Mult}_1(T, \| \ . \ \|)$ is dense in $\text{Mult}(T, \| \ . \ \|)$. Next, $\| \ . \ \|_0$ is a norm equal to $\| \ . \ \|_{\text{sp}}$. So, Properties (a) and (b) are satisfied.

Consider now a uniformly open subset D of E with respect to λ_T. Identifying E with $\mathcal{X}(T,\mathbb{L})(T)$, by Corollary 4.8.4, D is also uniformly open with respect to δ. Consequently, since T is compatible, the characteristic function u of D belongs to T and we have $\inf_{x \in D} |u(x)| = 1$, $\sup_{x \notin D} |u(x)| = 0$, which ends the proof. $\qquad\square$

Notations: Let $(A, \| \ . \ \|)$ be a $\mathbb{L}$-based algebra. We will denote by $\widetilde{A}$ the algebra of all bounded Lipschitz functions from the space $E = (\mathcal{X}(A,\mathbb{L}), \lambda_A)$ to $\mathbb{L}$ and we denote by $\| \ . \ \|^{\sim}$ the norm $\| \ ,. \ , \|^{\sim} = \max(\| . \|_0, \| . \|_{\widetilde{1}})$ where $\|f\|_{\widetilde{1}} = \sup\{\frac{|f(x)-f(y)|}{\lambda_A(x,y)} \ x, \ y \in E \ x \neq y\}$ for every $f \in \widetilde{A}$.

Theorem 4.8.6. *Let E be a complete ultrametric space equipped with its distance δ and let T be the $\mathbb{L}$-based algebra of all bounded Lipschitz functions from E to $\mathbb{L}$ equipped with the norm $\| . \| = \max(\| . \|_0, \| . \|_1)$. Then the algebras T and $\widetilde{T}$ are isomorphic and the norms $\| . \|$ and $\| . \|^{\sim}$ are equivalent. Moreover $|x - y| \geq \lambda_A(x,y) \ \forall x,y \in E$. Further, if E is bounded, then there exists a constant $M \geq 1$ such that $|x - y| \leq M\lambda_T(x,y) \ \forall x,y \in E$ and the distances δ and λ_A are equivalent on E.*

Proof. Identifying E with $\mathcal{X}(T,\mathbb{L})(T)$, by Theorem 4.8.3, we have $|b-a| \geq \lambda_T(a,b)$ for all a and b in E and consequently, if $f \in T^{\sim}$ then $f \in A$ and $\|f\|_1 \leq \|f\|_1^{\sim}$ for all $f \in T$. Furthermore, by Theorem 4.8.2, T is algebraically isomorphic to a sub-algebra of $T^{\sim}$ and, considering T as a sub-algebra of $T^{\sim}$, there exists a constant $c \geq 1$ such that $\|f\|_1^{\sim} \leq c\|f\|$ for all $f \in T$. We conclude that the algebras T and $T^{\sim}$ are isomorphic and for every $f \in T$ we finally have : $\|f\| \leq \|f\|^{\sim} \leq c\|f\|$, which proves that these norms are equivalent.

Let us now suppose that E is bounded, say $|x| \leq M \; \forall x \in E$, with $M \geq 1$, and show that $M\lambda_A(x,y) \geq |x-y| \; \forall x, \; y \in E$. Take $a,b \in E$ with $a \neq b$ and set $r = |a-b|$. Let u be the characteristic function of $B(a,r)$. Thus, b does not belong to $B(a,r)$. Let $\alpha \in \mathbb{L}$ be such that $|\alpha| = \min(r,1)$ and set $h = \alpha u$. Now, $\|u\|_1 = \frac{1}{r}$.

Suppose first $r = |a-b| \geq 1$. We have $\|u\| = \|u\|_0 = 1$ hence $\|h\| = 1$. Then $\lambda_A(a,b) \geq |h(a) - h(b)| = |u(a) - u(b)| = 1$.

Suppose now $r = |a-b| < 1$. We have $\|u\| = \|u\|_1 = \frac{1}{r}$ hence $\|h\| = 1$. Then $\lambda_A(a,b) \geq |h(a) - h(b)| = r|u(a) - u(b)| = r$. Consequently, $\lambda_A(a,b) \geq |a-b|$. Therefore, gathering the two inequalities, we get $\lambda_A(a,b) \geq \min(1, |a-b|)$ and since $M \geq 1$, $M\lambda_A(a,b) \geq \min(M, |a-b|)$. But $|a-b| \leq M$, therefore $\min(M, |a-b|) \geq |a-b|$ and hence $M\lambda_A(a,b) \geq |a-b|$, which by Theorem 4.8.3, completes the proof. $\qquad\square$

Remark 18. By Lemma 4.8.1, we have $\lambda_A(x,y) \leq 1 \; \forall x,y \in A$. Therefore, if E is not bounded with respect to the absolute value of $\mathbb{L}$, there exists no $M > 0$ such that $M\lambda_A(x,y) \geq |x-y| \; \forall x,y \in A$.

Chapter 5

Circular Filters and Tree Structure

5.1. Infraconnected sets

Infraconnected sets were introduced in order to provide a class of sets as wide as possible, aimed at playing the same role as connected sets do in complex analysis towards holomorphic functions. We will see that for many things, this class of sets is quite satisfactory. Many proofs of results given in this chapter can be found in [28, Chapter 1].

Definitions and notations: Throughout the chapter the valuation of $\mathbb{L}$ is supposed to be dense. Let $a \in \mathbb{L}$ and let r_1 and r_2 such that $0 < r_1 < r_2$. We will denote by $\Gamma(a, r_1, r_2)$ the annulus $\{x \in \mathbb{L} | \ r_1 < |x - a| \ < r_2\}$ and by $\Lambda(a, r_1, r_2)$ the annulus $\{x \in K | \ r_1 \leq |x - a| \ \leq r_2\}$.

We know that if $b \in d(a, r)$ then $d(b, r) = d(a, r)$. In the same way if $b \in d(a, r^-)$ then $d(b, r^-) = d(a, r^-)$. Moreover, given two disks T and T' such that $T \cap T' \neq \emptyset$ then either $T \subset T'$ or $T' \subset T$.

Of course the following three statements are equivalent:

(i) $d(a, r) = d(a, r^-)$,
(ii) $C(a, r) = \emptyset$,
(iii) $r \notin | \mathbb{L} |$.

Moreover, the disks $d(b, r^-)$ included in $C(a, r)$ (resp., in $d(a, r)$) are the disks $d(b, r^-)$ such that $b \in C(a, r)$ (resp., in $d(a, r)$). They are called *the classes* of $C(a, r)$ (resp., of $d(a, r)$).

Henceforth D will denote a subset of $\mathbb{L}$.

The closure of D is denoted by $\overline{D}$ and the interior of D is denoted by $\overset{\circ}{D}$.

We put $\mathrm{diam}(D) = \sup\{|x - y| \ |x \in D, y \in D\}$ and $\mathrm{diam}(D)$ is named the *diameter* of D.

If D is bounded of diameter R we denote by $\widetilde{D}$ the disk $d(a, R)$ for any $a \in D$. If D is not bounded we put $\widetilde{D} = \mathbb{L}$.

Given a point $a \in \mathbb{L}$ we put $\delta(a, D) = \inf\{|x - a| \, |x \in D\}$. Then $\delta(a, D)$ is named *the distance of a to D.*

Lemma 5.1.1. $\widetilde{D} \setminus \overline{D}$ *admits a unique partition of the form* $(T_i)_{i \in I}$, *where each* T_i *is a disk of the form* $d(a_i, r_i^-)$ *with* $r_i = \delta(a_i, D)$.

Definition: Such disks $d(a_i, r_i^-)$ are called *the holes of D.*

Example 1. The holes of a disk $d(a, r^-)$ with $r \in |\mathbb{L}|$ are the classes of $C(a, r)$.

Example 2. The unique hole of $\mathbb{L} \setminus d(0, 1^-)$ is $d(0, 1^-)$.

Example 3. The holes of $\mathbb{L} \setminus d(0, 1)$ are the disks $d(a, 1^-)$ with $a \in d(0, 1)$.

Definitions: A set D is said to be *infraconnected* if for every $a \in D$, the mapping I_a from D to $\mathbb{R}_+$ defined by $I_a(x) = |x - a|$ has an image whose closure in $\mathbb{R}_+$ is an interval. In other words, a set D is not infraconnected if and only if there exist a and $b \in D$ and an annulus $\Gamma(a, r_1, r_2)$ with $0 < r_1 < r_2 < |a - b|$ such that $\Gamma(a, r_1, r_2) \cap D = \emptyset$. In such a situation, $\Gamma(a, r_1, r_2)$ is called *an empty-annulus of D.*

D is said to be *strongly infraconnected* if for every hole $T = d(a, r^-)$ with $r \in |\mathbb{L}|$, there exists a sequence $(x_n)_{n \in \mathbb{N}}$ in D such that $|x_n - a| = |x_n - x_m| = r$ whenever $n \neq m$.

Lemma 5.1.2. *If D is infraconnected of diameter $R \in \mathbb{R}$ (resp., $+\infty$) then* $\overline{I_a(D)} = [0, R]$ *(resp.,* $\overline{I_a(D)} = [0, +\infty[$).

Lemma 5.1.3. *Let D be strongly infraconnected. Let $a \in D$ and let $r \in |\mathbb{L}|$ be such that $r < \mathrm{diam}(D)$. There exists a sequence $(x_n)_{n \in \mathbb{N}}$ in D such that* $|x_n - a| = |x_n - x_m| = r$ *whenever* $n \neq m$.

Corollary 5.1.4. *A strongly infraconnected set is infraconnected.*

Lemma 5.1.5. *Let D be infraconnected and let α belong to a hole T of diameter ρ. The closure of the set $\{|x - \alpha| \, |x \in D\}$ is an interval whose lower bound is ρ.*

Theorem 5.1.6. *Let A and B be two infraconnected subsets of $\mathbb{L}$ such that $A \cap B \neq \emptyset$. Then $A \cup B$ is infraconnected.*

Notation: Let $\mathcal{R}_D$ be the relation defined in D as $x \mathcal{R}_D y$ if there exists an infraconnected subset of D containing both x, y.

Corollary 5.1.7. *The relation $\mathcal{R}_D$ is an equivalence relation.*

Definition: The equivalence classes with respect to $\mathcal{R}_D$ are called *the infraconnected components*.

Examples: (1) $d(0, 1^-) \cup d(1, 1^-)$ is infraconnected. Its holes are the disks $d(\alpha, 1^-)$ with $|\alpha| = |\alpha - 1| = 1$.
(2) Let $r \in]0, 1[$ and let $D = d(0, 1^-) \cup d(1, r)$. Then D is not infraconnected, its infraconnected components are $d(0, 1^-)$ and $d(1, r)$. The holes of D are the disks $d(\alpha, 1^-)$ with $|\alpha| = |\alpha - 1| = 1$ and the disks $d(\alpha, |\alpha - 1|^-)$ with $r < |\alpha - 1| < 1$.

Definition and notation: We call *an empty-annulus of D* an annulus $\Gamma(a, r_1, r_2)$ such that

(i) $r_1 = \sup\{|x - a| \quad |x \in D, |x - a| < r_2\}$,
(ii) $r_2 = \inf\{ |x - a| \quad |x \in D, |x - a| > r_1\}$.

The set $d(a, r_1) \cap D$ will be denoted by $\mathcal{I}_D(\Gamma(a, r_1, r_2))$ while the set $(\mathbb{L} \setminus d(a, r_2^-)) \cap D$ will be denoted by $\mathcal{E}_D(\Gamma(a, r_1, r_2))$. When there is no risk of confusion about the set D we will just write $\mathcal{I}(\Gamma(a, r_1, r_2))$ (resp., $\mathcal{E}(\Gamma(a, r_1, r_2))$) instead of $\mathcal{I}_D(\Gamma(a, r_1, r_2))$ (resp., $\mathcal{E}_D(\Gamma(a, r_1, r_2))$).

Remark 1. By definition, D is not infraconnected if and only if it admits an empty annulus.

Remark 2. If a set D admits an empty annulus $\Gamma(a, r_1, r_2)$, then $\{\mathcal{I}(\Gamma(a, r_1, r_2)), \mathcal{E}(\Gamma(a, r_1, r_2))\}$ is a partition of D.

Examples: Let $r \in]0, 1[$, let $D = d(0, r) \cup d(1, 1^-)$ and let $D' = d(0, r^-) \cup d(1, r)$. Then $\Gamma(0, r, 1)$ is an empty annulus of D and also of D'. In the same way $\Gamma(1, r, 1)$ is also an empty annulus of D'.

Notation: Let $\mathcal{O}(D)$ be the set of empty annuli of D. Given F_1 and $F_2 \in \mathcal{O}(D)$, it is easily seen that $\mathcal{I}(F_1) \subset \mathcal{I}(F_2)$ is equivalent to $\mathcal{E}(F_1) \supset \mathcal{E}(F_2)$. We will denote by $\leq$ the relation defined on $\mathcal{O}(D)$ by $F_1 \leq F_2$ if $\mathcal{I}(F_1) \subset \mathcal{I}(F_2)$. It is easily seen that $\leq$ is an order relation on $\mathcal{O}(D)$.

Theorem 5.1.8. *Let A and B be infraconnected subsets such that $\widetilde{A} = \widetilde{B}$. Then $A \cup B$ is infraconnected.*

Proof. Let $\widetilde{A} = d(a, r)$ and $A \cup B = X$. Suppose that X is not infraconnected and let $\Gamma(c, r_1, r_2)$ be an empty annulus of X, with $c \in X$ and $0 < r_1 < r_2 \leq r$. Then there exist $a, b \in X$ such that $|c - a| \leq r_1$, $|c - b| \geq r_2$. Since both A and B are infraconnected, either $a \in A$, $b \in B$

or $a \in B$, $b \in\in A$. Thus, without loss of generality, we can suppose that $a \in A$, $b \in B$. But then, there exists no $x \in A \setminus d(0, r_2^-)$, as we just saw. Consequently, $\operatorname{diam}(A) < r_2$, a contradiction. That finishes proving that X is infraconnected. □

Notation: We will denote by $\leq$ the order relation defined on $\mathcal{O}(D)$ by $F_1 <\leq F_2$ if $\mathcal{I}_D(F_1) \subset \mathcal{I}_D(F_2)$.

The following Lemmas 5.1.9 and 5.1.10 are easily seen.

We will denote by $<$ the order relation defined by $F_1 < F_2$ if $F_1 \leq F_2$ and $F_1 \neq F_2$.

The following Lemmas 5.1.9 and 5.1.10 are easily seen.

Lemma 5.1.9. *Let F_1 and F_2 be two empty annuli of D. The following assertions are equivalent:*

(i) *F_1 and F_2 are not comparable with respect to the order $\leq$,*
(ii) *$\mathcal{I}(F_1) \subset \mathcal{E}(F_2)$,*
(iii) *$\mathcal{I}(F_2) \subset \mathcal{E}(F_1)$,*
(iv) *$\mathcal{I}(F_1) \cap \mathcal{I}(F_2) = \emptyset$.*

Lemma 5.1.10. *Let $F \in \mathcal{O}(D)$ and let $x \in \mathcal{I}(F)$ (resp., $x \in \mathcal{E}(F)$). The infraconnected component of x is included in $\mathcal{I}(F)$ (resp., in $\mathcal{E}(F)$). If $F' \in \mathcal{O}(D)$ is such that $F < F'$ then $\mathcal{I}(F') \cap \mathcal{E}(F) \neq \emptyset$.*

Corollary 5.1.11. *Let F be an empty annulus of D. The family of the empty annuli G of D such that $G \geq F$ is totally ordered.*

We will also need Lemma 5.1.12 in [23, Lemma 8.12; 28].

Lemma 5.1.12. *Let $a \in \widetilde{D}$, let ρ be the distance from a to D and let R be such that $\rho \leq R \leq \operatorname{diam}(D)$. For $j = 1, \ldots, q$ let $\alpha_j \in d(a, R)$ and let $r_j', r_j'' \in \mathbb{R}_+$ be such that $r_j' < R < r_j''$. Then $\bigcap_{j=0}^q (\Gamma(\alpha_j, r_j', r_j'') \cap D) \neq \emptyset$.*

Sets having finitely many infraconnected components are characterized in [23, 28].

Theorem 5.1.13. *Let $Y(D)$ be the set of empty annuli of D. Then D has finitely many infraconnected components if and only if it has finitely many empty annuli. Moreover, if so does D, then one of the infraconnected components is $A_0 = \bigcap_{F \in Y(D)} \mathcal{E}(F)$ while the others are of the form $A_i = \mathcal{I}(F_i) \cap \left(\bigcap_{F < F_i} \mathcal{E}(F) \right)$, with $F_i \in \mathcal{O}(D)$.*

Definitions: A bounded closed infraconnected subset A of $\mathbb{L}$ is said to be *affinoid* if it has finitely many holes and if $\text{diam}(A)$ and each diameter of any hole lies in $|L|$. More generally, a set A is said to be *affinoid* if it is a finite union of infraconnected affinoid sets.

From Theorem 5.1.13, Lemma 5.1.14 is easily proven.

Lemma 5.1.14. *A finite union and a finite intersection of affinoid subsets of $\mathbb{L}$ (which is not empty) is affinoid.*

Theorem 5.1.15. *Let A be a unital commutative ultrametric normed $\mathbb{L}$-algebra. Suppose that A has an element u such that $sp(u)$ admits an empty annulus $\Gamma(a, r_1, r_2)$. Then there exist ψ_1, $\psi_2 \in \text{Mult}(A, \| \, . \, \|)$ such that $\psi_1(t) = r_1$, $\psi_2(t) = r_2$.*

Proof. By definition, $r_1 = \sup\{|\lambda| \mid \lambda \in sp(u) \; |\lambda| < r_2\}$, and $r_2 = \inf\{|\lambda| \mid \lambda \in sp(u) \; |\lambda| > r_1\}$. Consequently, both r_1, r_2 lie in the closure of $\{\psi(u) \mid \psi \in \text{Mult}(A, \| \, . \, \|)\}$, which, is a closed subset of $\mathbb{R}$. Since $\text{Mult}(A, \| \, . \, \|)$ is compact, there exist ψ_1, $\psi_2 \in \text{Mult}(A, \| \, . \, \|)$ such that $\psi_1(t) = r_1$, $\psi_2(t) = r_2$. $\qquad\square$

5.2. Monotonous filters

Monotonous filters are indispensable to understand the behavior of rational functions, and all kinds of analytic functions. Circular filters which are closely linked to monotonous filters, will be introduced in Section 5.3. Most of proofs of results given in this chapter can be found in [28, Chapter 2].

Definitions and notations: Throughout Section 5.2, $\mathbb{L}$ is a complete ultrametric field whose valuation is dense and D is a subset of $\mathbb{L}$. According to classical definitions, we call *basis* of a filter $\mathcal{F}$ a subset $\mathcal{B}$ of $\mathcal{F}$ such that every element of $\mathcal{F}$ contains an element of $\mathcal{B}$. A filter $\mathcal{F}$ is said to be *secant with D* if the family of sets $\{B \cap D \mid B \in \mathcal{F}\}$ is a filter. When a filter $\mathcal{F}$ is secant with a subset D of $\mathbb{L}$, the filter $\{B \cap D \mid B \in \mathcal{F}\}$ is called *intersection of $\mathcal{F}$ with D*, and denoted by $\mathcal{F} \cap D$.

A filter $\mathcal{F}$ is said to be *secant with* another filter $\mathcal{G}$ if for every $X \in \mathcal{F}$ and $Y \in \mathcal{G}$, then $X \cap Y \neq \emptyset$ and the two filters $\mathcal{F}$, $\mathcal{G}$ are said to be *secant*. In such a case, the filter generated by the $X \cap Y$, $X \in \mathcal{F}$, $Y \in \mathcal{G}$ is called *intersection of $\mathcal{F}$ and $\mathcal{G}$*.

A filter $\mathcal{F}$ on $\mathbb{L}$ is said to be *thinner* than a filter $\mathcal{G}$ if every element of $\mathcal{G}$ belongs to $\mathcal{F}$. In such a case, $\mathcal{G}$ will also be said to be *less thin than $\mathcal{F}$*.

A sequence $(u_n)_{n\in\mathbb{N}}$ in $\mathbb{L}$ will be said to be *thinner* than a filter $\mathcal{G}$ if so is the filter defined by the sets $A_q = \{u_n | n \geq q\}$ ($q \in \mathbb{N}$). In such a case, $\mathcal{G}$ will also said to be *less thin than* the sequence $(u_n)_{n\in\mathbb{N}}$.

A sequence $(u_n)_{n\in\mathbb{N}}$ in $\mathbb{L}$ will be said to be *an increasing distances sequence* (resp., *a decreasing distances sequence*) if the sequence $|u_{n+1} - u_n|$ is strictly increasing (resp., decreasing) and has a limit $\ell \in \mathbb{R}_+^*$.

The sequence $(u_n)_{n\in\mathbb{N}}$ will be said to be *a monotonous distances sequence* if it is either an increasing distances sequence or a decreasing distances sequence.

A sequence $(u_n)_{n\in\mathbb{N}}$ in $\mathbb{L}$ will be said to be *an equal distances sequence* if $|u_n - u_m| = |u_m - u_q|$ whenever $n, m, q \in \mathbb{N}$ such that $n \neq m \neq q$.

In the same way, a sequence of holes $(T_n)_{n\in\mathbb{N}}$ of D with $T_n = d(a_n, r_n^-)$ and $T_m \cap T_n = \emptyset$ $\forall m, n \in \mathbb{N}$, will be called *an increasing distances* (resp., *a decreasing distances*, resp., *an equal distances*) *holes sequence*, if the sequence $(a_n)_{n\in\mathbb{N}}$ is an increasing distances (resp., a decreasing distances, resp., an equal distances) sequence. And an increasing distances (resp., a decreasing distances) will be called *a monotonous distance holes sequence*.

Let $a \in \widetilde{D}$ and $R \in \mathbb{R}_+^*$ be such that $\Gamma(a, r, R) \cap D \neq \emptyset$ whenever $r \in]0, R[$ (resp., $\Gamma(a, R, r) \cap D \neq \emptyset$ whenever $r > R$). We call *an increasing* (resp., *a decreasing) of center a and diameter R, on D* the filter $\mathcal{F}$ on D that admits for basis the family of sets $\Gamma(a, r, R) \cap D$ (resp., $\Gamma(a, R, r) \cap D$). For every sequence $(r_n)_{n\in\mathbb{N}}$ such that $r_n < r_{n+1}$ (resp., $r_n > r_{n+1}$) and $\lim_{n\to\infty} r_n = R$, it is seen that the sequence $\Gamma(a, r_n, R) \cap D$ (resp., $\Gamma(a, R, r_n) \cap D$) is a basis of $\mathcal{F}$. Such a basis will be called *a canonical basis*. We call *a decreasing filter with no center, of canonical basis $(D_n)_{n\in\mathbb{N}}$ and diameter $R > 0$, on D* a filter $\mathcal{F}$ on D that admits for basis a sequence $(D_n)_n \in \mathbb{N}$ in the form $D_n = d(a_n, r_n) \cap D$ with $D_{n+1} \subset D_n$, $r_{n+1} < r_n, \lim_{n\to\infty} r_n = R$, and $\bigcap_{n\in\mathbb{N}} d(a_n, r_n) = \emptyset$.

Given an increasing (resp., a decreasing) filter $\mathcal{F}$ on D of center a and diameter r we will denote by $\mathcal{P}_D(\mathcal{F})$ the set $\{x \in D | \, |x - a| \geq r\}$ (resp., the set $\{x \in D | \, |x - a| \leq r\}$ and by $\mathcal{B}_D(\mathcal{F})$ the set $\{x \in D | \, |x - a| < r\}$ (resp., the set $\{x \in D | \, |x - a| > r\}$. When there is no risk of confusion we will only write $\mathcal{P}(\mathcal{F})$ instead of $\mathcal{P}_D(\mathcal{F})$, and $\mathcal{B}(\mathcal{F})$ instead of $\mathcal{B}_D(\mathcal{F})$. Besides $\mathcal{B}_D(\mathcal{F})$ will be named *the body of $\mathcal{F}$* and $\mathcal{P}_D(\mathcal{F})$ will be named *the beach of $\mathcal{F}$*.

We call *a monotonous filter on D* a filter which is either an increasing filter or a decreasing filter (with or without a center). Given a monotonous filter $\mathcal{F}$ we will denote by $\text{diam}(\mathcal{F})$ its diameter.

The field $\mathbb{L}$ is said to be *spherically complete* if every decreasing filter on $\mathbb{L}$ has a center in $\mathbb{L}$. The field $\mathbb{C}_p$ for example is known not to be spherically complete [28, Theorem 5.23]. However, every algebraically closed complete ultrametric field admits a spherically complete algebraically closed extension (see, for example, [28, Chapter 7]).

Theorem 5.2.1. *Let $(u_n)_{n \in \mathbb{N}}$ be a bounded sequence in $\mathbb{L}$. Either we may extract a convergent subsequence or we may extract a monotonous distances subsequence or we may extract an equal distances subsequence from the sequence $(u_n)_{n \in \mathbb{N}}$.*

Lemma 5.2.2. *Let $(a_n)_n \in \mathbb{N}$ be an increasing distances (resp., a decreasing distances) sequence in D. There exists a unique increasing (resp., decreasing) filter $\mathcal{F}$ on D such that the sequence $(a_n)_n \in \mathbb{N}$ is thinner than $\mathcal{F}$. Moreover, given a monotonous filter $\mathcal{F}$ on D, there exists a unique monotonous filter $\mathcal{G}$ on $\mathbb{L}$ inducing $\mathcal{F}$.*

Lemma 5.2.3. *Let D be infraconnected. Let $\mathcal{F}$ be an increasing filter (resp., a decreasing filter) on $\mathbb{L}$, of center $\alpha \in \widetilde{D}$ and diameter $R \leq \mathrm{diam}(D)$ (resp., $R < \mathrm{diam}(D)$) such that α does not belong to a hole of diameter $\rho \geq R$ (resp., $\rho > R$). Then $\mathcal{F}$ is secant with D and induces on D an increasing filter (resp., a decreasing filter) of center α and diameter R, on D.*

Definitions: Let $\mathcal{F}$ be an increasing (resp., a decreasing) filter of center a and diameter R on D. Then $\mathcal{F}$ is said *to be pierced* if for every $r \in]0, R[$, (resp., $r < R$), $\Gamma(a, r, R)$ (resp., $\Gamma(a, R, r)$) contains some hole T_m of D.

A decreasing filter with no center $\mathcal{F}$ on D is said *to be pierced* if for every $m \in \mathbb{N}$, $\widetilde{D}_m \setminus \widetilde{D}_{m+1}$ contains some hole T_m of D.

Remarks. The definition of a pierced filter with no center also applies to a decreasing filter with a center and then is equivalent to the one given just above for such a filter.

If $\mathcal{F}$ is an increasing (resp., a decreasing) filter of center a, of diameter R, then $\mathcal{F}$ is pierced if and only if there exists a sequence of holes $(T_n)_{n \in \mathbb{N}}$ of D such that $\delta(a, T_n) < \delta(a, T_{n+1})$, (resp., $\delta(a, T_n) > \delta(a, T_{n+1})$), $\lim_{n \to \infty} \delta(a, T_n) = R$.

Given a Cauchy filter $\mathcal{F}$ on D, of limit a in $\mathbb{L}$, we will call *a canonical basis of $\mathcal{F}$* a sequence D_m in the form $d(a, r_m) \cap D$ with $0 < r_m < r_{m+1}$ and $\lim_{m \to \infty} r_m = 0$. The filter $\mathcal{F}$ is said *to be pierced* if for every $m \in \mathbb{N}$, $\widetilde{D}_m$ contains some hole of D.

Let $a \in \widetilde{D}$. Let $(T_{m,i})_{1 \leq i \leq s(m)_{m \in \mathbb{N}}}$ be a sequence of holes of D which satisfies $\delta(a, T_{m,i}) = d_m$ ($1 \leq i \leq h_m$), $d_m < d_{m+1}$ (resp., $d_m > d_{m+1}$), $\lim_{m \to \infty} d_m = S > 0$.

The sequence $(T_{m,i})_{1\leq i\leq s(m)_{m\in\mathbb{N}}}$ is called *a quasi-increasing* (resp., a *quasi-decreasing) distances holes sequence that runs the increasing* (resp., *decreasing) filter of center a, of diameter S.*

Now let $(T_{m,i})_{1\leq i\leq s(m)_{m\in\mathbb{N}}}$ be a sequence of holes of D that satisfies $\delta(T_{m+1,j}, T_{m,i}) = d_m$ $(1 \leq i \leq s(m),\ 1 \leq j \leq s(m+1))$, $d_m > d_{m+1}$, $\lim_{m\to\infty} d_m = R > 0$, where the filter $\mathcal{F}$ of basis $D_m = d(a_m, d_m) \cap D$ is a decreasing filter with no center. The sequence $(T_{m,i})_{1\leq i\leq s(m)_{m\in\mathbb{N}}}$ is called *a quasi-decreasing distances holes sequence that runs $\mathcal{F}$.*

In each case, the sequence $(d_m)_{m\in\mathbb{N}}$ is called *monotony* of the sequence $(T_{m,i})_{1\leq j\leq s(m+1)},\ _{m\in\mathbb{N}}$.

Summarizing these definitions, an increasing (resp., decreasing) distances holes sequence that runs an increasing (resp., decreasing) filter $\mathcal{F}$ will be just named *an increasing* (resp., *decreasing) distances holes sequence* and the filter $\mathcal{F}$ will be named *the increasing* (resp., *decreasing) filter associated to the sequence* $(T_{m,i})_{1\leq i\leq s(m)_{m\in\mathbb{N}}}$. The diameter of $\mathcal{F}$ will be called *the diameter of the sequence* $(T_{m,i})_{1\leq i\leq s(m)_{m\in\mathbb{N}}}$. If $\mathcal{F}$ has a center a, a will be named *the center of the sequence* $(T_{m,i})_{1\leq i\leq s(m)_{m\in\mathbb{N}}}$. If $\mathcal{F}$ has no center, the sequence $(T_{m,i})$ will be called *a decreasing distances holes sequence with no center.*

Let $(T_{m,i})_{1\leq i\leq s(m)_{m\in\mathbb{N}}}$ be a monotonous distances holes sequences and for every $(m,i)_{1\leq i\leq s(m)_{m\in\mathbb{N}}}$ let $\rho_{m,i} = \operatorname{diam}(T_{m,i})$. The number $\inf_{1\leq i\leq s(m)_{m\in\mathbb{N}}}\rho_{m,i}$ will be called *piercing of the sequence* $(T_{m,i})_{1\leq i\leq s(m)_{m\in\mathbb{N}}}$.

If a monotonous distances holes sequence has a piercing $\rho > 0$, it will be said to be *well pierced*. If a monotonous filter $\mathcal{F}$ is run by a well pierced monotonous holes sequence, $\mathcal{F}$ will be said to be well pierced.

In each case the sequence of circles $C(a, d_m)$ when $\mathcal{F}$ has center a (resp., $C(a_{m+1}, d_m)$ when $\mathcal{F}$ has no center) will be said *to run the filter $\mathcal{F}$*, and *to carry the monotonous distances holes sequence* $(T_{m,i})_{1\leq i\leq s(m)_{m\in\mathbb{N}}}$.

A monotonous distances holes sequences $(T_{m,i})_{1\leq i\leq s(m)_{m\in\mathbb{N}}}$ will be said to be *pointwise* if $s(m) = 1$ for all $m \in \mathbb{N}$.

Moreover, a sequence of holes $(T_m)_{m\in\mathbb{N}}$ of D will be called *a Cauchy sequence of holes of limit $a \in \mathbb{L}$* if $\lim_{m\to\infty}\delta(a, T_m) = 0$. Such a sequence will be said *to run the Cauchy filter of basis* $\{d(a,r) \cap D | r > 0\}$.

Let $(T_m)_{m\in\mathbb{N}}$ be a monotonous distances holes sequence or an equal distances holes sequence. We will call *superior gauge* of the sequence the number $\limsup_{m\to\infty}\operatorname{diam}(T_m)$.

As an obvious consequence of Theorem 5.2.1, we have Theorem 5.2.4.

Theorem 5.2.4. *Let $(T_n)_{n\in\mathbb{N}}$ be a bounded sequence of holes of D. Either we may extract from the sequence $(T_n)_{n\in\mathbb{N}}$ a subsequence which converges* *in* $\mathbb{L}$, *or we may extract a monotonous distances holes sequence or we may extract an equal distances holes sequence.*

Lemma 5.2.5 is also obvious.

Lemma 5.2.5. *Let* $(T_m)_{m \in \mathbb{N}}$ *be a monotonous distances holes sequence. Then the superior gauge* s *of the sequence satisfies* $s \leq r$.

Definition: Given a monotonous filter $\mathcal{F}$, we will call *superior gauge of* $\mathcal{F}$ the upper bound of all superior gauges of monotonous holes sequences running $\mathcal{F}$.

From the definition, Lemma 5.2.6 is immediate again.

Lemma 5.2.6. *Let* $\mathcal{F}$ *be an increasing (resp., decreasing) filter on* D *of center* a, *of diameter* r, *of superior gauge* s. *For every* $\epsilon > 0$, *there exists a hole* T *of* D *included in* $\Gamma(a, r, r+\epsilon)$ (*resp.,* $\Gamma(a, r-\epsilon, r)$) *such that* $\operatorname{diam}(T) > s - \epsilon$.

Let $\mathcal{F}$ *be a decreasing filter on* D *with no center, of diameter* r, *of superior gauge* s. *For every* $\epsilon > 0$, *there exists a hole* T *of* D *included in in a disk* $d(a, r + \epsilon)$ *such that* $\operatorname{diam}(T) > s - \epsilon$.

Notation: From now on, to the end of the chapter, γ denotes the homographic function $b + \frac{1}{x-a}$ with $a, b \in \mathbb{L}$.

Proposition 5.2.7. *Let* $\alpha \in \mathbb{L}$, $r > 0$ *and let* $a \in \mathbb{L}$ *be such that* $|a - \alpha| < r$. *Then* $\gamma(C(\alpha, r)) = C(b, \frac{-1}{r})$.

Proof. We may assume $b = 0$ and then the proof is immediate. $\square$

Corollary 5.2.8. *Let* $\alpha \in \mathbb{L}$, $r_1, r_2 \in]0, +\infty[$ *with* $|a - \alpha| < r_1 < r_2$. *Then* $\gamma(\Gamma(\alpha, r_1, r_2)) = \Gamma\left(b, \frac{1}{r_2}, \frac{1}{r_1}\right)$.

Corollary 5.2.9. *Let* $\mathcal{F}$ *be the an increasing (resp., a decreasing) filter of center* α *and diameter* $R > |a - \alpha|$, *on* $\mathbb{L} \setminus \{a\}$. *Then* $\gamma(\mathcal{F})$ *is the decreasing (resp., increasing) filter of center* b *and diameter* $\frac{1}{R}$.

Lemma 5.2.10. *Let* $\alpha \in \mathbb{L}$ *be such that* $|\alpha - a| \neq r$. *Then*

$$\gamma(C(\alpha, r)) = C\left(\gamma(\alpha), \frac{r}{|a - \alpha|^2}\right).$$

Corollary 5.2.11. *Let* $\alpha \in \mathbb{L}$ *and* $r, r' \in]0, +\infty[$ *be such that* $0 < r < r' < |a - \alpha|$. *Then we have* $\gamma(\Gamma(\alpha, r, r')) = \Gamma\left(\gamma(\alpha), \frac{r}{|a-\alpha|^2}, \frac{r'}{|a-\alpha|^2}\right)$, $\gamma(d(\alpha, r)) = d\left(\gamma(\alpha), \frac{r}{|a-\alpha|^2}\right)$, $\gamma(d(\alpha, r^-)) = d\left(\gamma(\alpha), \left(\frac{r}{|a-\alpha|^2}\right)^-\right)$.

Corollary 5.2.12. *Let $\mathcal{F}$ be the increasing* (resp., *decreasing) filter of center α and diameter R on $\mathbb{L} \setminus \{a\}$ with $|a - \alpha| > R$. Then $\gamma(\mathcal{F})$ is an increasing* (resp., *a decreasing) filter of center $\gamma(\alpha)$, of diameter $\frac{R}{|a-\alpha|^2}$ on $\mathbb{L} \setminus \{b\}$.*

Corollary 5.2.13. *Let $\mathcal{F}$ be a decreasing filter with no center, of canonical basis $(D_n)_{n\in\mathbb{N}}$ on $\mathbb{L} \setminus \{a\}$ such that $a \notin D_0$. Then $\gamma(\mathcal{F})$ is a decreasing filter with no center, of canonical basis $(\gamma(D_n))_{n\in\mathbb{N}}$ on $\mathbb{L} \setminus \{b\}$.*

Theorem 5.2.14. *We suppose $a \in D$. Let $D' = \gamma(D)$. Let $\mathcal{F}$ be a filter on D which is either a monotonous filter or a Cauchy filter. Then $\mathcal{F}$ is pierced if and only if $\gamma(\mathcal{F})$ is a pierced filter on D'.*

5.3. Circular filters

Definitions and notations: Throughout Chapter 5, the complete field $\mathbb{L}$ is supposed to have a dense valuation. D will denote a subset of $\mathbb{L}$. Let $a \in \widetilde{D}$, let $\rho = \delta(a, D)$ and let $R \in]0, +\infty[$ be such that $\rho \leq R \leq \mathrm{diam}(D)$. We call *circular filter of center a and diameter R on $\mathbb{L}$* the filter $\mathcal{F}$ which admits as a generating system the family of sets $\Gamma(\alpha, r', r'')$ with $\alpha \in d(a, R), r' < R < r''$, i.e., $\mathcal{F}$ is the filter which admits for basis the family of sets of the form $\bigcap_{i=1}^{q} \Gamma(\alpha_i, r'_i, r''_i))$ with $\alpha_i \in d(a, R), r'_i < R < r''_i$ $(1 \leq i \leq q,\ q \in \mathbb{N})$.

For reasons that will appear when characterizing the absolute values of $\mathbb{L}(x)$, a decreasing filter with no center, of canonical basis $(D_n)_{n\in\mathbb{N}}$ is also be called *a circular filter on $\mathbb{L}$ with no center, of canonical basis $(D_n)_{n\in\mathbb{N}}$*.

Finally the filter of neighborhoods of a point $a \in \mathbb{L}$ is called *circular filter of the neighborhoods of a on $\mathbb{L}$*. It will also be named *circular filter of center a and diameter 0* or *Cauchy circular filter of limit a*. A circular filter on $\mathbb{L}$ will be said to be *large* if it has diameter different from 0 and to be *punctual* if it is a cauchy circular filter.

Given a circular filter $\mathcal{F}$, its diameter will be denoted by $\mathrm{diam}(\mathcal{F})$ and the set of its centers is denoted by $\mathcal{Q}(\mathcal{F})$. So, if $\mathcal{F}$ is a circular filter of center a and diameter r, then $\mathcal{Q}(\mathcal{F}) = d(a, r)$.

Given a circular filter $\mathcal{F}$ on $\mathbb{L}$, an infraconnected affinoid subset B of $\mathbb{L}$ will be called $\mathcal{F}$*-affinoid* if it belongs to $\mathcal{F}$. Thus, in particular, if $\mathcal{F}$ has center a and diameter r, a $\mathcal{F}$ affinoid either is a disk $d(a, s)$, with $s \in |\mathbb{L}|$, or is of the form $d(a, r'') \setminus \bigcup_{i=1}^{q} d(a_i, r'^{-})$, with $r', r'' \in |\mathbb{L}|$, $r' < r < r''$, and $|a_i - a_j| = r\ \forall i \neq j$.

Lemma 5.3.1 is easily checked.

Lemma 5.3.1. *Let $\mathcal{F}$, $\mathcal{G}$ be two different circular filters. Then $\mathcal{F}$ is not secant with $\mathcal{G}$. Moreover, if $\mathcal{Q}(\mathcal{F}) \neq \emptyset$, then $\mathcal{Q}(\mathcal{F}) \neq \mathcal{Q}(\mathcal{G})$.*

Definitions and notations: We call *circular filter on D* the filter induced on D by a circular filter on $\mathbb{L}$ secant with D.

The set of circular filters on $\mathbb{L}$ secant with D will be denoted by $\Phi(D)$. In the same way, the set of large circular filters on $\mathbb{L}$ secant with D will be denoted by $\Phi'(D)$.

Thus, by definition, every circular filter on D is induced by a unique circular filter on $\mathbb{L}$. Given a circular filter $\mathcal{F}$ on D induced by a circular filter $\mathcal{G}$ on $\mathbb{L}$, we will denote by $\mathcal{Q}(\mathcal{F})$ the set of its centers in $\mathbb{L}$, i.e., $\mathcal{Q}(\mathcal{G})$.

Given a circular filter $\mathcal{F}$ on D, induced by a circular filter $\mathcal{G}$ on $\mathbb{L}$, we shall call $\mathcal{F}$-*affinoid* any $\mathcal{G}$-affinoid.

Lemma 5.3.2. *Let $\mathcal{F}$ be a circular filter. Then $\mathcal{F}$ admits a basis consisting of the family of all $\mathcal{F}$-affinoids. If $\mathcal{F}$ does not admit a countable basis, it has a center and its diameter belongs to $|\mathbb{L}|$. If $\mathcal{F}$ has no center and is secant with an affinoid subset E of $\mathbb{L}$ then E lies in $\mathcal{F}$. If $\mathcal{F}$ has center a and diameter r, then an affinoid set E lies in $\mathcal{F}$ if and only if satisfy $E \cap (\mathbb{L} \setminus d(a,r)) \neq \emptyset$, $E \cap d(b,r^-) \neq \emptyset \; \forall b \in d(a,r)$.*

Proof. By definition, a circular filter with no center has a countable basis, and of course so does a Cauchy circular filter. In both cases, it admits a basis consisting of a family of disk which are $\mathcal{F}$-affinoid sets. Now, consider a circular filter of center a and diameter r. Therefore, $\mathcal{F}$ admits for basis the family of annuli $d(a, r + \frac{1}{n}) \setminus (\bigcup_{i=1}^{q} d(a_i, (r - \frac{1}{n})^-)$ where the a_i are centers of $\mathcal{F}$ satisfying $|a_i - a_j| = r$. In particular, if $r \notin |\mathbb{L}|$ we have $q = 1$ and we obtain a basis of the form $\Gamma(a, r - \frac{1}{n}, r + \frac{1}{n})$ which is countable.

Now, suppose that $\mathcal{F}$ is secant with an affinoid subset E of $\mathbb{L}$. Let $(A_n)_{n \in \mathbb{N}}$ be a canonical basis of $\mathcal{F}$. Since each A_n admits common points with E, each is included in $\widetilde{E}$, and therefore it is included in E if and only if it contains no hole of E. But since $\mathcal{F}$ has no center, $\bigcap_{n=0}^{\infty} A_n = \emptyset$, hence there exists $q \in \mathbb{N}$ such that $A_n \subset E \; \forall n \geq q$, and therefore $E \in \mathcal{F}$.

Now suppose that $\mathcal{F}$ has center a and diameter r. If $E \in \mathcal{F}$, it obviously satisfies $E \cap (\mathbb{L} \setminus d(a,r)) \neq \emptyset$, $E \cap d(b,r^-) \neq \emptyset \; \forall b \in d(a,r)$. Now, suppose that E satisfies $E \cap (\mathbb{L} \setminus d(a,r)) \neq \emptyset$, $E \cap d(b,r^-) \neq \emptyset \; \forall b \in d(a,r)$. Since E has finitely many holes, on one hand there exists $s > r$ such that $\Gamma(a,r,s) \subset E$, and on the other hand, all classes of $d(a,r)$ are included in E, except finitely many: $d(b_j, r^-)$, $1 \leq j \leq n$. And for each $j = 1, \ldots, n$, there exists $r_j < r$ such that $\Gamma(b_j, r_j, r) \subset E$. Finally, E contains the set $d(a,s) \setminus \bigcup_{j=1}^{n} d(b_j, r_j)$ which obviously lies in $\mathcal{F}$. $\qquad\square$

Remarks. By Lemma 5.3.2, it is easily seen that there exist circular filters without countable basis if and only if the residue class field of $\mathbb{L}$ is not countable.

The following Proposition 5.3.3 is just a translation of the definitions.

Proposition 5.3.3. *Let $\mathcal{F}$ be an increasing filter (resp., a decreasing filter) of center a and diameter r, on D. Then the circular filter of center a and diameter r on $\mathbb{L}$ is secant with D and is the only circular filter on D less thin than $\mathcal{F}$.*

Conversely let $\mathcal{F}$ be a circular filter of center a and diameter r on D, secant with $d(a, r^-)$ (resp., $\mathbb{L} \setminus d(a, r)$). Then the increasing filter (resp., decreasing filter) of center a and diameter r on $\mathbb{L}$ is secant with D and less thin than $\mathcal{F}$.

Definition: Let $\mathcal{F}$ be an increasing filter (resp., a decreasing filter) of center a and diameter r, on D. Then the circular filter of center a and diameter r will be called *circular filter associated to $\mathcal{F}$.*

Proposition 5.3.4 is immediate (and is [28, Proposition 2.17]).

Proposition 5.3.4. *Let D be infraconnected, let $a \in \tilde{D}$ and let S be the closure of $\{|x - a| \,|x \in D\}$ in $\mathbb{R}$. For every $r \in S$ the circular filter $\mathcal{F}$ of center a and diameter r on $\mathbb{L}$ is secant with D.*

Lemma 5.3.5. *Let $\mathcal{F}$ be a circular filter secant with two disks $d(a, r)$ and $d(b, s)$. Then either $d(a, r) \subset d(b, s)$ or $d(b, s) \subset d(a, r)$.*

Proof. Let $R = \delta(d(a, r), d(b, s))$. Suppose that none of the two inclusions is satisfied. Then $d(a, r) \cap d(b, s) = \emptyset$. But since $\mathcal{F}$ is secant with $d(a, r)$, then diam$(\mathcal{F}) \leq r$, and therefore $\mathcal{F}$ is not secant with $C(a, R)$, in particular neither is it with $d(b, s)$. $\qquad\square$

Corollary 5.3.6. *Let D be infraconnected, and let $d(a, r^-)$ be a hole of D. Then, the circular filter of center a and diameter r on $\mathbb{L}$ is secant with D.*

Proposition 5.3.7. *Let $(a_n)_{n \in \mathbb{N}}$ be a sequence in L that is either a monotonous distances sequence or a constant distances sequence. Then there exists a unique circular filter on $\mathbb{L}$ less thin than the sequence (a_n).*

Proof. By [28, Proposition 2.18] we know that there exists a unique circular filter on $\mathbb{L}$ less thin than the sequence (a_n). By definition, this filter is secant with the set $\{a_n \mid n \in \mathbb{N}\}$, hence with D. $\qquad\square$

Corollary 5.3.8. *Let $(a_n)_{n \in \mathbb{N}}$ be a bounded sequence in $\mathbb{L}$. Then there exists a subsequence $(a_{n_t})_{t \in \mathbb{N}}$ and a unique circular filter $\mathcal{F}$ on $\mathbb{L}$ less thin than the subsequence $(a_{n_t})_{t \in \mathbb{N}}$.*

Proof. Since the sequence $(a_n)_{n \in \mathbb{N}}$ is bounded, by Theorem 5.2.1, we can extract either a monotonous distances subsequence or a constant distances subsequence, or a converging subsequence. In all cases, once such a subsequence is chosen, there exists a unique circular filter $\mathcal{F}$ on $\mathbb{L}$ less thin than the subsequence. $\qquad\square$

Lemma 5.3.9. *Let $(a_n)_{n \in \mathbb{N}}$, $(b_n)_{n \in \mathbb{N}}$ be two sequences such that $|a_n - b_n| <\leq t < r \; \forall n \in \mathbb{N}$. Suppose that the sequence $(a_n)_{n \in \mathbb{N}}$ is thinner than a circular filter $\mathcal{F}$ of diameter r. Then the sequence $(b_n)_{n \in \mathbb{N}}$ also is thinner than $\mathcal{F}$.*

Proof. We can find a $\mathcal{F}$-affinoid B, whose codiameter s is strictly superior to t. Then if a_n belongs to B, so does b_n. Now, when n is big enough, all a_n belong to B and hence so do all b_n. And since $\mathcal{F}$ admits a basis of $\mathcal{F}$-affinoids with a codiameter $s > t$, we see that the sequence $(b_n)_{n \in \mathbb{N}}$ is thinner than $\mathcal{F}$. $\qquad\square$

Remark. If $\mathcal{F}$ is the circular filter of center a and diameter r, it is not secant with $C(a, r)$ if and only if $r \notin |\mathbb{L}|$.

Definitions and notations: A circular filter $\mathcal{F}$ on $\mathbb{L}$ will be said to be *(a,r)-approaching* if it is secant with the circle $C(a, r)$, or if it is the circular filter of center a and diameter r,

A circular filter $\mathcal{F}$ on $\mathbb{L}$ will be said to be *D-bordering* if it is secant with both D and $\mathbb{L} \setminus \overline{D}$. In the same way, a circular filter on D will be said to be *D-bordering* if it is induced by a D-bordering circular filter on $\mathbb{L}$.

If D is bounded, $\mathcal{F}$ will be said to be *D-peripheral*, or to be *peripheral to D* if $\mathcal{Q}(\mathcal{F}) = \tilde{D}$.

Finally, a circular filter $\mathcal{F}$ will be said to be *strictly D-bordering* if either it is D-peripheral, or it is peripheral to some hole of D.

Proposition 5.3.10. *Let D be closed and infraconnected and let $\mathcal{F}$ be a circular filter on $\mathbb{L}$. If every element of $\mathcal{F}$ contains infinitely many holes of D, then $\mathcal{F}$ is D-bordering. Conversely, if $\mathcal{F}$ is D-bordering but not strictly D-bordering, then every element of $\mathcal{F}$ contains infinitely many holes of D and then, either D admits a monotonous pierced filter thinner than $\mathcal{F}$, or $\mathcal{F}$ has a center a and D admits an equal distances holes sequence $(T_n)_{n\mathbb{N}}$ such that $\delta(T_n, T_m) = \delta(T_n, a) = \operatorname{diam}(\mathcal{F})$ for all $m \neq n$.*

Proof. Suppose that every element of $\mathcal{F}$ contains infinitely many holes of D. Then $\mathcal{F}$ is obviously secant with $\mathbb{L} \setminus D$. Let $A \in \mathcal{F}$ be affinoid and let $s = \text{diam}(\mathcal{F})$. Suppose that $A \cap D = \emptyset$. If $D \subset \mathbb{L} \setminus \tilde{A}$, then every hole of D has empty intersection with A, a contradiction. But then, since D is infraconnected, there exists a hole $T = d(a, r^-)$ of A such that $D \subset T$. Since A is affinoid, r belongs to $|\mathbb{L}|$, hence there exists another affinoid $B \in \mathcal{F}$, with a hole $d(a, s^-)$, with $s > r$, thereby $\tilde{D} \cap B = \emptyset$, and finally every hole of D has empty intersection with B, a contradiction again. Thus, $\mathcal{F}$ is also secant with D, and therefore $\mathcal{F}$ is D-bordering.

Conversely, consider a D-bordering filter $\mathcal{F}$ which is not strictly D-bordering. Let $s = \text{diam}(\mathcal{F})$. Let $A \in \mathcal{F}$ be affinoid and let $a \in A \cap (\mathbb{L} \setminus D)$. Since D is closed, a belongs to a hole $T = d(a, r^-)$ of D. Since $A \cap D \neq \emptyset$, A is not included in $d(a, r^-)$. We will show that there exists $B \in \mathcal{F}$ such that $B \subset A$ and $T \cap B = \emptyset$. For any $r' \in]0, r[$ the set $B = A \setminus d(a, r')$ is an affinoid set that also belongs to $\mathcal{F}$. Since T is not included in B, then B must contain another hole of D. Thus, by induction we can construct a strictly decreasing sequence $(A_m)_{m \in \mathbb{N}}$ of affinoid sets such that each A_m contains a hole S_m which is not included in A_{m+1}. Thus, every element of $\mathcal{F}$ contains infinitely many holes of D. More precisely, suppose that D does not admit a monotonous pierced filter thinner than $\mathcal{F}$. If $\mathcal{F}$ had no center, it would be a not pierced decreasing filter, and therefore would admit a basis consisting of a sequence of disks $(D_n)_{n \in \mathbb{N}}$ included in D, a contradiction to the hypothesis: every element of $\mathcal{F}$ contains infinitely many holes of D. Consequently, $\mathcal{F}$ admits a center a. Let $r = \text{diam}(\mathcal{F})$. By Proposition 5.2.4, we can extract from the sequence $(S_m)_{m \in \mathbb{N}}$ a subsequence $(T_n)_{n \in \mathbb{N}}$ which is either a monotonous distances holes sequence, or an equal distances holes sequence. Consider the sequence $(B_n)_{n \in \mathbb{N}}$ defined as $B_n = \bigcup_{j=n}^{\infty} T_j$. This sequence is the basis of a filter thinner than $\mathcal{F}$. If the sequence $(T_n)_{n \in \mathbb{N}}$ is a monotonous distances holes sequence, then the sequence $(B_n)_{n \in \mathbb{N}}$ defined as $B_n = \bigcup_{j=n}^{\infty} T_j$ is the basis of a filter thinner than a monotonous filter $\mathcal{G}$ thinner than $\mathcal{F}$ and such that the filter induced by $\mathcal{G}$ on D is pierced, a contradiction to the hypothesis. Consequently, the sequence $(T_n)_{n \in \mathbb{N}}$ is an equal distances holes sequence of center b: $\delta(T_n, T_m) = l = \delta(b, T_n) \ \forall n \neq m$. If $l < r$, (resp., $l > r$), there exists an annulus $\Gamma(b, r', r'') \in \mathcal{F}$ such that $T_n \subset d(b, r')$ (resp., $T_n \subset K \setminus d(b, r'')$), and therefore $T_n \cap \Gamma(b, r', r'') = \emptyset$. But since the sequence (B_n) is a basis of a filter thinner than $\mathcal{F}$, each B_n has a non-empty intersection with $\Gamma(b, r', r'')$. Consequently, $l = r = \delta(a, T_n)$ $\forall n \in \mathbb{N}$. $\qquad \square$

Lemma 5.3.11. *Let D be an affinoid subset of $\mathbb{L}$. Then the set of D-bordering filters is finite.*

Proof. If D is infraconnected, it has finitely many holes. And if it is not infraconnected, it has finitely many infraconnected components $D_1, \ldots, D_q$, hence the set of its D-bordering filters is the union of the sets of D_j-bordering filters, $1 \leq j \leq q$. $\qquad \square$

Lemma 5.3.12. *Let $\mathcal{F}$ be a circular filter on $\mathbb{L}$ and let $a \in \mathbb{L}$. There exists a unique $r > 0$ such that $\mathcal{F}$ is (a, r)-approaching. Moreover, for every $s >$ $\mathrm{diam}(\mathcal{F})$ there exists a unique disk E of the form $d(b, s)$ which belongs to $\mathcal{F}$.*

Proof. Let $\rho = \mathrm{diam}(\mathcal{F})$. If a is a center of $\mathcal{F}$, then $\mathcal{F}$ is (a, ρ)-approaching. Suppose now a is not a center of $\mathcal{F}$. Then we can find a disk $D \in \mathcal{F}$ which does not contain a: let $r = \delta(a, D)$. Then $\mathcal{F}$ is secant with $C(a, r)$. Let $s > \mathrm{diam}(\mathcal{F})$. If $\mathcal{F}$ has a center a, then $\mathcal{F}$ is secant with $d(a, s)$. Now, suppose that $\mathcal{F}$ has no center. By definition of the circular filters with no center, there exists a disk $d(b, r) \in \mathcal{F}$ such that $r < s$. Then $d(b, s)$ obviously belongs to $\mathcal{F}$. Moreover, in both cases, any disk $d(b, s)$ other than $d(a, s)$ has empty intersection with $d(a, s)$, hence $\mathcal{F}$ is not secant with $d(b, s)$. $\qquad \square$

Lemma 5.3.13. *Let $\mathcal{F}$ be a filter admitting a basis $(A_i)_{i \in J}$ consisting of infraconnected affinoid sets. There exists at least one circular filter $\mathcal{G}$ secant with $\mathcal{F}$.*

Proof. Suppose first that $\bigcap_{i \in J} \widetilde{A}_i = \emptyset$. Then $(\widetilde{A}_i)_{i \in J}$ is a basis of a circular filter with no center $\mathcal{G}$ which actually is equal to $\mathcal{F}$. Suppose now that $\bigcap_{i \in J} \widetilde{A}_i \neq \emptyset$ and let $a \in \bigcap_{i \in J} \widetilde{A}_i$. Let $r = \inf_{i \in J} \mathrm{diam}(A_i)$. Then the circular filter of center a and diameter r is clearly secant with $\mathcal{F}$, which ends the proof. $\qquad \square$

Lemma 5.3.14. *Let $\mathcal{G}$ be a circular filter on $\mathbb{L}$ of center a and diameter r and let $\mathcal{F}$ be a filter secant with $\mathcal{G}$, admitting a basis $(A_i)_{i \in J}$ consisting of infraconnected affinoid sets. Either $\mathcal{F}$ is less thin than $\mathcal{G}$, or $\mathcal{F}$ is secant with another circular filter $\mathcal{G}'$.*

Proof. Since $\mathcal{F}$ is secant with $\mathcal{G}$, for each $i \in J$ we have $A_i \cap C(a, r) \neq \emptyset$ and $\mathrm{diam}(A_i) \geq r$. Suppose that $\mathcal{F}$ is not less thin than $\mathcal{G}$. Then there exists an affinoid infraconnected set $B \in \mathcal{G}$ which does not contain A_i, whenever $i \in J$. Let $s = \inf_{i \in J} \mathrm{diam}(A_i)$.

Suppose first that $\mathcal{G}$ has no center. Then we can assume that B is a disk $d(b, \lambda)$, with $\lambda \in]r, s]$. Let $t \in]r, s[\cap |\mathbb{L}|$. Since the A_i are infraconnected and affinoid, for each $i \in J$ and for every class D of $C(b, t)$ except at most for finitely many, D is included in A_i. Consequently, $\mathcal{F}$ is secant with the circular filter $\mathcal{G}'$ of center b and diameter t.

Suppose now that $\mathcal{G}$ has center a. Since B has finitely many holes, either $s > r$, or there exists a hole $T = d(b, \rho)$ of B such that $T \cap A_i \neq \emptyset$ $\forall i \in J$. Suppose first that $s > r$. Let $t \in]r, s[\cap |\mathbb{L}|$. Similarly to the previous case, since the A_i are infraconnected and affinoid, for each $i \in J$ and for every class D of $C(a, t)$ except at most for finitely many, D is included in A_i. Consequently, $\mathcal{F}$ is secant with the circular filter $\mathcal{G}'$ of center a and diameter t. Similarly, suppose now that there exists a hole $T = d(b, \rho)$ of B such that $T \cap A_i \neq \emptyset$ $\forall i \in J$. We can find $t \in]\rho, r[\cap |\mathbb{L}|$ and then all classes of $C(b, t)$ except at most finitely many are included in A_i whenever $i \in J$. Consequently, $\mathcal{F}$ is secant with the circular filter $\mathcal{G}'$ of center b and diameter t, which ends the proof.   □

5.4.   Tree structure and metric on circular filters

In this chapter, we will show that the set of circular filters is equipped with a tree structure as defined in Section 1.2 of Chapter 1, and that the diameter here is an increasing function with values in $\mathbb{R}$, defining distances associated to this structure, as shown in [26, Chapter 11]. The first remarks on that tree structure are due to Motzkin [38]. The field $\mathbb{L}$ is a complete ultrametric field equipped with a dense ultrametric absolute value.

Circular filters and monotonous filters are filters on the field $\mathbb{L}$ or on an infraconnected subset D of $\mathbb{L}$.

Definitions and notations: Given two circular filters $\mathcal{F}$ and $\mathcal{G}$, $\mathcal{F}$ is said *to surround* $\mathcal{G}$ if either $\mathcal{G}$ is secant with $\mathcal{Q}(\mathcal{F})$, or if $\mathcal{F} = \mathcal{G}$. Similarly, a circular filter $\mathcal{F}$ is said *to surround a monotonous filter* $\mathcal{G}$ if it surrounds the circular filter associated to $\mathcal{G}$. A monotonous filter $\mathcal{F}$ is said *to surround a circular filter* $\mathcal{G}$ if its associated circular filter surrounds $\mathcal{G}$, and $\mathcal{F}$ is said *to surround a monotonous filter* $\mathcal{G}$ if the circular filter associated to $\mathcal{F}$ surrounds the circular filter associated to $\mathcal{G}$.

We will denote by $\preceq$ the relation on the set of circular filters defined as $\mathcal{F} \preceq \mathcal{G}$ if $\mathcal{G}$ surrounds $\mathcal{F}$ and by $\prec$ the relation defined as $\mathcal{F} \prec \mathcal{G}$ if $\mathcal{F} \preceq \mathcal{G}$ and $\mathcal{F} \neq \mathcal{G}$.

By definition of the relation $\preceq$, Lemma 5.4.1 is then immediate.

Lemma 5.4.1. *Let $\mathcal{F}, \mathcal{G}$ be two circular filters such that $\mathcal{Q}(\mathcal{F}) \neq \emptyset$ and $\mathcal{Q}(\mathcal{G}) \neq \emptyset$. Then $\mathcal{F} \preceq \mathcal{G}$ if and only if $\mathcal{Q}(\mathcal{F}) \subset \mathcal{Q}(\mathcal{G})$.*

Corollary 5.4.2. *Let $\mathcal{F}, \mathcal{G}$ be two circular filters such that $\mathcal{F} \preceq \mathcal{G}$. Then $\mathrm{diam}(\mathcal{F}) \leq \mathrm{diam}(\mathcal{G})$. Moreover, $\mathcal{F} = \mathcal{G}$ if and only if $\mathrm{diam}(\mathcal{F}) = \mathrm{diam}(\mathcal{G})$.*

Theorem 5.4.3. *The relation $\preceq$ is an order relation on $\Phi(\mathbb{L})$ and $\prec$ is the strict order associated to this order relation.*

Proof. Indeed, the relation is reflexive by definition. Let $\mathcal{F}, \mathcal{G}, \mathcal{H}$ be 3 circular filters. Suppose that $\mathcal{F} \preceq \mathcal{G}$ and $\mathcal{G} \preceq \mathcal{F}$. If $\mathcal{F} \neq \mathcal{G}$, then by definition $\mathcal{Q}(\mathcal{F}) \neq \emptyset$, $\mathcal{Q}(\mathcal{G}) \neq \emptyset$, $\mathcal{F}$ is secant with $\mathcal{Q}(\mathcal{G})$ and $\mathcal{G}$ is secant with $\mathcal{Q}(\mathcal{F})$. Consequently, $\mathcal{Q}(\mathcal{F}) = \mathcal{Q}(\mathcal{G})$ and therefore by Corollary 5.4.2, $\mathcal{F} = \mathcal{G}$. Finally, suppose that $\mathcal{F} \preceq \mathcal{G}$ and $\mathcal{G} \preceq \mathcal{H}$, with $\mathcal{F} \neq \mathcal{G} \neq \mathcal{H}$. Then $\mathcal{F}$ is secant with $\mathcal{Q}(\mathcal{G})$, hence with $\mathcal{Q}(\mathcal{H})$ and therefore $\mathcal{F} \preceq \mathcal{H}$. $\qquad\square$

Theorem 5.4.4. *Let D be infraconnected and let $\mathcal{F} \in \Phi(D)$. Then $\mathcal{F}$ is a minimal element in $\Phi(D)$ if and only if:*
   either it is punctual,
   or it has no center,
   or $\mathcal{Q}(\mathcal{F}) \cap D = \emptyset$.

Proof. On one hand it is easily seen that if $\mathcal{F}$ is punctual, or has no center, then it is minimal in $\Phi(\mathbb{L})$, and therefore in $\Phi(D)$. Suppose now that $\mathcal{Q}(\mathcal{F}) \cap D = \emptyset$. Suppose that $\mathcal{G} \in \Phi(D)$ satisfies $\mathcal{G} \prec \mathcal{F}$. Then $\mathcal{G}$ is secant with $\mathcal{Q}(\mathcal{F})$, and has a diameter strictly inferior to the diameter r of $\mathcal{F}$, hence there exists a disk $d(b, s) \in \mathcal{G}$, with $s < r$. Consequently, $d(b, s) \subset \mathcal{Q}(\mathcal{F})$. Since $d(b, s) \cap D = \emptyset$, we see that $\mathcal{G}$ is not secant with D. $\qquad\square$

Now, let $\mathcal{F} \in \Phi(D)$ be minimal in $\Phi(D)$. Suppose that $\mathcal{F}$ is not punctual and has a center a, and let $r = \operatorname{diam}(\mathcal{F})$. If $\mathcal{Q}(\mathcal{F}) \cap D \neq \emptyset$, then for any $a \in \mathcal{Q}(\mathcal{F}) \cap D$, of course $\mathcal{F}$ surrounds the filter of neighborhoods of a, a contradiction to the hypothesis "$\mathcal{F}$ minimal in $\Phi(D)$". Thus, $\mathcal{Q}(\mathcal{F}) \cap D = \emptyset$.

Lemma 5.4.5. *Let $\mathcal{F}, \mathcal{G}$ be two circular filters such that $\mathcal{Q}(\mathcal{F}) \cap \mathcal{Q}(\mathcal{G}) \neq \emptyset$. Then $\mathcal{F}$ and $\mathcal{G}$ are comparable for $\preceq$.*

Proof. Since $\mathcal{Q}(\mathcal{F}) \cap \mathcal{Q}(\mathcal{G}) \neq \emptyset$, and since both $\mathcal{Q}(\mathcal{F})$, $\mathcal{Q}(\mathcal{G})$ are disks, we can suppose for instance $\mathcal{Q}(\mathcal{F}) \subset \mathcal{Q}(\mathcal{G})$. Then $\mathcal{F}$ is secant with $\mathcal{Q}(\mathcal{G})$, hence $\mathcal{F} \preceq \mathcal{G}$. $\qquad\square$

Theorem 5.4.6. *Let $\mathcal{F}$ be a circular filter on $\mathbb{L}$ and let $s > \operatorname{diam}(\mathcal{F})$. There exists a unique circular filter on $\mathbb{L}$, of diameter s, surrounding $\mathcal{F}$.*

Proof. By Lemma 5.3.12, there exists a unique disk $d(b, s)$ such that $\mathcal{F}$ is secant with this disk. Then the circular filter $\mathcal{G}$ of center b and diameter s obviously surrounds $\mathcal{F}$. Conversely, let $\mathcal{H}$ be another circular filter of diameter s, surrounding $\mathcal{H}$. Since $s > \operatorname{diam}(\mathcal{F})$, $\mathcal{F}$ is secant with $\mathcal{Q}(\mathcal{H})$ which is of the form $d(b, s)$, hence $\mathcal{H} = \mathcal{G}$. $\qquad\square$

Proposition 5.4.7. *Let* $\mathcal{F}$, $\mathcal{G}$ *be two circular filters surrounding a certain circular filter. Then* $\mathcal{F}$ *and* $\mathcal{G}$ *are comparable with respect to* $\preceq$.

Proof. Suppose that both $\mathcal{F}$, $\mathcal{G}$ surrounds $\mathcal{H}$. If $\mathcal{F} = \mathcal{H}$, then $\mathcal{G}$ just surrounds $\mathcal{F}$. Now suppose that both $\mathcal{F}$, $\mathcal{G}$ strictly surrounds $\mathcal{H}$. Then $\mathcal{H}$ is secant with both $\mathcal{Q}(\mathcal{F})$ and $\mathcal{Q}(\mathcal{G})$. But by Lemma 5.3.5, either $\mathcal{Q}(\mathcal{F}) \subset \mathcal{Q}(\mathcal{G})$ or $\mathcal{Q}(\mathcal{G}) \subset \mathcal{Q}(\mathcal{F})$, hence by Lemma 5.4.5, $\mathcal{F}$ and $\mathcal{G}$ are comparable. $\qquad\square$

By Proposition 5.3.4, we obtain Theorem 5.4.8

Theorem 5.4.8. *Let* D *be infraconnected, let* $\mathcal{F}$ *be a circular filter on* $\mathbb{L}$ *secant with* D *and let* $r \in]\mathrm{diam}(\mathcal{F}), \mathrm{diam}(D)[$. *The unique circular filter of diameter* r *surrounding* $\mathcal{F}$ *is secant with* D.

Proof. By Lemma 5.3.12, there exists a unique disk $d(a, r)$ $\mathcal{F}$ is secant with. By Proposition 5.3.4, the circular filter of center a and diameter r is secant with D, and on the other hand, by Theorem 5.4.6, this is the unique circular filter of diameter r surrounding $\mathcal{F}$. $\qquad\square$

Proposition 5.4.9. *Let* $\mathcal{F}$, $\mathcal{G}$ *be two circular filters on* $\mathbb{L}$ *which are not comparable for the relation* $\preceq$. *There exist disks* $F \in \mathcal{F}$ *and* $G \in \mathcal{G}$ *such that* $F \cap G = \emptyset$. *Moreover, given* $F' \in \mathcal{F}$, $G' \in \mathcal{G}$ *such that* $F' \cap G' = \emptyset$, *we have* $\delta(F, G) = \delta(F', G') > \max(\mathrm{diam}(\mathcal{F}), \mathrm{diam}(\mathcal{G}))$.

Proof. Suppose first that both $\mathcal{F}$, $\mathcal{G}$ have no center. Then $\mathcal{F}$ (resp., $\mathcal{G}$) admits a canonical basis $\mathcal{D}$ (resp., $\mathcal{E}$) consisting of a decreasing family of disks. But since the two filters are not secant, we can obviously find $F \in \mathcal{D}$ and $G \in \mathcal{E}$ such that $F \cap G = \emptyset$.

Suppose now that $\mathcal{F}$ has centers and let $d(a, r) = \mathcal{Q}(\mathcal{F})$. If all disks $G \in \mathcal{G}$ contain $d(a, r)$, then $\mathcal{G}$ surrounds $\mathcal{F}$, hence we can find a disk $G = d(b, u) \in \mathcal{G}$ which does not contain $d(a, r)$. Of course, if $G \subset d(a, r)$, then $\mathcal{F}$ surrounds $\mathcal{G}$, hence $G \cap d(a, r) = \emptyset$. Consequently, we have $|a - b| > r$ and $u < |a - b|$. Let $s \in]r, |a - b|[$. Then the disk $F = d(a, s)$ belongs to $\mathcal{F}$ and is such that $F \cap G = \emptyset$.

Since $\mathcal{F}$ is a filter, we have $F \cap F' \neq \emptyset$, hence either $F \subset F'$, or $F' \subset F$. In the same way, $G \cap G' \neq \emptyset$, hence either $G \subset G'$, or $G' \subset G$. Without loss of generality, we can assume that $F \subset F'$. If $G \subset G'$, then our claim is obvious. Suppose now that $G' \subset G$. If $F' \cap G \neq \emptyset$, then either $G \subset F'$, therefore $G \subset G'$, a contradiction, or $F' \subset G$, hence $F \subset G$, a contradiction again. Consequently, we have $F' \cap G = \emptyset$. Putting $l = \delta(F', G)$ we have $|a - b| = l$ $\forall a \in F'$, $b \in G$, hence in particular, $\delta(F, G) = \delta(F', G') = l$. Then we notice that $l > \mathrm{diam}(F) > \mathrm{diam}(\mathcal{F})$ and $l > \mathrm{diam}(G) > \mathrm{diam}(\mathcal{G})$. $\qquad\square$

Notations: Let $\mathcal{F}, \mathcal{G}$ be two circular filters which are not comparable for $\preceq$ and let $F \in \mathcal{F}$, $G \in \mathcal{G}$ be disks such that $F \cap G = \emptyset$. By Proposition 5.4.9, $\delta(F, G)$ does not depend on the choice of disks F, G satisfying these properties, so we can put $\pi(\mathcal{F}, \mathcal{G}) = \delta(F, G)$ with F, G disks such that $F \in \mathcal{F}, G \in \mathcal{G}$, $F \cap G = \emptyset$.

Theorem 5.4.10. *Let D be infraconnected and let $\mathcal{F}, \mathcal{G} \in \Phi(D)$. There exists* $\sup(\mathcal{F}, \mathcal{G}) \in \Phi(D)$ *and it is the unique circular filter of diameter* $\pi(\mathcal{F}, \mathcal{G})$ *which surrounds both $\mathcal{F}$, $\mathcal{G}$.*

Proof. The claim is trivial when the two filters are comparable for $\preceq$. So we assume they are not. Let $l = \pi(\mathcal{F}, \mathcal{G})$. By Theorem 5.4.6, there exists a unique circular filter $\mathcal{S} \in \Phi(L)$ of diameter l surrounding $\mathcal{F}$. Then $\mathcal{S}$ has centers. Let $\mathcal{Q}(\mathcal{S}) = d(a, l)$. Then by Proposition 5.4.9, $\mathcal{G}$ contains disks included in $d(a, l)$ and therefore is secant with $d(a, l)$, hence $\mathcal{S}$ surrounds $\mathcal{G}$. We will check that $\mathcal{S}$ is the smallest element of the set of filters on $\mathbb{L}$ surrounding both $\mathcal{F}$ and $\mathcal{G}$. Indeed, let $\mathcal{H}$ be a circular filter surrounding $\mathcal{F}$ and $\mathcal{G}$. Then both $\mathcal{F}$ and $\mathcal{G}$ are secant with $\mathcal{Q}(\mathcal{H})$. Let $d(b, s) = \mathcal{Q}(\mathcal{H})$. Consider disks $F = d(\alpha, \rho) \in \mathcal{F}$ and $G = d(\beta, \nu) \in \mathcal{G}$ such that $F \cap G = \emptyset$. Since $F \cap d(b, s) \neq \emptyset$ and $G \cap d(b, s) \neq \emptyset$, and since $F \cap G = \emptyset$, it is easily seen that both F, G are included in $d(b, s)$, hence $s \geq l$. But since both $\mathcal{S}$, $\mathcal{H}$ surround $\mathcal{F}$, by Proposition 5.4.7, they are comparable for $\preceq$. Then, since $l \leq s$, $\mathcal{H}$ surrounds $\mathcal{S}$, and therefore $\mathcal{S}$ is the smallest element of the set of filters on $\mathbb{L}$ surrounding both $\mathcal{F}$ and $\mathcal{G}$.

Now, suppose that both $\mathcal{F}$, $\mathcal{G}$ are secant with D. Then $d(\alpha, \rho) \cap D \neq \emptyset$, and $d(\beta, \nu) \cap D \neq \emptyset$, hence $\mathcal{S}$ which is the circular filter of center α and diameter $l = |\alpha - \beta|$ is secant with D because D is infraconnected. $\square$

Theorem 5.4.11. *Let D be infraconnected. $\Phi(D)$ is a tree with respect to the order $\preceq$ and the mapping* diam *is strictly increasing from $\Phi(D)$ to $\mathbb{R}_+$.*

Proof. This is an obvious consequence of Theorems 5.4.3, 5.4.4, 5.4.10 and Proposition 5.4.7. $\square$

Notation: Let $\mathcal{F}$, $\mathcal{G}$ be two circular filters and let $\mathcal{S} = \sup(\mathcal{F}, \mathcal{G})$. We put $\delta(\mathcal{F}, \mathcal{G}) = \max(\text{diam}(\mathcal{S}) - \text{diam}(\mathcal{F}), \ \text{diam}(\mathcal{S}) - \text{diam}(\mathcal{G}))$ and $\delta'(\mathcal{F}, \mathcal{G}) = 2\text{diam}(\mathcal{S}) - \text{diam}(\mathcal{F}) - \text{diam}(\mathcal{G})$.

Remarks. (1) Particularly, if $\mathcal{F} \preceq \mathcal{G}$, we have $\delta(\mathcal{F}, \mathcal{G}) = \text{diam}(\mathcal{G}) - \text{diam}(\mathcal{F})$.

(2) If $\mathcal{F}$ and $\mathcal{G}$ are two circular filters reduced to two points a and b respectively, then $\delta(\mathcal{F}, \mathcal{F}) = |a - b| = \delta(a, b)$. Thus, δ appears as a continuation of the classical distance on $\mathbb{L}$ to $\Phi(\mathbb{L})$.

According to Lemma 1.2.1 of Chapter 1, we can state Corollary 5.4.12

Corollary 5.4.12. δ *is the supremum distance associated to the mapping* diam *defined on the tree* $\Phi(\mathbb{L})$ *and* δ' *is the whole distance associated to the mapping* diam *defined on the tree* $\Phi(\mathbb{L})$ *and they satisfy*

$$\delta(\mathcal{F}, \mathcal{G}) \leq \delta'(\mathcal{F}, \mathcal{G}) \leq 2\delta(\mathcal{F}, \mathcal{G}).$$

Definition: We will call *metric-topology* or δ-*topology* the topology defined on $\Phi(D)$ by these two equivalent distances.

Lemma 5.4.13 is immediate and will be useful.

Lemma 5.4.13. *Let* $\mathcal{F}$ *be a circular filter and let* $B \in \mathcal{F}$ *be infraconnected and affinoid. Let* $\mathcal{G}_1, \ldots, \mathcal{G}_n$ *be the* B-*bordering filters and let* $\mu = \min_{1 \leq j \leq n}(\delta(\mathcal{G}, \mathcal{G}_j))$. *For every* $\mathcal{G} \in \Phi(K)$ *such that* $\delta(\mathcal{G}, \mathcal{F}) < \mu$, B *lies in* $\mathcal{G}$.

Proof.   Every disk $d(a, r)$ with $a \in B$ and $r < \mu$ is clearly included in B. Consequently, $\mathcal{G}'$ is secant with B. Now, suppose that $B \notin \mathcal{G}'$. Then $\mathcal{G}'$ is B-bordering, and therefore is one the $\mathcal{G}_j$ ($1 \leq j \leq n$), a contradiction to th hypothesis $\delta(\mathcal{G}, \mathcal{F}) < \mu$. Hence B lies in $\mathcal{G}'$. $\square$

Theorem 5.4.14. *Every bounded monotonous sequence of* $\Phi(\mathbb{L})$ *has a limit with respect to the* δ-*topology.*

Proof.   Let $(\mathcal{F}_n)_{n \in \mathbb{N}}$ be a bounded monotonous sequence of circular filters, and for each $n \in \mathbb{N}$, let $r_n = \text{diam}(\mathcal{F}_n)$. Without loss of generality, we can obviously assume that the sequence is strictly monotonous. Consequently, for each $n \in \mathbb{N}$, each filter $\mathcal{F}_n$ has a center. Let $l = \lim_{n \to \infty} r_n$. Suppose first that it is a decreasing sequence. For each $n \in \mathbb{N}$ we can find a center $a_n \notin d(a_{n+1}, r_{n+1})$. Thus, the sequence $(a_n)_{n \in \mathbb{N}}$ is such that the sequence $(|a_{n+1} - a_n|)_{n \in \mathbb{N}}$ is strictly decreasing. Hence by Proposition 5.3.7, there exists a unique circular filter $\mathcal{F}$ less thin than the sequence $(a_n)_{n \in \mathbb{N}}$ of radius l. And then, we check that the sequence $(\mathcal{F}_n)_{n \in \mathbb{N}}$ converges to $\mathcal{F}$ with respect to the metric topology. If the sequence is increasing, it is easily seen that the sequence converges to the circular filter of center a_1 and diameter l. $\square$

Theorem 5.4.15 ([26]). $\Phi(\mathbb{L})$ *is complete with respect to the* δ-*topology.*

Proof.   Let $(\mathcal{F}_n)_{n \in \mathbb{N}}$ be a Cauchy sequence with respect to the δ-topology. For every $m, n \in \mathbb{N}$, let $\mathcal{S}_{m,n}$ denote $\sup(\mathcal{F}_m, \mathcal{F}_n)$. Next, for each $n \in \mathbb{N}$, let $r_n = \text{diam}(\mathcal{F}_n)$, let $s_n = \sup_{m \geq n}\{\text{diam}(\mathcal{S}_{n,m})\}$. We will show that the sequence $d(a_n, s_n)_{n \in \mathbb{N}}$ is decreasing with respect to inclusion. Indeed, let $m, t > n$. Both $\mathcal{S}_{n,t}, \mathcal{S}_{n,m}$ are secant with $d(a_n, s_n)$ and surround $\mathcal{F}_n$, hence are comparable with respect to $\preceq$. Then, $\sup(\mathcal{S}_{n,t}, \mathcal{S}_{n,m})$ is secant with

$d(a_n, s_n)$, and surrounds $\mathcal{F}_n$, $\mathcal{F}_m$, $\mathcal{F}_t$, hence surround $\mathcal{S}_{m,t}$. Consequently, $\mathcal{S}_{m,t}$ is secant with $d(a_n, s_n)$. Now, let us fix $t > n$. This true for all $m \geq t$, hence $\text{diam}(\mathcal{S}_{m,t}) \leq s_n \; \forall m \geq t$, and therefore $s_m \leq s_n$. Thus, for every $n \in \mathbb{N}$, let let $\mathcal{G}_n$ be the circular filter of center a_n and diameter s_n. The sequence $(\mathcal{G}_n)_{n \in \mathbb{N}}$ is then decreasing with respect to the order $\preceq$ and therefore, by Theorem 5.4.14, has a limit $\mathcal{G}$ with respect to the δ-topologies. Let $s = \lim_{n \to \infty} s_n$. Since the sequence $(\mathcal{F}_n)_{n \in \mathbb{N}}$ is a Cauchy sequence, and since by definition of the distance δ, we have $\delta(\mathcal{F}_n, \mathcal{S}_{n,m}) \leq \delta(\mathcal{F}_n, \mathcal{F}_m)$, it is clear that $\lim_{n \to \infty} s_n - r_n = 0$. Consequently, $\lim_{n \to \infty} \delta(\mathcal{F}_n, \mathcal{G}) = 0$ and therefore the Cauchy sequence has limit $\mathcal{G}$. $\qquad \square$

Theorem 5.4.16. *Let B be a totally ordered subset of $\Phi(\mathbb{L})$. Then B admits an infimum $\mathcal{T}$ with respect to the order $\preceq$. Further, a subset A of $\Phi(\mathbb{L})$ admits a supremum with respect to the order $\preceq$ if and only if it is bounded with respect to δ.*

Proof. Let B be a totally ordered subset of $\Phi(\mathbb{L})$ and let $r = \inf\{\text{diam}(\mathcal{G}) \mid \mathcal{G} \in B\}$. We can see that all sequences $(\mathcal{G}_n)_{n \in \mathbb{N}}$ of B satisfying $\lim_{n \to \infty} \text{diam}(\mathcal{G}_n) = r$ are Cauchy sequences and admit the same limit $\mathcal{T}$. Therefore, $\mathcal{T}$ is clearly the infimum of B with respect to $\preceq$. Now, let A be a subset of $\Phi(\mathbb{L})$. If it admits a supremum with respect to the order $\preceq$, it is obviously bounded. Conversely, assume that A is bounded. We can clearly find a disk $d(b, s)$ such that all elements of A is secant with $d(b, s)$. Consequently, the circular filter of center b and diameter s surrounds all elements of A. Now, by Proposition 5.4.7, the set A^* of circular filters surrounding all elements of A is totally ordered and therefore admits an infimum $\mathcal{S}$ and we have $\mathcal{G} \preceq \mathcal{S}$ for every $\mathcal{G} \in A$. Consequently, $\mathcal{S}$ is the supremum of A. $\qquad \square$

Chapter 6

Rational Functions and Circular Filters

6.1. Rational functions and algebras $R(D)$

Notations: Recall that $\mathbb{K}$ denotes an algebraically closed complete ultra-metric field and we put $\mathbb{K}^* = \mathbb{K} \setminus \{0\}$. Next, D will be an infinite subset of $\mathbb{K}$.

Rational functions without poles in a set D of the field $\mathbb{K}$ are the only material we handle to define a kind of holomorphic functions in D. But first, we have to know perfectly the properties of rational functions, with regards to the ultrametric structure of $\mathbb{K}$.

Given a function f from D to $\mathbb{K}$, we put $\|f\|_D = \sup\{|f(x)| \mid x \in D\} \in [0, +\infty]$ and we denote by $R(D)$ (resp., $R_b(D)$) the $\mathbb{K}$-algebra of all rational functions $h \in \mathbb{K}(x)$ with no pole in D (resp., the $\mathbb{K}$-algebra of all rational functions $h \in \mathbb{K}(x)$ with no pole in D which are bounded in D).

Given a bounded closed subset of $\mathbb{K}$, we denote by $H(D)$ the completion of $R(D)$ with respect to the norm $\| \cdot \|_D$ of uniform convergence on D [19, 35, 23].

We will now define the antivaluation on $\mathbb{K}$ and on $\mathbb{K}[x]$. The antivaluation is the opposite of the classical valuation defined on these sets, but is easier to follow computations than the classical valuation [28]. Thus, throughout the book, given $a \in \mathbb{K}$, we put $\Psi(a) = \log(|a|)$ and given $P(x) = \sum_{n=0}^{q} a_n x^n \in \mathbb{K}[x]$, and $\mu \in \mathbb{R}$, we put $\Psi(P, \mu) = \sup_{0 \le j \le q} \Psi(a_n) + n\mu$. We then denote by $\nu^+(P, \mu)$ the biggest of the integers j such that $\Psi(a_j) + j\mu = \sup_{0 \le j \le q} \Psi(a_n) + n\mu$ and by $\nu^-(P, \mu)$ the smallest of the integers j such that $P(a_j) + j\mu = \sup_{0 \le j \le q} \Psi(a_n) + n\mu$.

Next, given $h = \frac{P}{q} \in \mathbb{K}(x)$ with $P, Q \in \mathbb{K}[x]$, we put $\Psi(h) = \Psi(P) - \Psi(Q)$.

By results of [23, Chapters 4, Lemmas 4.4, 4.6, 4.8, Theorem 4.11 and Corollaries 4.12 and 4.13], we have the following lemmas.

Lemma 6.1.1. *Let $P(x) \in \mathbb{K}[x]$. Let $r \in \mathbb{R}_+$ and let $a \in \mathbb{K}$ be such that $|a| \leq r$. Then $|P(x)|$ has a limit $\varphi_{a,r}(P)$ when $|x - a|$ approaches r but remains different from r. Moreover, if $P(x) = \sum_{j=0}^n \alpha_j (x - a)^j$, then $\varphi_{a,r}(P) = \max_{0 \leq j \leq n} |\alpha_j| r^j = \|P\|_{d(a,r)}$. Thus, $\varphi_{a,r}$ belongs to $\mathrm{Mult}(\mathbb{K}[x])$ and has continuation to $\mathbb{K}(x)$. Moreover, given $h \in \mathbb{K}(x)$, if $r \in |\mathbb{K}|$, then $\varphi_{a,r}(h) \in |\mathbb{K}|$.*

Corollary 6.1.2. *Let $P(x) = \sum_{j=0}^n \alpha_j x^j \in \mathbb{K}[x]$. Then $\Psi(P, \log r) = \log(\varphi_{0,r}(P))$.*

Lemma 6.1.3. *Let $h \in \mathbb{K}(x)$ and let $r \in \mathbb{R}_+$. For every $a \in d(0, r)$ we have $\lim_{\substack{|x-a| \to r \\ |x-a| \neq 0}} |h(x)| = \varphi_{a,r}(h)$. Let $x \in C(0, r)$. If h has no zeros (resp., no poles) in the class of x in $d(0, r)$ then $|h(x)| \geq \varphi_{0,r}(h)$ (resp., $|h(x)| \leq \varphi_{0,r}(h)$). If h has neither any zeros nor any poles in the class of x inside $d(0, r)$, then $|h(x)| = \varphi_{0,r}(h)$.*

Corollary 6.1.4. *Let $h \in \mathbb{K}(x) \setminus \{0\}$. We have $\Psi(h(x)) = \Psi(h, \Psi(x))$ for every $x \in \mathbb{K}$ such that h has no zero α satisfying $|x - \alpha| < |x|$ and no pole β satisfying $|x - \beta| < |x|$.*

Theorem 6.1.5 ([30]). *For every large circular filter $\mathcal{F}$ on $\mathbb{K}$, for every rational function $h(x) \in \mathbb{K}(x)$, $|h(x)|$ has a limit $\varphi_{\mathcal{F}}(h)$ along the filter $\mathcal{F}$. If $\mathcal{F}$ has center 0 and diameter r, then $\varphi_{\mathcal{F}}(h) = |h|(r)$.*

Lemma 6.1.6. *Let $h \in \mathbb{K}(x) \setminus \{0\}$. The function in μ $\Psi(h, .)$ is continuous and piecewise affine and has at each point μ a left-side derivative $\Psi'^l(h, \mu)$ and a right-side derivative $\Psi'^r(h, \mu)$. Moreover, if $h \in \mathbb{K}[x]$, then $\Psi'^l(h, \mu) = \nu^-(h, \mu)$, and $\Psi'^r(h, \mu) = \nu^+(h, \mu)$.*

Let $\nu \in \mathbb{R}$.

If $d(0, \omega^\nu)$ contains s zeros and t poles of h, (taking multiplicities into account), and if $C(0, \omega^\nu)$ contains neither any zero nor any pole, then $\Psi(h, .)$ has a derivative equal to $s - t$ at ν.

If $C(0, \omega^\nu)$ contains s zeros and t poles of h (taking multiplicities into account), then we have $\Psi'^r(h, \nu) - \Psi'^l(h, \nu) = s - t$. Moreover, if the function $\Psi(h, \mu)$ is not derivable at ν, then ν lies in $\Psi(\mathbb{K})$.

Lemma 6.1.7. *Let $h \in \mathbb{K}(x) \setminus \{0\}$ have s zeros and t poles in $d(0, r)$ (taking multiplicities into account), and have neither any zero nor any pole* *in* $\Gamma(0, r, r')$. *Then in* $\Gamma(0, r, r')$, $\Psi(h(x))$ *is of the form* $A + (s - t)\Psi(x)$, *with* $A = \log(\varphi_{0,r}(h))$.

Corollary 6.1.8. *Let* $h \in \mathbb{K}(x) \setminus \{0\}$. *If* $d(0, r_1)$ *contains* s *zeros and* t *poles of* h *(taking multiplicities into account) and if* $\Gamma(0, r_1, r_2)$ *contains neither any zero nor any pole of* h, *then* $|h(x)| = \left(\frac{|x|}{r_2}\right)^{s-t} \varphi_{0,r_2}(h) = \left(\frac{|x|}{r_1}\right)^{s-t} \varphi_{0,r_1}(h) \ \forall x \in \Gamma(0, r_1, r_2)$.

Theorem 6.1.9. *Let* $h \in \mathbb{K}(x) \setminus \{0\}$. *If* $r \in |\mathbb{K}|$, *then* $\varphi_{0,r}(h)$ *lies in* $|\mathbb{K}|$. *Conversely, if* h *has* s *zeros and* t *poles in* $d(0, r)$, *if* $s \neq t$ *with* $\varphi_{0,r}(h) \in |\mathbb{K}|$, *then* r *lies* $|\mathbb{K}|$.

Proof. The direct claim is an immediate consequence of Lemma 6.1.7. So we will show the converse claim and suppose that $\varphi_{0,\rho}(h) \in |\mathbb{K}|$ and that $s \neq t$. If h admits zeros or poles in $C(0, r)$ then of course r lies in $|\mathbb{K}|$. So, we assume that h has neither any zero nor any pole in $C(0, r)$. Therefore, we can find $\rho \in]0, r[\cap |\mathbb{K}|$ such that h has neither any zero nor any pole in $\Lambda(0, \rho, r)$, hence h admits exactly s zeros and t poles in $d(0, \rho)$. And by Lemmas 6.1.3 and 6.1.8 we have $\varphi_{0,r}(h) = \varphi_{0,\rho}(h)\left(\frac{r}{\rho}\right)^{s-t}$. But since $\rho \in |\mathbb{K}|$ and since $s - t \neq 0$, by the direct claim it is clear that $\frac{\rho^{s-t}}{\varphi_{0,r}(h)}$ also lies in $|\mathbb{K}|$, and so does r. $\qquad\square$

Notation: Henceforth, we denote by U the disk $d(0, 1)$ in $\mathbb{K}$. Given a disk $A = d(a, r)$ or $A = d(a, r^-)$, we put $\varphi_A = \varphi_{a,r}$.

Theorem 6.1.10. *Let* $h \in \mathbb{K}(x)$ *satisfy* $\deg(h) > 0$. *Then the set* $h^{-1}(U) = \{x \in \mathbb{K} \mid |h(x)| \leq 1\}$ *is affinoid.*

Proof. Let $D = h^{-1}(U)$ and let $h = \frac{P}{Q}$, with P, $Q \subset \mathbb{K}[x]$ relatively prime, and of course $\deg(P) > \deg(Q)$. We first notice that D is bounded. Let $\Gamma(a, l', l'')$, with $a \in D$, be an empty annulus of D. Suppose that h has no zeros and no poles in $d(a, l')$. There exists $l > l'$ such that h admits no zeros and no poles in all $d(a, l)$, hence by Lemma 6.1.2, $|h(x)|$ is constant in all $d(a, l)$, hence $|h(x)| = |h(a)| \leq 1 \ \forall x \in d(a, l)$, thereby $d(a, l) \subset D$, a contradiction. Thus, for each empty annulus F of D, $\mathcal{I}(F)$ contains at least a zero or a pole of h. Consequently, D admits finitely many empty annuli and therefore by Theorem 5.1.13, D has finitely many infraconnected components.

Let E be an infraconnected component of D, of diameter r, and let $a \in E$. Let t (resp., s) be the number of poles (resp., zeros) of h inside $\widetilde{E}$.

Since E is infraconnected, the circular filter of center a and diameter r is secant with E and thereby $\varphi_{a,r}(h) \leq 1$. It is also secant with $\mathbb{K} \setminus \tilde{E}$, hence $\varphi_{a,r}(h) \geq 1$, therefore $\varphi_{a,r}(h) = 1$. On the other hand, we have $s - t > 0$ because if $s - t \leq 0$, then by Lemma 6.1.7, there exists $r' > r$ such that $|h(x)| = \varphi_{a,r}(h) \; \forall x \in \Gamma(a, r, r')$, a contradiction to the hypothesis $D \subset d(a, r)$. Consequently, by Lemma 6.1.7, the equality $\varphi_{a,r}(h) = 1$ implies $r \in |\mathbb{K}|$.

Similarly, consider now a hole $T = d(b, \rho^-)$ of E. Let $l = \varphi_{b,\rho}(h)$. Since E is infraconnected, the circular filter of center b and diameter ρ is secant with E and thereby $\varphi_{b,\rho}(h) \leq 1$. It is also secant with T, hence $\varphi_{b,\rho}(h) \geq 1$, therefore $\varphi_{b,\rho}(h) = 1$. Then by Lemma 6.1.3, T contains at least one pole of h. Thus, E has finitely many holes. Moreover, since $|h(x)| > 1 \; \forall x \in T$, by Lemma 6.1.8, inside T the number of poles of h is strictly bigger than the number of zeros. Consequently by Lemma 6.1.7, ρ lies in $|\mathbb{K}|$. $\qquad \square$

Theorem 6.1.11. *Let D be affinoid. There exists $h \in R(D)$ such that $\deg(h) > 0$, $h(D) = U$, $D = h^{-1}(U)$.*

Proof. Suppose first that D is infraconnected. Then D is of the form $d(a, r_0) \setminus \bigcup_{j=1}^n d(b_j, r_j^-)$, with $|b_i - b_j| \geq \max(r_i, r_j)$, $\forall i \neq j$, $b_j \in d(a, r_0)$ $r_j \leq r_0 \; \forall j = 1, \ldots, n$, $r_j \in |\mathbb{K}| \; \forall j = 0, \ldots, n$. For each $j = 0, \ldots, n$, we can take $\lambda_j \in \mathbb{K}$ such that $|\lambda_j| = r_j$. Let $h(x) = \frac{x-a}{\lambda_0} + \sum_{j=1}^n \frac{\lambda_j}{x-b_j}$. Outside $d(b-j, r_j^-)$, we have $|\frac{\lambda_j}{x-b_j}| \leq 1$, inside $d(b-j, r_j^-)$, we have $|\frac{\lambda_j}{x-b_j}| > 1$. Inside $d(a, r_0)$ we have $|\frac{x-a}{\lambda_0}| \leq 1$, outside $d(a, r_0)$ we have $|\frac{x-a}{\lambda_0}| > 1$. Consequently, we check that $|h(x)| \leq 1$ if and only if $x \in D$, so we have constructed h when D is infraconnected. Moreover, by construction, we check that $\lim_{|x| \to \infty} |h(x)| = +\infty$, so $\deg(h) > 0$.

We now consider the general case and denote by $D_1, \ldots, D_q$ the infraconnected components of D. For each $i = 1, \ldots, q$, there exists $h_i \in R(D_i)$ such that $|h(x)| \leq 1$ if and only if $x \in D_i$. We then put $g = \sum_{i=1}^n \frac{1}{h_i}$ and $h = \frac{1}{g}$. We can easily check that $|g(x)| \geq 1$ if and only if $x \in D$. Moreover, we notice that for each $i = 1, \ldots, n$ we have $\lim_{|x| \to \infty} |h_i(x)| = +\infty$, hence $\lim_{|x| \to \infty} |g(x)| = 0$, therefore $\lim_{|x| \to \infty} |h(x)| = +\infty$. So, $\deg(h) > 0$ and h satisfies our claim. $\qquad \square$

According to [23, Theorem 9.3] we have Theorem 6.1.12.

Theorem 6.1.12. $R_b(D) = R(D)$ *if and only if D is closed and bounded. Moreover, if D is closed and bounded, $\| \cdot \|_D$ is a semi-multiplicative ultrametric norm of $\mathbb{K}$-algebra on $R_b(D)$.*

6.2. Pointwise topology on $\text{Mult}(\mathbb{K}[x])$

As we saw in Section 2.3 of Chapter 2, $\text{Mult}(\mathbb{K}[x])$ is equipped with the topology of pointwise convergence. On the other hand, according to Section 5.3 of Chapter 5, circular filters are equipped with a distance. Now, we will see that the circular filters characterize the multiplicative semi-norms on $\mathbb{K}[x]$. Thus, $\text{Mult}(\mathbb{K}[x])$ admits three topologies that we shall compare.

Notation: We denote by Ω the mapping from $\Phi(\mathbb{K})$ into $\text{Mult}(\mathbb{K}[x])$ defined as $\Omega(\mathcal{F}) = \varphi_{\mathcal{F}}$.

Throughout the section, D will denote a subset of $\mathbb{K}$ and $\mathcal{U}_D$ will denote the topology of uniform convergence on D.

The characterization of all absolute values on $\mathbb{K}(x)$ is given in [30, 32, 23]. According to [23, Theorems 4.14], we have this theorem.

Theorem 6.2.1. (Garandel–Guennebaud). *The mapping M from $\Phi(\mathbb{K})$ into $\text{Mult}(\mathbb{K}[x])$ defined as $\Omega(\mathcal{F}) = \varphi_{\mathcal{F}}$ is a bijection. Moreover, the restriction Ω' of Ω to $\Phi'(\mathbb{K})$ is a bijection from $\Phi'(\mathbb{K})$ onto the set of multiplicative norms on $\mathbb{K}[x]$, and therefore onto the set of multiplicative norms on $\mathbb{K}(x)$. Moreover, if $\mathcal{F}$ has center a and diameter r, then $\varphi_{\mathcal{F}}(h) = \varphi_{a,r}(h) \ \forall h \in \mathbb{K}(x)$.*

Definitions: A multiplicative semi-norm on $\mathbb{K}[x]$ defined by a punctual circular filter of limit a will be called *a punctual multiplicative semi-norm* and will be denoted by φ_a.

Remarks. The punctual multiplicative semi-norm φ_a is just of the form $\phi(P) = |P(a)|$, with $a \in \mathbb{K}$. In a field F which is complete but not algebraically closed, such as $\mathbb{Q}_p$, we have to consider the (algebraically closed) completion $\mathbb{G}$ of an algebraic closure. Then every element ψ of $\text{Mult}(\mathbb{G}[x])$ has a restriction to $F[x]$ which obviously belongs to $\text{Mult}(F[x])$. And conversely, every element of $\text{Mult}(F[x])$ admits extensions to $\text{Mult}(\mathbb{G}[x])$. Therefore, we have a surjection from the set of circular filters on $\mathbb{G}$ onto $\text{Mult}(F[x])$. But this surjection is not injective: if a and b are conjugate over F, the circular filters of centers a and b and of same diameter r define the same absolute values on $F[x]$.

Theorem 6.2.2. $\text{Mult}_1(\mathbb{K}[x]) = \text{Mult}_a(\mathbb{K}[x]) = \text{Mult}_m(\mathbb{K}[x])$ *is dense in* $\text{Mult}(\mathbb{K}[x])$ *with respect to the topology of pointwise convergence.*

Proof. It is well known that all maximal ideals of $\mathbb{K}[x]$ have codimension 1, hence $\text{Mult}_a(\mathbb{K}[x]) = \text{Mult}_m(\mathbb{K}[x])$. Now, let $\varphi_{\mathcal{F}} \in \text{Mult}(\mathbb{K}[x])$. Let $P_1, \ldots, P_n \in \mathbb{K}[x]$ and let $\epsilon > 0$. For each $j = 1, \ldots, n$, there exists $E_j \in \mathcal{F}$

such that $|P_j(x) - \varphi_{\mathcal{F}}(P_j)|_\infty \le \epsilon \; \forall x \in E_j$. Let $E = \cap_{j=1}^n E_j$. Then E lies in $\mathcal{F}$ and we have $|P_j(x) - \varphi_{\mathcal{F}}(P_j)|_\infty \le \epsilon \; \forall x \in E$, which shows that φ_a belongs to the neighborhood $\mathcal{W}(\varphi_{\mathcal{F}}, P_1, \ldots, P_n, \epsilon)$ for every $a \in E$. □

Theorem 6.2.3. *Let D be infraconnected, closed, bounded. Then* $\mathrm{Mult}(R(D), \| \cdot \|_D)$ *is sequentially compact with respect to the topology of pointwise convergence.*

Proof. Let $(\psi_n)_{n\in\mathbb{N}}$ be a sequence of $\mathrm{Mult}(R(D), \| \cdot \|_D)$. Since $\mathrm{Mult}(R(D), \| \cdot \|_D)$ is compact, the sequence $(\psi_n)_{n\in\mathbb{N}}$ admits a point of adherence $\varphi_{\mathcal{F}}$. Let $r = \mathrm{diam}(\mathcal{F})$. If $\mathcal{F}$ has no center, or if it has a center and if $r \notin |\mathbb{K}|$, by Lemma 5.3.2 of Chapter 5, it has a countable basis, hence there does exist a subsequence of the sequence $(\psi_n)_{n\in\mathbb{N}}$ converging to $\varphi_{\mathcal{F}}$. Now suppose that $\mathcal{F}$ has a center a and that $r \in |\mathbb{K}|$. Suppose that we can extract from the sequence (ψ_n) a subsequence $(\psi_{m_t})_{t\in\mathbb{N}}$, with $\psi_{m_t} = \varphi_{\mathcal{F}_t}$, $(t \in \mathbb{N})$, such that for a certain fixed $b \in d(a, r)$, each $\mathcal{F}_t$ is (b, r_t)-approaching with $r_t \neq r$. Consider now $x - b$. For convenience, we put $\phi_t = \psi_{m_t}$ $(t \in \mathbb{N})$. By hypothesis we can reextract from this sequence $(\phi_t)_{t\in\mathbb{N}}$ a new subsequence $(\phi_{t_q})_{q\in\mathbb{N}}$ such that $\lim_{q\to\infty} \phi_{t_q}(x - b) = \varphi_{\mathcal{F}}(x - b)$. This clearly proves that $\lim_{q\to\infty} r_{t_q} = r$, whereas $r_{t_q} \neq r$. Consequently, for every element A of $\mathcal{F}$ we can find a rank $s \in \mathbb{N}$ such that $\mathcal{F}_{t_s}$ is secant with A. Therefore, for every $\epsilon > 0$ and for every $f \in R(D)$ we can find a rank $s \in \mathbb{N}$ such that $|\phi_{t_s}(f) - \varphi_{\mathcal{F}}(f)|_\infty \le \epsilon$, hence the sequence (ϕ_{t_s}) converges to $\varphi_{\mathcal{F}}$.

Finally, suppose that we cannot find $b \in d(a, r)$ and a subsequence $(\psi_{m_t})_{t\in\mathbb{N}}$, with $\psi_{m_t} = \varphi_{\mathcal{F}_t}$, $(t \in \mathbb{N})$, such that for a certain fixed $b \in d(a, r)$, each $\mathcal{F}_t$ be (b, r_t)-approaching with b and diameter $r_t \neq r$. We first notice that there exists $r' > r$ such that none of the $\varphi_{\mathcal{F}_t}$ is secant with $\Gamma(a, r, r')$. In the same way, for every $c \in d(a, r)$, there exists $\rho(c) \in]0, r[$ such that none of the $\phi_{\mathcal{F}_t}$ is secant with $\Gamma(a, \rho, r)$. Let $(c_t)_{t\in\mathbb{N}}$ be a sequence in $d(a, r)$ such that $|c_j - c_k| = r \; \forall j \neq k$. We can easily construct a decreasing sequence of affinoid sets $(A_t)_{t\in\mathbb{N}}$ of the form $A_t = d(a, r') \backslash \left(\bigcup_{j=1}^t d(c_j, \rho(c_j)^-) \right)$. For each $t \in \mathbb{N}$, let $\epsilon_t = \min(\frac{1}{t}, r' - r, r - \rho(c_1), \ldots, r - \rho(c_t))$. By hypothesis, for each $t \in \mathbb{N}$, there exists $n_t \in \mathbb{N}$ such that $|\psi_{n_t}(x - c_t) - \varphi_{\mathcal{F}}(x - c_t)|_\infty < \epsilon_t$, hence, putting $\psi_{n_t} = \varphi_{\mathcal{F}_t}$, we can see that $\mathcal{F}_t$ is secant with $d(a, r')$ but is not secant secant with $\Gamma(a, r, r')$ and is not secant with $\bigcup_{j=1}^t d(c_j, r^-)$. Consequently, $\mathcal{F}_t$ is secant with a class of $d(a, r)$ different from all $d(c_j, r) \; \forall j = 1, \ldots, t$. So, we can construct a sequence $(b_t)_{t\in\mathbb{N}}$ such that $|b_j - a| = |b_j - b_k| = r \; \forall j \neq k$ and such that $\mathcal{F}_t$ is secant with $d(b_t, r^-)$. □

Now, let $f \in R(D)$. For each $t \in \mathbb{N}$, since $\mathcal{F}_t$ is secant with $d(b_t, r^-)$, we can find $a_t \in d(b_t, r^-)$ such that $|\psi_{n_t}(f) - |f(a_t)|\|_\infty < \epsilon_t$. Then the sequence

$(a_t)_{t \in \mathbb{N}}$ is thinner than $\mathcal{F}$, and therefore $\lim_{t \to \infty} |f(a_t)| = \varphi_{\mathcal{F}}(f)$. But of course $\lim_{t \to \infty} |f(a_t)| = \lim_{t \to \infty} \psi_{n_t}(f)$, so the sequence (ψ_{n_t}) converges to $\varphi_{\mathcal{F}}$ again.

Notations: In Section 5.4 of Chapter 5, the set of circular filters was equipped with an order relation $\preceq$ that makes it a tree, and we have denoted by $\prec$ the strict order associated to $\preceq$. On the other hand, we can apply to $\mathrm{Mult}(\mathbb{K}[x])$ the usual order $\leq$ on functions with values in $\mathbb{R}$ as $\varphi_{\mathcal{F}} \leq \varphi_{\mathcal{G}}$ if $\varphi_{\mathcal{F}}(P) \leq \varphi_{\mathcal{G}}(P) \ \forall P \in \mathbb{K}[x]$, and we denote by $<$ the strict order associated to $\leq$, as $\varphi_{\mathcal{F}} < \varphi_{\mathcal{G}}$ if $\varphi_{\mathcal{F}}(P) \leq \varphi_{\mathcal{G}}(P) \ \forall P \in \mathbb{K}[x]$ and $\varphi_{\mathcal{F}} \neq \varphi_{\mathcal{G}}$.

Theorem 6.2.4. *Let $\varphi_{\mathcal{F}}, \varphi_{\mathcal{G}} \in \mathrm{Mult}(\mathbb{K}[x])$. Then $\varphi_{\mathcal{F}} < \varphi_{\mathcal{G}}$ if and only if $\mathcal{F} \prec \mathcal{G}$.*

Proof. Let $\mathcal{F}, \mathcal{G} \in \Phi(\mathbb{K})$. Suppose first that they satisfy $\mathcal{F} \prec \mathcal{G}$ and let $s = \mathrm{diam}(\mathcal{G})$. By Corollary 5.4.2 of Chapter 5, $\mathcal{F}$ is secant with a disk $d(a, r)$ strictly included in $d(b, s)$. Let $P \in \mathbb{K}[x]$. We have $\varphi_{\mathcal{F}}(P) \leq \|P\|_{d(a,r)} \leq \|P\|_{d(a,s)} = \varphi_{\mathcal{G}}(P) \ \forall P \in \mathbb{K}[x]$. Now, let $Q(x) = x - a$. Then $\varphi_{\mathcal{F}}(Q) \leq \|Q\|_{d(a,r)} = r$ and $\varphi_{\mathcal{G}}(Q) = \|Q\|_{d(a,s)} = s$, hence $\varphi_{\mathcal{F}}(Q) < \varphi_{\mathcal{G}}(Q)$. Suppose that $\mathcal{F}$ and $\mathcal{G}$ are not comparable with respect to $\preceq$. By Proposition 5.4.9 of Chapter 5, there exist disks $F = d(a, r) \in \mathcal{F}$ and $G = d(b, s) \in \mathcal{G}$ such that $F \cap G = \emptyset$ and $\delta(F, G) > \max(\mathrm{diam}(\mathcal{F}), \mathrm{diam}(\mathcal{G}))$. Consider $P(x) = x - a$ and $Q(x) = x - b$. Then $\varphi_{\mathcal{F}}(P) \leq \|P\|_F = r$ and $\varphi_{\mathcal{G}}(P) = \|P\|_G = |b - a|$, hence $\varphi_{\mathcal{F}}(P) < \varphi_{\mathcal{G}}(P)$. In the same way, $\varphi_{\mathcal{G}}(Q) \leq \|P\|_G = s$ and $\varphi_{\mathcal{F}}(Q) = \|P\|_F = |b - a|$, hence $\varphi_{\mathcal{G}}(Q) < \varphi_{\mathcal{F}}(Q)$. Thus, neither $\varphi_{\mathcal{F}} \leq \varphi_{\mathcal{G}}$ nor $\varphi_{\mathcal{G}} \leq \varphi_{\mathcal{F}}$ is true whenever $\mathcal{F}$ and $\mathcal{G}$ are not comparable. This finishes proving that $\varphi_{\mathcal{F}} < \varphi_{\mathcal{G}}$ if and only if $\mathcal{F} \prec \mathcal{G}$. $\square$

Corollary 6.2.5. *Ω is a strictly increasing bijection from $\Phi(\mathbb{K})$ onto $\mathrm{Mult}(\mathbb{K}[x])$ and Ω^{-1} is a strictly increasing bijection from $\mathrm{Mult}(\mathbb{K}[x])$ onto $\Phi(\mathbb{K})$.*

Corollary 6.2.6. *Let D be closed. The restriction of M to the set of $\mathcal{F} \in \Phi(\mathbb{K})$ which are not converging to a point of $\mathbb{K} \setminus D$ is a bijection from this set onto $\mathrm{Mult}(R(D))$.*

According to [30, Theorem 10.4] we have Theorem 6.2.7.

Theorem 6.2.7. (Garandel). *Let $\mathcal{F}$ be a large circular filter on $\mathbb{K}$. Then $\varphi_{\mathcal{F}}$ belongs to $\mathrm{Mult}(R(D), \mathcal{U}_D)$ if and only if $\mathcal{F}$ is secant with D.*

Corollary 6.2.8. *The mapping from $\Phi(D)$ to $\mathrm{Mult}(R(D))$ which associates to each circular filter $\mathcal{F}$ secant with D the multiplicative semi-norm $\varphi_{\mathcal{F}}$, is a bijection from $\Phi(D)$ onto $\mathrm{Mult}(R(D), \mathcal{U}_D)$.*

Definitions and notations: According to Section 1.2 of Chapter 1, the function diam defines two equivalent distances on the set of circular filters, in particular $\delta(\mathcal{F}, \mathcal{G}) = \text{diam}(\sup(\mathcal{F}, \mathcal{G}) - \min(\text{diam}(\mathcal{F}), \text{diam}(\mathcal{G}))$. Henceforth, we will apply this distance to $\text{Mult}(\mathbb{K}[x])$ and to $\text{Mult}(R(D), \mathcal{U}_D)$ as well as to $\Phi(\mathbb{K})$ or $\Phi(D)$ by putting $\delta(\varphi_{\mathcal{F}}, \varphi_{\mathcal{G}}) = \delta(\mathcal{F}, \mathcal{G})$. Thus, $\text{Mult}(\mathbb{K}[x])$ and $\text{Mult}(R(D), \mathcal{U}_D)$ are equipped with a metric topology defined by δ.

Given $h = \frac{P}{Q} \in \mathbb{K}(x)$, with $P, Q \in \mathbb{K}[x]$, we put $\deg(h) = \deg(P) - \deg(Q)$, and $\deg(h)$ will be called *the algebraic degree* of h.

Theorem 6.2.9. *Let $\varphi_{\mathcal{F}} \in \text{Mult}(\mathbb{K}(x))$. Let $f \in \mathbb{K}(x)$, and let $\epsilon > 0$. There exists an affinoid subset E of $\mathbb{K}$ of diameter $l > \text{diam}(\mathcal{F})$, which belongs to $\mathcal{F}$, such that $| |f(x)| - \varphi_{\mathcal{F}}(f)|_\infty \le \epsilon, \ \forall x \in E$.*

Proof. If $\mathcal{F}$ has no center, there exists a disk $d(a, l) \in \mathcal{F}$, with $r \in |\mathbb{K}|$, containing neither zeros nor poles of f, therefore by Lemma 6.1.3, $|f(x)|$ is a constant equal to $\varphi_{\mathcal{F}}(f)$ in $d(a, r)$, so our claim is obvious. Now, suppose that $\mathcal{F}$ is the circular filter of center a and diameter r. Let $A_1, \dots, A_q$ be the classes of $d(a, r)$ containing at least one zero or one pole of f. By Lemma 6.1.3, $|f(x)|$ is a constant equal to $\varphi_{\mathcal{F}}(f)$ in $d(a, r) \setminus \left(\bigcup_{j=1}^q A_j \right)$. Consider a class $A_j = d(a_j, r^-)$, and let s_j (resp., t_j) be the number of zeros (resp., poles) of f inside A_j, and let s_0 (resp., t_0) be the number of zeros (resp., poles) of f in all $d(a, r)$. Let $\rho \in]0, r[\cap|\mathbb{K}|$ be such that $\left| \left(\frac{r}{\rho} \right)^{s_j - t_j} - 1 \right) | \varphi_{\mathcal{F}}(f) \le \epsilon \ \forall j = 0, \dots, q$. Let $l = \frac{r^2}{\rho}$ and let $E = d(a, l) \setminus \bigcup_{j=1}^q d(a_j, \rho^-)$. By Lemmas 6.1.7 and Corollary 6.1.8, we can check that the inequality $| |f(x)| - \varphi_{\mathcal{F}}(f)|_\infty \le \epsilon$ holds in all E. Since $\rho < r$, E is an affinoid set which belongs to $\mathcal{F}$. Moreover, by definition, $l > \text{diam}(\mathcal{F})$. $\qquad\square$

Theorem 6.2.10. *Let D be a closed bounded set and $\varphi_{\mathcal{F}} \in \text{Mult}(R(D), \| \cdot \|_D)$. There exists a basis of neighborhoods of $\varphi_{\mathcal{F}}$ in $\text{Mult}(R(D), \| \cdot \|_D)$, with respect to the topology of pointwise convergence, consisting of the family of sets of the form $\text{Mult}(R(E \cap D), \| \cdot \|_{E \cap D})$, where E is a $\mathcal{F}$-affinoid.*

Proof. Let $f_1, \dots, f_q \in R(D)$ and consider a neighborhood W of $\varphi_{\mathcal{F}}$: we will show that it contains a neighborhood of the form $\text{Mult}(R(E \cap D), \| \cdot \|_{E \cap D})$, where E is a $\mathcal{F}$-affinoid. Indeed, by Lemma 6.2.9, there exist affinoid sets $E_j \in \mathcal{F}$, $(1 \le j \le q)$ such that $| |f_j(x)| - \varphi_{\mathcal{F}}(f_j)|_\infty \le \epsilon \ \forall x \in E_j$ $(1 \le j \le q)$. Then $\bigcap_{j=1}^q E_j$ is an infraconnected affinoid set which belongs to $\mathcal{F}$. We can put $E = \bigcap_{j=1}^q E_j$. So, we have an affinoid set F such that $| |f_j(x)| - \varphi_{\mathcal{F}}(f_j)|_\infty \le \epsilon \ \forall x \in E_j$ $(1 \le j \le q)$, and therefore $| |\psi(f_j)| - \varphi_{\mathcal{F}}(f_j)|_\infty \leq \epsilon \ \forall x \in E_j \ (1 \leq j \leq q), \ \forall \psi \in$ Mult$(R(E \cap D), \| \ . \ \|_{E \cap D})$. Now, by Lemma 5.3.2 of Chapter 5, F contains a $\mathcal{F}$-affinoid E. Then Mult$(R(E), \| \ . \ \|_E)$ is a neighborhood of $\varphi_{\mathcal{F}}$ in Mult$(\mathbb{K}(x))$ and Mult$(R(E \cap D), \| \ . \ \|_{E \cap D})$ is a neighborhood of $\varphi_{\mathcal{F}}$ in Mult$(R(D), \| \ . \ \|_D)$. $\qquad\square$

Corollary 6.2.11. *Let $\varphi_{\mathcal{F}} \in$ Mult$(\mathbb{K}[x])$. There exists a basis of neighborhoods of $\varphi_{\mathcal{F}}$ in Mult$(\mathbb{K}[x])$ consisting of the family of sets of the form* Mult$(R(E), \| \ . \ \|_E)$*, where E is a $\mathcal{F}$-affinoid.*

Corollary 6.2.12. Mult$(\mathbb{K}[x])$ *is locally compact and locally sequentially compact.*

Proof. Let $\varphi_{\mathcal{F}} \in$ Mult$(\mathbb{K}[x])$ and let E be an affinoid set such that Mult$(R(E), \| \ . \ \|_E)$ is a neighborhood of $\varphi_{\mathcal{F}}$ in Mult$(\mathbb{K}[x])$. Then Mult$(R(E), \| \ . \ \|_E)$ is a compact neighborhood of $\varphi_{\mathcal{F}}$. Moreover, it is a sequentially compact neighborhood. $\qquad\square$

Remark 1. In general, the metric Θ studied in Section 2.1 of Chapter 2, defines a topology that is not equivalent to the pointwise topology on Mult$(A, \| \ . \ \|)$ as the following example shows.

Let $\mathbb{L} = \mathbb{K}$ and let A be the Banach $\mathbb{K}$-algebra $H(d(0,1))$ provided with the norm of uniform convergence on $d(0,1)$. Let $r \in |\mathbb{K}| \cap]0,1[$ and for each $n \in \mathbb{N}^*$, set $r_n = \sqrt[n]{r}$. We know that A is the algebra of analytic elements on $d(0,1)$ [1, 35] and for every $s \in]0,1]$, the mapping defined on A as $\psi(f) = \lim_{|x| \to s, |x| \neq s} |f(x)|$ belongs to Mult$(A, \| \ . \ \|)$ [3, 30, 32]. Then it is well known and easily checked that the norm of uniform convergence $\phi = \| \ . \ \|$ on $d(0,1)$ admits the sets $\{\psi_s \mid s \leq 1\}$ as a basis of neighborhoods with respect to the pointwise topology [28, 30].

Now, let ϕ_n be the element of Mult$(A, \| \ . \ \|)$ defined by $\phi_n(f) = \lim_{|x| \to r_n, |x| \neq r_n} |f(x)|$. Then the sequence $(\phi_n)_{n \in \mathbb{N}}$ tends to $\| \ . \ \|$ when n goes to $+\infty$, with respect to the pointwise convergence on Mult$(A, \| \ . \ \|)$.

However, let $P_n(x) = x^n$, $n \in \mathbb{N}^*$ and consider $\phi(P_n) - \phi_n(P_n) = 1 - (r_n)^n = 1 - r$. Since P_n lies in A_0 for every $n \in \mathbb{N}$, we have $\Theta(\phi, \phi_n) \geq 1 - r$ and hence ϕ_n does not tend to ϕ with respect to the metric Θ.

6.3. Topologies on Mult$(\mathbb{K}[x])$

In Section 5.4 of Chapter 5, we defined the δ-topology on $\Phi(\mathbb{K})$ and we showed that $\Phi(\mathbb{K})$ is complete with respect to this metric. In Section 6.2, we saw that $\Phi(\mathbb{K})$ is in bijection with Mult$(\mathbb{K}[x])$ which is equipped with the topology of pointwise convergence. Henceforth, thanks to this one to one correspondence, we will consider that both $\Phi(\mathbb{K})$ and $\text{Mult}(\mathbb{K}[x])$ are equipped with both topologies.

Definition and notations: We will denote by $\Sigma(D)$ the set of $\varphi_{\mathcal{F}} \in \text{Mult}(\mathbb{K}[x])$ such that $\mathcal{F}$ is D-bordering, and by $\Sigma_0(D)$ the set of $\varphi_{\mathcal{F}} \in \text{Mult}(\mathbb{K}[x])$ such that $\mathcal{F}$ is strictly D-bordering.

Theorem 6.3.1. *Let D be closed and bounded. Then the boundary of* $\text{Mult}(R(D), \| \cdot \|_D)$ *inside* $\text{Mult}(\mathbb{K}[x])$, *with respect to the topology of pointwise convergence, is equal to* $\Sigma(D)$.

Proof. Let $\mathcal{F}$ be D-bordering. Let W be a neighborhood of $\varphi_{\mathcal{F}}$ in $\text{Mult}(\mathbb{K}[x])$. By Theorem 6.2.10, there exists an affinoid set $E \in \mathcal{F}$ such that $\text{Mult}(R(E), \| \cdot \|_E) \subset W$. Since $E \cap (\mathbb{K} \setminus D) \neq \emptyset$, for all $a \in E \cap (\mathbb{K} \setminus D)$, φ_a belongs to $\text{Mult}(R(E), \| \cdot \|_E)$, which shows that $\varphi_{\mathcal{F}}$ belongs to the closure of $\text{Mult}(\mathbb{K}[x]) \setminus \text{Mult}(R(D), \| \cdot \|_D)$. Hence,

$$\Sigma(D) \subset \text{Mult}(\mathbb{K}[x]) \setminus \text{Mult}(R(D), \| \cdot \|_D).$$

Conversely, let $\varphi_{\mathcal{F}} \in \text{Mult}(R(D), \| \cdot \|_D)$ belong to the closure of $\text{Mult}(\mathbb{K}[x]) \setminus \text{Mult}(R(D), \| \cdot \|_D)$ in $\text{Mult}(\mathbb{K}[x])$. Let $E \in \mathcal{F}$ be an affinoid set. Then $\text{Mult}(R(E), \| \cdot \|_E)$ is a neighborhood of $\varphi_{\mathcal{F}}$, hence it contains an open neighborhood W of $\varphi_{\mathcal{F}}$. Therefore,

$$W \cap (\text{Mult}(\mathbb{K}[x]) \setminus Mult(R(D), \| \cdot \|_D)) \neq \emptyset.$$

Moreover, $W \cap (\text{Mult}(\mathbb{K}[x]) \setminus \text{Mult}(R(D), \| \cdot \|_D))$ is open in $\text{Mult}(\mathbb{K}[x])$ because so are W and $\text{Mult}(\mathbb{K}[x]) \setminus \text{Mult}(R(D), \| \cdot \|_D)$. Since $\text{Mult}_a(\mathbb{K}[x])$ is dense inside $\text{Mult}(\mathbb{K}[x])$, there exists $\varphi_a \in (\text{Mult}_a(\mathbb{K}[x]) \setminus \text{Mult}(R(D), \| \cdot \|_D)) \cap \text{Mult}(R(E), \| \cdot \|_E)$. Consequently, a lies in E, which proves that $\mathcal{F}$ is secant with $\mathbb{K} \setminus D$. Therefore, the boundary of $\text{Mult}(R(D), \| \cdot \|_D)$ inside $\text{Mult}(\mathbb{K}[x])$ is included in $\Sigma(D)$, and finally these two sets are equal. $\qquad \square$

Theorem 6.3.2. *Let D be closed, bounded and infraconnected. Then $\Sigma_0(D)$ is included in the boundary of* $\text{Mult}(R(D), \| \cdot \|_D)$ *inside* $\text{Mult}(\mathbb{K}[x])$, *with respect to the δ-topology. Moreover, the boundary of* $\text{Mult}(R(D), \| \cdot \|_D)$ *inside* $\text{Mult}(\mathbb{K}[x])$, *with respect to the δ-topology, is equal to $\Sigma(D)$ if and only if for every $\varphi_{\mathcal{G}} \in \Sigma(D) \setminus \Sigma_0(D)$, there exists either a monotonous distances holes sequence or an equal distances holes sequence which is thinner than $\mathcal{G}$, whose superior gauge is equal to its diameter.*

Proof. Let $\tilde{D} = d(a, r)$, with $a \in D$. Let $\varphi_{\mathcal{G}} \in \Sigma(D)$. Suppose first that $\mathcal{G}$ is the peripheral of D, hence $\varphi_{\mathcal{G}}$ lies in the closure of $\{\varphi_{a,s} \mid s > r\}$ for both topologies, hence it lies in the closure of $\mathrm{Mult}(\mathbb{K}[x]) \setminus \mathrm{Mult}(R(D), \| \cdot \|_D)$ for both topologies. In the same way, $\varphi_{\mathcal{G}}$ lies in the closure of $\{\varphi_{a,s} \mid 0 \leq s \leq r\}$ for both topologies. But since D is infraconnected, for every $s \in [0, r]$, the circular filter of center a and diameter s is secant with D, and then, by Corollary 6.2.8, $\varphi_{a,s}$ belongs to $\mathrm{Mult}(R(D), \| \cdot \|_D)$. Consequently, $\varphi_{\mathcal{G}}$ lies in the closure of $\mathrm{Mult}(R(D), \| \cdot \|_D)$ for both topologies. This shows that $\varphi_{\mathcal{G}}$ belongs to the boundary of $\mathrm{Mult}(R(D), \| \cdot \|_D)$ inside $\mathrm{Mult}(\mathbb{K}[x])$ with respect to the δ-topology.

Similarly, suppose now that $\mathcal{G}$ is the peripheral of a hole $T = d(b, l^-)$ of D. Then $\varphi_{\mathcal{G}}$ lies in the closure of $\{\varphi_{b,s} \mid s < l\}$ for both topologies, hence it lies in the closure of $\mathrm{Mult}(\mathbb{K}[x]) \setminus \mathrm{Mult}(R(D), \| \cdot \|_D)$ for both topologies. In the same way, $\varphi_{\mathcal{G}}$ lies in the closure of $\{\varphi_{a,s} \mid l \leq s \leq r\}$ for both topologies. But since D is infraconnected, for every $s \in [l, r]$, the circular filter of center b and diameter s is secant with D, and then, by Corollary 6.2.8, $\varphi_{b,s}$ belongs to $\mathrm{Mult}(R(D), \| \cdot \|_D)$. Consequently, $\varphi_{\mathcal{G}}$ lies in the closure of $\mathrm{Mult}(R(D), \| \cdot \|_D)$ for both topologies and therefore belongs to the boundary of $\mathrm{Mult}(R(D), \| \cdot \|_D)$ inside $\mathrm{Mult}(\mathbb{K}[x])$ with respect to the δ-topology.

Now, let $\varphi_{\mathcal{G}} \in \Sigma(D) \setminus \Sigma_0(D)$. Let $r = \mathrm{diam}(\mathcal{G})$. Suppose first that there exists a monotonous holes sequence, or an equal distances holes sequence, thinner than $\mathcal{G}$, whose superior gauge is equal to its diameter. Let $(D_n)_{n \in \mathbb{N}}$ be a basis of $\mathcal{G}$ consisting of affinoid sets. Then of course $\lim_{n \to \infty} \mathrm{diam}(D_n) = r$. Thus, there exists a monotonous distances holes sequence or an equal distances holes sequence, $(T_n)_{n \in \mathbb{N}}$ such that $\lim_{n \to \infty} \mathrm{diam}(T_n) = r$, and $T_n \subset D_n$ $\forall n \in \mathbb{N}$. For each $n \in \mathbb{N}$, let $T_n = d(b_n, r_n^-)$, let $s_n = \mathrm{diam}(D_n)$ and $\lambda_n = \delta(\varphi_{\mathcal{G}}, \varphi_{b_n, r_n})$. Assume first that the monotonous sequence (T_n) is decreasing. Since the circular filter $\mathcal{F}_n$ peripheral of T_n is secant with D_n, we have $\mathrm{diam}(\sup(\mathcal{F}_n, \mathcal{G})) \leq s_n$. Consequently, by definition of the distance δ, we can check that $\lambda_n \leq \max(s_n - r_n, s_n - r)$, and finally $\lim_{n \to \infty} \lambda_n = 0$. Hence $\mathcal{G}$ is a point of adherence of the sequence $(\mathcal{F}_n)$ with respect to the δ-topology, and therefore belongs to the closure of $\mathrm{Mult}(\mathbb{K}[x]) \setminus \mathrm{Mult}(R(D), \| \cdot \|_D)$ inside $\mathrm{Mult}(\mathbb{K}[x])$. Since it belongs to $\mathrm{Mult}(R(D), \| \cdot \|_D)$, it belongs the boundary of $\mathrm{Mult}(R(D), \| \cdot \|_D)$ inside $\mathrm{Mult}(\mathbb{K}[x])$ with respect to the δ-topology. Assume now that the sequence (T_n) is an increasing holes sequence, or an equal distances holes sequence. Then each filter $\mathcal{F}_n$ is surrounded by $\mathcal{G}$. Consequently, $\lambda_n = r - r_n$, and therefore the conclusion is immediate.

Finally, suppose now that there exist neither any monotonous holes sequence nor any equal distances holes sequence whose superior gauge is equal to its diameter. Then, we can find $\lambda < r$ and an affinoid set $B \in \mathcal{G}$ such that every hole of D included in B has a diameter $\rho \leq \lambda$. Let $B = d(a, s_0) \setminus \bigcup_{j=1}^{q} d(a_j, s_j^-)$ and let $\varphi_{\mathcal{F}} \in \mathrm{Mult}(\mathbb{K}[x]) \setminus \mathrm{Mult}(R(D), \| \cdot \|_D)$. Either $\mathcal{F}$ is secant with $K \setminus (d(a, r_0) \bigcup (\bigcup_{\leq j \leq q} d(a_j, r_j^-)))$, and then we check that $\delta(\mathcal{G}, \mathcal{F}) \geq \min_{0 \leq j \leq q} |r - s_j|_\infty$, or it is secant with a hole T_k included in B, and then $\delta(\mathcal{G}, \mathcal{F}) \geq r - \lambda$. Thus we have proven that for every $\varphi_{\mathcal{F}} \in \mathrm{Mult}(\mathbb{K}[x]) \setminus \mathrm{Mult}(R(D), \| \cdot \|_D)$, we have

$$\delta(\mathcal{G}, \mathcal{F}) \geq \min(\min_{0 \leq j \leq q} |r - s_j|_\infty, |r - \lambda|_\infty).$$

This finishes proving that $\varphi_{\mathcal{G}}$ does not belong to the boundary of $\mathrm{Mult}(R(D), \| \cdot \|_D)$ inside $\mathrm{Mult}(\mathbb{K}[x])$ with respect to the δ-topology. $\square$

Theorem 6.3.3. *Let D be a closed bounded set. The topology of pointwise convergence is weaker than the δ-topology on $\mathrm{Mult}(R(D), \| \cdot \|_D)$. If at least one large circular filter on $\mathbb{K}$ is secant with D, then $\mathrm{Mult}(R(D), \| \cdot \|_D)$ is not compact for the δ-topology. However, on a totally ordered subset of $\mathrm{Mult}(R(D), \| \cdot \|_D)$ the two topologies are equal.*

Proof. Let $\varphi_{\mathcal{F}} \in \mathrm{Mult}(R(D), \| \cdot \|_D)$. First we will show that every neighborhood of $\varphi_{\mathcal{F}}$, with respect to the pointwise convergence topology, which is of the form $\mathrm{Mult}(R(E \cap D), \| \cdot \|_{E \cap D})$, where E is an affinoid set, contains a neighborhood of $\varphi_{\mathcal{F}}$ with respect to the δ-topology. Let $r = \mathrm{diam}(\mathcal{F})$, and let l be the maximum of the diameters of holes of E. For every $\mathcal{G} \in \mathrm{Mult}(R(D), \| \cdot \|_D)$ such that $\delta(\mathcal{F}, \mathcal{G}) \leq r - l$, then $\mathcal{G}$ is secant with E, and therefore lies in $\mathrm{Mult}(R(E \cap D), \| \cdot \|_{E \cap D})$. This shows that $\mathrm{Mult}(R(E \cap D), \| \cdot \|_{E \cap D})$ is a neighborhood of $\varphi_{\mathcal{F}}$ with respect to the δ-topology. But by Theorem 6.2.10, the family of neighborhoods of $\varphi_{\mathcal{F}}$, for the pointwise convergence topology inside $\mathrm{Mult}(R(D), \| \cdot \|_D)$, of the form $\mathrm{Mult}(R(E \cap D), \| \cdot \|_{E \cap D})$, with E an affinoid set, forms a basis of neighborhoods of $\varphi_{\mathcal{F}}$ for the pointwise convergence topology: this shows that every neighborhood of $\varphi_{\mathcal{F}}$ for the pointwise convergence topology is a neighborhood of $\varphi_{\mathcal{F}}$ for the δ-topology, and therefore, the pointwise convergence topology is weaker than the δ-topology.

Suppose that at least one large circular filter $\mathcal{F}$ is secant with D. There exists a sequence (a_n) in D which is thinner than $\mathcal{F}$, and we know that the sequence φ_{a_n} converges to $\varphi_{\mathcal{F}}$ with respect to the topology of pointwise convergence. On the other hand, the sequence satisfies $\lim_{n \to \infty} |a_{n+1} - a_n| = \mathrm{diam}(\mathcal{F})$. But since $\delta(\varphi_a, \varphi_b) = |a - b|$, the sequence

φ_{a_n} satisfies $\lim_{n\to\infty} \delta(\varphi_{a_{n+1}}, \varphi_{a_n}) = \text{diam}(\mathcal{F})$, and therefore it has no limit in $\text{Mult}(R(D), \| \cdot \|_D)$. Consequently, if D admits at least one large circular filter, the pointwise convergence topology is strictly weaker than the δ-topology.

Suppose now that a subset S of $\text{Mult}(R(D), \| \cdot \|_D)$ is totally ordered with respect to the order $\preceq$. Let $\varphi_{\mathcal{F}} \in \text{Mult}(R(D), \| \cdot \|_D)$ and consider a neighborhood of $\varphi_{\mathcal{F}}$ with respect to the δ-topology, of the form

$$B = \{\psi \in \text{Mult}(R(D), \| \cdot \|_D) \mid \delta(\psi, \varphi_{\mathcal{F}}) \leq \epsilon\}.$$

Let $r = \text{diam}(\mathcal{F})$. If $\mathcal{F} = \inf(S)$, we denote by E the unique disk in $\mathbb{K}$ of diameter $r + \epsilon$, such that $\mathcal{F}$ is secant with E, and we put $s = r - \epsilon$. Else, since there exists $\mathcal{F}_0 \in S$ such that $\mathcal{F}_0 \prec \mathcal{F}$, we put $r_0 = \text{diam}(\mathcal{F}_0)$ and consider a disk $d(b, s)$ of diameter $s \in]\max(r_0, r - \epsilon), r[$ such that $\mathcal{F}_0$ is secant with $d(b, s)$ and we denote by E the annulus $\Gamma(b, s, r + \epsilon)$. Now, in all cases, consider $\varphi_{\mathcal{G}} \in S \cap (\text{Mult}(R(E \cap D), \| \cdot \|_{E\cap D}))$. Thus, $\mathcal{G}$ is secant with $d(b, r + \epsilon)$, hence $\text{diam}(\mathcal{G}) \leq r + \epsilon$. If $\mathcal{F} \prec \mathcal{G}$ then $r \leq \text{diam}(\mathrm{G}) \leq r + \epsilon$, hence of course $\mathcal{G}$ lies in B. Now suppose $\mathcal{G} \prec \mathcal{F}$, hence $\mathcal{F}_0 \prec \mathcal{G} \prec \mathcal{F}$, hence $s \leq \text{diam}(\mathcal{G}) \leq r$, and therefore $\mathcal{G}$ lies in B again. This proves that $(\text{Mult}(R(E\cap D), \| \cdot \|_{E\cap D})) \cap S$ is included in B, which obviously shows that the two topologies are equal on S. $\qquad\square$

Corollary 6.3.4. *The topology of pointwise convergence is weaker than the δ-topology on $\text{Mult}(\mathbb{K}[x])$. However, on a totally ordered subset of $\text{Mult}(\mathbb{K}[x])$ the two topologies are equal.*

Remarks. Since the topology of pointwise convergence is weaker than the δ-topology, the boundary of $\text{Mult}(R(D), \| \cdot \|_D)$ inside $\text{Mult}(\mathbb{K}[x])$, with respect to the δ-topology is obviously included in the boundary of $\text{Mult}(R(D), \| \cdot \|_D)$ inside $\text{Mult}(\mathbb{K}[x])$, with respect to the topology of pointwise convergence. The equivalence does not hold in the general case.

By Theorem 5.4.15 of Chapter 5, the following Theorem 6.3.5 is immediate.

Theorem 6.3.5. *Let D be a closed and bounded subset of $\mathbb{K}$. Then $\text{Mult}(R(D), \| \cdot \|_D)$ is complete with respect to the distance δ.*

Proposition 6.3.6. *Let $(\varphi_{\mathcal{F}_n})_{n\in N}$, $(\varphi_{\mathcal{G}_n})_{n\in N}$ be sequences in $\text{Mult}(R(D), \| \cdot \|_D)$ such that the sequence $(\varphi_{\mathcal{F}_n})_{n\in N}$ converges to a limit $\varphi_{\mathcal{T}} \in \text{Mult}(R(D), \| \cdot \|_D)$ with respect to the topology of pointwise convergence. Suppose that $\lim_{n\to\infty} \delta(\mathcal{F}_n, \mathcal{G}_n) = 0$. Then the sequence $(\varphi_{\mathcal{G}_n})_{n\in N}$ also converges to $\mathcal{T}$ with respect to the topology of pointwise convergence.*

Proof. Let $r = \text{diam}(\mathcal{T})$, $B \in \mathcal{T}$ be a $\mathcal{T}$-affinoid, let $B = d(a, r'') \setminus \bigcup_{i=1}^{q} d(a_i, r')$ where the a_i are centers of $\mathcal{T}$ satisfying $|a_i - a_j| = r \forall i \neq j$.

Let $\epsilon = \frac{1}{2} \min(r'' - r, r - r')$, let $s'' =]r, r + \epsilon[\cap |K|$, $s' \in]r - \epsilon, r[\cap |K|$ and let $A = d(a, s'') \setminus \bigcup_{i=1}^{q} d(a_i, s')$. According to hypotheses, we can find a rank N such that $\mathcal{F}_n$ is secant with A and $\delta(\mathcal{F}_n, \mathcal{G}_n) < \epsilon \, \forall n \geq N$. Then, by Lemma 5.4.13 of Chapter 5, each $\mathcal{G}_n$ is secant with B for all $n \geq N$. Thus, given any $\mathcal{F}$-affinoid B, there exists a rank $N \in \mathbb{N}$ such that all $\varphi_{\mathcal{G}_n}$ belong to $\text{Mult}(R(B), \| \cdot \|_B)$ whenever $n \geq N$. $\square$

Remarks. We can now make a short comparison between the three topologies defined on $\text{Mult}(H(D), \| \cdot \|_D)$ in the particular and very simple case when D is the disk $d(0, 1)$.

In Section 2.1 of Chapter 2, given a normed $\mathbb{K}$-algebra we introduced a distance Θ on $\text{Mult}(A, \| \cdot \|)$. We can now ask how to compare Θ and δ. The question is quite complicated in the general case. In a particular case, we can see that Θ and δ satisfy $\Theta(\phi, \psi) \geq \delta(\phi, \psi)$ but define not equivalent topologies. Let D be the disk $d(0, 1)$ and let $\phi = \varphi_{\mathcal{F}}$, $\psi = \varphi_{\mathcal{G}}$, suppose that $\text{diam}(\mathcal{F}) = r$, $\text{diam}(\mathcal{G}) = s$ with $r \leq s$ and for simplicity, suppose that 0 is a center of $\mathcal{F}$ and that $\mathcal{G}$ is secant with a circle $C(0, t)$, with $r \leq t$. Then we can check that $\delta(\phi, \psi) = t - r$. On the other hand, since x belongs to the unit ball of $R(D)$ and since $\phi(x) = r$, $\psi(x) = t$, we have $\left| \phi, \psi \right|_\infty \geq \left| t - r \right|_\infty$, therefore, $\Theta(\phi, \psi) \geq \delta(\phi, \psi)$.

Now, suppose $t = 1$ and let us fix $n \in \mathbb{N}$ and consider $\psi(x^n) - \phi(x^n) = 1 - r^n$ hence $\Theta(\phi, \psi) \geq 1 - r^n$. That holds for all $n \in \mathbb{N}$ and hence $\Theta(\phi, \psi) \geq 1$. But $\delta(\mathcal{F}, \mathcal{G}) = 1 - r$. That holds for every $r < 1$ and hence the ratio $\frac{\Theta(\phi, \psi)}{\delta(\phi, \psi)}$ is unbounded on $\text{Mult}(H(D), \| \cdot \|_D)$, which proves that Θ and δ are not equivalent.

However, on this example, consider the sequence $(\phi_n)_{n \in \mathbb{N}}$ defined as $\phi_n = \varphi_{d(0, r, n)}$, where $(r_n)_{n \in \mathbb{N}}$ is a sequence of $]0, 1[$ of limit 1. The sequence (ϕ_n) is strictly ordered for the order of circular filters, hence the pointwise topology is equivalent to the δ-topology on this sequence, but the Θ-topology is not equivalent with both.

In conclusion, by Corollary 6.3.4 and by this remark, the δ-topology is not equivalent to the pointwise topology and the Θ-topology is neither equivalent to the δ-topology nor to the pointwise topology.

Theorem 6.3.7. ([11]). *Let S be a locally compact subset of $\text{Mult}(\mathbb{K}[x])$ with respect to the pointwise convergence topology and let T be the set* of circular filters $\mathcal{H}$ such that $\varphi_{\mathcal{H}} \in S$. The following statements are equivalent:

(i) S is connected with respect to the topology of pointwise convergence.
(ii) S is arcwise connected with respect to the topology of pointwise convergence.
(iii) S is connected with respect to the δ-topology.
(iv) S is arcwise connected with respect to the δ-topology.
(v) There exists no annulus A together with filters $\mathcal{F}$, $\mathcal{G} \in T$ such that $\mathcal{F}$ be secant with $\mathcal{I}(A)$, $\mathcal{G}$ be secant with $\mathcal{E}(A)$, and none of circular filters $\mathcal{H} \in T$ be secant with A.

Proof. First, (iv) trivially implies (iii) and (ii) trivially implies (i). Next, by Corollary 6.3.4, it is obviously seen that (iv) implies (ii) and that (iii) implies (i). Consequently, (iv) implies (i). We can also easily check that (i) implies (v). Indeed, suppose that (v) is not true. Thus there exists an annulus $A = \Gamma(a, r, s)$ together with $\mathcal{F}$, $\mathcal{G} \in T$ such that $\mathcal{F}$ is secant with $\mathcal{I}(A)$, $\mathcal{G}$ is secant with $\mathcal{E}(A)$, but none of the circular filters $\mathcal{H}$ are secant with A whenever $\mathcal{H} \in T$. Let E be the set of $\varphi_{\mathcal{H}}$, $\mathcal{H} \in T$ such that $\mathcal{H}$ is secant with $d(a, r)$, and let F be the set of $\varphi_{\mathcal{H}}$, $\mathcal{H} \in T$ such that $\mathcal{H}$ is secant with $\mathbb{K} \setminus d(a, s^-)$. Then E, F are two subsets of S closed in $\mathrm{Mult}(\mathbb{K}[x])$ for the topology of pointwise convergence, and make a partition of S, therefore S is not connected with respect to the topology of pointwise convergence. Consequently, (i) implies (v). Thus, it only remains us to prove that (v) implies (iv). The mapping diam defined on $\Phi(\mathbb{K})$ is strictly increasing by Corollary 5.4.2 of Chapter 5. Let $\mathcal{F}$, $\mathcal{G} \in T$ be such $\mathcal{G}$ surrounds $\mathcal{F}$, let $r = \mathrm{diam}(\mathcal{F})$, $s = \mathrm{diam}(\mathcal{G})$. Let $l \in [r, s]$. Let $\mathcal{R}$ be the unique circular filter of diameter l surrounding $\mathcal{F}$. By Theorem 6.2.11, every neighborhood of $\varphi_{\mathcal{R}}$, with respect to the topology of pointwise convergence, contains a set of the form $\mathrm{Mult}(R(E), \| . \|_E)$, with E an affinoid subset of $\mathbb{K}$, and such a set contains an annulus of the form $\Gamma(a, l - \epsilon, l)$. But by hypothesis, there exists $\mathcal{B} \in T$ secant with $\Gamma(a, l - \epsilon, l)$. Consequently, every neighborhood of $\varphi_{\mathcal{R}}$ with respect to the topology of pointwise convergence contains elements of S, and therefore (since S is locally compact), $\mathcal{R}$ belongs to T. Consequently, the mapping diam satisfies $\{\mathrm{diam}(\mathcal{H}) \mid \mathcal{F} \preceq \mathcal{H} \preceq \mathcal{G}\} = [\mathrm{diam}(\mathcal{F}), \mathrm{diam}(\mathcal{G})]$. Hence we can apply Theorem 1.2.4 of Chapter 1 which shows that S is arcwise connected with respect to the δ-topology. This finishes proving that (v) implies (iv). $\qquad \square$

Corollary 6.3.8. *Let D be a closed bounded set. The following statements are equivalent*

(i) $\mathrm{Mult}(R(D), \| \, . \, \|_D)$ *is connected with respect to the topology of pointwise convergence.*

(ii) $\mathrm{Mult}(R(D), \| \, . \, \|_D)$ *is arcwise connected with respect to the topology of pointwise convergence.*

(iii) $\mathrm{Mult}(R(D), \| \, . \, \|_D)$ *is connected with respect to the δ-topology.*

(iv) $\mathrm{Mult}(R(D), \| \, . \, \|_D)$ *is arcwise connected with respect to the δ-topology.*

(v) D *is infraconnected.*

Notation: Let D be a closed bounded infraconnected subset of $\mathbb{K}$ and let $f \in H(D)$. We denote by ${}^*f$ the mapping from $\mathrm{Mult}(H(D), \| \, . \, \|)$ to $\mathbb{K}$ defined by ${}^*f(\phi) = \phi(f)$.

Theorem 6.3.9. *Let D be a closed bounded infraconnected subset of $\mathbb{K}$ and let $f \in H(D)$. Then ${}^*f$ is continuous with respect to the δ-topology on $\mathrm{Mult}(H(D), \| \, . \, \|)$ and the absolute value on $\mathbb{K}$.*

Proof. Let $\phi, \psi \in \mathrm{Mult}(H(D), \| \, . \, \|)$. We know that ϕ is of the form $\varphi_{\mathcal{F}}$ and ψ is of the form $\varphi_{\mathcal{G}}$. When ϕ and ψ are close enough, one surrounds the other, hence when $\varphi_{\mathcal{F}}$ tends to $\varphi_{\mathcal{G}}$, then $\varphi_{\mathcal{F}}(f)$ tends to $\varphi_{\mathcal{G}}(f)$, which ends the proof. $\qquad \square$

Chapter 7

Analytic Elements and T-Filters

7.1. Analytic elements

Let us briefly recall that analytic elements where introduced by Marc Krasner in order to define a general notion of holomorphic functions in sets which are not just disks: indeed, when two disks have a non-empty intersection, this which has the biggest radius contains the other. As a consequence, it is hopeless to cover a set (which is not a disk) with a chained family of disks. But, according to Runge's Theorem, complex holomorphic functions may be viewed as uniform limits of rational functions. Marc Krasner adopted this point of view to defining holomorphic functions in a set D [35].

Definition and notation: Throughout this chapter and all the next ones, $\mathbb{K}$ is an algebraically closed field complete for an ultrametric absolute value and D is an infinite subset of $\mathbb{K}$.

By definition, the set $H(D)$ is equipped with the topology of uniform convergence on D for which it is complete and every $f \in H(D)$ defines a function on D which is the uniform limit of a sequence $(h_n)_{n \in \mathbb{N}}$ in $R(D)$. Thus, given another set D' containing D, the restriction to D of elements of $H(D')$ enables us to consider that $H(D')$ is included in $H(D)$. Here, we will only consider bounded analytic elements, and we will denote by $H_b(D)$ the set of the elements $f \in H(D)$ bounded on D. Then $H_b(D)$ is clearly a $\mathbb{K}$-vector subspace of $H(D)$ and is closed in $H(D)$. Moreover, $\| \cdot \|_D$ is a $\mathbb{K}$-algebra norm on $R_b(D)$, hence its continuation to $H_b(D)$ is a $\mathbb{K}$-algebra norm that makes $H_b(D)$ a Banach $\mathbb{K}$-algebra. If D is unbounded, we will denote by $H_0(D)$ the set of the $f \in H(D)$ such that $\lim_{\substack{|x| \to +\infty \\ x \in D}} f(x) = 0$.

According to Theorem 2.5.6 of Chapter 2 and Corollary 6.2.8 of Chapter 6, it is obviously seen that each element $\varphi_{\mathcal{F}}$ of $\mathrm{Mult}(R(D), \| \cdot \|_D)$

has a unique continuation to $\mathrm{Mult}(H(D), \|\cdot\|_D)$ and will be denoted by $\varphi_{\mathcal{F}}$ again. Particularly, when $\mathcal{F}$ is the circular filter of center a and diameter r, we will also denote by $\varphi_{a,r}$ the continuation of $\varphi_{a,r}$ to $\mathrm{Mult}(H(D), \|\cdot\|_D)$. Thus, we identify $\mathrm{Mult}(H(D), \|\cdot\|_D)$ to $\mathrm{Mult}(R(D), \|\cdot\|_D)$, and therefore all properties already proven in $\mathrm{Mult}(R(D), \|\cdot\|_D)$ also hold in $\mathrm{Mult}(H(D), \|\cdot\|_D)$.

Theorem 7.1.1 is well known [28, Chapter 11].

Theorem 7.1.1. *$H_b(D)$ is a Banach $\mathbb{K}$-subalgebra of $\mathbb{K}^D$. The following four conditions are equivalent:*

(i) *$H_b(D) = H(D)$,*
(ii) *$H(D)$ is a topological $\mathbb{K}$-vector space,,*
(iii) *$(H(D), \|\cdot\|_D)$ is a $\mathbb{K}$-Banach algebra,,*
(iv) *D is closed and bounded.*

If these conditions are satisfied, then $\|\cdot\|_D$ is a semi-multiplicative norm.

Corollary 7.1.2. *Let D be a closed and bounded subset of $\mathbb{K}$. The norm $\|\cdot\|_D$ is the spectral norm of $H(D)$.*

The following two lemmas are classical and easy.

Lemma 7.1.3. *$R(D)$ is a full $\mathbb{K}$-subalgebra of $H(D)$.*

Lemma 7.1.4. *Let $f \in H(D)$ be such that $\inf_{x \in D} |f(x)| > 0$. Then f^{-1} belongs to $H(D)$. Moreover if D is closed and bounded, an element $g \in h(D)$ belongs to $H(D)$ if and only if $\inf_{x \in D} |g(x)| > 0$.*

Let $g, h \in H(D)$ satisfy $|g(x)| = 1$ for all $x \in D$ and $\|h-1\|_D > \|g-1\|_D$. Then we have $\|hg - 1\|_D = \|h - 1\|_D$

Lemma 7.1.5. *Let $a \in \mathbb{K} \setminus D$ and let $f \in R(D)$ be such that $|f(a)| > \|f\|_D$. Then $\frac{1}{x-a}$ belongs to the closure of $\mathbb{K}[f, x]$ in $H(D)$.*

Proof. Let A be the closure of $\mathbb{K}[f, x]$ in $H(D)$. Since A is a Banach $\mathbb{K}$-algebra, $f(a) - f$ is invertible in A, hence of course in $H(D)$. But by Lemma 7.1.4, its inverse g actually belongs to $R(D)$. Now, since a is clearly a zero of $f(a) - f$, it is a pole of g. Consequently, by Lemma 1.1.8 of Chapter 1, $\frac{1}{x-a}$ belongs to $\mathbb{K}[g, x]$ which is included in A. $\square$

Lemma 7.1.6. *Let $\alpha \in \overline{D} \setminus D$ and let $f \in H(D)$ be such that there exists $n \in \mathbb{N}$ such that $(x - \alpha)^n f \in H(D \cup \{\alpha\})$. There exists a unique $q \in \mathbb{N}$ such that $(x - \alpha)^q f \in H(D \cup \{\alpha\})$ and such that the value of $(x - \alpha)^q f$ at α is not zero.*

Definitions and notations: In the hypothesis of Lemma 7.1.6 of Chapter 7, f is said to be *meromorphic at α and to admit α as a pole of order q*.

Let $D \subset \mathbb{K}$ be infraconnected closed and bounded, and let T be a hole of D. Let $f \in H(D)$. Then f will be said to be *meromorphic in T* if there exist finitely many points $(a_i)_{(1 \leq i \leq n)}$ in T such that f has continuation to an element of $H((D \cup T) \setminus \{a_i| \ 1 \leq i \leq n\})$.

Let f be meromorphic in T, and belong to $H(D \cup T) \setminus \{a_i| \ 1 \leq i \leq n\})$. For each $i = 1, \ldots, n$ if $f \notin H((D \cup T) \setminus \{a_h \ |h \neq i\})$ then a_i is a pole of f as an element of $H((D \cup T) \setminus \{a_j| 1 \leq j \leq n\})$. Let q_i be its order. Then a_i will be called *a pole of f of order q_i in T*. The polynomial $P(x) = \prod_{i=1}^{n}(x - a_i)^{q_i}$ will be called *the polynomial of the poles of f in T*.

Let $d(b_i, r_i^-))_{i \in I}$ be the family of holes of D. We will denote by $H'(D, (b_i)_{i \in I})$ the $\mathbb{K}$-subvector space of $H(D)$ consisting of the $f \in H(\overline{D} \setminus (\bigcup_{i \in I} d(b_i, r_i^-))$ admitting each b_i as a pole of order 0 or 1 and having no other poles in any hole $d(b_i, r_i^-)$.

Lemma 7.1.7 is immediate.

Lemma 7.1.7. *Let D be closed bounded and infraconnected, let T be a hole of D, and let $f \in H(D)$. If f is meromorphic in T, the polynomial P of the poles of f in T satisfies $Pf \in H(D \cup T)$. Conversely, if there exists a polynomial Q such that Qf lies in $H(D \cup T)$ then f is meromorphic in T.*

Theorem 7.1.8 is in [28, Theorem 14.6].

Theorem 7.1.8. *Let $r \in \mathbb{R}_+^*$, let $\mathcal{F}$ be the circular filter of center 0 and diameter r on $\mathbb{K}$, and let $E = d(0, r)$. Then $H(E)$ is the set of the power series $f(x) = \sum_{n=0}^{\infty} a_n x^n$ such that $\lim_{n \to \infty} |a_n| r^n = 0$ and we have $\|f\|_E = \max_{n \in \mathbb{N}} |a_n| r^n = \varphi_{\mathcal{F}}(f) = \|f\|_{C(0,r)}$. For every $\alpha \in E$, $H(E)$ is also equal to the set of the series $f(x) = \sum_{n=0}^{\infty} b_n(x - \alpha)^n$ such that $\lim_{n \to \infty} |b_n| r^n = 0$.*

Let $B = \mathbb{K} \setminus d(0, r^-)$. Then $H(B)$ is the set of the Laurent series $f(x) = \sum_{n=0}^{\infty} \frac{a_n}{x^n}$ such that $\lim_{n \to \infty} |a_n| r^{-n} = 0$ and we have $\|f\|_B = \max_{n \in \mathbb{N}} |a_n| r^{-n} = \varphi_{\mathcal{F}}(f) = \|f\|_{C(0,r)}$.

For every $\alpha \in d(0, r^-)$, $H(B)$ is also equal to the set of the series $f(x) = \sum_{n=0}^{\infty} \frac{b_n}{(x-\alpha)^n}$ such that $\lim_{n \to \infty} |b_n| r^{-n} = 0$.

By Theorem 7.1.8, we have Theorem 7.1.9.

Theorem 7.1.9. *Let $\alpha \in D$ and $r \in \mathbb{R}_+^*$ be such that $d(\alpha, r) \subset D$. Let $f \in H(D)$. In $d(\alpha, r)$, $f(x)$ is equal to a power series of the form $\sum_{n=0}^{\infty} a_n(x-\alpha)^n$ such that $\lim_{n \to \infty} |a_n| r^n = 0$. If $f(\alpha) = 0$ and if $f(x)$ is not identically zero in $d(\alpha, r)$, then there exists a unique integer $q \in \mathbb{N}^*$ such that $a_n = 0$ for* *every $n < q$ and $a_q \neq 0$ and α is an isolated zero of f in $d(\alpha, r)$. Moreover, there exists $g \in H(D)$ such that $f(x) = (x - \alpha)^q g(x)$.*

As a consequence of Theorems 7.1.8 and 7.1.9, we can deduce Theorem 7.1.10.

Theorem 7.1.10. *The characteristic of $\mathbb{K}$ is supposed equal to 0. Let $f \in H(d(0, r^-))$ and for every $j = 0, \ldots, n$ let $a_j = \frac{f^{(j)}(0)}{j!}$. Then f is equal to the power series $\sum_{n=0}^{\infty} a_n x^n$, and $\|f\|_{d(0,r^-)} = \sup_{n \in \mathbb{N}} |a_n|$. Moreover, for every $q \in \mathbb{N}$, f is of the form $\sum_{j=0}^{q} a_j x^j + x^n g$, with $g \in H(M)$ and $\|f\|_{d(0,r^-)} = \max(|a_0|, |a_1|, \ldots |a_{n-1}| r^{n-1}, \|x^n g\|_{d(0,r^-)})$.*

We must also recall [28, Theorem 18.12].

Theorem 7.1.11. *If $\mathbb{K}$ has characteristic 0, then an element $f \in H(d(0, r))$ has a derivative identically equal to 0 if and only if it is equal to a constant.*

If $\mathbb{K}$ has a characteristic $p \neq 0$, then an element $f \in H(d(0, r))$ has a derivative identically equal to 0 if and only if there exists $g \in H(d(0, r))$ such that $f(x) = (g(x))^p$.

Definitions and notations: Let $f \in H(D)$, let $\alpha \in \overset{\circ}{D}$, let $r > 0$ be such that $d(\alpha, r) \subset D$ and let $f(x) = \sum_{n=q}^{\infty} b_n (x - \alpha)^n$ whenever $x \in d(\alpha, r)$, with $b_q(\alpha) \neq 0$, and $q > 0$. Then α is called *a zero of multiplicity order q*, or more simply, *a zero of order q*. In the same way, q will be named *the multiplicity order of α*.

An element $f \in H(D)$ is said to be *quasi-invertible* if it factorizes in the form Pg with $P \in \mathbb{K}[x]$ all zeros of which lie $\overset{\circ}{D}$, and $g \in H(D)$ invertible in $H(D)$.

Theorem 7.1.12. *Let D be closed and bounded and let $f \in H(D)$ be quasi-invertible. There exists $h \in R(D)$ such that $|f(x)| = |h(x)| \; \forall x \in D$.*

Proof. Since f is quasi-invertible, there exists a polynomial P whose zeros lie in D, and an invertible element g in $H(D)$ such that $f = Pg$. Let $m = \inf_{x \in D} |g(x)|$. Since g is invertible, we have $m = \frac{1}{\|\frac{1}{g}\|_D}$, hence $m > 0$, and then we can find $l \in R(D)$ such that $\|g - l\|_D < m$. Consequently, we have $|g(x)| = |l(x)| \; \forall x \in D$ and we obtain h by putting $h = Pl$. Moreover, if $r \in |\mathbb{K}|$, then $\varphi_{0,r}(h) \in |\mathbb{K}|$. Conversely, if $s \neq t$ and if $\varphi_{0,r}(h) \in |\mathbb{K}|$, then $r \in |\mathbb{K}|$. $\square$

The following Theorem 7.1.13 is [28, Theorem 15.1] and is stated in [34] for quasi-connected sets.

Theorem 7.1.13 (M. Krasner). *Let D be bounded (resp., unbounded) and infraconnected and let $f \in H_b(D)$. There exists a unique sequence of holes $(T_n)_{n \in \mathbb{N}^*}$ of D and a unique sequence $(f_n)_{n \in \mathbb{N}}$ in $H(D)$ such that $f_0 \in H(\widetilde{D})$ (resp., $f_0 \in \mathbb{K}$), $f_n \in H_0(\mathbb{K} \setminus T_n)$ $(n > 0)$, $\lim_{n \to \infty} \|f_n\|_D = 0$ and*

(i) $f = \sum_{n=0}^{\infty} f_n$ *and* $\|f\|_D = \sup_{n \in \mathbb{N}} \|f_n\|_D$.

Moreover for every hole $T_n = d(a_n, r_n^-)$, we have

(ii) $\|f_n\|_D = \|f_n\|_{\mathbb{K} \setminus T_n} = \varphi_{a_n, r_n}(f_n) \leq \varphi_{a_n, r_n}(f) \leq \|f\|_D$.

If D is bounded and if $\widetilde{D} = d(a, r)$ we have

(iii) $\|f_0\|_D = \|f_0\|_{\widetilde{D}} = \varphi_{a, r}(f_0) \leq \varphi_{a, r}(f) \leq \|f\|_D$.

If D is not bounded then $|f_0| = \lim_{\substack{|x| \to \infty \\ x \in D}} |f(x)| \leq \|f\|_D$.

Let $D' = \widetilde{D} \setminus (\bigcup_{n=1}^{\infty} T_n)$. Then f belongs to $H(D')$ (resp., $H_b(D')$) and its decomposition in $H(D')$ is given again by (i) and then f satisfies $\|f\|_{D'} = \|f\|_D$.

Corollary 7.1.14. *Let D be closed and infraconnected. Let $(T_i)_{i \in I}$ be the family of holes of D. Let J be a subset of D and let $L = I \setminus J$. Let $E = D \bigcup (\bigcup_{i \in J} T_i)$ and let $F = D \bigcup (\bigcup_{i \in L} T_i)$. Then we have $H(D) = H_0(E) \oplus H(F)$, and for each $g \in H_0(E)$, $h \in H(F)$, we have $\|g + h\|_D = \max(\|g\|_E, \|h\|_F)$.*

Theorem 7.1.15. *Let D be closed and infraconnected. Then $\Sigma_0(D)$ is a minimal boundary for $(H(D), \| \cdot \|_D)$.*

Proof. By Theorem 7.1.13, it is a boundary for $(H(D), \| \cdot \|_D)$. Conversely, let B be a boundary included in $\mathcal{S}_0(D)$. Let $\widetilde{D} = d(a, r)$. Considering $(x - a)$ we can see that $\varphi_{a, r}$ must belong to B because if $\phi \in \Sigma_0(D) \setminus \{\varphi_{a, r}\}$ we have $\phi(x - a) < r = \|x - a\|_D$. In the same way, let $T = d(\alpha, \rho^-)$ and let $h = \frac{1}{x - \alpha}$. Then we can see that for every $\phi \in \mathcal{S}_0(D) \setminus \{\varphi_{\alpha, \rho}\}$ we have $\phi(h) < \frac{1}{\rho} = \|h\|_D$. Consequently, $\mathcal{S}_0(D) = B$ and therefore is a minimal boundary for $(H(D), \| \cdot \|_D)$. $\square$

Lemma 7.1.16. *Let D be infraconnected and suppose that $H(D)$ contains a dense Luroth $\mathbb{K}$-algebra $\mathbb{K}[h, x]$. Then each hole of D contains at least one pole of h.*

Proof. Let B be the closure of $\mathbb{K}[h, x]$ in $H(D)$ and let $T_1, \ldots, T_q$ be the holes of D containing at least one hole of D and let $D' = \widetilde{D} \setminus \bigcup_{j=1}^{q} T_j$. Clearly $\mathbb{K}[h, x]$ is included in $H(D')$. But since each hole of D' is a hole of D, and since $\widetilde{D'} = \widetilde{D}$, by Corollary 7.1.15 $H(D')$ is a closed $\mathbb{K}$-subalgebra of $H(D)$. Consequently $D = D'$. $\square$

The Mittag–Leffler theorem suggests some new definitions.

Definitions and notations: Let $f \in H_b(D)$. We consider the series $\sum_{n=0}^{\infty} f_n$ obtained in Theorem 7.1.13, whose sum is equal to f in $H(D)$, with $f_0 \in H(\widetilde{D}), f_n \in H(\mathbb{K} \setminus T_n) \setminus \{0\}$ whereas the T_n are holes of D. Each T_n will be called a f-*hole* and f_n will be called the *Mittag–Leffler term of* f *associated to* T_n, whereas f_0 will be called the *principal term* of f.

For each f-hole T of D, the Mittag–Leffler term of f associated to T will be denoted by $\overline{f}_T$ whereas the principal term of f will be denoted by $\overline{f}_0$.

The series $\sum_{n=0}^{\infty} f_n$ will be called *the Mittag–Leffler series of* f *on the infraconnected set* D.

More generally, let E be an infraconnected set and $f \in H(E)$. According to [28, Theorem 11.5], f is of the form $g + h$ with $g \in R(\mathbb{K} \setminus (\overline{E} \setminus E))$, and $h \in H_b(\overline{E})$, and such a decomposition is unique, with respect to an additive constant. For every hole T of $\overline{E}$, we will denote by $\overline{f}_T$ the Mittag–Leffler term of h associated to T, and $\overline{f}_T$ will still be named *the Mittag–Leffler term of* f *associated to* T.

Corollary 7.1.17. *Let D be infraconnected. Let $f \in H_b(D)$ let $(T_n)_{n \in \mathbb{N}^*}$ be the sequence of the f-holes, with $T_n = d(a_n, \rho_n^-)$, let $f_0 = \overline{f}_0$, and $f_n = \overline{f}_{T_n}$ for every $n \in \mathbb{N}^*$. Let $\widetilde{D} = d(a, s)$, (resp., $\widetilde{D} = \mathbb{K}$). There exists $q \in \mathbb{N}$ such that $\|f\|_D = \|f_q\|_D$. If $q \geq 1$ then $\|f\|_D = \varphi_{a_q, r_q}(f) = \varphi_{a_q, r_q}(f_q)$. If $q = 0$ and if D is bounded (resp., is not bounded), then $\|f\|_D = \varphi_{a,s}(f) = \varphi_{a,s}(f_0)$ (resp., $\|f\|_D = |f_0|$). Further, given a hole T of D, if f belongs to $H_b(D)$, and if g belongs to $H_0(\mathbb{K} \setminus T)$ and satisfies $f - g \in H(D \cup T)$, then $\overline{f}_T$ is equal to g.*

Corollary 7.1.18. *Let D be infraconnected. Let $f \in H_b(D)$. There exists a large circular filter $\mathcal{F}$ with center $\alpha \in \widetilde{D}$ secant with D such that $\varphi_{\mathcal{F}}(f) = \|f\|_D$. If D is bounded, there exists a D-bordering filter $\mathcal{F}$ such that $\varphi_{\mathcal{F}}(f) = \|f\|_D$.*

Corollary 7.1.19. *Let $f \in H(d(0, 1^-))$ and let $(d(\alpha_m, 1^-))_{m \in \mathbb{N}^*}$ be the family of the f-holes. Then f is of the form*

$$(1) \quad \sum_{n=0}^{\infty} a_{n,0} x^n + \sum_{m,n \in \mathbb{N}^*} \frac{a_{n,m}}{(x - \alpha_m)^n}$$

with $\lim_{n\to\infty} a_n = 0$, $\lim_{n\to\infty} |a_{n,m}| = 0$ *whenever* $m \in \mathbb{N}^*$ *and* $\lim_{m\to\infty}$ $(\sup_{n\in\mathbb{N}^*} |a_{n,m}|) = 0$. *Moreover* f *satisfies* $\|f\|_{d(0,1-)} = \sup_{m,n\in\mathbb{N}^*} |a_{n,m}|$.

Conversely, every function of the form (1), *with the* α_m *satisfying* $|\alpha_m| = |\alpha_j - \alpha_m| = 1$ *whenever* $m \neq j$, *belongs to* $H(d(0,1^-))$. *The norm* $\| \cdot \|_{d(0,1-)}$ *is multiplicative and equal to* $\varphi_{0,1}$.

Theorem 7.1.20 is proven as [28, Corollary 15.6].

Theorem 7.1.20. *Let* $r_1, r_2 \in \mathbb{R}_+$ *satisfy* $0 < r_1 < r_2$. *Then* $H(\Lambda(0, r_1, r_2))$ *is equal to the set of Laurent series* $\sum_{-\infty}^{+\infty} a_n x^n$ *with* $\lim_{n\to-\infty} |a_n| r_1^n = \lim_{n\to\infty} |a_n| r_2^n = 0$ *and we have* $\| \sum_{-\infty}^{+\infty} a_n x^n \|_{\Lambda(0,r_1,r_2)} = \max\left(\sup_{n\geq 0} |a_n| r_1^n, \sup_{n<0} |a_n| r_2^n\right)$.

Notations: Let D be a closed bounded infraconnected containing the circle $C(0, r)$ and let $f \in H(D)$ and let $f(x) = \sum_{-\infty}^{+\infty} a_n x^n$. Generalizing notations introduced with rational functions, we denote by $\nu^+(f, \log(r))$ the biggest of the integers l such that $|a_l| r^l = \sup_{n\in\mathbb{Z}} |a_n| r^n$ and by $\nu^-(f, \log(r))$ the smallest of the integers l such that $|a_l| r^l = \sup_{n\in\mathbb{Z}} |a_n| r^n$.

Let D be closed bounded infraconnected. For every $a \in D$ we will denote by $\mathcal{I}(a)$ the ideal of the $f \in H(D)$ such that $f(a) = 0$.

Theorem 7.1.21. *Let* D *be closed bounded infraconnected. The mapping* ϕ *from* D *into the set of ideals of* $H(D)$ *defined by* $\phi(a) = \mathcal{I}(a)$ *is a bijection from* D *onto* $\text{Max}_1(H(D))$.

Finally, we can notice that analytic elements define uniformly continuous functions, in a closed bounded set.

Theorem 7.1.22. *Let* D *be closed and bounded and let* $f \in H(D)$. *Then* f *is uniformly continuous on* D.

Proof. Let $h \in R(D)$. Since h has finitely many poles, we can find $r_0 > 0$ such that for all $a \in D$, $d(a, r_0)$ contains no poles of h. Let $M = \|h\|_D$. Thus, by Corollary 7.1.19, for all $a \in D$, we have

$$\sup_{k\geq 1} \left| \frac{h^{(k)}(a)}{k!} \right| r_0^k \leq \sup_{k\geq 0} \left| \frac{h^{(k)}(a)}{k!} \right| r_0^k \leq \|h\|_{d(a,r_0)} \leq M.$$

Consequently, for every $r \in]0, r_0]$, we have

$$\sup_{k\geq 1} \left| \frac{h^{(k)}(a)}{k!} \right| r^k \leq \sup_{k\geq 1} \left(\left| \frac{h^{(k)}(a)}{k!} \right| r_0^k \left(\frac{r}{r_0}\right)^k \right) \leq M \frac{r}{r_0}.$$

Thus, we can see that $h(d(a,r)) \subset d(h(a), M\frac{r}{r_0})$, which proves that h is uniformly continuous on D. Then, so is every element of $H(D)$, as uniform limits of uniformly continuous functions on D. □

7.2. Properties of the function Ψ for analytic elements

Throughout this chapter D is infraconnected.

The function $\Psi(f, \mu)$ was defined for rational functions in Section 6.3 of Chapter 6. Here we will generalize that function to analytic elements. Its interest is to transform the multiplicative property of the norm $|\,.\,|$ into an additive property. Overall, Ψ is piecewise affine. Long ago, such a function was first defined in classical works such as the valuation function of an analytic element [1, 23] denoted by $v(f, \mu)$. However, the function $v(f, \mu)$ has the inconvenient of being contravariant: $\mu = -\log(|x|)$ and $v(f, -\log(|x|)) = -\log(|f|(r))$. Here we will change both senses of variation: $\Psi(f, \mu) = -v(f, -\mu)$.

Among applications, we can show that a set E is infraconnected if and only if for all $f \in H(E)$, $f(E)$ is infraconnected and that an analytic element converges along a monotonous filter $\mathcal{F}$ if and only if f' is vanishing along $\mathcal{F}$.

Notations: For every $a \in \tilde{D}$, we put $z(a) = \log(\delta(a, D))$ if $\delta(a, D) > 0$ and $z(a) = -\infty$ if $\delta(a, D) = 0$. We denote by S the diameter of D, with $S = +\infty$ if D is not bounded.

Let $a \in \tilde{D}$ and let $\mathcal{F}$ be a circular filter of center a and diameter $r \in [\delta(a, D), S] \cap \mathbb{R}$. Then $\mathcal{F}$ is secant with D and then defines an element $_D\varphi_{\mathcal{F}}$ of $\mathrm{Mult}(H(D), \mathcal{U}_D)$.

Consider a disk $d(a, r^-)$. We denote by $\mathcal{A}(d(a, r,^-))$ the algebra of power series $\sum_{n=0}^{\infty} a_n(x - a)^n$ converging in $d(a, r^-))$. Consider an annulus $\Gamma(a, r, t)$. We denote by $\mathcal{A}(\Gamma(a, r, t))$ the algebra of Laurent series $\sum_{-\infty}^{\infty} a_n(x - a)^n$ converging in $\Gamma(a, r, t)$. Consider a set $\mathbb{K} \setminus d(a, r)$. We denote by $\mathcal{A}(\mathbb{K} \setminus d(a, r)$ the algebra of series $\sum_{-\infty}^{0} a_n(x - a)^n$ converging in $\mathbb{K} \setminus d(a, r)$.

For every $f \in H(D)$ such that $_D\varphi_{\mathcal{F}}(f) \neq 0$ we put $\Psi_a(f, \log r) = \log\left(_D\varphi_{\mathcal{F}}(f)\right)$. Next, for an $f \in H(D)$ such that $_D\varphi_{\mathcal{F}}(f) = 0$ we put $\Psi_a(f, \log r) = -\infty$.

When $a = 0$ for simplicity we just put $\Psi(f, \mu) = \Psi_0(f, \mu)$.

In the same way, consider an annulus $\Gamma(0, r, t)$ and $f \in \mathcal{A}(\Gamma(0, r, t))$. Then for any $s \in]r, t[$, f belongs to $H(C(0, s))$, so we can put consider $\Psi(f, \log(s)) = \log(|f|(s))$. If we consider $f \in \mathcal{A}(a, r^-)$, so much the more, we can consider $\Psi_a(f, \ell)$ for each $\ell < \log(r)$.

Remarks. Let $f(x) = \sum_{-\infty}^{+\infty} a_n x^n \in H(C(0,r))$ for some $r > 0$. Then we have $\Psi(f, \log r) = \log\big(_{C(0,r)}\varphi_{0,r}(f)\big) = \log\|f\|_{C(0,r)} = \sup_{n\in\mathbb{Z}} \Psi(a_n) + n \log r$.

Throughout the chapter, D is a subset of $\mathbb{K}$. Here we shall go back to the basic properties of analytic elements which are transmitted by rational functions.

Proposition 7.2.1 is classical and given in [28, Proposition 20.3].

Proposition 7.2.1. *Let $a \in \tilde{D}$, let $r \in [\delta(a, D), \mathrm{diam}(D)]$, and let $f \in H(D)$. If $\varphi_{a,r}(f) \neq 0$ the equality $|f(x)| = \varphi_{a,r}(f)$ holds in all classes of $C(a, r)$ except maybe in finitely many ones.*

Proposition 7.2.2. *Let $a \in \tilde{D}$, let $\mu \in [\zeta(a), \log(S)] \cap \mathbb{R}$ and let $f, g \in H(D)$. Then $\Psi_a(f + g, \mu) \leq \max(\Psi_a(f, \mu), \Psi_a(g, \mu))$ and when $\Psi_a(f, \mu) > \Psi_a(g, \mu)$, then $\Psi_a(f + g, \mu) = \Psi_a(f, \mu)$. Moreover, $\Psi_a(fg, \mu) = \Psi_a(f, \mu) + \Psi_a(g, \mu)$.*

Let $r, t \in]0, +\infty[$ be such that $r \leq t$. Let $f \in H(D)$ be such that $\Psi_a(f, \mu)$ is bounded in $[\log(r), \log(t)]$. Then $\Psi_a(f, \mu)$ is continuous and piecewise affine in $[\log(r), \log(t)]$. Further, there exists $h \in R(D)$ such that $\Psi_a(f, \mu) = \Psi_a(h, \mu) \ \forall \mu \in [\log(r), \log(t)]$.

Inside $D \cap \Gamma(a, r, t)$, the relation $\Psi(f(x)) = \Psi_a(f, \Psi(x - a))$ holds in all classes of all circles $C(a, s)$, except maybe in finitely many classes of finitely many circles $C(a, s)$.

Moreover, if $\Gamma(a, r, t) \subset D$, the function $\Psi_a(f, \mu)$ is convex in $[\log r, \log t]$.

Proof. Without loss of generality, we can assume $a = 0$. The first statements concerning operations and inequalities come directly from those of multiplicative semi-norms $_D\varphi$. Now, suppose that $\Psi(f, \mu)$ is bounded in $[\log(r), \log(t)]$, hence there exists $\epsilon > 0$ such that $\Psi(f, \mu) > \log \epsilon \ \forall \mu \in [\log r, \log t]$.

Let $h \in R(D)$ satisfy $\|f - h\|_D < \epsilon$. Particularly, for every circular filter $\mathcal{F}$ secant with D, we have $_D\varphi_{\mathcal{F}}(f - h) < \epsilon$ and particularly $_D\varphi_{a,\rho}(f - h) < \epsilon \ \forall \rho \in [r, t]$ i.e., $\Psi(f - h, \mu) < \log(\epsilon) < \Psi(f, \mu) \ \forall \mu \in [\log r, \log t]$. Consequently, $\Psi(f, \mu) = \Psi(h, \mu) \ \forall \mu \in [\log r, \log t]$. Now, the function $\Psi(h, \mu)$ is continuous, piecewise in $[\log r, \log t]$ and so is $\Psi(f, \mu)$. Moreover, if $\Gamma(a, r, t) \subset D$, the function $\Psi(h, \mu)$ is convex in $[\log r, \log t]$, hence so is $\Psi(f, \mu)$.

By Lemma 7.2.1, the relation $\Psi(h(x)) = \Psi_a(h, \Psi(x - a))$ holds in all classes of all circles $C(a, s)$, except maybe in finitely many classes of finitely many circles $C(a, s)$. Therefore, the same relation holds for f. $\qquad\square$

Proposition 7.2.3. *Let $a \in D$ and let $f \in H(D)$ satisfy $f(a) \neq 0$. There exists $\mu_o \in \mathbb{R}$ such that $\Psi_a(f, \mu) = \Psi(f(a))$ whenever $\mu \leq \mu_o$. Let $r \in \mathbb{R}_+^*$, let $G = C(0, r)$ and let f and $g \in H(G)$ satisfy $\|f - g\|_G < \|f\|_G$. Then we have $\nu^+(f, \log r) = \nu^+(g, \log r)$, $\nu^-(f, \log r) = \nu^-(g, \log r)$.*

Proof.   Indeed let us take $r > 0$ such that $|f(x) - f(a)| < |f(a)|$ whenever $x \in d(a, r) \cap D$ hence $|f(x)| = |f(a)|$ whenever $x \in d(a, r) \cap D$ and therefore $\Psi_a(f, \mu) = \Psi(f(a))$ whenever $\mu \leq \log(r)$.

Let $f(x) = \sum_{-\infty}^{+\infty} a_n x^n$ and let $g(x) = \sum_{-\infty}^{+\infty} b_n x^n$. From the hypothesis we see that $\|f\|_G = \|g\|_G$. By [28, Corollary 14.9], we have

(1) $\sup_{n \in \mathbb{Z}} |a_n| r^n = \|f\|_G = \sup_{n \in \mathbb{Z}} |b_n| r^n$
    and $\|f - g\|_G = \sup_{n \in \mathbb{Z}} |a_n - b_n| r^n$.

Let $s = \nu^-(f, \log r)$ and let $t = \nu^+(f, \log r)$. We see that $|a_s - b_s| r^s \leq \|f - g\|_G < \|f\|_G = |a_s| r^s$ hence

(2) $|b_s| = |a_s|$.

In the same way we have

(3) $|a_t| = |b_t|$.

Now for every $n < s$ and for every $n > t$ we have $|a_n| r^n < |a_s| r^s = \|f\|_G$ hence $|b_n| r^n < \|f\|_7$. Finally by (1), (2), (3) we see that $\nu^-(g, \log r) = s$, $\nu^+(g, \log r) = t$. $\square$

We can now derive Corollary 7.2.4

Corollary 7.2.4. *Let $f(x) \in H(\Gamma(0, r_1, r_2))$ (resp., $f(x) \in H(\Lambda(0, r_1, r_2))$) (with $0 < r_1 < r_2$) and let $\sum_{-\infty}^{+\infty} a_n x^n$ be its Laurent series. The function $\mu \to \Psi(f, \mu)$ is bounded in $]\log r_1, \log r_2[$ (resp., in $[\log(r_1), \log(r_2)]$) and equal to $\sup_{n \in \mathbb{Z}}(\Psi(a_n) + n\mu)$. Next, we have $\Psi(f(x)) \leq \Psi(f, \Psi(x))$ whenever $x \in \Gamma(0, r_1, r_2)$ (resp., whenever $x \in \Lambda(0, r_1, r_2)$) and the equality holds in all of $\Gamma(0, r_1, r_2)$ (resp., un all of $\Lambda(0, r_1, r_2)$) except in finitely many classes of finitely many circles $C(0, r)$ ($r_1 < r < r_2$) (resp., $r_1 \leq r \leq r_2$). The right side derivative (resp., the left side derivative) of the function $\Psi(f, .)$ at μ is equal to $\nu^+(f, \mu)$ (resp., to $\nu^-(f, \mu)$). Moreover, if the function in $\mu\Psi(f, \mu)$ is not derivable at μ, then μ lies in $\Psi(\mathbb{K})$.*

Further, the function $\Psi(f, .)$ is convex in $]\log r_1, \log r_2[$ (resp., in $[\log r_1, \log r_2]$). Next, given another $g \in H(\Gamma(0, r_1, r_2))$, (resp., $g \in H(\Lambda(0, r_1, r_2))$) the functions ν^+ and ν^- satisfy $\nu^+(fg, \mu) = \nu^+(f, \mu) + \nu^+(g, \mu)$, $\nu^-(fg, \mu) = \nu^-(f, \mu) + \nu^-(g, \mu)$. Further, the function $\nu^+(f, .)$ is *continuous on the right and the function* $\nu^-(f,.)$ *is continuous on the left at each point* μ. *They are continuous at* μ *if and only if they are equal.*

Proposition 7.2.5. *Let* $a \in \widetilde{D}$ *and let* $f \in H(D)$. *If* $f(a) \neq 0$, *there exists* $s > 0$ *such that* $\Psi(f, \mu) = \Psi(f(a)) \,\forall \mu \leq s$. *Let* $b \in D$ *be such that* $|a - b| = r$ *and* $d(b, r^-) \subset D$, *Then we have* $\Psi_b(f, \mu) = \Psi_a(f, \mu) \,\forall \mu \leq \Psi(b - a)$.

Proof. Since $f(a) \neq 0$, the first statement is immediate since $|f(x)|$ is a constant inside a disk of center a. Next, the relation $\Psi_a(f, \mu) = \Psi_b(f, \mu)$ when $\Psi(a - b) \leq \mu$ is true for every $f \in R(D)$, hence by (2), is obviously generalized to every $f \in H(D)$. $\qquad\qquad\qquad\qquad\qquad\qquad\qquad\qquad \square$

Proposition 7.2.6. *Let* $\mu \in \mathbb{R}$ *and let* $f(x) = \sum_{-\infty}^{+\infty} a_n x^n \in H(C(0, \theta^\mu))$. *Then* $\Psi(f, \mu)$ *is equal to* $\sup_{n \in \mathbb{Z}} \Psi(a_n) + n\mu$ *and we have* $\Psi(f(x)) \leq \Psi(f, \mu)$ *for all* $x \in C(0, \theta^\mu)$. *Moreover, the equality holds in every class except in finitely many classes where* f *admits zeros. Further, if* $\nu^+(f, \mu) = \nu^-(f, \mu)$, *then* $\Psi(f(x)) = \Psi(f, \mu)$ *whenever* $x \in C(0, \theta^\mu)$.

If $h \in H(C(0, \theta^\mu))$ *satisfies* $\Psi(f - h, \mu) < \Psi(f, \mu)$, *then* $\nu^+(f, \mu) = \nu^+(h, \mu)$ *and* $\nu^-(f, \mu) = \nu^-(h, \mu)$.

Proof. Let $G = C(0, \theta^\mu)$, let $s = \nu^-(f, \mu)$ and let $t = \nu^+(f, \mu)$. By the Remark above $\Psi(f, \mu)$ is obviously equal to $\sup_{n \in \mathbb{Z}}(\Psi(a_n) + n\mu)$. Let $x \in C(0, \theta^\mu)$. The inequality $\Psi(f(x)) \leq \Psi(f, \mu)$ is true because $\Psi(f, \mu) = \log \|f\|_G \geq \Psi(f(x))$. Finally by Proposition 7.2.2, the equality holds in all the classes except in finitely many. If $\nu^+(f, \mu) = \nu^-(f, \mu)$ then $\Psi(a_s x^s) = \Psi(a_s) + s\mu > \Psi(a_n x^n)$ whenever $n \neq s$ hence $\Psi(f(x)) = \Psi(f, \mu)$.

Now, let $h \in H(G)$ satisfy $\Psi(f - h, \mu) < \Psi(f, \mu)$ and let $h(x) = \sum_{-\infty}^{+\infty} b_n x^n$. We have $\Psi(a_n - b_n) + n\mu < \Psi(a_s) + s\mu$ whenever $n \in \mathbb{Z}$ hence $\Psi(b_s) = \Psi(a_s)$, $\Psi(b_t) = \Psi(a_t)$, $\Psi(b_n) + n\mu < \Psi(a_s) + s\mu$ whenever $n < s$ and $n > t$ and $\Psi(b_n) + n\mu \leq \Psi(a_s) + s\mu$ whenever $n \in [s, t]$, hence finally $\nu^+(h, \mu) = \nu^+(f, \mu)$ and $\nu^-(h, \mu) = \nu^-(f, \mu)$. $\qquad\qquad \square$

Corollary 7.2.7. *Let* $f(x) = \sum_{-\infty}^{+\infty} a_n x^n \in A(\Gamma(0, r_1, r_2))$ *(with* $0 < r_1 < r_2$). *The function* $\mu \to \Psi(f, \mu)$ *defined in* $]\log r_1, \log r_2[$ *is equal to* $\sup_{n \in \mathbb{Z}}(\Psi(a_n) + n\mu)$. *Next, we have* $\Psi(f(x)) \leq \Psi(f, \Psi(x))$ *whenever* $x \in \Gamma(0, r_1, r_2)$ *and the equality holds in all of* $\Gamma(0, r_1, r_2)$ *except in finitely many classes of each circle* $C(0, r)$ $(r_1 < r < r_2)$. *The right side derivative (resp., the left side derivative) of the function* $\Psi(f, .)$ *at* μ *is equal to* $\nu^+(f, \mu)$ *(resp., to* $\nu^-(f, \mu))$. *Moreover, if the function in* μ $\Psi(f, \mu)$ *is not derivable at* μ, *then* μ *lies in* $\Psi(\mathbb{K})$.

Further, the function $\Psi(f,.)$ *is convex in* $]\log r_1, \log r_2[$. *Next, given another* $g \in \mathcal{A}(\Gamma(0, r_1, r_2))$ *the functions* ν^+ *and* ν^- *satisfy*

$$\nu^+(fg, \mu) = \nu^+(f, \mu) + \nu^+(g, \mu), \quad \nu^-(fg, \mu) = \nu^-(f, \mu) + \nu^-(g, \mu).$$

Moreover, the function $\nu^+(f,.)$ *is continuous on the right and the function* $\nu^-(f,.)$ *is continuous on the left at each point* μ. *They are continuous at* μ *if and only if they are equal.*

Proof. All statements hold in all annuli $\Gamma(0, r', r'')$ with $r_1 < r' < r'' < r_2$ because the restriction of f to $\Gamma(0, r', r'')$ belongs to $H(\Gamma(0, r', r''))$. $\quad\square$

Proposition 7.2.8. *Let* $\mu \in \mathbb{R}$ *and let* $f, g \in H(C(0, \theta^\mu))$. *Then* $\nu^+(fg, \mu) = \nu^+(f, \mu) + \nu^+(g, \mu)$ *and* $\nu^-(fg, \mu) = \nu^-(f, \mu) + \nu^-(g, \mu)$.

Proof. By Proposition 7.2.3, the relations are obvious when f and $g \in R(C(0, \theta^\mu))$ because there is an annulus $\Gamma(0, r_1, r_2) \supset C(0, \theta^\mu)$ such that $f, g \in R(\Gamma(0, r_1, r_2))$. Now we may obviously extend them to $H(C(0, \theta^\mu))$ by taking h and $\ell \in R(C(0, \theta^\mu))$ such that $\Psi(f - h, \mu) < \Psi(f, \mu)$ and $\Psi(g - \ell, \mu) < \Psi(g, \mu)$. $\quad\square$

Proposition 7.2.9. *Let* r_1, $r_2 \in \mathbb{R}$ *and let* f, $g \in \mathcal{A}(\Gamma(0, r_1, r_2))$ *having no zero in* $\Gamma(0, r_1, r_2)$ *and satisfying* $\nu(f, \mu) \neq \nu(g, \mu)$, $\forall \mu \in]\log r_1, \log r_2[$. *Then both* $\nu^+(f + g, \mu)$ *and* $\nu^-(f + g, \mu)$ *are equal either to* $\nu(f, \mu)$ *or to* $\nu(g, \mu)$.

Proof. Let $\mu_j = \log(r_j)$, $j = 1, 2$. Since both f, g have no zero in $\Gamma(0, r_1, r_2)$, $\nu(f, \mu)$ is a constant integer s and $\nu(g, \mu)$ is a constant integer $t \neq s$. Consequently, $\Psi(f, \mu)$ is of the form $a + s\mu$, $\Psi(g, \mu)$ is of the form $b + t\mu$, therefore the two functions in μ can coincide at most at one point in $[\mu_1, \mu_2]$. So, by Proposition 7.2.2, we have $\Psi(f + g, \mu) = \max(\Psi(f, \mu), \Psi(g, \mu))$ for all $\mu \in [\log(r_1), \log(r_2)]$ except maybe at all point. But then, by continuity, the equality holds in all $[\log(r_1), \log(r_2)]$.

Let us fix $\mu_0 \in]\mu_1, \mu_2[$. Suppose $\Psi(f + g, \mu) = \Psi(f, \mu)$ in a neighborhood $]\mu_1, \mu_2[$ of μ_0. Then of course, $\nu(f+g, \mu_0) = \nu(f, \mu_0)$. Suppose now that $\Psi(f + g, \mu) = \Psi(f, \mu)$ in a left neighborhood $]\mu_1, \mu_0]$ of μ_0 and $\Psi(f+g, \mu) = \Psi(g, \mu)$ in a right neighborhood $[\mu_0, \mu_2[$ of μ_0, which implies $\Psi(f, \mu) > \Psi(g, \mu)$ $\forall \mu \in]\mu_1, \mu_0[$ and $\Psi(f, \mu) < \Psi(g, \mu)$ $\forall \mu \in]\mu_0, \mu_2[$. Then we have $\nu(f + g, \mu) = \nu(f, \mu) \forall \mu \in]\mu_1, \mu_0[$ and $\nu(f + g, \mu) = \nu(g, \mu) \forall \mu \in]\mu_0, \mu_2[$. Consequently, since ν^+ is continuous on the left and ν^- is continuous on the right, we can check that both $\nu^+(f + g, \mu_0)$ and $\nu^-(f + g, \mu_0)$ are equal either to $\nu(f\mu_0)$ or to $\nu(g, \mu_0)$. $\quad\square$

Theorem 7.2.10. *Let $f \in \mathcal{A}(\mathbb{K} \setminus d(0, R))$. There exists $q \in \mathbb{N}$ such that*

$$\lim_{r \to +\infty} |f|(r)r^q = +\infty.$$

Proof. Let $s \in]R, +\infty[$ be such that $\nu^+(f, \log s) = \nu^-(f, \log s)$ and let $\tau = \nu^+(f, \log s)$. Thus, $\Psi(f, \mu)$ has a derivative at $\log s$ equal to τ. Consequently, since by Proposition 7.2.2, $\Psi(f, \mu)$ is convex, we have $\Psi(f, \mu) - \Psi(f, \log s) \geq \tau(\mu - \log s)$. Therefore,

$$\lim_{\mu \to +\infty} [\Psi(f, \mu) + (1 - \tau)\mu] = +\infty,$$

i.e., $\lim_{r \to +\infty} |f|(r)r^{(1-\tau)} = +\infty$. Finally we can take $q = \max(0, 1 - \tau)$. □

Definitions: Let $f \in H(D)$. Then f is said to be *quasi-minorated* if for every sequence $(a_n)_{n \in \mathbb{N}}$ of D such that $f(a_n) \neq 0 \ \forall n \in \mathbb{N}$, $\lim_{n \to \infty} f(a_n) = 0$ either we may extract a subsequence that converges in $\mathbb{K}$ or we may extract a subsequence $(a_{n_q})_{q \in \mathbb{N}}$ such that $\lim_{q \to \infty} |a_{n_q}| = +\infty$.

Let $\mathcal{F}$ be an increasing filter of center a and diameter r. Then f is said to be *strictly vanishing along* $\mathcal{F}$ if $\Psi_a(f, \mu) > -\infty \ \forall \mu < \log(r)$ and $\Psi_a(f, \log(r)) = \infty$.

Now, let $\mathcal{F}$ be a decreasing filter of center a and diameter r. A symmetric definition applies: f is said to be *strictly vanishing along* $\mathcal{F}$ if $\Psi_a(f, \mu) > -\infty \ \forall \mu > \log(r)$ and $\Psi_a(f, \log(r)) = \infty$.

Finally, $\mathcal{F}$ be a decreasing filter of diameter r, with no center in $\mathbb{K}$. We can just take a center a in a spherical completion of $\mathbb{K}$ and then get to the same definition. However, another method consists of taking a decreasing distances sequence $(a_n)_{n \in \mathbb{N}}$ thinner than $\mathcal{F}$. Consider the sequence $(r_n)_{n \in \mathbb{N}}$ defined as $r_n = |a_n - a_{n+1}|$. Then the we have $\Psi_{a_n}(f, \mu) = \Psi_{a_{n+1}}(f, \mu) \ \forall \mu > \log(r_n)$ and hence we can define $\Psi_{\mathcal{F}}(f, \mu) = \Psi_{a_n}(f, \mu) \ \forall \mu > \log(r_n)$. □

Proposition 7.2.11. *Let D be bounded and let $f \in H(D)$ be not identically zero. Then f is not quasi-minorated if and only if there exists a monotonous filter $\mathcal{F}$ such that f is strictly vanishing along $\mathcal{F}$.*

Proof. On one hand, if f is strictly vanishing along a monotonous filter $\mathcal{F}$, then obviously, f is not quasi-minorated. On the other hand, suppose that f is not quasi-minorated and let $(a_n)_{n \in \mathbb{N}}$ be a sequence such that $f(a_n) \neq 0 \ \forall n \in \mathbb{N}$, $\lim_{n \to \infty} f(a_n) = 0$ and such that one can't extract any converging subsequence from that sequence. Then, one can extract from that sequence a monotonous distances subsequence which, by definition, is thinner than a monotonous filter $\mathcal{F}$ which defines $\varphi_{\mathcal{F}}$ and hence we have $\varphi_{\mathcal{F}}(f) = 0$. Thus, f is vanishing along $\mathcal{F}$. Now, suppose that f is not strictly vanishing along $\mathcal{F}$.

Suppose for instance that $\mathcal{F}$ is increasing. For convenience suppose $a_1 = 0$. Let $s = \operatorname{diam}(\mathcal{F})$. There exists $r \in\,]0, s]$ such that $\Psi(f, \mu) > -\infty \ \forall \mu < \log(r)$ and $\Psi(f, \log(r)) = -\infty)$, hence f is strictly vanishing along the increasing filter of center 0 and diameter r.

Suppose now that $\mathcal{F}$ is decreasing. A symmetric proof holds when $\mathcal{F}$ is decreasing and with a center. If $\mathcal{F}$ has no center, we can take a center in a spherically complete extension and make the same proof. We can also consider the second definition of an element strictly decreasing along a decreasing filter with no center and use the definition of $\Psi_{\mathcal{F}}$. $\qquad\square$

7.3. Holomorphic properties on infraconnected sets

Throughout the chapter, D is a subset of $\mathbb{K}$. Here we shall go back to the basic properties of analytic elements which are transmitted by rational functions.

Theorems 7.2.1 and 7.2.3 are proven in [28, Chapter 22].

Theorem 7.3.1. *Let $f \in H(C(0, r))$ (resp., $f \in H(d(0, r))$). The number of zeros of f in $C(0, r)$ (resp., in $d(0, r)$) is equal to $\nu^+(f, \log r) - \nu^-(f, \log r)$, (resp., $\nu^+(f, \log r)$), taking multiplicities into account.*

Corollary 7.3.2. *Let $f \in H(C(0, r))$ have t zeros in $C(0, r)$. Let $q = \nu^+(f, \log r) - \nu^-(f, \log r)$. Then $r = \sqrt[t]{\left|\dfrac{a_q}{a_{q+t}}\right|}$.*

Theorem 7.3.3. *Let $f = \sum_{-\infty}^{+\infty} a_n x^n \in H(C(0, r))$ have no zero in $C(0, r)$. Let $t = \nu^+(f, \log r)$. Then we have $\nu^+(f, \log r) = \nu^-(f, \log r)$ and $\Psi(f(x)) = \Psi(f, \log r) = \Psi(a_t) + t \log r$ whenever $x \in C(0, r)$.*

Corollary 7.3.4. *Let $f \in H(\Gamma(0, r, s))$ have no zero in $\Gamma(0, r, s)$ and be equal to a Laurent series $\sum_{-\infty}^{+\infty} a_n x^n \ \forall x \in \Gamma(0, r, s)$. Assume that there exists an integer $m \in \mathbb{Z}^*$ such that $|a_m| \rho^m > |a_n| \rho^n \ \forall n \neq m, \ \forall \rho \in\,]r, s[$. If $m > 0$, we have $f(\Gamma(0, r, s)) = \Gamma(0, |a_m| r^m, |a_m| s^m)$, and if $m < 0$, we have $f(\Gamma(0, r, s)) = \Gamma(0, |a_m| s^m, |a_m| r^m)$.*

Theorem 7.3.5. *Let $f(x) = \sum_{n=0}^{\infty} a_n x^n \in H(d(0, r))$. The number of zeros of f in $d(0, r)$ is $\nu^+(f, \log(r))$ (taking multiplicity into account).*

Theorem 7.3.6 will be useful when considering restricted power series.

Theorem 7.3.6. *Let $f(x) = \sum_{n=0}^{\infty} a_n x^n \in H(U)$. Then f is invertible in $H(U)$ if and only if $|a_0| > \sup_{n>0} |a_n|$. And f is irreducible in $H(U)$ if and only if $|a_0| \leq |a_1|, \ |a_1| > \sup_{n>1} |a_n|$.*

Proof. If $|a_0| > \sup_{n>0} |a_n|$, by Theorem 7.3.6, f is obviously invertible in $H(U)$. Else, by Lemma 7.3.1, f admits at least one zero, and therefore is not invertible. Precisely, let q be the unique integer such that $|a_q| = \sup_{n \in \mathbb{N}} |a_n|$. If $q = 1$, then by Theorem 7.2.2, f has exactly one zero α, and since it is quasi-invertible, it is of the form $(x - \alpha)g$, with g invertible, hence f is irreducible. If $q > 1$, since f admits q zeros, f factorizes in the form $\prod_{j=1}^{q}(x - a_j)g(x)$, whereas $g \in H(U)$ has no zeros, and therefore is invertible and hence f is not irreducible. □

Proposition 7.3.7. *Let D, D' be closed bounded subsets of $\mathbb{K}$, and let $f \in H(D)$ be such that $f(D) \subset D'$. Let $g \in H(D')$. Then $g \circ f$ belongs to $H(D)$.*

Homomorphisms between $\mathbb{K}$-algebras $H(D)$ are characterized in [28, Proposition 11.6 and Theorem 11.7]. Here, for convenience, we will restrict the statement to the case of closed bounded sets.

Proposition 7.3.8. *Let D, D', D'' be closed bounded subsets of $\mathbb{K}$ and let $\gamma \in H(D')$ satisfy $\gamma(D') \subset D$. Let ϕ_γ be the mapping from $H(D)$ into $H(D')$ defined as $\phi_\gamma(f) = f \circ \gamma$. Then ϕ_γ is a $\mathbb{K}$-algebra homomorphism from $H(D)$ into $H(D')$ continuous with respect to the topology of uniform convergence on D for $H(D)$ and on D' for $H(D')$. If $\gamma \in H(D')$ and $h \in H(D'')$ satisfy $\gamma(D') = D$ and $h(D'') = D'$, then $\phi_h \circ \phi_\gamma = \phi_{\gamma \circ h}$.*

If γ is a bijection from D' onto D and if $\overset{-1}{\gamma} \in H(D)$ then ϕ_γ is a $\mathbb{K}$-algebra isomorphism from $H(D)$ onto $H(D')$ bicontinuous with respect to the topology of uniform convergence on D for $H(D)$ and on D' for $H(D')$, satisfying $\left(\phi_\gamma\right)^{-1} = \phi_{\gamma^{-1}}$.

By Propositions 7.3.8, we can easily deduce Corollary 7.3.9.

Corollary 7.3.9. *Let ϕ_γ be an isomorphism from $H(D)$ onto $H(D')$, with $\gamma \in H(D')$. Then γ is a bijection from D' onto D such that $\overset{-1}{\gamma} \in H(D)$.*

Definition: Let D be open. An element f of $H(D)$ is said to be *strictly injective* if it is injective and such that f' has no zero in D.

According to [28, Theorems 27.1 and 27.2], we have these statements.

Theorem 7.3.10. *Let $\mathbb{K}$ have characteristic zero. Let $f \in H(D)$ be injective in D. Then f is strictly injective.*

Theorem 7.3.11. *Let $a \in K$, $r \in \mathbb{R}_+$, let $f(x) = \sum_{n=0}^{\infty} a_n(x - a)^n \in H(d(a, r))$, and let $s = \sup_{n \geq 1} |a_n| r^n$ be > 0. Then the following statements are equivalent:*

(α) $|a_1| > |a_n|r^{n-1}$ whenever $n > 1$,

(β) $|f(x) - f(y)| = |x - y||a_1|$ whenever $x, y \in d(a, r)$,

(γ) f is strictly injective in $d(a, r)$.

Moreover when conditions $(\alpha), (\beta), (\gamma)$ *are satisfied, then we have* $s = |a_1|r$ *and* $|f'(x)| = |a_1|$ *whenever* $x \in d(a, r)$.

Theorem 7.3.12 is proven in [28, Theorem 26.1 and Corollary 26.2].

Theorem 7.3.12. *Let* $f(x) = \sum_{n=0}^{\infty} a_n(x - a)^n \in H(d(a, r))$ *and let* $s = \sup_{n \geq 1} |a_n| r^n$. *Then* $f(d(a, r)) = d(a_0, s)$, $f(d(a, r^-)) = d(a_0, s^-)$, *and* $\Psi_a(f - a_0, \log r) = \log s$.

Theorem 7.3.13 is proven in [28, Chapter 27, Corollary 27.6].

Theorem 7.3.13. *Let* $f \in H(d(a, r))$ *be strictly injective and let* $d(b, s) = f(d(a, r))$. *Then* f^{-1} *belongs to* $H(d(b, s))$.

Theorem 7.3.14 is proven in [28, Chapter 11 (Proposition 11.15)].

Theorem 7.3.14. *Let* D *have an empty annulus* A. *Let* w_1, w_2 *be the functions defined on* D *by* $w_1(x) = 1, w_2(x) = 0$ *if* $x \in \mathcal{I}(A)$ *and* $w_1(x) = 0, w_2(x) = 1$ *if* $x \in \mathcal{E}(A)$. *Then* w_1 *and* w_2 *belong to* $H(D)$.

As an immediate consequence, we have Corollaries 7.3.15 and 7.3.16.

Corollary 7.3.15. *If* D *has infinitely many infraconnected components then* $H(D)$ *is not Noetherian.*

Corollary 7.3.16. *If* D *has finitely many infraconnected components* $D_1, \ldots, D_q$. *Then* $H(D) = H(D_1) \times \cdots \times H(D_q)$.

Theorem 7.3.17. *Let* D *be a closed bounded infraconnected subset of* $\mathbb{K}$ *and let* $f \in H(D)$. *Then* $f(D)$ *is infraconnected.*

Proof. Let $R = \text{diam}(D)$ (with eventually $D = +\infty$). Let $E = f(D)$ and suppose that E is not infraconnected. Then E admits an empty annulus $\Gamma(a, r', r'')$. Without loss of generality, we can suppose $a = 0$. Let $A = \mathcal{I}(\Gamma(a, r', r''))$ and $B = \mathcal{E}(\Gamma(a, r', r''))$. By Theorem 7.3.14, the characteristic function u of A and the characteristic function u of B belong to $H(E)$. In D, we have $u \circ f(x) = 1$ whenever $x \in D$ such that $f(x) \in A$ and $u \circ f(x) = 0$ whenever $x \in D$ such that $f(x) \in B$. Let $g = u \circ f$. Then g_* is continuous in $\text{Mult}((H(D), \| . \|)$ and hence $g_*(\varphi)$ only takes values 1 and 0. Suppose first that there exists $s > 0$ such that $g_*(0,s\varphi) = 0$ and let $t = \inf\{s \;_{0,s}\varphi(g) = 0\}$. Since by Theorem 6.3.9, g_* is continuous on $\text{Mult}(H(D), \| . \|)$, $g_*(0,t\varphi)$ should be equal to both 0 and 1, a contradiction.

Suppose now that $g_*(0,s\varphi) = 1 \,\forall s \leq R$. Let $b \in D$ be such that $f(b) \in B$ and let $q = |b|$. Then we have $_{b,q}\varphi(g) =_{0,q} \varphi(g) = 1$, but $_{b,s}\varphi(g)$ is obviously null when s is small enough. So we can put $t = \inf\{s \ _{b,s}\varphi(g) = 1\}$ and by continuity we have $_{b,t}\varphi(g) = 0\}$, a contradiction again. Consequently, the hypothesis "E not infraconnected" is wrong and this proves the claim. $\qquad\square$

Theorem 7.3.18. *Let D be closed and bounded. If an ideal contains a quasi-invertible element, then it is generated by a polynomial whose zeros belong to $\overset{\circ}{D}$.*

Proof. Let $\mathcal{J}$ be an ideal of $H(D)$ that contains a quasi-invertible element f, and let $\mathcal{J}_0$ be the set of polynomials that belong to $\mathcal{J}$. Then $\mathcal{J}_0$ is an ideal of $\mathbb{K}[x]$ and hence is a principal ideal, generated by a polynomial T whose zeros belong to D. By hypothesis f factorizes in $H(D)$ in the form Pg with g invertible in $H(D)$ and $P(x) \in \mathbb{K}[x]$, all the zeros of P lying inside $\overset{\circ}{D}$. Since fg^{-1} belongs to $\mathcal{J}$, obviously P belongs to $\mathcal{J}_0$. Hence T divides P and then all the zeros of T lie in $D \cap \overset{\circ}{\overline{D}}$. We will show that $\mathcal{J} = TH(D)$.

It is clearly seen that $\mathcal{J}_0$ is an ideal of $\mathbb{K}[x]$, hence there exists $T(x) \in \mathbb{K}[x]$ such that $\mathcal{J}_0 = T(x)\mathbb{K}[x]$.

Let $\alpha_1, \ldots, \alpha_q$ be the zeros of T. Now we suppose that there exists some $h \in \mathcal{J} \setminus TH(D)$. Then we can find $r > 0$ such that $d(\alpha_i, r) \subset D$, whenever $i = 1, \ldots, q$. Let $G = \bigcup_{i=1}^{q} d(\alpha_i, r)$ and let $D' = D \cup G$. Since the zeros of T lie in G there exists $\lambda > 0$ such that $|T(x)| \geq \lambda$ whenever $x \in D \setminus G$. Now since D is closed and bounded, there exists $b \in \mathbb{K}$ such that $\|b\ell\|_D < \lambda$. We put $\phi = T + b\ell$. Clearly outside G, we have $|\phi(x)| \geq \lambda$. Next in each disk $d(\alpha_i, r)$, ϕ has finitely many zeros, hence in D', ϕ has finitely many zeros, all of them in G. Hence it factorizes in the form $Q(x)W(x)$ with $W \in H(D'), W(x) \neq 0$ whenever $x \in D'$ and Q a polynomial whose zeros belong to G. Since W has no zero in G, $|W(x)|$ has a strictly positive lower bound in G and another non-zero lower bound in $D' \setminus G$ because $|Q(x)|$ is obviously bounded in D'. Finally W has a non-zero lower bound in D', therefore it is invertible in $H(D')$. Hence Q belongs to $\mathcal{J}$. But then T divides Q in $\mathbb{K}[x]$. Since $T + b\ell$ is equal to WQ, then T divides $T + b\ell$, and ℓ, and h too. This contradicts the hypothesis $h \in \mathcal{J} \setminus TH(D)$, and finishes proving that T generates $\mathcal{J}$. $\qquad\square$

Proposition 7.3.19. *Let $\mathcal{F}$ be a pierced filter on D, let $(T_n)_{n\in\mathbb{N}}$ be a sequence of holes of D that runs $\mathcal{F}$ and let $D^* = \mathbb{K} \setminus (\bigcup_{n=0}^{\infty} T_n)$. Let $g_1, \ldots, g_q \in H_b(D^*)$ be vanishing along $\mathcal{F}$, with g_1 properly vanishing.*

For every $x \in D^*$ let $S(x) = \sup_{1 \leq i \leq q} |g_i(x)|$ and let F be the ideal generated by $g_1, \ldots, g_q$ in $H_b(D^*)$, and let $\overline{F}$ be its closure in $H_b(D)$.

There exists a sequence $(z_n)_{n \in \mathbb{N}}$ in D, thinner than F such that $g_1(z_n) \neq 0$ and an element $T \in \overline{F}$ such that $\lim_{n \to \infty} \frac{|T(z_n)|}{S(z_n)} = +\infty$.

Proof. Without loss of generality we may assume F to be a decreasing filter or a Cauchy filter. Indeed if F is an increasing filter of center α of diameter R, let $T(b, \rho)$ be a hole of D included in $d(\alpha, R^-)$, let $\gamma(x) = \frac{1}{x-b}$, and let $D' = \gamma(D)$. Then D' admits a decreasing pierced filter F', image of F, by γ. Hence we will assume F to be a decreasing pierced filter or a Cauchy pierced filter.

Without loss of generality we may clearly assume $D = D^*$. Since the g_j are bounded, we may obviously assume $\|g_j\|_D \leq 1$ whenever $j = 1, \ldots, q$. Let $R = \text{diam}(F)$ and let $(x_m)_{m \in \mathbb{N}}$ be a sequence in D thinner than F such that $g_1(a_m) \neq 0$ whenever $m \in \mathbb{N}$, with $|x_{m+2} - x_{m+1}| < |x_{m+1} - x_m|$. Since F is pierced, there exists a subsequence $(x_{m_q})_{q \in \mathbb{N}}$ of the sequence (x_m) together with a sequence of holes $(T_q)_{q \in \mathbb{N}}$ of D such that

$$T_q \subset d(x_{m_q+1}, d_{m_q}) \setminus d(x_{m_q+2}, d_{m_q+1}).$$

Hence without loss of generality we may assume that we have a sequence of holes $(T_m)_{m \in \mathbb{N}}$ of D such that $T_m \subset d(x_{m+1}, d_m) \setminus d(x_{m+2}, d_{m+1})$.

We put $D_m = d(x_{m+1}, d_m) \cap D$ and $A_n = D_{2n+1} \setminus D_{2n+3}$. For each n, let $u_n \in A_n$ be such that $|g_1(u_n)| \geq \|g_1\|_{A_n} \left(\frac{n}{n+1}\right)$. For each $j = 1, \ldots, t$, let $M_n^j = \|g_j\|_{A_n}$ and let $M_n = \max_{1 \leq j \leq t} M_n^j$. Since $g_1(x_m) \neq 0$ we have $M_n > 0$ whenever $n \in \mathbb{N}$, and since $\|g_j\|_D \leq 1$ for all j, we have $M_n \leq 1$ whenever $n \in \mathbb{N}$.

We will construct a sequence (U_n) in $H_b(D)$ satisfying

(1) $|U_n(x)| \leq \frac{1}{n+1}$ whenever $x \in D \setminus A_n$.

(2) $\sqrt{M_n}\left(\frac{n+1}{n}\right) > \|g_1 U_n\|_{A_n} > \sqrt{M_n}$.

For every $n \in \mathbb{N}$, let $T_n = d(\beta_n, \rho_n^-)$, $u_n = x_{2n+2}$, $a_n = \beta_{n+1}$, $b_n = \beta_{n+2}$, $c_n = \beta_{2n+3}$ and let $\epsilon_n \in d(0, \frac{1}{n})$. Let us fix $n \in \mathbb{N}$. It is seen that $|u_n - a_n| > |u_n - b_n|$, hence there exists $q_n \in \mathbb{N}$ such that

(3) $|\epsilon_n| \left|\frac{u_n - a_n}{u_n - b_n}\right|^{q_n} g(u_n) > \sqrt{M_n}$

and of course there exists q_n' such that

(4) $\left(\frac{d_{2n+1}}{d_{2n+2}}\right)^{q_n} \left(\frac{d_{2n+3}}{d_{2n+2}}\right)^{q_n'} < 1$

We put $h_n(x) = \epsilon_n \left(\frac{x-a_n}{x-b_n}\right)^{q_n} \left(\frac{x-c_n}{x-b_n}\right)^{q'_n}$.

Then by (4) we see that:

when $|x - c_n| > d_{2n+1}$ we have $|h_n(x)| = |\epsilon_n| < \frac{1}{n}$

when $|x - c_n| \le d_{2n+3}$ we have $|x - a_n| = |a_n - c_n| = d_{2n+1}$ and $|x - b_n| = |b_n - c_n| = d_{2n+2}$ hence $|h_n(x)| \le |\epsilon_n| \left(\frac{d_{2n+1}}{d_{2n+2}}\right)^{q_n} \left(\frac{d_{2n+3}}{d_{2n+2}}\right)^{q'_n} < \frac{1}{n}$.

But now we notice that x belongs to $D \setminus A_n$ if and only if x satisfies: either $|x - c_n| > d_{2n+1}$ or $|x - c_n| \le d_{2n+3}$, hence we have proven that $|h_n(x)| < \frac{1}{n}$ whenever $x \in D \setminus A_n$. This shows h_n satisfies (1).

When $x \in A_n$, i.e., when $d_{2n+3} < |x - c_n| < d_{2n+1}$, we see that $\|g_1 h_n\|_{A_n} \ge |g_1(u_n)h_n(u_n)|$ hence by (3) we have $\|g_1 h_n\|_{A_n} \ge \sqrt{M_n}$. Hence there trivially exists $\lambda_n \in d(0,1)$ such that $\left(\frac{n+1}{n}\right)\sqrt{M_n} > |\lambda_n g_1 h_n|_{A_n} > \sqrt{M_n}$.

Now we put $U_n = \lambda_n h_n$, and we see that U_n satisfies (1) and (2). In particular we have $\|g_1 U_n\|_D \le \max\left(\sqrt{M_n}\left(\frac{n+1}{n}\right), \frac{\|g_1\|_D}{n+1}\right)$ hence $\lim_{n \to \infty} \|g_1 U_n\|_D = 0$. Let $T = \sum_{n=0}^{\infty} g_1 U_n$. By definition T belongs to $\overline{F}$ because for every $t \in \mathbb{N}$, $g \sum_{n=0}^{t} U_n$ belongs to F.

By (2) there exists a sequence $(z_n))_{n \in \mathbb{N}}$ in D satisfying $z_n \in A_n$ and

(5) $\sqrt{M_n} < |g_1(z_n)U(z_n)| < M_n\left(\frac{n+1}{n}\right)$

hence we have

(6) $|U_n(z_n)| > \frac{\sqrt{M_n}}{|g_1(z_n)|} \ge \frac{1}{\sqrt{M_n}}$ because $|g_1(z_n)| \le M_n$.

Besides when $j \ne n$, z_n belongs to $D \setminus A_j$ hence by (1) and (6) we have $|U_j(z_n)| < \frac{1}{j+1} < \frac{1}{\sqrt{M_n}} < |U_n(z_n)|$ whenever $j \ne n$. Hence we see that $|T(z_n)| = |g_1(z_n)U_n(z_n)|$ whenever $n \in \mathbb{N}$. But then, by (5) we see that $\frac{|T(z_n)|}{S(z_n)} = \frac{|T(z_n)|}{M_n} > \frac{1}{\sqrt{M_n}}$. Thus, we have $\lim_{n \to \infty} \frac{|T(z_n)|}{S(z_n)} = +\infty$ and this finishes the proof of Proposition 7.3.19. $\square$

Notation: For any integer $n \in \mathbb{N}$ we will denote by $\mathcal{Q}_n(D)$ the set of the quasi-invertible elements $f \in H(D)$ that have exactly n zeros, taking multiplicity into account, and by $\mathcal{Q}(D)$ the set $\bigcup_{n=0}^{\infty} \mathcal{Q}_n(D)$.

7.4. *T*-filters and *T*-sequences

The behavior of analytic elements is linked to the existence of certain pierced filters, called *T*-filters [17, 20, 23]. This has a strong implication on ultrametric spectral theory.

Notations: In all of this chapter, D denotes an infraconnected closed set of diameter $S \in \overline{\mathbb{R}}_+$. Let $a \in \tilde{D}$, let $r \in]\delta(a, D), S] \cap |\mathbb{K}|$ and let $(T_i)_{i \in I}$ be the set of the holes of D included in $C(a, r)$. We will denote by $\mathcal{T}(D, a, r)$ the set $\bigcup_{i \in I} T_i$.

For every $q \in \mathbb{N}$ we will denote by $\mathcal{S}(D, a, r, q)$ the set of the monic polynomials P of degree q whose zeros lie in $\mathcal{T}(D, a, r)$.

Let $a \in \tilde{D}$. Let $r \in [\delta(a, D), \mathrm{diam}(D)] \cap |\mathbb{K}|$ and let $q \in \mathbb{N}$. We put

$$\gamma_D(a, r, q) = r^q \inf_{P \in \mathcal{S}(D,a,r,q)} \left\| \frac{1}{P} \right\|_D.$$

For every $q \in \mathbb{N}$ we will denote by $\mathcal{Q}(D, a, r, q)$ the set of the monic polynomials P of degree q whose zeros lie in $\mathcal{T}(D, a, r)$.

Lemma 7.4.1. *Let* $D = d(a, r^-) \setminus d(a, \rho^-)$, *with* $0 < \rho 0 < r$. *Then*

$$\gamma_D(a, r, q) = \left(\frac{r}{\rho} \right)^q.$$

Proof. Indeed, for each monic polynomial P of degree q, having all its zeros in $d(a, \rho^-)$ we have $\left\| \frac{1}{P} \right\|_D = \frac{1}{\rho^q}$. □

Lemma 7.4.2. *Let* $a \in \tilde{D}$, *let* $r \in [\delta(a, D), \mathrm{diam}(D)] \cap |\mathbb{K}|$, *let* $q \in \mathbb{N}$ *and let* $P \in \mathcal{Q}(D, a, r, q)$. *Let* $(T_j)_{1 \leq j \leq \ell}$ *be the holes of* D *included in* $C(a, r)$ *which contain at least one zero of* P. *For every* $j = 1, \ldots, \ell$ *let* $T_j = d(a_j, r_j^-)$, *and let* t_j *be the number of zeros of* P *in* T_j *(taking multiplicities into account). Then we have*

$$\left\| \frac{1}{P} \right\|_D = \left\| \frac{1}{P} \right\|_{D \cap C(a,r)} = \frac{1}{\inf_{1 \leq j \leq \ell} \left(r_j^{t_j} \prod_{\substack{1 \leq m \leq \ell \\ m \neq j}} |a_m - a_j|^{t_m} \right)}.$$

Proof. Let $D' = \mathbb{K} \setminus (\bigcup_{j=1}^{\ell} T_j)$, let $\lambda = \frac{1}{\min_{1 \leq j \leq \ell} {}_D\varphi_j(P)}$. It is seen that $\frac{1}{P}$ belongs to $H_0(D')$ and then by Theorem 7.1.13, we have $\left\| \frac{1}{P} \right\|_{D'} = \lambda$. But then, we see that $\left\| \frac{1}{P} \right\|_{D'} \geq \left\| \frac{1}{P} \right\|_D \geq \left\| \frac{1}{P} \right\|_{D \cap C(a,r)} \geq \lambda$ and therefore we have

(1) $\left\| \frac{1}{P} \right\|_D = \lambda$. Now, it is easily seen that ${}_D\varphi_j(x - a) = r_j$ when $\alpha \in T_j$, and that ${}_D\varphi_j(x - a) = |\alpha - a_j|$ when $\alpha \notin T_j$ so that we have ${}_D\varphi_j(P) = r_j^{t_j} \prod_{\substack{1 \leq m \leq \ell \\ m \neq j}} |a_m - a_j|^{t_m}$. Finally by (1) the conclusion is clear. □

Notation: Let $a \in \tilde{D}$. Let $r \in [\delta(a, D), \mathrm{diam}(D)] \cap |\mathbb{K}|$ and let $q \in \mathbb{N}$. We put $\gamma_D(a, r, q) = r^q \inf_{P \in \mathcal{Q}(D, a, r, q)} \|\frac{1}{P}\|_D$.

Let D be bounded, of diameter r and included in a disk $d(a, r^-)$. Let $W(d, q)$ be the set of monic polynomials whose zeros lie in $d(a, r^-) \setminus D$. We denote by $\vartheta(D, q)$ the number $r^q \inf_{P \in W(D, q)} \|\frac{1}{P}\|_D$.

The following Lemma 7.4.3 is shown in [23, Proposition 34.1].

Lemma 7.4.3. *Let D be bounded, of diameter r, and included in a disk $d(a, r^-)$. Let ψ be defined as $\psi(x) = \alpha x + \beta$, and let $D' = \psi(D)$. Then for every $q \in \mathbb{N}$ we have $\vartheta(D', q) = \vartheta(D, q)$.*

Corollary 7.4.4. *Let $a \in \tilde{D}$, let $r \in [\delta(a, D), \mathrm{diam}(D)] \cap |\mathbb{K}|$ and let $q \in \mathbb{N}$. Let $(T_i)_{i \in I}$ be the set of the holes of D included in $C(a, r)$ with $T_i = d(a_i, r_i^-)$. Then we have*

$$\gamma_D(a, r, q)$$

$$= \frac{r^q}{\sup\left\{ \inf_{1 \leq j \leq \ell} \left(r_{i_j}^{t_j} \prod_{\substack{1 \leq m \leq \ell \\ m \neq i_j}} |a_m - a_{i_j}|^{t_m} \right) \middle| 1 \leq \ell \leq q, \ (i_1, \ldots, i_\ell) \in I^\ell, \ \sum_{j=1}^\ell t_j = q \right\}}.$$

Moreover, for every $\epsilon > 0$, there exist classes G_j, $(1 \leq j \leq t)$ of $C(a, r)$, and integers q_j, $(1 \leq j \leq t)$ such that $\sum_{j=1}^t q_j = q$, and $\max_{1 \leq j \leq t} \vartheta(G_j \cap D, q_j) \leq \gamma_D(a, r, q) + \epsilon$.

Proposition 7.4.5. *Let $a \in D$ and $r \in [\delta(a, D)], \mathrm{diam}(D)] \cap |\mathbb{K}|$. Let $f \in R(D)$ and let s (resp., t) be the number of zeros (resp., of the poles) of f inside $C(a, r)$ (taking multiplicities into account). We suppose $s \leq t$ and put $q = t - s$. If $C(a, r) \cap D \neq \emptyset$ then we have $\|f\|_{D \cap C(a, r)} \geq {}_D\varphi_{a, r}(f)\gamma_D(a, r, q)$.*

Proof. Let $f = \frac{P}{Q}g$ where P and Q are monic polynomials whose zeros lie in $C(a, r)$ while $g \in R(D)$ has neither any zero nor any pole in $C(a, r)$. We know that $|g(x)| = {}_D\varphi_{a, r}(g)$ for all $x \in C(a, r)$. Hence without loss of generality we may assume $g = 1$. If $P = 1$ the inequality we want to prove is obvious. Thus, the inequality is already proven when $s = 0(\forall t)$.

Now, given $m, n \in \mathbb{N}$, with $m \leq n$, we assume the inequality proven when $s \leq m$, $t \leq n$, and $s \leq t$ and will prove it when $s = m + 1$, $t = n + 1$. Indeed, suppose $s = m + 1$, $t = n + 1$. Let E be the set of the couples (ξ, η) such that ξ is a zero of P, and η is a zero of Q. Now let $(\alpha, \beta) \in E$ be such that $|\alpha - \beta| = \inf\{|\xi - \eta| \mid (\xi, \eta) \in E\}$. Let $u(x) = \frac{x - \alpha}{x - \beta}$ and let $f(x) = h(x)u(x)$. We will show

(1) $\|f\|_{D \cap C(a, r)} \geq \|h\|_{D \cap C(a, r)}$.

We notice

(2) $_D\varphi_{a,r}(u) = 1$.

First we suppose that $\|h\|_{D \cap C(a,r)} = _D\varphi_{a,r}(h)$. We have $_D\varphi_{a,r}(f) = _D\varphi_{a,r}(h)$ hence $\|f\|_{D \cap C(a,r)} \geq _D\varphi_{a,r}(f) = \|h\|_{D \cap C(a,r)}$.

Now we suppose that $_D\varphi_{a,r}(h) < \|h\|_{D \cap C(a,r)}$. Let $D' = D \cap C(a,r)$. It is seen that $_D\varphi_{a,r}(h) = _{D'}\varphi_{D'}(h)$ and then, there exists a hole $T = d(b, \rho^-)$ of $D \cap C(a,r)$ which contains at least one zero of Q, such that $_{D \cap C(a,r)}\varphi_{b,\rho}(h) = \|h\|_{D \cap C(a,r)}$. Actually T is a hole of D included in $C(a,r)$ because it contains a pole of h. Hence we may assume that b is just a pole of h. We will show

(3) $_D\varphi_{b,\rho}(u) \geq 1$.

By definition of (α, β) we have

(4) $|\alpha - b| \geq |\alpha - \beta|$.

If $\alpha \in d(b, \rho^-)$ then β also belongs to $d(b, \rho^-)$ and we have $_D\varphi_{b,\rho}(u) = 1$.
If $\alpha \notin d(b, \rho^-)$:

> either $|b - \beta| > |\alpha - \beta|$ while $\rho \leq |b - \beta|$ and then we have $_D\varphi_{b,\rho}(u) = 1$,
> or $|b - \beta| = |\alpha - \beta|$ while $|b - \alpha| = |\beta - \alpha|$ and then we have $_D\varphi_{b,\rho}(u) = 1$ again,
> or $|b - \beta| < |\alpha - \beta|$ and then we have $_D\varphi_{b,\rho}(u) > 1$.

Thus, (3) is now proven in all the cases. Hence we have

$$\|f\|_{D \cap C(a,r)} \geq _D\varphi_{b,\rho}(h) = \|h\|_{D \cap C(a,r)}$$

and this finishes showing (1). But now h clearly has m zeros and n poles in $C(a,r)$. Hence it satisfies the inequality

$$\|f\|_{D \cap C(a,r)} \geq _D\varphi_{a,r}(h)\gamma_D(a, r, q)$$

and therefore by (1) and (2) we have

$$\|f\|_{D \cap C(a,r)} \geq _D\varphi_{a,r}(h)_D\varphi_{a,r}(u)\gamma_D(a, r, q).$$

We have now proven the inequality when $s = m+1$, $t = n+1$. Therefore, we can check that the inequality announced in Proposition 7.4.5 is proven for every couple (s, t). Since the inequality is true for $(0, t - s)$, it is true for $(1, t - s + 1), \dots, (s, t)$. This ends the proof of Proposition 7.4.5. $\qquad\square$

Definitions and notations: Let $\mathcal{F}$ be an increasing filter on D of center $a \in \tilde{D}$ and diameter S. The filter $\mathcal{F}$ will be called *an increasing T-filter* if there exists a sequence of circles $(\Sigma_m)_{m \in \mathbb{N}}$ with $\Sigma_m = C(a, d_m)$ $(m \in \mathbb{N})$ such that $\lim_{m \to \infty} d_m = S$ and $d_m < d_{m+1}$ together with a sequence of natural integers $(q_m)_{m \in \mathbb{N}}$ satisfying

$$\lim_{m \to \infty} \gamma_D(a, d_m, q_m) \prod_{j=1}^{m-1} \left(\frac{d_j}{d_m}\right)^{q_j} = 0. \tag{T.1}$$

Given an increasing filter $\mathcal{F}$ of center a and diameter S, we denote by $\mathcal{B}(\mathcal{F})$ the set $\{x \in D \mid |x - a| < S\}$ and by $\mathcal{P}(\mathcal{F})$ the set $D \setminus \mathcal{B}(\mathcal{F})$.

Given a decreasing filter $\mathcal{F}$ of center a and diameter S, we denote by $\mathcal{B}(\mathcal{F})$ the set $\{x \in D \mid |x - a| > S\}$ and by $\mathcal{P}(\mathcal{F})$ the set $D \setminus \mathcal{B}(\mathcal{F})$. Given an decreasing filter $\mathcal{F}$ with no center and diameter S, we put $\mathcal{B}(\mathcal{F}) = D$.

A decreasing filter $\mathcal{F}$ will be called *a decreasing T-filter* if it admits a basis $(D_m)_{m \in \mathbb{N}}$ with $D_m = d(a_m, d_m) \cap D \setminus (\bigcap_{m \in \mathbb{N}} d(a_m, r_m))$ together with a sequence of natural integers $(q_m)_{m \in \mathbb{N}}$ such that, putting $\Sigma_m = C(a_m, d_m)$, the sequences $(a_m)_{m \in \mathbb{N}}, (d_m)_{m \in \mathbb{N}}, (q_m)_{m \in \mathbb{N}}$ satisfy

$$\lim_{m \to \infty} \gamma_D(a, d_m, q_m) \prod_{j=1}^{m-1} \left(\frac{d_m}{d_j}\right)^{q_j} = 0. \tag{T.2}$$

In particular, this definition holds when $\mathcal{F}$ is a decreasing filter of center a and diameter S, and then we can take $a_m = a$ for every $m \in \mathbb{N}$.

Another way to define T-filters consists of introducing T-sequences. We will call *a weighted sequence* a sequence $(T_{m,i}, q_{m,i})_{\substack{1 \le i \le s(m) \\ m \in \mathbb{N}}}$ with $(T_{m,i})_{\substack{1 \le i \le s(m) \\ m \in \mathbb{N}}}$ a monotonous distances holes sequence and $(q_{m,i})_{\substack{1 \le i \le s(m) \\ m \in \mathbb{N}}}$ a sequence of integers.

If a weighted sequence $(T_{m,i}, q_{m,i})$ $(1 \le i \le k(m), m \in \mathbb{N})$ is increasing (resp., decreasing) we call *monotony* the sequence $(d_m)_{m \in \mathbb{N}}$ defined by $d_m = \delta(T_{m,1}, T_{m-1,1})$ (resp., $d_m = \delta(T_{m,1}, T_{m+1,1})$) and we call *piercing* the sequence $(\rho_{m,i})_{1 \le i \le k(m), m \in \mathbb{N}}$ defined by $\rho_{m,i} = \text{diam}(T_{m,i})$.

A weighted sequence $(T_{m,i}, q_{m,i})_{\substack{1 \le i \le s(m) \\ m \in \mathbb{N}}}$ will be said to be *idempotent* if $q_{m,i} = 0$ or 1 for all (m, i), $(1 \le i \le s(m), m \in \mathbb{N})$.

All definitions given about monotonous distances holes sequences will apply in the same way to weighted sequences: a weighted sequence $(T_{m,i}, q_{m,i})_{\substack{1 \le i \le s(m) \\ m \in \mathbb{N}}}$ will be said "to be something" (or "to satisfy a certain property") if "so is" (or "so does") the monotonous distances holes sequence $(T_{m,i})_{\substack{1 \le i \le s(m) \\ m \in \mathbb{N}}}$.

In particular, the diameter, the monotony, the piercing of the monotonous distances holes sequence $(T_{m,i})_{\substack{1 \le i \le s(m) \\ m \in \mathbb{N}}}$ will be called, respectively, *the diameter, the monotony, the piercing* of the weighted sequence $(T_{m,i}, q_{m,i})_{\substack{1 \le i \le s(m) \\ m \in \mathbb{N}}}$.

Now let $(T_{m,i}, q_{m,i})_{\substack{1 \le i \le s(m) \\ m \in \mathbb{N}}}$ be an increasing (resp., decreasing) weighted sequence of monotony $(d_m)_{m \in \mathbb{N}}$, associated to $\mathcal{F}$ and for every (i, m), $(1 \le i \le s(m), m \in \mathbb{N})$, let $T_{m,i} = d(a_{m,i}, \rho_{m,i}^-)$ and let $q_m = \sum_{i=1}^{s(m)} q_{m,i}$.

The weighted sequence will be said to be a *T-sequence* if it satisfies

$$\lim_{m \to \infty} \left(\sup_{1 \le j \le s(m)} \left[\left(\frac{d_m}{\rho_{m,j}} \right)^{q_{m,j}} \prod_{\substack{i \ne j \\ 1 \le i \le s(m)}} \left(\frac{d_m}{|a_{m,i} - a_{m,j}|} \right)^{q_{m,i}} \right] \prod_{n=1}^{m-1} \left(\frac{d_n}{d_m} \right)^{q_n} \right) = 0$$

(T.3)

resp.,

$$\lim_{m \to \infty} \left(\sup_{1 \le j \le s(m)} \left[\left(\frac{d_m}{\rho_{m,j}} \right)^{q_{m,j}} \prod_{\substack{i \ne j \\ 1 \le i \le s(m)}} \left(\frac{d_m}{|a_{m,i} - a_{m,j}|} \right)^{q_{m,i}} \right] \prod_{n=1}^{m-1} \left(\frac{d_m}{d_n} \right)^{q_n} \right) = 0).$$

(T.4)

We will call *subsidence* of a T-sequence $(T_{m,i}, q_{m,i})$ defined as above, the number

$$\sup_{m \in \mathbb{N}} \left(\sup_{1 \le j \le s(m)} \left[\log \left(\frac{d_m}{\rho_{m,j}} \right)^{q_{m,j}} + \sum_{\substack{i \ne j \\ 1 \le i \le s(m)}} \log \left(\frac{d_m}{|a_{m,i} - a_{m,j}|} \right)^{q_{m,i}} \right] \right.$$
$$\left. - \sum_{n=1}^{m-1} \left| \log \left(\frac{d_n}{d_m} \right)^{q_n} \right|_\infty \right)$$

Given a T-sequence $(T_{m,i}, q_{m,i})_{\substack{1 \le i \le s(m) \\ m \in \mathbb{N}}}$, a monotonous filter $\mathcal{F}$ will be said *to admit the T-sequence* $(T_{m,i}, q_{m,i})_{\substack{1 \le i \le s(m) \\ m \in \mathbb{N}}}$ if it is associated to the monotonous distances holes sequence $(T_{m,i})_{\substack{1 \le i \le s(m) \\ m \in \mathbb{N}}}$.

Remark. Given a T-sequence $(T_{m,i}, q_{m,i})_{\substack{1 \le i \le s(m) \\ m \in \mathbb{N}}}$, for every $t \in \mathbb{N}$, the weighted sequence $(T_{m,i}, q_{m,i})_{\substack{1 \le i \le s(m) \\ m \ge t}}$ is a T-sequence again.

Lemma 7.4.6. *Let* $(T_{m,i}, q_{m,i})_{\substack{1 \le i \le s(m) \\ m \in \mathbb{N}}}$ *be an increasing (resp., decreasing) weighted sequence, let* $q_m = \sum_{i=1}^{s(m)} q_{m,i}$ *and let* $(C(a_m, d_m))_{m \in \mathbb{N}}$ *be a sequence of circles that runs the weighted sequence. The weighted sequence is a T-sequence if and only if there exists a sequence of monic polynomials*

$(Q_m)_{m\in\mathbb{N}}$ *such that for each* $(m,i)_{1\leq i\leq s(m)}$, Q_m *admits exactly a zero of order* $q_{m,i}$ *in* $T_{m,i}$ *and has no other zero in* $\mathbb{K}$, *satisfying further*

$$\lim_{m\to\infty}\left(\left(\varphi_{a_m,d_m}(Q_m)\|\frac{1}{Q_m}\|_{C(a_m,d_m)\cap D}\right)\prod_{n=1}^{m-1}\left(\frac{d_n}{d_m}\right)^{q_n}\right)=0\ \left(resp.,\right.$$

$$\lim_{m\to\infty}\left(\left(\varphi_{a_m,d_m}(Q_m)\|\frac{1}{Q_m}\|_{C(a_m,d_m)\cap D}\right)\prod_{n=1}^{m-1}\left(\frac{d_m}{d_n}\right)^{q_n}\right)=0.\right)$$

Lemma 7.4.7. *Let* $\mathcal{F}$ *be a monotonous filter on* D. *Then* $\mathcal{F}$ *is a T-filter if and only if there exists a T-sequence associated to* $\mathcal{F}$.

Proposition 7.4.8. *Let* $\mathcal{F}$ *be an increasing (resp., decreasing) T-filter of center* α *and diameter* r. *Let* $h(x)=\frac{1}{x-a}$, *let* $D'=h(D)$ *and* $\mathcal{F}'=h(\mathcal{F})$. *If* $|a-\alpha|<r$, $\mathcal{F}'$ *is a decreasing (resp., increasing) T-filter of center* α *and diameter* $\frac{1}{r}$. *If* $|a-\alpha|\geq r$, $\mathcal{F}'$ *is an increasing (resp., decreasing) T-filter of center* $h(\alpha)$ *and diameter* $\frac{r}{|a-\alpha|^2}$.

Proposition 7.4.9. *Let* D *admit a T-sequence* $(T_{m,i},q_{m,i})_{1\leq i\leq k(m),\ m\in\mathbb{N}}$, *and let* $(V_{m,j})_{1\leq j\leq l(m),\ m\in\mathbb{N}}$ *be a monotonous distance holes sequence of an infraconnected set* $E\subset D$ *such that for every* (m,i), $1\leq i\leq k(m)$, $m\in\mathbb{N}$, $T_{m,i}$ *is included in certain hole* $V_{m,j}$ *of* E. *For every* $(m,j)(1\leq j\leq l(m))$, *we denote by* $\mathcal{T}_{m,j}$ *the set of the* (m,i) *such that* $T_{m,i}$ *is included in* $V_{m,j}$ *and we put* $s_{m,j}=\sum_{(m,i)\in\mathcal{T}_{m,j}}q_{m,i}$ *when* $\mathcal{T}_{m,j}\neq\emptyset$, *and* $s_{m,j}=0$ *when* $\mathcal{T}_{m,j}=\emptyset$. *Then the weighted sequence* $(V_{m,j},s_{m,j})_{1\leq j\leq l(m),\ m\in\mathbb{N}}$ *is a T-sequence of* E.

Lemma 7.4.10 is elementary.

Lemma 7.4.10. *Let* $(d_m)_{m\in\mathbb{N}}$ *be a strictly monotonous sequence of limit* r. *Then* $\lim_{m\to\infty}\sum_{j=0}^{m-1}|\log d_m-\log d_j|_\infty=+\infty$ *if and only if* $\sum_{j=0}^\infty|\log r-\log d_j|_\infty=+\infty$.

Theorem 7.4.11 is an immediate application of Lemma 7.4.10.

Theorem 7.4.11. *Let* $(T_m)_{m\in\mathbb{N}}$ *be a well pierced monotonous distances holes sequence of monotony* $(d_m)_{m\in\mathbb{N}}$ *of diameter* r. *There exists an idempotent T-sequence* $(T_m,u_n)_{m\in\mathbb{N}}$ *if and only if* $\sum_{j=0}^\infty|\log r-\log d_j|_\infty=+\infty$.

Lemma 7.4.12 is elementary.

Lemma 7.4.12. *Let* E *be a set which is not countable and let* f *be a function from* D *into* $\mathbb{R}_+$. *There exists a sequence* $(x_m)_{m\in\mathbb{N}}$ *in* E *and* $\lambda>0$ *such that* $f(x_m)\geq\lambda$ *for all* $m\in\mathbb{N}$.

Theorem 7.4.13, roughly, was proven separately and simultaneously by Motzkin and Robba [38, 40], and by the author in March 1969 [23, 20] (however, Motzkin–Robba's claim was not stated in terms of T-filters, but in terms of sequences of holes that look like T-sequences [43]).

Theorem 7.4.13. *Let $\mathcal{F}$ be a monotonous filter on D. Let $f \in H(D)$ be strictly vanishing along $\mathcal{F}$. Then $\mathcal{F}$ is a T-filter.*

Proof. Let us suppose $\mathcal{F}$ to be increasing, of center a and diameter S. With no loss of generality we may assume $a = 0$. Let $\lambda = \log S$. There exists $\xi < \lambda$ such that $\Psi(f, \mu) > -\infty$ whenever $\mu \in [\xi, \lambda]$ whereas $\Psi(f, \lambda) = -\infty$. Let r satisfy $\log(r) = |\xi|$. Hence the function $\Psi(f, .)$ is bounded in every interval $[\eta, \xi]$ and therefore the equality $\Psi(f(x)) = \Psi(f, \Psi(x))$ holds in all of $D \cap \Gamma(0, r, S)$ but inside finitely many classes of circles $C_m = C(0, r_m)$ with $r_m < r_{m+1}, \lim_{m \to \infty} r_m = S$.

We fix $m \in \mathbb{N}$ and take r', r'' satisfying $r \leq r' < r_m < r'' < S$. If $D \cap C_m \neq \emptyset$ we put $\theta_m = \|f\|_{D \cap C_m}$ and if $D \cap C_m = \emptyset$ we put $\theta_m = {}_D\varphi_{0,r_m}(f)$. Since $\Psi(f, \mu)$ is bounded in $[\log r', \log r'']$ by a constant M we may find $h \in R(D)$ such that $\log(\|h - f\|_D) < M$. Hence we have

(1) $\quad {}_D\varphi_{0,r_m}(f) = {}_D\varphi_{0,r_m}(h)$

and if $D \cap C_m \neq \emptyset$ we have

(2) $\quad \|h\|_{C_m \cap D_m} = \|f\|_{C_m \cap D_m} = \theta_m.$

Let $(T_{m,i})_{1 \leq i \leq s(m)}$ be the holes of D inside C_m which contain at least as many poles as many zeros and, for each one, let $q_{m,i}$ be the difference between the number of the poles and the number of the zeros (taking multiplicities into account). Let $q_m = \sum_{i=1}^{s(m)} q_{m,i}$. Then we know that

(3) $\quad \Psi'^l(h, \log r_m) - \Psi'^r(h, \log r_m) \geq q_m.$

By Relation (1) we have

(4) $\quad \theta_m \geq \gamma_D(0, r_m, q_m) {}_D\varphi_{0,r_m}(h)$

when $D \cap C_m \neq \emptyset$ and $\theta_m = {}_D\varphi_{0,r_m}(h)$ when $D \cap C_m = \emptyset$.

Hence Relation (4) is true anyway. In terms of valuations (4) is equivalent to $\log \theta_m \geq \log \gamma_D(0, r_m, q_m) + \Psi(f, \log r_m)$. Now by (3) we see that $\Psi(f, \log r_m) \leq \Psi(f, \log r_{m-1}) - q_{m-1}(\log r_m - \log r_{m-1})$.

Hence by induction we can easily obtain $\Psi(f, \log r_m) \geq \Psi(f, \log r_1) + \sum_{j=1}^{m-1} q_j(\log r_m - \log r_j)$ and finally $-\log \theta_m \leq \log \gamma_D(0, r_m, q_m) + \sum_{j=1}^{m-1} q_j(\log r_m - \log r_j)$. Since f is vanishing along $\mathcal{F}$, we have $\lim_{m \to \infty}(-\log \theta_m) = +\infty$ hence

$$\lim_{m \to \infty} \left(-\log \gamma_D(0, r_m, q_m) + \sum_{j=1}^{m-1} q_j(\log r_m - \log r_j) \right) = +\infty.$$

This just shows $\mathcal{F}$ to be an increasing T-filter. A symmetric reasoning is made when $\mathcal{F}$ is a decreasing filter equipped with a center.

Now let $\mathcal{F}$ be decreasing with no center. Let $\widehat{\mathbb{K}}$ be a spherical completion of $\mathbb{K}$. Then in $\widehat{\mathbb{K}}$ we denote by $(d(\alpha_j, \rho_j))_{j \in J}$ the family of the holes of D and we put $d(a, r) = \widetilde{D}$ and $\widehat{D} = \widehat{d}(a, r) \setminus (\bigcup_{j \in J} \widehat{d}(\alpha_j, \rho_j^-))$.

In $\widehat{\mathbb{K}}$, $\mathcal{F}$ has a center a. Then the filter $\widehat{\mathcal{F}}$ of center a and diameter S on $\widehat{D}$ is a T-filter because f belongs to the algebra $H_{\widehat{\mathbb{K}}}(\widehat{D})$ of analytic elements on $\widehat{D}$, with coefficients in $\mathbb{K}$ and is strictly vanishing along $\widehat{\mathcal{F}}$. Hence there exists a decreasing T-sequence $(\widehat{T}_{m,i}, m_{q,i})_{\substack{1 \leq i \leq s(m) \\ m \in \mathbb{N}}}$ of center a and diameter R that runs $\widehat{\mathcal{F}}$. But for each (m, i), $\widehat{T}_{m,i} \cap \mathbb{K}$ is a hole $T_{m,i}$ of D, and then, the weighted sequence $(T_{m,i}, m_{q,i})_{\substack{1 \leq i \leq s(m) \\ m \in \mathbb{N}}}$ is a T-sequence of D. □

Corollary 7.4.14. *Let $f \in H(D)$ be such that $f(a) \neq 0$ and $_D\varphi_a(r) = 0$. Then f is strictly vanishing along an increasing T-filter of center a and diameter $s \in]0, r]$.*

In [23, Chapter 37] it is shown that given a monotonous filter $\mathcal{F}$ on an infraconnected set D, there exist elements $H(D)$ strictly vanishing along $\mathcal{F}$ if and only if $\mathcal{F}$ is a T-filter. Moreover, by [23, Theorem 37.2], we have this theorem.

Theorem 7.4.15. *Let $\mathcal{F}$ be a T-filter on D, let $(T_{m,i}, m_{q,i})_{\substack{1 \leq i \leq s(m) \\ m \in \mathbb{N}}}$ be a T-sequence associated to $\mathcal{F}$. Let $D' = \mathbb{K} \setminus (\bigcup_{\substack{1 \leq i \leq s(m) \\ m \in \mathbb{N}}} T_{m,i})$, and for every (m, i), let $a_{m,i} \in T_{m,i}$. Let $\mathcal{F}'$ be the T-filter on D' associated to the T-sequence $(T_{m,i}, q_{m,i})$. Let $\alpha \in \mathcal{B}(\mathcal{F}')$. There exists $g \in H(D')$ satisfying these properties:*

(i) *g is meromorphic in $T_{m,i}$, admits $a_{m,i}$ as a pole of order at most $q_{m,i}$ and has no other pole in $T_{m,i}$,*

(ii) *g is strictly vanishing along $\mathcal{F}'$ and equal to zero in $\mathcal{P}_{D'}(\mathcal{F}')$,*

(iii) *for every circular filter $\mathcal{G}$ different from $\mathcal{F}$ and secant with $\mathcal{B}(\mathcal{F}')$,*
    $_{D'}\varphi_{\mathcal{G}}(g)$ is different from 0,

(iv) $g(\alpha) \neq 0$.

Proof. Without loss of generality we may obviously assume $D' = D$ hence $\mathcal{F}' = \mathcal{F}$. Besides we may assume $\mathcal{F}$ to be decreasing because an increasing filter of center a can be obtained by an inversion of center a of a decreasing filter of center a.

Actually the biggest difficulty happens when $\mathcal{F}$ has no center. Hence we do not suppose $\mathcal{F}$ to have a center (although it might have one). Indeed, here we have to construct the element f as a limit of rational functions with coefficients in $\mathbb{K}$. If we take an origin in $\widehat{\mathbb{K}}$, we work with a variable in $\widehat{\mathbb{K}}$ so that it is not clear how to obtain rational functions with coefficients in $\mathbb{K}$. For this reason we will perform a sequence of change of origin. (However, the proof becomes much easier to understand by assuming that $\mathcal{F}$ has center 0 and avoiding these changes of variable).

Let R be the diameter of $\mathcal{F}$ and let (d_m) be the monotony of the T-sequence $(T_{m,i}, q_{m,i})$. For every $m \in \mathbb{N}$, let $q_m = \sum_{i=1}^{s(m)} q_{m,i}$, and let $C_m = C(b_m, d_m)$. Let $G_m = \widetilde{C_m} \cap D$, let $D_m = G_m \setminus \mathcal{P}(\mathcal{F})$ and let $b_m = a_{m+1,1}$. The sequence (D_m) is a canonical basis of $\mathcal{F}$. We will construct f satisfying

(1) $\Psi_{b_m}(f, \mu) > -\infty$ whenever $\mu > \log R$,

(2) $\|f\|_{G_m} < \frac{1}{m}$.

Then f will clearly be strictly vanishing along $\mathcal{F}$ and equal to zero in all of $\mathcal{P}(\mathcal{F})$. Let us suppose we have defined increasing sequences of integers $(h(n))_{n\in\mathbb{N}}$ and $(\ell(n))_{n\in\mathbb{N}}$ satisfying $h(n) < \ell(n) < \ell(n) + 1 = h(n+1)$.

Now, for every $m \in \mathbb{N}$, let $Q_m = \prod_{i=1}^{s(m)}(x - a_{m,i})^{q_{m,i}}$. Let $\lambda_m = (d_m)^{q_m}\|\frac{1}{Q_m}\|_{C_m \cap D}$ and for every $n \in \mathbb{N}$ let $E_n = \prod_{m=h(n)}^{\ell(n)} Q_m$. Since the weighted sequence $(T_{m,i}, q_{m,i})$ is a T-sequence we have

(3) $\lim_{m\to\infty} \lambda_m \prod_{j=1}^{m-1}\left(\frac{d_m}{d_j}\right)^{q_j} = 0$.

Let $a_n \in G_{h(n+2)}$ and let $X_n = x - a_n$. Clearly we have $|X_n| \leq d_{h(n)}$ if and only if $x \in \widetilde{G}_{h(n)}$. We notice that as $a_n \in \Lambda_{h(n+2)}, (x \in D_{h(n+1)} \setminus D_{h(n+2)})$ is equivalent to $(d_{h(n+2)} < |X_n| \leq d_{h(n+1)}, X_n \in A_n)$. In particular $(x \in D_m \setminus D_{m+1})$ is equivalent to $(d_{m+1} < |X_n| \leq d_m, X_n \in A_n)$ whenever

$m \leq \ell(n+1)$. We set $E_n(x) = U_n(X_n)$ and develop $U_n : U_n(X_n) = A_{0,n} + \cdots + A_{\tau(n),n} X_n^{\tau(n)}$ with

(4) $\tau(n) = \deg(U_n) = \deg(E_n) = \sum_{m=h(n)}^{\ell(n)} q_m.$

Before going on, we notice this property. For every integer $e \leq \tau(n)$, we denote by $j(e)$ the unique integer such that

$$q_{h(n)} + \cdots + q_{j(e)-1} \leq \tau(n) - e < q_{h(n)} + \cdots + q_{j(e)}.$$

Now putting $\beta(e) = \tau(n) - e - (q_{h(n)} + \cdots + q_{j(e)-1})$, we have $\beta(e) \geq 0$ and

(5) $|A_{e,n}| \leq (d_{h(n)})^{q_{h(n)}} \cdots (d_{j(e)-1})^{q_{j(e)-1}} (d_{j(e)})^{\beta(e)}.$

Indeed this comes from the fact that $A_{e,n}$ is a sum of products of $\tau(n) - e$ zeros of U_n (taking multiplicities into account). Hence $|A_{e,n}|$ is bounded by the product of the $\tau(n) - e$ "biggest" zeros of U_n. If β_m is a zero of Q_m, then $\beta_m - a_m$ is a zero of U_m and then satisfies $|\beta_m - a_m| = d_m$. Thus, we obtain (5).

We are now going to introduce a sequence of rational functions F_n which will be involved in the construction of f. Let $(\sigma(n))_{n \in \mathbb{N}}$ and $(\tau(n))_{n \in \mathbb{N}}$ be sequences of integers satisfying $0 < \sigma(n) < \tau(n)$

(6) $q_{h(n)} + \cdots + q_{t(n)-1} < \tau(n) - \sigma(n) \leq q_{h(n)} + \cdots + q_{t(n)}$, and let $\alpha_n = \tau(n) - \sigma(n) - (q_{h(n)} + \cdots + q_{t(n)-1}).$

Let $S_n(X_n) = A_{\sigma(n)+1,n} X_n^{\sigma(n)+1} + \cdots + A_{\tau(n),n} X_n^{\tau(n)}$, $G_n(X_n) = \frac{S_n(X_n)}{U_n(X_n)}$ $F_n(x) = G_n(X_n)$ and $f_n = \prod_{i=1}^{n} F_i$.

Let us put $\phi(n) = \tau(n) - \sigma(n)$, $n \in \mathbb{N}$. We will prove that the sequence f_n converges in $H(D)$. The biggest problem consists of finding for $|1 - F_n(x)|$ a good upper bound on $D \setminus D_{h(n)}$ while $|F_n(x)|$ is equal to 1. We will show

(7) $|1 - F_n(x)| \leq \left(\frac{d_{h(n)}}{|x - a_n|} \right)^{\phi(n)} \leq \left(\frac{d_{h(n)}}{d_{h(n-1)}} \right)^{\phi(n)}$ whenever $x \in D \setminus D_{h(n)}$.

We will prove (7). Let e_1, e_2 be such that $0 \leq e_1 < e_2 < \tau(n)$. When $|X_n| > d_{h(n)}$ we have

(8) $|X_n|^{e_1} (d_{h(n)})^{q_{h(n)}} \cdots (d_{j(e_1)-1})^{q_{j(e_1)-1}} . (d_{j(e_1)})^{\beta(e_1)}$

$< |X_n|^{e_2} . (d_{h(n)})^{q_{h(n)}} \cdots (d_{j(e_2)-1})^{q_{j(e_2)-1}} (d_{j(e_2)-1})^{\beta(e_2)}.$

(Indeed by hypothesis the d_m are less than $|X_n|$. Since the sum of the powers of $|X_n|$ and that of the d_m are equal in the two members, the bigger member is the right one, which has the bigger power in X_n). Hence by (8) and (5) we see that when $|X_n| > d_{h(n)}$, for every $e \leq \sigma(n)$ we have

$$|X_n|^e |A_{e,n}| < |X_n|^e (d_{\ell(n)})^{q_{h(n)}} \cdots (d_{j(e)-1})^{q_{j(e)-1}} (d_{j(e)})^{\beta(e)}$$

$$\leq |X_n|^{\sigma(n)} (d_{h(n)})^{q_{h(n)}} \cdots (d_{t(n)-1})^{q_{t(n)-1}} (d_{t(n)})^{\alpha_n}.$$

This shows that $|(U_n - S_n)(X_n)| \leq \sup_{0 \leq e \leq \sigma(n)} |A_{e,n}| |X_n|^\ell \leq |X_n|^{\sigma(n)}$ $(d_{h(n)})^{q_{h(n)}} \cdots (d_{t(n)-1})^{q_{t(n)-1}} (d_{t(n)})^{\alpha_n}$ whenever $|X_n| > d_{h(n)}$. But as $d_m < d_{h(n)}$ whenever $m = h(n) + 1, \ldots, t(n)$, we see that $|(U_n - S_n)(X_n)| \leq |X_n|^{\sigma(n)} (d_{h(n)})^{\tau(n)-\sigma(n)}$.
On the other hand, we have $|U_n(X_n)| = |X_n|^{\tau(n)}$ whenever $|X_n| > d_{h(n)}$ because the zeros of U_n do belong to $d(0, d_{h(n)})$.

So, we have finally proven $|1 - G_n(X_n)| \leq \left(\frac{d_{h(n)}}{|X_n|}\right)^{\phi(n)}$ whenever $|X_n| > d_{h(n)}$ and this completes the proof of Relation (7).

Now we will give $|S_n(X_n)|$ an upper bound when $|X_n| \leq d_{h(n)}$. By (3) we know that

$$|X_n|^e |A_{e,n}| \leq |X_n|^e (d_{h(n)})^{q_{h(n)}} \cdots (d_{j(e)-1})^{q_{j(e)-1}} . (d_{j(e)})^{\beta(e)}.$$

But all the terms in S_n have an index $e \geq \sigma(n) + 1$. Hence we have

$$|X_n|^e |A_{e,n}| \geq |X_n|^{\sigma(n)+1} |X_n|^{e-\sigma(n)-1} d_{h(n)}^{q_{h(n)}} \cdots (d_{j(e)-1})^{q_{j(e)-1}} . (d_{j(e)})^{\beta(e)}.$$

Now, as $|X_n| \leq d_{h(n)}$ and $d_m < d_{h(n)}$ for all $m = h(n) + 1, \ldots, j(e)$ we see that $|X_n|^e |A_{e,n}| \leq |X_n|^{\sigma(n)+1} (d_{h(n)})^{\tau(n)-\sigma(n)-1}$ for every $e \geq \sigma(n) + 1$. Finally when $|X_n| \leq d_{h(n)}$ we have

$$|S_n(X_n)| \leq |X_n|^{\sigma(n)+1} (d_{h(n)})^{\tau(n)-\sigma(n)-1}.$$

So much the more, we have Relation (9)

$$(9) \quad |S_n(X_n)| \leq |X_n|^{\sigma(n)} (d_{h(n)})^{\tau(n)-\sigma(n)}$$

whenever $|X_n| \leq d_{h(n)}$.

We are now going to give $|U_n(X_n)|$ a lower bound when $|X_n| \leq d_{h(n)}$, with $X_n + a_n \in D$. Let $A_n = \{X_n \mid X_n + a_n \in D\}$. For $m = h(n), \ldots, \ell(n)$, let $V_m(X_n) = Q_m(x)$. Then $U_n(X_n) = \prod_{m=h(n)}^{\ell(n)} V_m(X_m)$. We consider the $V_j(X_n)$ when $d_{m+1} < |X_n| \leq d_m$ with $h(n) \leq m \leq \ell(n)$ and $h(n) \leq j \leq \ell(n)$. Obviously we have

(10) $|V_j(X_n)| = |X_n|^{q_j}$ when $j > m$ and

(11) $|V_j(X_n)| = (d_j)^{q_j}$ when $j < m$.

Moreover, we have

(12) $\quad |V_m(X_n)| \geq \frac{(d_m)^{q_m}}{\lambda_m}$

because $|V_m(X_n)| = (d_m)^{q_m}$ when $|X_n| < d_m$ and

$|V_m(X_n)| \geq \frac{(d_m)^{q_m}}{\lambda_m}$ when $|X_n| = d_m \quad (X_n \in A_n)$.

We will study $|F_n(x)|$ when $x \in D_{h(n)} \setminus D_{h(n+1)}$, i.e., $d_{h(n+1)} < |X_n| \leq d_{h(n)}$, $X_n \in A_n$. By (10)–(12) we obtain

(13) $\quad |U_n(X_n)| \geq \frac{|X_n|^{q_{\ell(n)}+\cdots+q_{m+1}}(d_{h(n)})^{q_{h(n)}}\cdots(d_m)^{q_m}}{\lambda_m}$,

whenever $d_{m+1} < |X_n| \leq d_m$ (with $h(n) \leq m \leq \ell(n)$).

Then by (9) and (12) we see that

$$|G_n(X_n)| \leq \frac{|X_n|^{\sigma(n)}(d_{h(n)})^{\tau(n)-\sigma(n)}\lambda_m}{|X_n|^{q_{\ell(n)}+\cdots+q_{m+1}}(d_{h(n)})^{q_{h(n)}}\cdots(d_m)^{q_m}}.$$

This may also be written

$$|G_n(X_n)| \leq \frac{\lambda_m|X_n|^{\tau(n)}|X_n|^{q_{h(n)}+\cdots+q_m}d_{h(n)}^{\tau(n)-\sigma(n)}}{|X_n|^{q_{h(n)}+\cdots+q_{\ell(n)}}|X_n|^{\tau(n)-\sigma(n)}(d_{h(n)})^{q_{h(n)}}\cdots(d_m)^{q_m}}.$$

Since the diameter of $\mathcal{F}$ is equal to R, and since Since $\tau(n) = \sum_{j=h(n)}^{\ell(n)} q_j$ we have

$$|G_n(X_n)| \leq \lambda_m \prod_{j=h(n)}^{m} \left(\frac{|X_n|}{d_j}\right)^{q_j} \left(\frac{d_{h(n)}}{|X_n|}\right)^{\tau(n)-\sigma(n)}$$

$$\leq \lambda_m \prod_{j=h(n)}^{m} \left(\frac{|X_n|}{d_j}\right)^{q_j} \left(\frac{d_{h(n)}}{R}\right)^{\tau(n)-\sigma(n)}.$$

In particular when $d_{m+1} < |X_n| \leq d_m$ we obtain

$|G_n(X_n)| \leq \lambda_m \prod_{j=h(n)}^{m} \left(\frac{d_m}{d_j}\right)^{q_j} \left(\frac{d_{h(n)}}{R}\right)^{\tau(n)-\sigma(n)}.$

Finally, we have proven Relation (14)

(14) $\quad |F_n(x)| \leq \lambda_m \prod_{j=h(n)}^{m} \left(\frac{d_m}{d_j}\right)^{q_j} \left(\frac{d_{h(n)}}{R}\right)^{\tau(n)-\sigma(n)}$,

whenever $x \in D_m \setminus D_{m+1}$ with $h(n) \leq m \leq \ell(n)$.

Now we consider $|F_n(x)|$ when $x \in D_{h(n+1)} \setminus D_{h(n+2)}$, i.e., $d_{h(n+2)} < |X_n| \leq d_{h(n+1)}$, $X_n \in A_n$. It is seen that for every $j = h(n), \ldots, \ell(n)$ we have $|V_j(X_n)| = (d_j)^{q_j}$ and then we have

(15) $\quad |U_n(X_n)| = \prod_{j=h(n)}^{\ell(n)} (d_j)^{q_j}$.

Now let us suppose $d_{m+1} < |X_n| \leq d_m$ with $h(n) \leq m \leq \ell(n)$. By (9) we have again $|S_n(X_n)| \leq |X_n|^{\sigma(n)} (d_{h(n)})^{\tau(n)-\sigma(n)}$, hence by (15) we see that $|G_n(X_n)| \leq 1$ whenever $|X_n| \leq d_{h(n+1)}$ hence

(16) $\quad |F_n(X)| \leq 1$ whenever $x \in G_{h(n+1)}$.

Besides, since $|X_n| \leq d_m$ we have $|S_n(X_n)| \leq d_m^{\sigma(n)} d_{h(n)}^{\tau(n)-\sigma(n)}$. Hence by (15) we have

$$|G_n(X_n)| \leq \frac{d_m^{\sigma(n)} d_{h(n)}^{\tau(n)-\sigma(n)}}{\prod_{j=h(n)}^{\ell(n)} d_j^{q_j}} = \prod_{j=h(n)}^{\ell(n)} \left(\frac{d_m}{d_j}\right)^{q_j} \left(\frac{d_{h(n)}}{d_m}\right)^{\tau(n)-\sigma(n)}$$

$$\leq \prod_{j=h(n)}^{\ell(n)} \left(\frac{d_m}{d_j}\right)^{q_j} \left(\frac{d_{h(n)}}{R}\right)^{\tau(n)-\sigma(n)}.$$

Finally we have proven (17).

(17) $\quad |F_n(x)| \leq \left(\prod_{j=h(n)}^{\ell(n)} \left(\frac{d_m}{d_j}\right)^{q_j}\right) \left(\frac{d_{h(n)}}{R}\right)^{\tau(n)-\sigma(n)}$

whenever $x \in D_m \setminus D_{m+1}$, for every $m = h(n+1), \ldots, \ell(n+1)$.

We are now able to construct the F_n just by defining the sequences $h(n)$ and $\sigma(n)$. For convenience we will use $\phi(n) = \ell(n) - \sigma(n)$, and then it is equivalent to define the sequences $h(n)$ and $\ell(n)$. Let us suppose these sequences $(h(n))_{n\in\mathbb{N}}$ and $(\phi(n))_{n\in\mathbb{N}}$ are already defined up to the rank N, satisfying for every $n \leq N$ the following relations (18)–(20)

(18) $\quad \left(\frac{d_{h(n-1)+1}}{d_{h(n-1)}}\right)^{\phi(n)} < \frac{1}{n}$,

(19) $\quad \left(\frac{d_{h(n)}}{s}\right)^{\phi(n)} < 2$,

(20) $\quad \lambda_m \prod_{j=h(n-1)}^{m} \left(\frac{d_m}{d_j}\right)^{q_j} < \frac{1}{4n}$ whenever $m \geq h(n)$.

Thus, we know that $\ell(N-1) = h(N) - 1$, and we want to choose $\phi(N+1)$ and $\ell(N)$ satisfying (18)–(20). Of course, we can find $\phi(N+1)$ big enough to satisfy

(21) $\left(\frac{d_{h(N)+1}}{d_{h(N)}}\right)^{\phi(N+1)} < \frac{1}{N+1}.$

Next, as $\lim_{m\to\infty} d_m = R$ there does exist a rank $M > N$ such that

(22) $\left(\frac{d_m}{R}\right)^{\phi(N+1)} < 2$ whenever $m \geq M.$

Finally by hypothesis we have $\lim_{m\to\infty} \lambda_m \prod_{j=h(N)}^{m} \left(\frac{d_m}{d_j}\right)^{q_j} = 0$, hence there exists a rank $M' \geq M$ such that

(23) $\lambda_m \prod_{j=h(N)}^{m} \left(\frac{d_m}{d_j}\right)^{q_j} < \frac{1}{8(N+1)}$ whenever $m \geq M'.$

We take $h(N+1) = M'$ and then by (21)–(23) we see that (18)–(20) are satisfied at the rank $N+1$. In order to begin the recurrence we can take $h(0) = 1, \sigma(0) = 0, \ell(1) = 1$ and then the sequences are defined for all $n \in \mathbb{N}$.

As it was announced above, we put $f_n = \prod_{i=1}^{n} F_i$ and we will show that the sequence $(f_n)_{n\in\mathbb{N}}$ converges in $H(D)$ to the element f announced in the theorem.

Let $n > 0$. We will study $|f_n(x)|$ in these 3 cases:

(α) $x \in D \setminus G_{h(n)}$, (β) $x \in G_{h(n+1)}$, (γ) $x \in G_{h(n)} \setminus G_{h(n+1)}.$

(α) By (17) we have $|F_n(x)| \leq 1$ hence $|f_n(x)| \leq |f_{n-1}(x)| \leq 1.$

(β) By (16) we notice that for every $j \leq n$, we have $|F_j(x)| \leq 1$. Next, by (17) we have $|F_n(x)| \leq \prod_{j=h(n)}^{\ell(n)} \left(\frac{d_{h(n+1)}}{d_j}\right)^{q_j} \left(\frac{d_{h(n)}}{s}\right)^{\phi(n)}$ and then by (19) and (20) we see that $|F_n(x)| \leq \frac{1}{2(n+1)}$. So, by (α) applied to f_{n-1}, we obtain (24) $|f_n(x)| \leq \frac{1}{2(n+1)}.$

(γ) Let $m \in \mathbb{N}$ be such that $x \in D_m \setminus D_{m+1}$. (Obviously $m \in [h(n), \ell(n)].$) We know that $|f_n(x)| \leq |F_{n-1}(x)F_n(x)|$, and by (17) we have

$$|F_{n-1}(x)F_n(x)| \leq \lambda_m \prod_{j=h(n-1)}^{m} \left(\frac{d_m}{d_j}\right)^{q_j} \left(\frac{d_{h(n-1)}}{R}\right)^{\phi(n-1)} \left(\frac{d_{h(n)}}{R}\right)^{\phi(n)}$$

hence by (19) and (20) we see that $|F_{n-1}(x)F_n(x)| \leq \frac{1}{n}$, hence $|f_n(x)| \leq \frac{1}{n}.$

Thus, after studying the cases (α), (β), (γ), we have proven that $\|f_n\|_D \leq \|f_0\|_D$ and that

(25) $\|f_n\|_{D_{h(n)}} \leq \frac{1}{n}.$

Hence we have obtained

(26) $\|f_{n+1} - f_n\|_{D_{h(n)}} \leq \frac{1}{n}$.

Now when $x \in D \setminus D_{h(n)}$ by (7) we have $|1 - F_n(x)| \leq \left(\frac{d_{h(n)}}{d_{h(n-1)}}\right)^{\phi(n)}$ and therefore $|1 - F_n(x)| \leq \left(\frac{d_{h(n-1)+1}}{d_{h(n-1)}}\right)^{\phi(n)}$. Hence by Relation (18) we have

(27) $|1 - F_n(x)| < \frac{1}{n}$ whenever $x \in D \setminus D_{h(n)}$.

Finally by (26) and (27) we have $\|f_{n+1} - f_n\|_D \leq \frac{1}{n} \sup(\|f_1\|_D, 1)$ and this shows that the sequence converges to a limit f in $H(D)$.

By construction it is clear that for each hole $T_{m,i}$, the sequence $\left((x - a_{m,i})^{q_{m,i}} f_n\right)_{n \in \mathbb{N}}$ converges in $H(D \cup T_{m,i})$, hence f admits $a_{m,i}$ as a unique pole of order $\leq q_{m,i}$. Next, by (16) and (25) it is seen that $\|f_j\|_{D_{h(n)}} \leq \frac{1}{n}$ whenever $j \geq n$, hence $\|f\|_{D_{h(n)}} \leq \frac{1}{n}$ and therefore f is vanishing along $\mathcal{F}$ whereas $f(x)$ is equal to zero in all of $\mathcal{P}(\mathcal{F})$. Thus, (i) is satisfied.

We check that (ii) and (iii) are satisfied. Indeed by (7) we have $|F_j(x)| = 1$ whenever $j > n$ and $x \in \mathbb{K} \setminus D_{h(n)}$, hence we have $|f_n(x)| = |f_j(x)| = |f(x)|$ whenever $j > n$ and $x \in \mathbb{K} \setminus D_{h(n)}$. This shows that Statements (ii) and (iii) are satisfied.

Finally, if $f(\alpha) = 0$, since D' is open, f factorizes in $H(D')$ in the form $(x - \alpha)^q g(x)$ with $g(\alpha) \neq 0$, and then it is immediately seen that g also satisfies (i), (ii), (iii), and further, satisfies (iv). This finishes the proof of Theorem 7.4.15.   □

Corollary 7.4.16. *Let $\mathcal{F}$ be a monotonous filter on D. There exist elements of $H(D)$ strictly vanishing along $\mathcal{F}$ if and only if $\mathcal{F}$ is a T-filter.*

The following purely arithmetical Lemma 7.4.17 is proven in [23] and in [43].

Lemma 7.4.17. *Let $(T_{m,i}, q_{m,i})_{1 \leq i \leq k_m}$ be a T-sequence of diameter r, of piercing $\rho > 0$, and let $A \in]0, +\infty[$. There exists a T-sequence $(T_{m,i}, u_{m,i})_{\substack{1 \leq i \leq k_m \\ m \in \mathbb{N}}}$ with $u_{m,i} \leq q_{m,i}$ (whenever $i = 1, \ldots, k_m, m \in \mathbb{N}$) whose subsidence ν satisfies $\nu \leq A + 3 \log(\frac{r}{\rho})$.*

Lemma 7.4.18. *Let $(T_{m,i}, q_{m,i})_{1 \leq i \leq k_m}$ be a T-sequence of diameter r, of piercing $\rho > 0$, and let $A \in]0, +\infty[$. There exists a T-sequence $(T_{m,i}, u_{m,i})_{\substack{1 \leq i \leq k_m \\ m \in \mathbb{N}}}$ with $u_{m,i} \leq q_{m,i}$ (whenever $i = 1, \ldots, k_m, m \in \mathbb{N}$) whose subsidence ν satisfies $\nu \leq A + 3 \log(\frac{r}{\rho})$.*

Proof. For every $i = 1, \ldots, k_m, m \in \mathbb{N}$, we put $\rho_{m,i} = \text{diam}(T_{m,i})$, and take $b_{m,i} \in T_{m,i}$. Let $(d_m)_{m \in \mathbb{N}}$ be the monotony of the T-sequence $(T_{m,i}, q_{m,i})$, let λ be its subsidence, and for each $m \in \mathbb{N}$, let $q_m = \sum_{j=1}^{k_m} q_{m,j}$, let
$$e_m = \max_{1 \leq i \leq k_m} \left(\frac{d_m}{\rho_{m,i}}\right)^{q_{m,i}} \prod_{\substack{j \neq i \\ 1 \leq j \leq k_m}} \left(\frac{d_m}{|b_{m,i} - b_{m,j}|}\right)^{q_{m,j}},$$
and let $\theta_m = \log e_m - \sum_{j=1}^{m} q_j |\log d_m - \log d_j|_\infty$. Then we have

(1) $\lim_{m \to +\infty} \theta_m = -\infty$, and

(2) $\lambda = \sup_{m \in \mathbb{N}} \theta_m$.

For every couple $(m, i)_{\substack{1 \leq i \leq k_m \\ m \in \mathbb{N}}}$ we put $t_{m,i} = \text{Int}(\frac{A}{\lambda} q_{m,i})$ and $u_{m,i} = \frac{A}{\lambda} q_{m,i} - t_{m,i}$. By Lemma 7.4.17, there exists a family of integers $(v_{m,i})_{1 \leq i \leq k_m}$ all equal to 0 or 1, satisfying

(3) either $0 \leq v_{m,i} - u_{m,i} < 1$ or $v_{m,i} = 0$.

(4) $0 \leq \sum_{i=1}^{k_m} v_{m,i} - \sum_{i=1}^{k_m} u_{m,i} < 1$

and

(5)
$$\max_{1 \leq i \leq k_m} \left(\sum_{\substack{j \neq i \\ 1 \leq j \leq k_m}} v_{m,j} \log\left(\frac{d_m}{|b_{m,j} - b_{m,i}|}\right) \right)$$
$$\leq \max_{1 \leq i \leq k_m} \left(\sum_{\substack{j \neq i \\ 1 \leq j \leq k_m}} u_{m,j} \log\left(\frac{d_m}{|b_{m,j} - b_{m,i}|}\right) \right) + 2 \log\left(\frac{d_m}{\rho}\right).$$

Let $s_{m,i} = t_{m,i} + v_{m,i}$ $(1 \leq i \leq k_m, m \in \mathbb{N})$, let $s_m = \sum_{i=1}^{k_m} s_{m,i}$ and let
$$e'_m = \max_{1 \leq i \leq k_m} \left[\left(\frac{d_m}{\rho_{m,i}}\right)^{s_{m,i}} \prod_{\substack{j \neq i \\ 1 \leq j \leq k_m}} \left(\frac{d_m}{|b_{m,j} - b_{m,i}|}\right)^{s_{m,j}} \right].$$

By (3) we notice that $(t_{m,i} + v_{m,i}) \log\left(\frac{d_m}{\rho_{m,i}}\right)$
$$\leq (t_{m,i} + u_{m,i}) \log\left(\frac{d_m}{\rho_{m,i}}\right) + \log\left(\frac{d_m}{\rho}\right)$$

and then by (5) we have
$$\log e'_m \leq \max_{1 \leq i \leq k_m} \left[(t_{m,i} + u_{m,i}) \log\left(\frac{d_m}{\rho_{m,i}}\right) \right.$$
$$\left. + \sum_{\substack{j \neq i \\ 1 \leq j \leq k_m}} (t_{m,j} + u_{m,j}) \log\left(\frac{d_m}{|b_{m,j} - b_{m,i}|}\right) + 3 \log\left(\frac{d_m}{\rho}\right) \right].$$

Hence we obtain

(6) $\log e'_m \leq \frac{A}{\lambda} \log e_m + 3 \log\left(\frac{d_m}{\rho}\right).$

We will check that the weighted sequence $(T_{m,i}, s_{m,i})_{\substack{1 \leq i \leq k_m \\ m \in \mathbb{N}}}$ is a T-sequence of subsidence $\leq A + 3 \log\left(\frac{r}{\rho}\right)$. Indeed, by (4) we have

$$s_m \geq \sum_{i=1}^{k_m} t_{m,i} + \sum_{i=1}^{k_m} u_{m,i} = \frac{A}{\lambda} q_m, \quad \text{hence} \quad \sum_{j=1}^{m} s_j \left| \log\left(\frac{d_m}{d_j}\right) \right|_\infty$$

$$\geq \frac{A}{\lambda} \sum_{j=1}^{m} q_j \left| \log\left(\frac{d_m}{d_j}\right) \right|_\infty$$

and then by (6) we obtain

(7) $\log e_m - \sum_{j=1}^{m} s_j \left| \log\left(\frac{d_m}{d_j}\right) \right|_\infty \leq \frac{A}{\lambda} \left(\log e_m - \sum_{j=1}^{m} q_j \left| \log\left(\frac{d_m}{d_j}\right) \right|_\infty \right) + 3 \log\left(\frac{r}{\rho}\right)$

Now by (1) it is clear that $\lim_{m \to \infty} \left(\log e'_m - \sum_{j=1}^{m} s_j \left| \log\left(\frac{d_m}{d_j}\right) \right|_\infty \right) = -\infty$ and therefore the weighted sequence we deal with is a T-sequence. Besides by hypothesis we have $\log e_m - \sum_{j=1}^{m} q_j \left| \log\left(\frac{d_m}{d_j}\right) \right|_\infty \leq \lambda$, hence by (7) the subsidence of the T-sequence $(T_{m,i}, s_{m,j})$ is clearly bounded by $A + 3 \log\left(\frac{r}{\rho}\right)$.

□

Lemma 7.4.19. is immediate.

Lemma 7.4.19. *Let* $\alpha_1, \ldots, \alpha_q \in \mathbb{K} \setminus \{0\}$ *and let* $Q(x) = \prod_{j=1}^{q}\left(1 - \frac{x}{\alpha_j}\right) = \sum_{j=0}^{q} a_j x^j.$ *Let* $m = \min_{1 \leq j \leq q} |\alpha_j|.$ *Then* $|a_j| \leq \frac{1}{m^j}$ *whenever* $j = 0, \ldots, q.$

Proposition 7.4.20. *Let* D *admit an increasing (resp., a decreasing) T-sequence* $(T_{m_i}, q_{m_i})_{1 \leq i \leq k_m, m \in \mathbb{N}}$ *of piercing* $\rho > 0$, *of center* α *and diameter* r.

Let $D' = \mathbb{K} \setminus \left(\bigcup_{1 \leq i \leq k_m, m \in \mathbb{N}} T_{m,i} \right).$ *Let* $\mathcal{F}$ *be the T-filter on D' associated to this T-sequence. For each* (m, i) $(1 \leq i \leq k_m, m \in \mathbb{N})$, *let* $b_{m,i} \in T_{m,i}$ *and let* $\epsilon \in]0, +\infty[$. *There exists* $\phi_\epsilon \in \mathcal{I}_0(\mathcal{F})$, *meromorphic on each hole* $T_{m,i}$, *satisfying further:*

(i) $|\phi_\epsilon(x) - 1| \leq \epsilon \ \forall x \in D \cap d(\alpha, r(1 - \epsilon)$ *(resp.,* $\forall x \in D \cap d(\alpha, \frac{r}{1-\epsilon})^-)$,
(ii) $\|\Phi_\epsilon\|_{D'} \leq \left(\frac{r}{\rho}\right)^3 + r^3 \epsilon,$
(iii) $b_{m,i}$ *is a pole of* ϕ_ϵ *of order* $u_{m,i} \leq q_{m,i}$ *and* ϕ_ϵ *has no pole different from* $b_{m,i}$ *in* $T_{m,i}$ *whenever* $i = 1, \ldots, k_m$, $m \in \mathbb{N}.$

Proof. We will follow the proof of Proposition 7.4.15 step to step, assuming further the sequence $(T_{m,i}, q_{m,i})_{\substack{1 \le i \le k_m \\ m \in \mathbb{N}}}$ to have piercing $\rho > 0$ and making ϕ_ϵ satisfy the strong conditions (a) and (b). Without loss of generality we may clearly assume $\alpha = 0$. If the Theorem is proven when the T-sequence is increasing, it is immediately generalized to the case when it is decreasing by considering the set $E = \{\frac{1}{x} \mid x \in D \setminus \{0\}\}$. So we will assume the T-sequence to be increasing without loss of generality.

For each $(m, i)_{\substack{1 \le i \le k_m \\ m \in \mathbb{N}}}$, we put $\rho_{m,i} = \text{diam}(T_{m,i})$ and for each $m \in \mathbb{N}$ we denote by $C(0, d_m)$ the circle of center 0 that contains the holes $(T_{m,i})_{1 \le i \le k_m}$. Given any $N \in \mathbb{N}$, we know that the family $(T_{m,i}, q_{m,i})_{\substack{1 \le i \le k_m \\ m \ge N}}$ also is a T-sequence. Therefore we may assume $d_m \ge r(1 - \epsilon)$ whenever $m \in \mathbb{N}$ without loss of generality. Let $q \in \mathbb{N}$ satisfy $(\mathcal{U}_0)$ $(\frac{d_0}{d_1})^q \le \epsilon$.

By Lemma 7.4.18, there clearly exists a T-sequence $(T_{m,i}, u_{m,i})_{\substack{1 \le i \le k_m \\ m > q}}$ whose subsidence λ is inferior or equal to $\log((\frac{r}{\rho})^3 + r^3 \epsilon)$ with $u_{m,i} \le q_{m,i}$ for all (m, i). For every $m \in \mathbb{N}$ we put $u_m = \sum_{i=1}^{k_m} u_{m,i}$, and

$$e_m = \sup_{1 \le i \le k_m} \left(\frac{d_m}{\rho_{m,i}}\right)^{u_{m,i}} \prod_{\substack{j \ne i \\ 1 \le j \le k_m}} \left(\frac{d_m}{|b_{m,j} - b_{m,i}|}\right)^{u_{m,j}}.$$

Since $(T_{m,i}, u_{m,i})_{\substack{1 \le i \le k_m \\ m \in \mathbb{N}}}$ is a T-sequence, for every $h \in \mathbb{N}$ we have

(1) $\quad \lim_{m \to \infty} \left(e_m \prod_{j=h}^m \left(\frac{d_j}{d_m}\right)^{u_j}\right) = 0$,

and for all $m \ge q$ we have

(2) $\quad \log\left(e_m \prod_{j=q}^m \left(\frac{d_j}{d_m}\right)^{u_j}\right) \le \log\left(\left(\frac{r}{\rho}\right)^3 + r^3 \epsilon\right)$.

By induction we will construct sequences of integers $s(n), \ell(n), w(n)$ satisfying:

$$(\mathcal{U}_n) \quad \left(\frac{d_{\ell(n)}}{d_{s(n)}}\right)^{w(n)} < \frac{\epsilon}{n+1},$$

$$(\mathcal{V}_n) \left(\frac{r}{d_{s(n-1)}}\right)^{w(n-1)} \left(\frac{r}{d_{s(n)}}\right)^{w(n)} e_m \prod_{j=s(n-1)+1}^m \left(\frac{d_j}{d_m}\right)^{u_j} < \frac{\epsilon}{n+1},$$

whenever $m \ge s(n)$.

Let us suppose we have already defined these three sequences up to the rank n. We will construct them a the rank $n + 1$ in this way.

First we choose $\ell(n+1)$ such that $e_m \prod_{j=\ell(n)}^m \left(\frac{d_j}{d_m}\right)^{u_j} < \frac{\epsilon}{n+1}$ for all $m \geq \ell(n+1)$. Then we can choose $w(n+1)$ such that $\left(\frac{d_{\ell(n+1)}}{d_{\ell(n+1)+1}}\right)^{w(n+1)} < \frac{\epsilon}{n+2}$. By (1) we can choose $s(n+1) > \max(\ell(n+1), s(n)+1+w(n))$ satisfying $(\mathcal{V}_{n+1})$.

Now it just remains to define $s(n), \ell(n), w(n)$ for $n = 1$ and $n = 2$, satisfying $(\mathcal{U}_2)$ and $(\mathcal{V}_2)$ in order to define the sequences for all $n \in \mathbb{N}$. We take $\ell(1) > q$, satisfying $e_m \prod_{j=q+1}^m \left(\frac{d_j}{d_m}\right)^{u_j} < \frac{\epsilon}{2}$ whenever $m \geq \ell(1)$. Then we can choose $w(1)$ such that $\left(\frac{d_{\ell(1)}}{d_{\ell(1)+1}}\right)^{w(1)} < \frac{\epsilon}{2}$. Next we take $s(1) > \ell(1)$ satisfying

$$(3) \quad \left(\frac{r}{d_{s(1)}}\right)^{w(1)} \prod_{j=q+1}^m \left(\frac{d_j}{d_m}\right)^{u_j} < \left(\frac{\epsilon}{2}\right) \quad \text{whenever } m \geq s(1).$$

By (1), we can choose $\ell(2)$ such that $e_m \prod_{j=\ell(2)}^m \left(\frac{d_j}{d_m}\right)^{u_j} < \frac{\epsilon}{3}$ whenever $m > \ell(2)$.

Hence we can choose $w(2)$ satisfying $(\mathcal{U}_2)$ and finally $s(2) > \ell(2)$ satisfying $(\mathcal{V}_2)$. Thus, the sequences $s(n), \ell(n), w(n)$ are now defined for all n, satisfying $(\mathcal{U}_n)$ and $(\mathcal{V}_n)$, for all $n \geq 2$.

For every $m \in \mathbb{N}$ we put $Q_m := \prod_{i=1}^{k_m} \left(1 - \frac{x}{b_{m,i}}\right)^{u_{m,i}}$ and for each $n \in \mathbb{N}^*$ we put $H_n(x) := \prod_{m=s(n)+1}^{s(n+1)} Q_m$ and $t(n) = \deg(H_n)$. We can develop $H_n(x)$ in the form $\sum_{h=0}^{t(n)} a_{n,h} x^h$ (with $a_{n,0} = 1$). Since $t(n) = \sum_{m=s(n)+1}^{s(n+1)} u_m$ and since $s(n+1) > s(n) + w(n)$, it is seen that $t(n) > w(n)$. Now we put $G_n(x) := \sum_{h=0}^{w(n)} a_{n,h} x^h$, and let $R_n(x) := \frac{G_n(x)}{H_n(x)}$. Thus, R_n is defined when $n > 0$. It only remains to define R_0. Let $P(x) = \prod_{m=1}^q \left(1 - \frac{x}{b_{m,1}}\right)$ and let $H_0(x) = P(x) \prod_{m=q+1}^{s(1)} Q_m(x)$. We can develop $H_0(x) = \sum_{h=0}^{t(0)} a_{0,h} x^h$. We put $G_0(x) = \sum_{h=0}^q a_{0,h} x^h$ and $R_0(x) = \frac{G_0(x)}{H_0(x)}$. By Lemma 7.4.19, we notice the relation (X_n) $|G_n(x)| \leq \max\left(1, \left(\frac{|x|}{d_{s(n)}}\right)^{w(n)}\right)$ whenever $x \in D$, whenever $n \in \mathbb{N}^*$.

Next, we check that $(\mathcal{Q}_0)$ $|R_0(x) - 1| \leq \epsilon$ whenever $x \in D \cap d(0, d_0)$. Indeed we have

$$|R_0(x) - 1| = \left| \frac{\sum_{h=q+1}^{t(0)} a_{0,h} x^h}{H_0(x)} \right|.$$

But $|H(x)| \geq 1$ for all $x \in D \cap d(0, d_0)$ because H_0 has no zero in $d(0, d_1)$. Besides, by Lemma 7.4.19, we see that

$$\Big| \sum_{h=q+1}^{t(0)} a_{0,h} x^h \Big| \leq \max_{q+1 \leq h \leq t(0)} \frac{|x|^h}{(d_1)^h} \leq \max_{q+1 \leq h \leq t(0)} \Big(\frac{d_0}{d_1} \Big)^h = \Big(\frac{d_0}{d_1} \Big)^{q+1}$$

Hence finally by $(\mathcal{U}_0)$ we obtain $|R_0(x) - 1| \leq \epsilon$ for every $x \in D \cap d(0, d_1)$. In the same way we will prove the relations $(\mathcal{Q}_n)$ $\quad |R_n(x) - 1| \leq \frac{\epsilon}{n+1}$ for all $x \in D \cap d(0, d_{\ell(n)})$.

Indeed as $H_n(x)$ has no zero in $d(0, (d_{s(n)+1})^-)$, we have $(\mathcal{T}_n)|H_n(x)| = 1$ for all $x \in d(0, d_{\ell(n)}) \cap D'$.

Next, by applying Lemma 7.4.19 to H_n we have $|a_{n,h}| \leq \frac{1}{(d_{s(n)+1})^h}$ whenever $h = 0, \dots, t(n)$, and therefore when $x \in d(0, d_{\ell(n)})$ we obtain

$$|H_n(x) - G_n(x)| = \Big| \sum_{h=w(n)+1}^{t(n)} a_{n,h} x^h \Big| \leq \max_{w(n)+1 \leq h \leq t(n)} \Big(\frac{d_{\ell(n)}}{d_{s(n)+1}} \Big)^h$$

$$= \Big(\frac{d_{\ell(n)}}{d_{s(n)+1}} \Big)^{w(n)+1} \leq \Big(\frac{d_{\ell(n)}}{d_{s(n)}} \Big)^{w(n)} \Big(\frac{d_{s(n)}}{d_{s(n)+1}} \Big)^{w(n)}.$$

But by $(\mathcal{U}_n)$ we have $\Big(\frac{d_{\ell(n)}}{d_{s(n)}} \Big)^{w(n)} \leq \frac{\epsilon}{n+1}$ and therefore by $(\mathcal{T}_n)$, we have proven $(\mathcal{Q}_n)$ $(n \in \mathbb{N})$. Besides, by $(\mathcal{X}_n)$, we notice that $|G_n(x)| \leq 1$ whenever $x \in d(0, d_{s(n)})$. So, we have $(\mathcal{S}_n)$ $\quad |R_n(x)| \leq 1$ for all $x \in D \cap d(0, d_{s(n)})$. Now we put $f_n(x) = \prod_{j=0}^{n} R_j(x)$. We will prove Relations $(\mathcal{R}_{n,k})$, $k = 1, \dots, n$ $n \in \mathbb{N}$.

$(\mathcal{R}_{n,k})|f_n(x)| \leq \frac{\epsilon}{k+1}$ whenever $x \in D \setminus d(0, d_{s(k)})$, whenever $k = 1, \dots, n$.

Let us suppose Relations $(\mathcal{R}_{n,k})$ when $n \leq N$ are already proven and let us show them for $n = N+1$. Since $(\mathcal{R}_{N,k})$ is satisfied, it is seen that it directly implies $(\mathcal{R}_{N+1,k})$ for $k \leq N$. Hence it just remains to prove $(\mathcal{R}_{N+1,N+1})$.

First we suppose $x \in D \setminus d(0, d_{s(N+2)})$. Then we have

$$|H_{N+1}(x)| = \prod_{j=s(N+1)+1}^{s(N+2)} \Big(\frac{|x|}{d_j} \Big)^{u_j} \geq \prod_{j=s(N+1)+1}^{s(N+2)} \Big(\frac{d_{s(N+2)}}{d_j} \Big)^{u_j}$$

and therefore by Relation $(\mathcal{X}_{N+1})$ we obtain

$$|R_{N+1}(x)| \leq \Big(\frac{r}{d_{s(N+1)}} \Big)^{w(N+1)} \prod_{j=s(N+1)+1}^{s(N+2)} \Big(\frac{d_j}{d_{s(N+2)}} \Big)^{u_j}.$$

Thus, by (2) we have $|R_{N+1}(x)| \leq \frac{\epsilon}{N+2}$ and therefore by $(\mathcal{R}_{N,N})$ Relation $(\alpha)| \frac{\Pi_{N+1}(x)| \leq \epsilon}{N+2}$ holds for all $x \in D$ such that $|x| > d_{s(N+2)}$.

We now suppose $d_{s(N+1)} \leq |x| \leq d_{s(N+2)}$. Actually, for convenience and more generally, we suppose $d_{s(N+1)} \leq |x| < d_{s(N+2)+1}$. For example, let $d_m \leq |x| < d_{m+1}$ with $s(N+1) \leq m < s(N+2)$. It is seen that

$$\frac{1}{|H_{N+1}(x)|} \leq e_m \prod_{j=s(N+1)+1}^{m} \left(\frac{d_j}{|x|}\right)^{u_j},$$

and

$$|R_{N+1}(x)| \leq e_m \left(\frac{r}{d_{s(N+1)}}\right)^{w(N+1)} \prod_{j=s(N+1)+1}^{m} \left(\frac{d_j}{|x|}\right)^{u_j}.$$

Then we have

$$|R_n(x)R_{N+1}(x)| \leq \left(\frac{r}{d_{s(N)}}\right)^{w(N)} \left(\prod_{j=s(N)+1}^{s(N+1)} \left(\frac{d_j}{d_m}\right)^{u_j}\right) e_m \left(\frac{r}{d_{s(N+1)}}\right)^{w(N+1)}$$

$$\times \left(\prod_{j=s(N+1)+1}^{m} \left(\frac{d_j}{d_m}\right)^{u_j}\right)$$

$$= e_m \left(\frac{r}{d_{s(N)}}\right)^{w(N)} \left(\frac{r}{d_{s(N+1)}}\right)^{w(N+1)} \prod_{j=s(N)+1}^{m} \left(\frac{d_j}{d_m}\right)^{u_j}.$$

But then by $(\mathcal{V}_{N+1})$ this is inferior or equal to $\frac{\epsilon}{N+2}$ by $(\mathcal{V}_{N+1})$ because $m > s(N+1)$. Hence, as $|f_{N-1}(x)| < 1$, we have proven that Relation (α) finally holds for all $x \in D$ such that $|x| \geq d_{s(N+1)}$. This just proves Relation $\mathcal{R}_{N+1,N+1}$.

Thus, it just remains to establish $\mathcal{R}_{1,1}$ in order to start the recurrence. This means $|R_0(x)R_1(x)| \leq \frac{\epsilon}{2}$ whenever $x \in D \setminus d(0, d_{s(1)})$ and this is also equivalent to

$$\left|\left(\frac{G_0(x)}{P(x)}\right)\left(\frac{1}{\prod_{m=q+1}^{s(1)} Q_m(x)}\right)\left(\frac{G_1(x)}{\prod_{m=s(1)+1}^{\infty} Q_m(x)}\right)\right| \leq \frac{\epsilon}{2},$$

whenever $x \in D$ such that $|x| > d_{s(1)}$. By construction, the coefficients of G_0 are the coefficients of H_0 of same index when this index runs from 0 to q. But by construction H_0 admits a unique zero in each circle $C(0, d_m)$ for $1 \leq m \leq q$ and no other zero in $\mathbb{K}$. Then it is easily seen that we have

(4) $\left|\dfrac{G_0(x)}{P(x)}\right| \leq \dfrac{1}{\rho}$ whenever $x \in D$,

and in particular

(5) $\left|\dfrac{G_0(x)}{P(x)}\right| = 1$ whenever $x \in D \setminus d(0, d_q)$.

Hence finally we have

$$|R_0(x)R_1(x)| = \frac{|G_1(x)|}{\left|\prod_{m=q+1}^{s(1)} Q_m(x)\right|}$$

whenever $x \in D \setminus d(0, d_q)$.

We first consider the case $|x| > d_{s(2)}$. Then we have

$$\frac{1}{\left|\prod_{j=q+1}^{s(2)} Q_j(x)\right|} \leq \prod_{j=q+1}^{s(2)} \left(\frac{d_j}{d_{s(2)}}\right)^{u_j},$$

hence

(6) $\dfrac{|G_1(x)|}{\left|\prod_{m=q+1}^{s(2)} Q_m(x)\right|} \leq \left(\dfrac{r}{d_{s(1)}}\right)^{w(1)} \prod_{j=q+1}^{s(2)} \left(\dfrac{d_j}{d_{s(2)}}\right)^{u_j}.$

Now it just remains to consider $x \in D$ when $d_{s(1)} \leq |x| \leq d_{s(2)}$. Say $d_m \leq |x| < d_{m+1}$, with $s(1) \leq m \leq s(2)$. We see that

$$\frac{1}{\left|\prod_{m=q+1}^{s(2)} Q_m(x)\right|} \leq \left(\prod_{j=q+1}^{m} \left(\frac{d_j}{d_m}\right)^{u_j}\right) e_m$$

hence

(7) $\dfrac{|G_1(x)|}{\left|\prod\limits_{m=q+1}^{s(2)} Q_m(x)\right|} \leq e_m \left(\dfrac{r}{d_{s(1)}}\right)^{w(1)} \prod_{j=q+1}^{m} \left(\dfrac{d_j}{d_m}\right)^{u_j}.$

Thus, by (3), (6), (7) we have $|R_0(x)R_1(x)| \leq \frac{\epsilon}{2}$ for all $x \in D$ such that $|x| > d_{s(1)}$. This proves $(\mathcal{R}_{1,1})$ and finishes proving all the relations $(\mathcal{R}_{n,k})$. Now by Relations $(\mathcal{Q}_n)$ and $(\mathcal{R}_{n,k})$ it is easily seen that the sequence $(f_n)_{n \in \mathbb{N}}$ converges in $H(D)$ to an element ϕ_ϵ satisfying

(8) $|\phi_\epsilon(x) - R_0(x)| \leq \epsilon$ for all $x \in D \cap d(0, d_0)$ and

$(\mathcal{G}_n)$ $|\phi_\epsilon(x)| = |f_n(x)|$ whenever $x \in D \cap d(0, d_{s(n)})$.

By construction, it is seen that ϕ_ϵ does satisfy Condition (iii) in Proposition 7.4.20. Next, by Relations $(\mathcal{Q}_n)$ and (8), we obtain

(9) $|\phi_\epsilon(x) - 1| \leq \epsilon$ whenever $x \in D \cap d(0, d_0)$

and therefore $|\phi_\epsilon(x) - 1| \leq \epsilon$ whenever $x \in D \cap d(0, r(1-\epsilon))$, which is just Condition (i) in Proposition 7.4.20.

Now we only have to check Condition (ii). By Relations $(\mathcal{G}_n)$ and $(\mathcal{R}_{n,k})$ true for $n \geq 1$, we have $|\phi_\epsilon(x)| < 1$ when $|x| > d_{s(1)}$ and therefore it just remains to show $|\phi_\epsilon(x)| \leq (\frac{r}{\rho})^3 + r^3\epsilon$ when $x \in D' \cap d(0, d_{s(1)})$. Actually by Relations $(\mathcal{S}_n)$ and then by (4) it is seen that $|H_n(x)| \leq \frac{r}{\rho}$ whenever $x \in D' \cap d(0, d_q)$ and therefore $|\phi_\epsilon(x)| \leq r^3(\epsilon + \frac{1}{\rho^3})$ whenever $x \in D' \cap d(0, d_q)$.

Now let $x \in D'$ satisfy $d_q \leq |x| \leq d_{s(1)}$. We still have $|R_n(x)| = 1$ for all $n \geq 1$ and then by (5) we obtain

$$|R_0(x)| = \frac{1}{\left|\prod_{m=q+1}^{s(1)} Q_m(x)\right|}.$$

For example, let $x \in D'$ satisfy $d_m \leq |x| < d_m$ with $q + 1 \leq m \leq s(1)$. Then $|R_0(x)| \leq e_m \prod_{j=q+1}^{m} \left(\frac{d_j}{d_m}\right)^{u_j}$, and therefore by (3) we see that $|R_0(x)| \leq \frac{r^3}{\rho^3} + r^3\epsilon$. This finishes showing that ϕ_ϵ satisfies (b), and this ends the proof of Proposition 7.4.20. $\qquad\square$

Corollary 7.4.21. *Let D have an increasing (resp., a decreasing) T-sequence $t(T_{m,i}, q_{m,i})_{\substack{1 \leq i \leq k_m \\ m \in \mathbb{N}}}$ of piercing $\rho > 0$, of center α and diameter r. Let $D' = \mathbb{K} \setminus \left(\bigcup_{\substack{1 \leq i \leq k_m \\ m \in \mathbb{N}}} T_{m,i}\right)$. Let $\mathcal{F}$ be the T-filter on D' associated to this T-sequence. For each (m, i) $(1 \leq i \leq k_m,\ m \in \mathbb{N})$, let $b_{m,i} \in T_{m,i}$. Let $\epsilon \in\ ]0, +\infty[$. There exists $\psi_\epsilon \in \mathcal{I}_0(\mathcal{F})$, meromorphic on each hole $T_{m,i}$, satisfying further $\psi_\epsilon(x) = 1$ whenever $x \in \mathcal{P}(\mathcal{F})$ and*

(a) $|\psi_\epsilon(x)| \leq \epsilon$ *whenever* $x \in D \cap d(\alpha, r(1-\epsilon))$ $\left(\text{resp.}, x \in D \setminus d\left(\alpha, \left(\frac{r}{1-\epsilon}\right)^-\right)\right)$

(b) $\|\psi_\epsilon\|_D \leq \left(\frac{r}{\rho}\right)^3 + r^3\epsilon$

(c) $b_{m,i}$ *is a pole of ψ_ϵ of order $u_{m,i} \leq q_{m,i}$.*

Proof. In Proposition 7.4.20, we just take $\psi_\epsilon = 1 - \phi_\epsilon$. $\qquad\square$

7.5. Examples and counter-examples

Throughout the chapter, D is a closed bounded infraconnected subset of $\mathbb{K}$. We give examples of T-filters and T-sequences. We construct a closed infraconnected set whose interior is empty, which admits no T-filter.

We remember that a monotonous distances holes sequence $(T_{m,i})_{\substack{1 \leq i \leq h(m) \\ m \in \mathbb{N}}}$ is said to be *simple* if $h(m) = 1$ whenever $m \in \mathbb{N}$. A simple monotonous distances holes sequence will be denoted by $(T_m)_{m \in \mathbb{N}}$.

When a monotonous distances holes sequence is simple, it is much easier to determine whether there exists a T-sequence $(T_m, q_m)_{m \in \mathbb{N}}$. Indeed if $(\sigma_m)_{m \in \mathbb{N}}$ is a sequence of circles that run the sequence $(T_m)_{m \in \mathbb{N}}$, then putting $\sigma_m = C(a_m, d_m)$ and $\rho_m = \text{diam}(T_m)$ we have $\gamma_D(a_m, d_m, q) = \left(\frac{d_m}{\rho_m} \right)^q$ for every $q \in \mathbb{N}$. Hence a weighted sequence $(T_m, q_m)_{m \in \mathbb{N}}$ is a T-sequence if and only if

$$\lim_{m \to \infty} \left(-q_m |\log d_m - \log \rho_m|_\infty + \sum_{j=0}^{m-1} q_j |\log d_m - \log d_j|_\infty \right) = +\infty.$$

Notations: In Theorems 7.5.1, 7.5.6 and Corollaries 7.5.4, 7.5.5, $(T_m)_{m \in \mathbb{N}}$ is a simple monotonous distances holes sequence of monotony $(d_m)_{m \in \mathbb{N}}$ and diameter r with $\rho_m = \text{diam}(T_m)$. Moreover, denoting by σ_m the circle $C(a_m, d_m), (m \in \mathbb{N})$, the sequence $(\sigma_m)_{m \in \mathbb{N}}$ is a sequence of circles that runs the sequence $(T_m)_{m \in \mathbb{N}}$.

Theorem 7.5.1. *If $d_m = \rho_m$ is satisfied for infinitely many m, then there exists a sequence $(q_m)_{m \in \mathbb{N}}$ such that $(T_m, q_m)_{m \in \mathbb{N}}$ is a T-sequence.*

Proof. Indeed without loss of generality we may assume $d_m = \rho_m$ for every $m \in \mathbb{N}$, just by considering the subsequence of the T_m such that $d_m = \rho_m$. Then we have $\gamma_D(a_m, d_m, q_m) = 1$ whenever q_m, whence we can take a sequence q_m such that $\sum_{j=0}^{\infty} q_j |\log d_m - \log d_j|_\infty = +\infty$, which ends the proof. $\qquad \square$

Lemma 7.5.2. *Let $(u_n)_{n \in \mathbb{N}}$ be a strictly decreasing sequence of limit 0 in $]0, \infty[$, and let $(q_n)\ n \in \mathbb{N}$ be sequence of natural integers such that $\sum_{n \in \mathbb{N}} q_n u_n = +\infty$. Then we have $\lim_{n \to \infty} \sum_{j=0}^{n} q_j (u_j - u_n) = +\infty$.*

Proof. Let $B \in \mathbb{R}_+$, and let $N \in \mathbb{N}$ be such that $\sum_{j=0}^{N} q_j u_j \geq 2B$. Let $q - \sum_{j=0}^{N} q_j$, and let $t \in \mathbb{N}$ be such that $t > N$ and $u_t < \frac{B}{q}$. Now let $m \in \mathbb{N}$ satisfy $m \geq t$. We have

$$\sum_{j=0}^{m} q_j (u_j - u_m) \geq \sum_{j=0}^{N} q_j (u_j - u_m) \geq \sum_{j=0}^{N} q_j (u_j - u_t) = \sum_{j=0}^{N} q_j u_j - q u_t.$$

But we have $q u_t \leq B$, hence $\sum_{j=0}^{m} q_j (u_j - u_m) \geq B$. This ends the proof. $\square$

In Proposition 7.5.3, the implication "(1) implies (2)" is due to Motzkin and Robba [40].

Proposition 7.5.3. *Let $(u_n)_{n\in\mathbb{N}}$ be a strictly decreasing sequence of limit 0 in $]0,\infty[$, and let $(\theta_n)_n \in \mathbb{N}$ be a sequence in $]0,+\infty[$. If there exists a sequence of natural integers $(q_n)_n \in \mathbb{N}$ such that*

$$(1) \quad \lim_{n\to\infty} \sum_{j=0}^{n} q_j(u_j - u_n) - q_n\theta_n = +\infty,$$

then we have

$$(2) \quad \sum_{n=0}^{\infty} \frac{u_n}{\theta_n} = +\infty.$$

Conversely, if (2) is satisfied and if the sequences $(u_n)_{n\in\mathbb{N}}$, $(\theta_n)_{n\in\mathbb{N}}$ satisfy at least one of these 2 additional conditions:

(α) the sequence $(\theta_n)_{n\in\mathbb{N}}$ is bounded,
(β) $\sum_{n=0}^{\infty} u_n < +\infty$,

then there exists a sequence of natural integers $(q_n)_{n\in\mathbb{N}}$ satisfying (1).

Proof. First we suppose the existence of a sequence of integers $(q_n)_{n\in\mathbb{N}}$ satisfying (1) and will deduce that $\sum_{m=0}^{\infty} \frac{u_m}{\theta_m} = +\infty$. For each $m \in \mathbb{N}$, we put $\phi_m = \sum_{j=1}^{m} q_j(u_j - u_m) - q_m\theta_m$. By hypothesis we have $\lim_{m\to+\infty} \phi_m = +\infty$. Since $\lim_{m\to\infty} \phi_m = +\infty$ there exists $N \in \mathbb{N}$ such that $q_m\theta_m < \sum_{j=1}^{m-1} q_j(u_j - u_m)$ whenever $m > N$ and then with greater reason we have

$$(3) \quad q_m\theta_m < \sum_{j=1}^{m-1} q_j u_j.$$

Now since $q_m\theta_m = \sum_{j=1}^{m} q_j(u_j - u_m) - \phi_m$, we have $q_{m+1}\theta_{m+1} = \sum_{j=1}^{m} q_j(u_j - u_{m+1}) - \phi_{m+1} = \sum_{j=1}^{m-1} q_j(u_j - u_m) + q_m(u_m - u_{m+1}) - \phi_{m+1}$ and therefore $q_{m+1}\theta_{m+1} < \sum_{j=1}^{m-1} q_j u_j + q_m u_m - \phi_{m+1}$.

Now we can write

$$\sum_{j=1}^{m} q_j u_j = \left(\sum_{j=1}^{m-1} q_j u_j\right)\left[1 + \frac{q_m u_m}{\sum_{j=1}^{m-1} q_j u_j}\right]$$

and then by (3) we have

$$(4) \quad \sum_{j=1}^{m} q_j u_j < \left(\sum_{j=1}^{m-1} q_j u_j\right)\left[1 + \frac{u_m}{\theta_m}\right].$$

Since $\lim_{m \to \infty} \phi_m = +\infty$ there exists $N \in \mathbb{N}$ such that $\phi_m > 0$ whenever $m > N$. By induction on m, from (4) we have

$$\sum_{j=1}^{m} q_j u_j < \left(\sum_{j=1}^{N} q_j u_j \right) \left(\prod_{j=N}^{m} \left(1 + \frac{u_j}{\theta_j} \right) \right),$$

and therefore we obtain

(5) $\quad q_m \theta_m < \left(\sum_{j=1}^{N} q_j u_j \right) \left(\prod_{j=N}^{m} \left(1 + \frac{u_j}{\theta_j} \right) \right) - \phi_m \forall m > N.$

Since $\theta_m \geq 0$ whenever $m \in \mathbb{N}$ and since $\lim_{m \to \infty} \phi_m = +\infty$ one sees that $\lim_{m \to \infty} \prod_{j=N}^{\infty} \left(1 + \frac{u_j}{\theta_j} \right) = +\infty$ and therefore the series $\sum_{m=0}^{\infty} \frac{u_m}{\theta_m}$ diverges.

Conversely, now we suppose that (2) is satisfied together with one of the hypothesis α), β) and we will show that there exists a sequence of integers $(q_n)_{n \in \mathbb{N}}$ satisfying (1).

First, we assume

(6) $\quad \theta_n \leq \lambda$ for every $n \in \mathbb{N}$.

Then we put $q_n = \mathrm{int}(\frac{1}{\theta_n} + 1)$. By (2), we see

(7) $\quad \sum_{n=0}^{\infty} q_n u_n = +\infty,$

and by (6) we have

(8) $\quad q_n \theta_n \leq \lambda + 1$ for all $n \in \mathbb{N}$.

But by Lemma 7.5.2, (7) shows that $\lim_{n \to +\infty} \sum_{j=0}^{n} q_j(u_j - u_n) = +\infty$, hence by (8) we finally obtain

(9) $\quad \lim_{n \to +\infty} \sum_{j=0}^{n} q_j(u_j - u_n) - q_n \theta_n = +\infty.$

Now, we stop assuming the sequence $(\theta_n)_{n \in \mathbb{N}}$ to be bounded, but we suppose

(10) $\quad \sum_{n=0}^{\infty} u_n < +\infty.$

Let J be the subset of the $n \in \mathbb{N}$ such that $\theta_n > 1$. Clearly by (10) we have $\sum_{n \in J} \frac{u_n}{\theta_n} < +\infty$, hence $\sum_{n \in \mathbb{N} \setminus J} \frac{u_n}{\theta_n} = +\infty$. Therefore we can consider an increasing bijection ϕ from $\mathbb{N}$ onto $\mathbb{N} \setminus J$ such that $(u_{\phi(s)})_{s \in \mathbb{N}}$, $(\theta_{\phi(s)})_{s \in \mathbb{N}}$ satisfy $\sum_{s=0}^{\infty} \frac{u_{\phi(s)}}{\theta_{\phi(s)}} = +\infty$. So, we define a sequence of integers $(q_{\phi(s)})_{s \in \mathbb{N}}$ satisfying $\lim_{s \to +\infty} \sum_{j=0}^{s} q_{\phi(j)}(u_{\phi(j)} - u_{\phi(s)}) - q_{\phi(s)} \theta_{\phi(s)} = +\infty$. Finally, we put $q_n = 0$ for every $n \in J$, and then the sequence $(q_n)_{n \in \mathbb{N}}$ is defined for every $n \in \mathbb{N}$, and clearly satisfies (9) again. This finishes the proof of Proposition 7.5.3. $\qquad \square$

Corollary 7.5.4. *If the sequence $(T_m, q_m)_{m \in \mathbb{N}}$ is a T-sequence, then we have*

$$\sum_{m=0}^{+\infty} \frac{|\log r - \log d_m|_\infty}{\log d_m - \log \rho_m} = +\infty.$$

Conversely, if

$$\sum_{m=0}^{+\infty} \frac{|\log r - \log d_m|_\infty}{\log d_m - \log \rho_m} = +\infty$$

and if $\sum_{m=0}^{\infty} |\log r - \log d_m|_\infty < +\infty$ and $\rho_m < d_m$ for all m, then there exists a sequence $(q_m)_{m \in \mathbb{N}}$ such that the weighted sequence $(T_m, q_m)_{m \in \mathbb{N}}$ is a T-sequence.

Proof. Indeed, we put $u_n = |\log r - \log d_n|_\infty$, and $\theta_n = \log r - \log \rho_n$, $(n \in \mathbb{N})$, and we apply Lemma 7.5.3 to these sequences. □

Proposition 7.5.5. *If $\sum_{m=0}^{\infty} |\log r - \log d_m|_\infty < +\infty$ and if $\liminf_{n \to \infty} (\log d_m - \log \rho_m) > 0$, then there exists no sequence of integers $(q_n)_{n \in \mathbb{N}}$ such that $(T_m, q_m)_{m \in \mathbb{N}}$ is a T-sequence.*

Proof. Indeed, as above, we put $u_n = |\log r - \log d_n|_\infty$, and $\theta_n = \log r - \log \rho_n$, $(n \in \mathbb{N})$. It is easily seen that

$$\sum_{m=0}^{+\infty} \frac{|\log r - \log d_m|_\infty}{\log d_m - \log \rho_m} < \infty,$$

hence by Corollary 7.5.4, there exists no T sequence of the form $(T_m, q_m)_{m \in \mathbb{N}}$. □

Remark 1. In Proposition 7.5.3, we would like to have a genuine reciprocal to the first claim, without assuming one of the additional conditions (α), or (β). Actually, if the sequence $(\theta_n)_{n \in \mathbb{N}}$ is not bounded, and if $\sum_{n=0}^{\infty} u_n = +\infty$, it is far from clear whether (2) implies the existence of a sequence $(q_n)_{n \in \mathbb{N}}$ satisfying (1). However, this example shows that there is little hope to generalize the reciprocal.

For every $n \in \mathbb{N}$, we put $u_n = \frac{1}{\sqrt{n}}$ and $\theta_n = 3\sqrt{n+1}$. Then (2) is clearly satisfied. Now, we can check that the sequence $(q_n)_{n \in \mathbb{N}}$ defined as $q_n = 1$ whenever $n \in \mathbb{N}^*$ does not satisfy (1). Indeed we have $\sum_{j=1}^{n} u_j \le 2\sqrt{n+1}$, hence

$$\sum_{j=1}^{n} u_j - u_n - \theta_n < \sum_{j=1}^{n} u_j - \theta_n \le -\sqrt{n+1}.$$

Remark 2. In particular, if $\sum_{m=0}^{\infty} |\log r - \log d_m|_\infty < +\infty$ and if $\limsup_{m\to\infty} \left(\frac{\rho_m}{r}\right) < 1$ then there is no T-sequence $(T_m, q_m)_{m\in\mathbb{N}}$.

Proposition 7.5.6. *We have* $\lim_{m\to\infty} \sum_{j=0}^{m-1} |\log d_m - \log d_j|_\infty = +\infty$ *if and only if* $\sum_{j=0}^{\infty} |\log r - \log d_j|_\infty = +\infty$.

Proof. We just have to show that $\sum_{j=0}^{\infty} |\log r - \log d_j|_\infty = +\infty$ implies $\lim_{m\to\infty} \sum_{j=0}^{m-1} |\log d_m - \log d_j|_\infty = +\infty$. We put $u_m = |\log r - \log d_m|_\infty$ and then, by Lemma 7.5.2, we have $|\log d_m - \log d_j|_\infty = u_m - u_j$ because both $\log d_m - \log d_j$, $\log r - \log d_m$ have the same sign for $j < m$. Hence the conclusion is clear. $\qquad\square$

Remark. The condition $\liminf_{m\to\infty} \rho_m > 0$ is not required provided the sequence $(d_m)_{m\in\mathbb{N}}$ satisfies good conditions.

Proposition 7.5.7. *There exist simple T-sequences $(T_m, q_m)_{m\in\mathbb{N}}$ with*

$$\lim_{m\to\infty} \operatorname{diam}(T_m) = 0.$$

Proof. Let d_m satisfy $\sum_{j=0}^{\infty} |\log r - \log d_j|_\infty = +\infty$ and let ρ_m satisfy

$$\log \rho_m = -\frac{1}{2} \sum_{j=0}^{m} |\log d_m - \log d_j|_\infty.$$

By Theorem 7.5.6, we know that $\lim_{m\to\infty} \sum_{j=0}^{m-1} |\log d_m - \log d_j|_\infty = +\infty$ hence $\lim_{m\to\infty} \rho_m = 0$. However, it is seen that

$$\lim_{m\to\infty} \left(-(\log d_m - \log \rho_m) + \sum_{j=0}^{m-1} |\log d_m - \log d_j|_\infty \right) = +\infty,$$

hence the weighted sequence $(T_m, 1)_{m\in\mathbb{N}}$ is a T-sequence. $\qquad\square$

Theorem 7.5.8. *If $\mathbb{K}$ is separable, there exist closed infraconnected sets whose interiors are empty, which admit no T-filter.*

Proof. We construct a set $D \subset d(0,1)$, by defining its holes first. Since $\mathbb{K}$ is separable we can take an injective sequence $S = (a_s)_{s\in\mathbb{N}}$ dense in $d(0,1)$. Let $\rho_s = p^{(2^s)}$ and let $\tau_s = d(a_s, \rho_s^-)$ $(s \in \mathbb{N})$. We can easily construct a sequence of natural integers $(s(n))_{n\in\mathbb{N}}$ satisfying

(1) $s(n) > s(n-1)$
(2) $\tau_{s(n)} \cap \tau_{s(j)} = \emptyset$ whenever $j \neq n$
(3) $a_i \in \bigcup_{j=0}^{n} \tau_{s(j)}$ whenever $i \neq 0, \ldots, s(n)$.

Indeed let $s_0 = 0$, assume the $s(j)$ already obtained up to $j = n$, satisfying (1), (2), (3) and let us define $s(n + 1)$. We take for $s(n + 1)$ the lowest integer m such that $\tau_m \cap \left(\bigcup_{j=0}^{n} \tau_{s(j)} \right) = \emptyset$. Such an integer $s(n + 1)$ is easily seen to exist because we can find $m \in \mathbb{N}$ such that the distance from a_m to $\bigcup_{j=0}^{n} \tau_{s(j)}$ is strictly superior to ρ_m. Then (1) and (2) are now satisfied up to the rank $n + 1$. Let $i \in \mathbb{N} \cap [s(n) + 1, s(n + 1) - 1]$. Since τ_i has a not empty intersection with one $\tau_{s(j)}$ for a certain $j \leq n$, and since the sequence $(\rho_s)_{s \in \mathbb{N}}$ is strictly decreasing, obviously we have $\tau_i \subset \tau_{s(j)}$. Then (3) is clearly satisfied for $i \leq s(n + 1) - 1$, and then also for $i \leq s(n + 1)$ because $a_{s(n+1)} \in \tau_{s(n+1)}$.

We put $a_n = \alpha_{s(n)}$, $T_n = \tau_{s(n)}$, $r_n = \rho_{s(n)}$. Clearly $\bigcup_{n=0}^{\infty} T_n$ is dense in $d(0, 1)$ because by (3) it contains all the a_s. We fix $N \in \mathbb{N}$. We will first show that for every $\epsilon > 0$, there exists $N' > N$ such that $r_N < \epsilon$ and $\tau_{N'} \subset C(a_N, r_N)$. Indeed $C(a_N, r_N)$ is not included in any T_n because otherwise, we should have $T_N \subset T_n$. Since $r_N \in |\mathbb{K}|$, there exist infinitely many T_m included in $C(a_N, r_N)$, one of them has a radius $r_m < \epsilon$, and we may call N' this m. By induction we shortly define a convergent subsequence of the sequence $(a_n)_{n \in \mathbb{N}}$ whose terms belong to $C(a_N, r_N)$. Indeed $C(a_{N'}, r_{N'})$ contains a $T_{N''}$ such that $r_{N''} < \epsilon^2$ whereas $T_{N''} \subset C(a_{N'}, r_{N'}) \subset C(a_N, r_N)$, and so on.

Let D be the set of the limits of the convergent subsequences of the sequence $(a_n)_{n \in \mathbb{N}}$. By definition D is closed and included in $d(0, 1)$. Moreover the T_n appear to be the holes of D. Indeed D obviously has an empty intersection with each T_n because $T_n \cap T_m = \emptyset$ for every $n \neq m$. Next, each T_n is included in $d(0, 1)$ and $C(a_n, r_n)$ contains points of D, hence $\widetilde{D} = d(0, 1)$. But then the distance from T_n to D is just r_n and therefore T_n is a hole of D. Finally we check that D has no hole other than the T_n. Indeed let $\alpha \in d(0, 1) \setminus \bigcup_{n=1}^{\infty} T_n$. Since $\bigcup_{n \in \mathbb{N}^*} T_n$ is dense in $d(0, 1)$, there exists a subsequence of the sequence $(T_n)_{n \in \mathbb{N}^*}$ which converges to α, hence α is the limit of a subsequence of the sequence $(a_n)_{n \in \mathbb{N}}$, hence $\alpha \in D$. Thus, if $\alpha \in d(0, 1) \setminus D$, α belongs to a certain T_N and then this finishes proving that every hole of D is a T_n.

Now it is easily seen that D is infraconnected. Indeed let $\alpha \in D$ and let $r \in |\mathbb{K}|$. Since $\bigcup_{n \in \mathbb{N}^*} T_n$ is dense in $d(0, 1)$, there exists $N \in \mathbb{N}$ such that $T_N \cap C(\alpha, r) \neq \emptyset$. Since α belongs to T_N, T_N is obviously included in $C(\alpha, r)$, hence we have $r_N < r$, and we know that $C(a_N, r_N)$ contains points of D which obviously belong to $C(\alpha, r)$, hence $C(\alpha, r) \cap D \neq \emptyset$. We check that D has empty interior because by definition every point $\alpha \in D$ is the limit of a subsequence of the sequence $(a_n)_{n \in \mathbb{N}}$ no term of which belong to D.

Now we will show that D has no T-filter. Indeed assume D to have a T-filter $\mathcal{F}$, of diameter r. We know that even if $\mathcal{F}$ is decreasing with no center in $\mathbb{K}$, it admits a center α in $\widehat{\mathbb{K}}$. With no loss of generality we may assume $\mathcal{F}$ to be decreasing. Then there exists a T-sequence $(U_m, u_m)_{m \in \mathbb{N}^*}$ associated to $\mathcal{F}$ where each U_m is a hole of D, whose diameter is denoted by σ_m, and where u_m belongs to $\mathbb{N}^*$, whereas the distance d_m from U_m to α satisfies $d_{m+1} \leq d_m, \lim_{m \to \infty} d_m = r$ and $d_m < r$ whenever $m \in \mathbb{N}^*$. By relation (4) in the definition of the T-sequences it is easily checked that our T-sequence $(U_m, u_m)_{m \in \mathbb{N}^*}$ must satisfy

(4) $\lim_{n \to \infty} \left[\left(\frac{d_n}{\sigma_n} \right)^{u_n} \prod_{j=1}^{n-1} \left(\frac{d_n}{d_j} \right)^{u_j} \right] = 0.$

We put $\lambda_n = \log \sigma_n$. Then by (4) we have

$$\lim_{n \to \infty} \left(u_n \left(\log r + \lambda_n \right) + \left(\sum_{j=1}^{n-1} u_j \right) \left(\log r - \log d_1 \right) \right) = +\infty$$

hence there obviously exists a positive constant C such that

(5) $u_n \left(\log r + \lambda_n \right) \leq C \sum_{j=1}^{n-1} u_j.$

Since $\lim_{m \to \infty} r_m = 0$ and since the sequence $(U_n)_{n \in \mathbb{N}}$ appears to be a reordered subsequence of the sequence T_n it is clearly seen that $\lim_{n \to \infty} \lambda_n = +\infty$. Then there exists a positive constant B such that $u_n (\log r + \lambda_n) \leq B \lambda_n$ whenever $n \in \mathbb{N}^*$ and then by (5) there exists a positive constant A such that

(6) $u_n \lambda_n \leq A \sum_{j=1}^{n-1} u_j.$

By applying (6) at the rank $n - 1$ we have

$$u_{n-1} \leq \frac{A}{\lambda_{n-1}} \sum_{j=1}^{n-2} u_j$$

hence by (6) at the rank n we obtain

$$u_n \leq \frac{A}{\lambda_n} \left(\sum_{j=1}^{n-2} u_j \right) \left(1 + \frac{A}{\lambda_{n-1}} \right)$$

and then by an immediate downing induction we have

$$u_n \leq \frac{A u_1}{\lambda_n} \left(\prod_{j=1}^{n-1} \left(1 + \frac{A}{\lambda_j} \right) \right).$$

We now consider the sequence $(w_n)_{n \in \mathbb{N}}$ defined as

$$w_n = \log\Big(\frac{1}{\lambda_n} \prod_{j=1}^{n-1}\Big(1 + \frac{A}{\lambda_j}\Big)\Big) = -\log \lambda_n + \sum_{j=1}^{n-1} \log\Big(1 + \frac{A}{\lambda_j}\Big).$$

It is seen that

(7) $w_n \leq -\log \lambda_n + A \sum_{j=1}^{n-1} \frac{1}{\lambda_j}.$

Now, by construction of D, every U_n is a certain T_m, hence every λ_j is of the form 2^m while the sequence $n \to \lambda_n$ is injective, hence

$$\sum_{j=1}^{n-1} \frac{1}{\lambda_j} \leq \sum_{m \in \mathbb{N}^*} \frac{1}{2^m} = 1.$$

Therefore, by (7) we see that $\lim_{n \to \infty} w_n = -\infty$, hence $u_n = 0$ when n is big enough. This contradicts the hypothesis $u_n \in \mathbb{N}^*$ and finishes proving that D has no T-filter and this ends the proof of Theorem 7.5.8. □

Lemma 7.5.9. *Let E be a set which is not countable and let f be a function from D into $\mathbb{R}_+$. There exists a sequence $(x_m)_{m \in \mathbb{N}}$ in E and $\lambda > 0$ such that $f(x_m) \geq \lambda$ for all $m \in \mathbb{N}$.*

Proof.   We assume that such a λ and such a sequence $(x_m)_{m \in \mathbb{N}}$ do not exist. Hence for every $q \in \mathbb{N}^*$, the set A_q of the $x \in E$ such that $f(x) \geq \frac{1}{q}$ is finite. But E is clearly equal to $\bigcup_{q=1}^{\infty} A_q$ and therefore E is countable. □

Theorem 7.5.10. *Let $\mathbb{K}$ be strongly valued. Let $a \in \widetilde{D}$ be such that $\delta(a, D) < \mathrm{diam}(D)$. We assume that for every $r \in]\delta(a, D), \mathrm{diam}(D)[\cap|\mathbb{K}|$, each class of $C(a, r)$ contains at least one hole of D. Then for each $r \in]\delta(a, D), \mathrm{diam}(D)]$ there exists on D an increasing idempotent T-sequence and a decreasing idempotent T-sequence of center a and diameter r.*

Proof.   Let $r \in]\delta(a, D), \mathrm{diam}(D)]$. We will construct an increasing idempotent T-sequence $(T_n)_{n \in \mathbb{N}}$ of D, of center a and diameter r.

Let $J =]\delta(a, D), \mathrm{diam}(D)[\cap|\mathbb{K}|$, and let $(r_n)_{n \in \mathbb{N}}$, $(r'_n)_{n \in \mathbb{N}}$ be sequences of J satisfying $r_n < r'_n < r_{n+1}$, $\lim_{n \to \infty} r_n = r$. By hypothesis, either the residue class field $\mathcal{K}$ of $\mathbb{K}$ is not countable, or $|\mathbb{K}|$ is not.

First, we suppose that $|\mathbb{K}|$ is not countable. Let n be fixed. By Lemma 7.5.9, there exists $\rho_n > 0$ together with an infinite family of circles $C(a, r)$, with $r_n < r < r'_n$, every one of them contains at least one hole of diameter bigger than ρ_n. Let $q_n \in \mathbb{N}$ satisfy

(1) $\Big(\frac{r_n}{r'_n}\Big)^{q_n} \frac{r'_{n+1}}{\rho_{n+1}} < \frac{1}{n+1}.$

Thus, we have defined the sequences $(\rho_n)_{n\in\mathbb{N}}$, $(q_n)_{n\in\mathbb{N}}$. Now, for each $n \in \mathbb{N}$, we put $t_n = \sum_{h=1}^{n} q_h$. From above, we can clearly find q_n circles $C(a, d_m)_{t_n \le m < t_{n+1}}$, with $r_n < d_m < d_{m+1} < r'_n$, which all contain at least one hole T_m of diameter bigger than ρ_n. Then by definition, T_m is included in $C(a, d_m)$ while $T_h \cap C(a, d_m) = \emptyset$ whenever $h \ne m$. Besides when $t_n \le m < t_{n+1}$, by hypothesis we have $\mathrm{diam}(T_m) \ge \rho_n$. Thus, by (1) the weighted sequence $(T_m, 1)$ is seen to satisfy

$$\frac{d_m}{\mathrm{diam}(T_m)} \prod_{j=1}^{m} \frac{d_j}{d_m} \le \frac{1}{n} \quad \text{whenever } t_n \le m < t_{n+1}.$$

This shows that we have

$$\lim_{m\to\infty} \frac{d_m}{\mathrm{diam}(T_m)} \prod \frac{d_j}{d_m} = 0$$

and therefore the weighted sequence $(T_m, 1)$ is an idempotent T-sequence.

Now we suppose that the residue class field $\mathcal{K}$ is not countable. By Lemma 7.5.9, for each $n \in \mathbb{N}$, there exist $\rho_n > 0$ together with an infinite family of classes of $C(a, r_n)$, such that each contains at least one hole of diameter bigger than ρ_n. In the same way as the previous case, for each $n \in \mathbb{N}$, we take $q_n \in \mathbb{N}$ satisfying (1) again and then, we have defined the sequences $(\rho_n)_{n\in\mathbb{N}}$, $(q_n)_{n\in\mathbb{N}}$. For each $n \in \mathbb{N}$, we put $t_n = \sum_{h=1}^{n} q_h$. Then in $C(a, r_n)$, we can clearly find q_n different classes such that each contains at least one hole of diameter bigger than ρ_n. Let $(T_m)_{t_n \le m < t_{n+1}}$ be these holes. Thus, by definition, we have

(2) $\delta(T_j, T_h) = r_n$ whenever h, j such that $t_n \le h < t_{n+1}, t_n \le j < t_{n+1}$, $h \ne j$.

By (2), it is easily checked that $\gamma_D(a, r_n, q_n) \le \frac{r_n}{\rho_n}$. Therefore the weighted sequence $(T_m, 1)_{m\in\mathbb{N}}$ is an idempotent T-sequence because by (1) we have

$$\gamma_D(a, r_n, q_n) \prod_{j=1}^{n} \left(\frac{r_j}{r_n}\right)^{q_j} \le \frac{1}{n}.$$

Symmetrically, we can prove the existence of a decreasing idempotent T-sequence of center a and diameter r: in the proof, we just have to replace (1) by

(1)' $\left(\frac{r'_n}{r_n}\right)^{q_n} \frac{r_{n+1}}{\rho_{n+1}} < \frac{1}{n+1}$.

This finishes the proof of Theorem 7.5.10. □

Lemma 7.5.11. *Put $s = \mathrm{diam}(D)$ and let D admit a partition of the form $(\mathcal{B}(\mathcal{F}_i))_{i \in I}$ where each $\mathcal{F}_i$ is an increasing T-filter, and assume that D has no other T-filter. Let $\mathcal{G}$ be the D-peripheral circular filter. Then $\mathcal{I}(\mathcal{F}_i) = \mathcal{I}(\mathcal{G})\ \forall i \in I$.*

Proof. Let $i \in I$ be fixed, let a_i be a center of $\mathcal{F}_i$, and let $r_i = \mathrm{diam}(\mathcal{F}_i)$. Let $f \in \mathcal{I}(\mathcal{F}_i)$, so we have $\varphi_{a_i, r_i}(f) = 0$. Suppose there exists $r > r_i$ such that $\varphi_{a_i, r}(f) > 0$. Let $u = \inf\{r \mid \varphi_{a_i, r}(f) > 0\}$. Then f is strictly vanishing along the decreasing filter of center a_i and diameter u, so this filter is a T-filter, a contradiction to the hypothesis. Consequently, we have $\varphi_{a_i, s}(f) = 0$, and therefore $\mathcal{I}(\mathcal{F}_i) \subset \mathcal{I}(\mathcal{G})$.

Now, let $f \in \{\mathcal{I}(\mathcal{G})$, so we have $\varphi_{a_i, r}(f) = 0$. Suppose that $\varphi_{a_i, r_i}(f) > 0$. Since $\varphi_{a_i, r}(f) = 0$, we can consider $u = \inf\{r \mid \varphi_{a_i, r}(f) = 0\}$. Then f is strictly vanishing along the increasing filter of center a_i and diameter u, so this filter is a T-filter, a contradiction to the hypothesis. Consequently, we have $\varphi_{a_i, r_i}(f) = 0$, and therefore $\mathcal{I}(\mathcal{F}_i) = \mathcal{I}(\mathcal{G})$. $\qquad\square$

Corollary 7.5.12. *Let $a \in D$ and $r > 0$ such that $D \cap d(a, r^-)$ admit a partition of the form $(\mathcal{B}(\mathcal{F}_i))_{i \in I}$ where each $\mathcal{F}_i$ is an increasing T-filter,, and assume that every other T-filter on $D \cap d(a, r^-)$ surrounds all the $\mathcal{F}_i$. Then, $\mathcal{I}(\mathcal{F}_i) = \mathcal{I}(\mathcal{F}_j)\ \forall i, j \in I$.*

Definitions and notations: If $\mathcal{F}$ is a circular filter on D, of center a and diameter r, we will denote by $\mathcal{Q}(\mathcal{F}, D)$ the set $d(a, r) \cap D$. If $\mathcal{F}$ is a circular filter with no center, we put $\mathcal{Q}(\mathcal{F}, D) = \emptyset$.

Given a monotonous filter $\mathcal{F}$, we will denote by $tcf\mathcal{F}$ the unique circular filter less thin than $\mathcal{F}$.

A circular filter $\mathcal{F}$ will be said to *surround* a circular filter $\mathcal{G}$ (resp., a monotonous filter $\mathcal{G}$) if $\mathcal{G}$ is secant with $\mathcal{Q}(\mathcal{F}, D)$, or if $\mathcal{G} = \mathcal{F}$ (resp., $\mathcal{G} = tcf\mathcal{F}$).

A monotonous filter $\mathcal{F}$ will be said to *surround* a circular filter (resp., a monotonous filter) $\mathcal{G}$ if $\check{\mathcal{F}}$ surrounds $\mathcal{G}$ (resp., $tcf\mathcal{F}$ surrounds $tcf\mathcal{G}$).

A monotonous filter or a circular filter $\mathcal{F}$ will be said to *strictly surround* a monotonous or a circular filter $\mathcal{G}$ if $\mathcal{F}$ surrounds $\mathcal{G}$ and if $\mathrm{diam}(\mathcal{F}) \neq \mathrm{diam}(\mathcal{G})$.

A partition $(A_i)_{i \in I}$ of D will be said to be *T-optimal* if for each $i \in I$, there exists an increasing T-filter $\mathcal{F}_i$ on D which is not surrounded by any increasing T-filter different from $\mathcal{F}_i$, such that $A_i = \mathcal{B}(\mathcal{F}_i)$.

If D admits a T-optimal partition $\mathcal{B}(\mathcal{F}_i)_{i \in I}$, this partition will be said to be *T-specific* if for every decreasing T-filter $\mathcal{G}$ on D there exists $h \in I$ such that $\mathcal{G}$ is secant with $\mathcal{B}(\mathcal{F}_h)$).

A circular filter $\mathcal{F}$ on D will be said to be *distinguished* if it satisfies the following two statements:

(i) Either $\mathcal{Q}(\mathcal{F}, D)$ is empty, or it admits a T-specific partition.
(ii) Either $\mathcal{Q}(\mathcal{F}, D) = D$, or D admits a decreasing T-filter thinner than $\mathcal{F}$.

Let $\mathcal{F}$ be a distinguished circular filter on D such that $\mathcal{Q}(\mathcal{F}, D) \neq \emptyset$ and let $\mathcal{B}(\mathcal{F}_i)_{i \in I}$ be the T-specific partition of $\mathcal{Q}(\mathcal{F}, D)$. The family of the increasing T-filters $(\mathcal{F}_i)_{i \in I}$ will be called *the T-family of $\mathcal{F}$*, and the $\mathcal{F}_i$ $(i \in I)$ will be called *the increasing T-filters of $\mathcal{F}$*.

A distinguished circular filter $\mathcal{F}$ on D will be said to be *regular* if either $\mathcal{Q}(\mathcal{F}, D) = \emptyset$, or $\mathcal{Q}(\mathcal{F}, D) \neq \emptyset$, and all the increasing T-filters of $\mathcal{F}$ have the same diameter.

A distinguished circular filter $\mathcal{F}$ on D that is not regular will be said to be *irregular*.

Theorem 7.5.14 will be useful in strongly valued fields. We first need to notice some basic results. We need to notice this basic lemma.

Lemma 7.5.13. *Let L be a metric space whose distance is denoted by σ and let $(A_i)_{i \in I}$ be a family of closed subsets of L. If there exists $r > 0$ such that $\sigma(A_i, A_j) \geq r$ for every $i, j \in I$, such that $i \neq j$, then $\bigcup_{i \in I} A_i$ is closed in L.*

Theorem 7.5.14. *If $\mathbb{K}$ is strongly valued, every distinguished circular filter is regular. If $\mathbb{K}$ is weakly valued, there exist closed bounded infraconnected sets with an irregular distinguished circular filter (in particular, we can construct a closed bounded infraconnected set with an irregular distinguished circular filter of diameter 1 included in $d(0, 1^-)$).*

Proof. First we suppose $\mathbb{K}$ strongly valued and we suppose that D admits an irregular distinguished circular filter $\mathcal{F}$ of diameter t. Hence $\mathcal{Q}(\mathcal{F}, D)$ admits a T-specific partition $\mathcal{B}(\mathcal{F}_i)_{i \in I}$ with a certain $\mathcal{F}_h$ of diameter $s \in]0, t[$. Let a be a center of $\mathcal{F}_h$. For every $\ell \in]s, t[\cap|\mathbb{K}|$, for every $b \in C(a, \ell) \cap D$, b belongs to a certain set of the form $\mathcal{B}(\mathcal{F}_j)$ included in $\mathcal{B}(a, \ell)$ because by hypothesis we have $\mathcal{B}(\mathcal{F}_j) \cap \mathcal{B}(\mathcal{F}_h) = \emptyset$. Hence $d(b, \ell^-)$ does contain holes of D. This way, we see that for every $\ell \in]s, t[\cap|\mathbb{K}|$, every class of $\mathcal{B}(a, \ell)$ contains at least one hole of D. Therefore, by Theorem 7.5.10, D admits an increasing and a decreasing T-filter of center a and diameter ℓ, a contradiction to the hypothesis "$\mathcal{F}$ distinguished".

Now we suppose $\mathbb{K}$ weakly valued and we will construct an infraconnected set E with an irregular distinguished circular filter.

First, we will construct a model of sets $B(u)$ for $u \in]0,1[$ that will be used in the construction of E. Let $(a_n)_{n\in\mathbb{N}}$ be a sequence in $\mathbb{K}$ satisfying for all $n \in \mathbb{N}^*$

(1) $\frac{u}{\sqrt{n+1}} < v(a_n) \leq \frac{u}{\sqrt{n}}$.

We notice that the sequence $v(a_n)_{n\in\mathbb{N}}$ is strictly decreasing, and obviously positive. Next, for every $n \in \mathbb{N}$, we put $\rho_n = \omega^{\frac{-u\sqrt{n}}{2}}$, $T_n = d(a_n, \rho_n^-)$, and $N(u) = d(0, 1^-) \setminus \left(\bigcup_{n=0}^{\infty} T_n \right)$. Thus, by definition, the only holes of $N(u)$ which are included in $d(0, 1^-)$ are the T_n and then $N(u)$ admits a simple pierced filter $\mathcal{T}$ of center 0 and diameter 1. We are going to check that $\mathcal{T}$ is a T-filter. For each $n \in \mathbb{N}$, we put $d_n = |a_n|$. By (1) we check

(2) $\sum_{j=1}^{n-1} \log\left(\frac{d_n}{d_j}\right) \geq \sum_{j=1}^{n-1} \frac{u}{\sqrt{j+1}} - \frac{nu}{\sqrt{n}} \geq u(\sqrt{n} - 2\sqrt{2})$.

Hence we have

$$\sum_{j=1}^{n-1} \log\left(\frac{d_n}{d_j}\right) - \log\left(\frac{d_n}{\rho_n}\right) \geq u(\sqrt{n} - 2\sqrt{2}) - \frac{u}{\sqrt{n}} + \frac{u\sqrt{n}}{\sqrt{2}}$$

and then by (2) it is seen that

$$\lim_{n\to\infty} \left(\sum_{j=1}^{n-1} \log\left(\frac{d_n}{d_j}\right) - \log\left(\frac{d_n}{\rho_n}\right) \right) = +\infty.$$

Since $\mathcal{T}$ is a simple increasing filter, this is sufficient to prove that it is a T-filter.

Now for $j = 1, \ldots, q$ let $e_j \in T_{n_j}$ with the T_{n_j} not necessarily all different, and let $P(x) = \prod_{j=1}^{q}(x - e_j)$. We want to consider $\left\| \frac{1}{P} \right\|_{<B(u)}$.

First, we assume that at least one index ℓ satisfies $n_\ell \geq q$, and then clearly we have

$$\log\left\| \frac{1}{P} \right\|_{N(u)} \geq \left\| \frac{1}{x - e_\ell} \right\|_{N(u)} = -\log(\rho_{n_\ell}) \geq \frac{u\sqrt{q}}{2}$$

Now we assume that $n_j < q$ for all $j = 1, \ldots, q$. Then when $|x| \leq |a_q|$ we have $|P(x)| \leq |a_q|^q$ hence $\log\left\| \frac{1}{P} \right\|_{N(u)} \geq uq \log(d_q) \geq u\sqrt{q}$. Thus, in both cases we have proven $\log\left\| \frac{1}{P} \right\|_{N(u)} \geq \frac{u\sqrt{q}}{2}$. As a consequence we have

(3) $\vartheta(B(u), q) \geq \frac{u\sqrt{q}}{2}$.

Now since $|\mathbb{K}|$ is countable, there exists a bijection b from $|\mathbb{K}|$ onto $\mathbb{N}$. For every $r \in |\mathbb{K}|$ we put $\phi(r) = 4^{b(r)}$, and $B(r) = b(r)^{\phi(r)}$. It is seen that B is injective. The set E we want to construct should satisfy $\log(\gamma_E(0, r, q)) \geq B(r)\sqrt{q}$ for every $r \in]\frac{1}{2}, 1[\cap|\mathbb{K}|$ and $q \in \mathbb{N}$. We fix $r \in]0, \frac{1}{2}[\cap|\mathbb{K}|$. Since $\mathbb{K}$ is countable we can find a sequence $(\beta_n)_{n \in \mathbb{N}^*}$ in $C(0, r)$ such that the family of disks $(d(\beta_n, r^-))_{n \in \mathbb{N}}$ makes a partition of $C(0, r)$.

For each $n \in \mathbb{N}^*$ we put $\psi_{n,r}(x) = \beta_n(1 + x)$. For every $u \in]0, 1[$ it is seen that the set $\psi_{n,r}(B(u))$ is an infraconnected subset of $d(\beta_n, r^-)$ which admits an increasing T-filter of center β_n and diameter r.

Now we put $A_{n,r} = \psi_{n,r}(B(2^{n+2}B(r)))$, $(n \in \mathbb{N})$, $E_r = \bigcup_{n \in \mathbb{N}} A_{n,r}$, and finally $I =]\frac{1}{2}, 1[\cap|\mathbb{K}|$, and $E = \bigcup_{r \in I} E_r$. Then E is obviously bounded. Besides, by Lemma 7.5.13, it is clearly closed because $\delta(A_{r,n}, A_{r',n'}) \geq \frac{1}{2}$ for all $(r, n) \neq (r', n')$.

Finally we check that E is infraconnected. Let $a \in E$, and let (n, r) be the couple such that $a \in A_{r,n}$. If $u \in]0, r[\cap|\mathbb{K}|$, there exists $x \in A_{n,r}$ such that $|x - a| = u$. If $u \in [r, 1[\cap|\mathbb{K}|$, then any $x \in A_{n,u}$ satisfies $|x - a| = u$. We fix $r \in I$, $q \in \mathbb{N}$ and $\varepsilon > 0$. We will study $\gamma_E(0, r, q)$. By Corollary 7.4.4 we can find indices $n_1, \ldots, n_k \in \mathbb{N}$ together with strictly positive integers $q_1, \ldots q_k$ satisfying

(4) $\sum_{j=1}^{k} q_j = q$, and

(5) $\log(\gamma_E(0, r, q)) + \varepsilon \geq \log(\vartheta(A_{r,n_j,j}))$

By Lemma 7.4.3, we know that for every $s \in \mathbb{N}$ we have $\vartheta(A_{r,n}, s) = \vartheta(B(2^{n+2}B(r)), s)$. Hence by (3) we have

(6) $\vartheta(A_{r,n}, s) \geq 2^{n+1}B(r)\sqrt{s}$.

Hence by (5), (6), we obtain

(7) $\log(\gamma_E(0, r, q)) + \varepsilon \geq 2^{n_j+1}\sqrt{q_j}B(r)$ whenever $j = 1, \ldots, k$.

From (7) we can deduce

$$\sum_{j=1}^{k} 2^{-n_j-1}\left(\frac{\log(\gamma_E(0, r, q)) + \varepsilon}{B(r)}\right)^2 \geq \sum_{j=1}^{k} q_j = q.$$

But since $n_i \neq n_j$ whenever $i \neq j$, we have $\sum_{j=1}^{k} 2^{-n_j-1} < \sum_{n=0}^{\infty} 2^{-n-1} = 1$. So we obtain $q \leq \left(\frac{\log(\gamma_E(0,r,q))+\varepsilon}{B(r)}\right)^2$. Actually ε was chosen arbitrarily, hence we have

(8) $q \leq \left(\frac{\log(\gamma_E(0,r,q))}{B(r)}\right)^2$.

Now, we will show that the circular filter $\mathcal{G}$ of center 0 and diameter 1 is an irregular distinguished circular filter on E. For each couple $(n, r) \in \mathbb{N} \times I$ we denote by $\mathcal{F}_{n,r}$ the increasing T-filter of $A_{n,r}$, that obviously defines a T-filter on E. Clearly the family $(\mathcal{B}(\mathcal{F}_{n,r}))_{(n,r) \in \mathbb{N} \times I}$ makes a partition of E. Hence, to prove that $\mathcal{G}$ is an irregular distinguished circular filter, it is sufficient to show that this partition is T-specific, i.e., none of the $\mathcal{F}_{n,r}$ is surrounded by any T-filter $\mathcal{F}' \neq \mathcal{F}_{n,r}$.

Indeed suppose that certain $\mathcal{F}_{n,\rho}$ is surrounded by a T-filter $\mathcal{F}' \neq \mathcal{F}_{n,\rho}$ and let $r = \mathrm{diam}(\mathcal{F}')$. If $\mathcal{F}'$ is decreasing then $\mathcal{P}(\mathcal{F}') = d(0, r)$, hence 0 is a center of $\mathcal{F}'$. If $\mathcal{F}'$ is increasing and different from $\mathcal{F}_{n,\rho}$, then $\mathcal{B}(\mathcal{F}')$ contains $\mathcal{B}(\mathcal{F}_{n,\rho})$ and has a diameter $\rho < r$, hence 0 is a center of $\mathcal{F}'$. Thus, we have checked that $\mathcal{F}'$ is a T-filter of center 0 and diameter r.

First we will suppose $\mathcal{F}'$ increasing. Hence E admits an increasing T-sequence of center 0 and diameter r. Let $C(0, r_m)$ be a sequence of circles carrying this T-sequence. So, there exists a sequence of strictly positive integers $(q_m)_{m \in \mathbb{N}}$ such that

$$\lim_{m \to \infty} (-\log(\gamma_E(0, r_m, q_m))) + \sum_{j=0}^{m-1} q_j \log\left(\frac{r_m}{r_j}\right) = +\infty.$$

Let $a = -\log r_0$. Trivially we have $a \sum_{j=0}^{m-1} q_j \geq \log(\gamma_E(0, r_m, q_m))$ and therefore by (8) we obtain

(9) $a \sum_{j=0}^{m-1} q_j \geq \sqrt{q_m}\, B(r_m)$.

For every $m \in \mathbb{N}$ we put $h(m) = \sup_{0 \leq j < m} q_j$. By (9) we have $am\, h(m) \geq \sqrt{q_m}\, B(r_m)$ hence

(10) $q_m \leq \left(\frac{am\, h(m)}{B(r_m)}\right)^2$.

Let $m \in \mathbb{N}$ be fixed. Let $w_0, \ldots, w_s$ be the ranks such that $h(w_j) = q_{w_j}$ with $w_0 = h(m)$, $w_1 = h(m-1)$ and $w(s) = h(0)$. By (10) each $h(w_j)$ satisfies

$$h(w_j) \leq \left(\frac{2a\, w_j\, h(w_{j-1})}{B(r_{w_j})}\right)^2 \quad (1 \leq j \leq s).$$

For convenience we put $\sigma(m) = B(r_m)$ $(m \in \mathbb{N})$, and $\theta(n) = 2^n$ $(n \in \mathbb{N})$. By induction on Relations (10) we shortly obtain:

(11) $q_m \leq \left(\frac{m}{\sigma(m)}\right)^2 \prod_{j=1}^{s} \left(\frac{2a\, w_j}{\sigma(w_j)}\right)^{\theta(j+1)}$.

Now we notice that $w_j < m$ for every $j = 1, \ldots, s$ and that $\sum_{j=0}^{s} 2^{j+1} \leq 2^m$ because $s \leq m$. So by (11) we have

(12)　$q_m \leq \dfrac{(am)^{\theta(m+2)}}{\sigma(m)^2}$　for every $m \in \mathbb{N}$.

Let $(\tau_m)_{m \in \mathbb{N}}$ be the sequence defined in $\mathbb{N}$ as $\tau_m = b(r_m)$. Since the mapping b is injective, so is the sequence $(\tau_m)_{m \in \mathbb{N}}$. Therefore, as it takes values in $\mathbb{N}$, it is seen that the inequality $\tau_m \geq m$ holds for infinitely many indices. Hence there exists a subsequence $(\tau_{m_t})_{t \in \mathbb{N}}$ such that $\tau_{m_t} \geq m_t$ whenever $t \in \mathbb{N}$. For convenience we put $\nu(t) = m_t$. By (12) we obtain

$$q_{\nu(t)} \leq \frac{(a\nu(t))^{\theta(\nu(t)+2)}}{\tau(\nu(t))^2} \leq \frac{(a\nu(t))^{\theta(\nu(t)+2)}}{(\nu(t))^{2\phi(\nu(t))}}.$$

Since the sequence $(\nu(t))_{t \in \mathbb{N}}$ goes to $+\infty$, it is seen that $q_{\nu(t)} = 0$ when t is big enough and then this contradicts the definition of the q_m. So this finishes proving that E does not admit any increasing T-sequence of center 0. Symmetrically, we can show that E does not admit any decreasing T-sequence of center 0. Hence we have proven that the distinguished circular filter is irregular.　$\square$

Remark. $\mathbb{C}_p$ is weakly valued. Indeed, the valuation group of $\mathbb{C}_p$ is isomorphic to $\mathbb{Q}$ and by [28, Theorem 5.11], the residue class field of $\mathbb{C}_p$ is an algebraic closure of $\mathbb{F}_p$, and therefore is countable.

7.6.　Characteristic property of T-filters

Throughout this chapter, D is infraconnected.

We will prove that there exist elements strictly vanishing along a monotonous filter $\mathcal{F}$ if and only if $\mathcal{F}$ is a T-filter. It is quite easy to show that a pierced filter admitting strictly vanishing elements is a T-filter (Proposition 7.6.1). The big problem consists of proving that given any T-filter, there do exist analytic elements strictly vanishing along it (Theorem 7.6.2, [20, 23]).

Proposition 7.6.1, roughly, was proven separately and simultaneously by E. Motzkin and Ph. Robba [40], and by the author in March 1969 [20, 23] (however, Motzkin–Robba's claim was not stated in terms of T-filters, but in terms of sequences of holes that look like T-sequences).

Proposition 7.6.1. *Let $\mathcal{F}$ be a monotonous filter on D. Let $f \in H(D)$ be strictly vanishing along $\mathcal{F}$. Then $\mathcal{F}$ is a T-filter.*

Proof.　Let us suppose $\mathcal{F}$ to be increasing, of center a and diameter S. With no loss of generality we may assume $a = 0$. Let $\lambda = \log S$. There exists $\nu < \lambda$

such that $\Psi(f, \mu) > -\infty \ \forall \mu \in]\nu, \lambda[$ while $\Psi(f, \lambda) = -\infty$. Let $\log(r) = \nu$. Hence the function $\Psi(f, .)$ is bounded in every interval $[\nu, \xi]$ with $\xi \in]\nu, \lambda[$ and therefore by Proposition 7.2.2, the equality $\Psi(f(x)) = \Psi(f, \Psi(x))$ holds in all of $D \cap \Gamma(0, r, S)$ but inside finitely many classes of circles $C_m = C(0, r_m)$ with $r_m < r_{m+1}, \lim_{m \to \infty} r_m = S$.

We fix $m \in \mathbb{N}$ and take r', r'' satisfying $r \leq r' < r_m < r'' < S$. If $D \cap C_m \neq \emptyset$ we put $\theta_m = \|f\|_{D \cap C_m}$ and if $D \cap C_m = \emptyset$ we put $\theta_m = {}_D\varphi_{0,r_m}(f)$. Since $\Psi(f, \mu)$ is bounded in $[\log r', \log r'']$ by a constant M we may find $h \in R(D)$ such that $\log(t\|h - f\|_D) < -M$. Hence we have

(1) ${}_D\varphi_{0,r_m}(f) = {}_D\varphi_{0,r_m}(h)$

and if $D \cap C_m \neq \emptyset$ we have

(2) $\|h\|_{C_m \cap D_m} = \|f\|_{C_m \cap D_m} = \theta_m.$

Let $(T_{m,i})_{1 \leq i \leq s(m)}$ be the holes of D inside C_m which contain at least as many poles as many zeros and, for each one, let $q_{m,i}$ be the difference between the number of the poles and the number of the zeros (taking multiplicities into account). Let $q_m = \sum_{i=1}^{s(m)} q_{m,i}$. Then we know that

(3) $\Psi'^l(h, \log r_m) - \Psi'^r(h, \log r_m) \geq q_m.$

By Proposition 7.4.5 and by Relation (1) we have

(4) $\theta_m \geq \gamma_D(0, r_m, q_m) {}_D\varphi_{0,r_m}(h)$

when $D \cap C_m \neq \emptyset$ and $\theta_m = {}_D\varphi_{0,r_m}(h)$ when $D \cap C_m = \emptyset$.

Hence Relation (4) is true anyway. In terms of valuations (4) is equivalent to $\log \theta_m \geq \log \gamma_D(0, r_m, q_m) + \Psi(f, \log r_m)$. Now by (3) we see that $\Psi(f, \log r_m) \geq \Psi(f, \log r_{m-1}) - q_{m-1}(\log r_m - \log r_{m-1})$. Hence by induction we can easily obtain

$$\Psi(f, \log r_m) \geq \Psi(f, \log r_1) - \sum_{j=1}^{m-1} q_j(\log r_m - \log r_j),$$

and finally $-\log \theta_m \leq \log \gamma_D(0, r_m, q_m) + \sum_{j=1}^{m-1} q_j(\log r_m - \log r_j)$. Since f is vanishing along $\mathcal{F}$, we have $\lim_{m \to \infty}(\log \theta_m) = -\infty$ hence $\lim_{m \to \infty} \log \gamma_D(0, r_m, q_m) - \sum_{j=1}^{m-1} q_j(\log r_m - \log r_j) = -\infty$. This just shows $\mathcal{F}$ to be an increasing T-filter. A symmetric reasoning is made when $\mathcal{F}$ is a decreasing filter equipped with a center.

Now let $\mathcal{F}$ be decreasing with no center. Then in $\widehat{\mathbb{K}}$ we denote by $(d(\alpha_j, \rho_j))_{j \in J}$ the family of the holes of D and we put $d(a, r) = \widetilde{D}$ if D

is bounded. If D is bounded (resp., unbounded), in $\widehat{\mathbb{K}}$ we put

$$\widehat{D} = \widehat{d}(a, r) \setminus \left(\bigcup_{j \in J} \widehat{d}(\alpha_j, \rho_j^-) \right) \quad \left(\text{resp.,} \ \widehat{D} = \widehat{\mathbb{K}} \setminus \left(\bigcup_{j \in J} \widehat{d}(\alpha_j, \rho_j^-) \right) \right).$$

In $\widehat{\mathbb{K}}$, $\mathcal{F}$ has a center a. Then the filter $\widehat{\mathcal{F}}$ of center a and diameter S on $\widehat{D}$ is a T-filter because f belongs to $H_{\widehat{\mathbb{K}}}(\widehat{D})$ and is strictly vanishing along $\widehat{\mathcal{F}}$. Hence there exists a decreasing T-sequence $(\widehat{T}_{m,i}, \ m_{q,i})_{\substack{1 \leq i \leq s(m) \\ m \in \mathbb{N}}}$ of center a and diameter R that runs $\widehat{\mathcal{F}}$. But for each (m, i), $\widehat{T}_{m,i} \cap \mathbb{K}$ is a hole $T_{m,i}$ of D, and then, by Remark in Chapter 7.6. the weighted sequence $(T_{m,i}, \ m_{q,i})_{\substack{1 \leq i \leq s(m) \\ m \in \mathbb{N}}}$ is a T-sequence of D. $\qquad \square$

Corollary 7.6.2. *Let $\mathcal{F}$ be a monotonous filter on D. There exist elements of $H(D)$ strictly vanishing along $\mathcal{F}$ if and only if $\mathcal{F}$ is a T-filter.*

7.7. Applications of T-filters

We will apply T-filters for characterizing the main algebraic properties of the algebras $H(D)$: Noetherian algebras $H(D)$, existence of divisors of zero. We will also characterize the property for a set E in $\mathbb{K}$ to be analytic (i.e., if an analytic element is equal to zero inside a disk, it is identically zero everywhere) [20, 38, 40].

Notation: Throughout the chapter, D is an infraconnected closed and bounded subset of $\mathbb{K}$.

By Propositions 7.2.11, 7.6.1 and 7.6.2, Lemma 7.7.1 is immediate:

Proposition 7.7.1. *Let $b \in D$, $l > 0$ and let $f \in H(D)$ satisfy $f(b) \neq 0$ and $_D\varphi_{b,l} = 0$. There exists an increasing T-filter $\mathcal{F}$ of center b and diameter $t \in]0, l[$ such that f is strictly vanishing along $\mathcal{F}$ and satisfies $_D\varphi_{b,s}(f) > 0$ for every $s \in]0, t[$.*

Let $a \in \widetilde{D}$ and let $r, s \in \mathbb{R}$ satisfy $\delta(a, D) \leq r \leq s \leq \text{diam}(D)$. There exists $f \in H(D)$ satisfying $_D\varphi_{a,s}(f) > 0$, $_D\varphi_{a,r}(f) = 0$, (resp., $_D\varphi_{a,r}(f) > 0$, $_D\varphi_{a,s}(f) = 0$) if and only if there exists an increasing (resp., a decreasing) T-filter of center a and diameter $t \in]r, s]$, (resp., $t \in [r, s[$). Moreover, if f satisfies $_D\varphi_{a,s}(f) > 0$, $_D\varphi_{a,r}(f) = 0$, (resp., $_D\varphi_{a,s}(f) = 0$, $_D\varphi_{a,r}(f) > 0$) then there exists an increasing (resp., a decreasing) T-filter $\mathcal{F}$ of center a and diameter $t \in]r, s]$, (resp., $t \in [r, s[$) such that f is strictly vanishing along $\mathcal{F}$.

Theorem 7.7.2. *Let $f \in H(D) \setminus \{0\}$. Then f is not quasi-minorated if and only if there exists a T-filter $\mathcal{F}$ on D such that f is strictly vanishing along $\mathcal{F}$.*

Proof. Suppose $f \in H(D)$ is not quasi-minorated. Then by Proposition 7.2.11, f is strictly vanishing along a monotonous filter. And by Theorem 7.4.13, this filter is a T-filter. ☐

Corollary 7.7.3. *Let D be open A non-zero element of $H(D)$ is quasi-invertible if and only if it is not strictly vanishing along a T-filter.*

Corollary 7.7.4. *Every non-zero element of $H(D)$ is quasi-minorated if and only if D has no T-filter.*

Theorem 7.7.5. *Every non-zero element of $H(D)$ is quasi-invertible if and only if D is open, with no T-filter.*

Proof. Suppose that D is open with no T-filter. By Theorem 7.7.1, every non-identically zero element is quasi-minorated. Therefore the set of its zeros is finite. But then, since D is open, it is quasi-invertible. Now, suppose that D ha a T-filter: there exist elements that are not quasi-minorated and hence, are not quasi-invertible. Finally, suppose that D is not open and let a be a point of D that us not interior to D. Without loss of generality, we can suppose that $a = 0$. Then any quasi-minorated element vanishing at 0 is not quasi-invertible. ☐

Theorem 7.7.6. *Let A be an infraconnected subset of D not reduced to a single point and let $b \in D \setminus A$. Let $f \in H(D)$ satisfy $f(u) = 0$ for all $u \in A$ and $f(b) \neq 0$. Then, f is strictly vanishing along a T-filter $\mathcal{F}$ on D such that $A \subset \mathcal{P}(\mathcal{F})$ and $b \in \mathcal{B}(\mathcal{F})$.*

Proof. Let $a \in A$, let $r = \mathrm{diam}(A)$, and let $\ell = |a - b|$. Since A is not reduced to a single point, we have $r > 0$.

First suppose $\ell \leq r$. Since A is infraconnected, we have $_D\varphi_{b,\ell}(f) = {}_A\varphi_{b,\ell}(f) = 0$. But since $f(b) \neq 0$, there does exists $t \in]0, \ell]$ such that $_D\varphi_{b,s}(f) > 0$ for all $s \in]0, t[$, and $_D\varphi_{b,t}(f) = 0$. Hence, f is strictly vanishing along the increasing filter $\mathcal{F}$ of center b and diameter t, and therefore A is included in $\mathcal{P}(\mathcal{F})$.

Now, suppose $\ell > r$. If $_D\varphi_{a,\ell}(f) = 0$ then $_D\varphi_{b,\ell}(f) = 0$ hence there exists an increasing T-filter $\mathcal{F}$ of center b and diameter $t \in]0, \ell]$, so we have $A \subset \mathcal{P}(\mathcal{F})$. Hence it only remains to consider the case $_D\varphi_{a,\ell}(f) > 0$. But then, there exists $t \in [r, \ell[$ such that $_D\varphi_{a,s}(f) > 0$ for all $s \in]t, \ell[$ and

$_D\varphi_{a,t}(f) = 0$. Hence f is strictly vanishing along a decreasing T-filter $\mathcal{F}$ of center a and diameter t, and therefore A is included in $\mathcal{P}(\mathcal{F})$. $\quad\square$

Corollary 7.7.7. *Let D have no T-filter. Let $a \in D$ and let $r \in \mathbb{R}_+^*$. Let $f \in H(D)$ satisfy $f(x) = 0$ whenever $x \in d(a,r) \cap D$. Then f is identically 0.*

Theorem 7.7.8. *Let D have a unique T-filter $\mathcal{F}$ and let $f \in H(D)$, $f \neq 0$, be vanishing along $\mathcal{F}$. Then f is strictly vanishing along $\mathcal{F}$ and satisfies $f(x) = 0$ whenever $x \in \mathcal{P}(\mathcal{F})$.*

Proof. Since f is not quasi-minorated, by Theorem 7.7.1, there exists a T-filter $\mathcal{G}$ such that f is strictly vanishing along $\mathcal{G}$. But $\mathcal{F}$ is the unique T-filter of D, hence $\mathcal{F} = \mathcal{G}$. Now let $D' = \mathcal{P}(\mathcal{F})$ and let $\tilde{f}$ be the restriction of f to D'. Let $\mathcal{G}$ be the circular filter on D less thin than $\mathcal{F}$. Then $\mathcal{G}$ is secant with D', and we have $_{D'}\phi_{\mathcal{G}}(\tilde{f}) = 0$. Hence $\tilde{f}$ is not quasi-minorated in $H(D')$ and then $\tilde{f}$ is either strictly vanishing along a T-filter of D', or identically zero. But since D has no T-filter other than $\mathcal{F}$, D' has no T-filter hence by Corollary 7.7.7, we have $f(x) = 0$ whenever $x \in D'$.

Analytic sets were introduced by Krasner, Motzkin and Robba. They were characterized in [20, 40]. $\quad\square$

Theorem 7.7.9. *Let E be a subset of $\mathbb{K}$. Then E is analytic if and only if E is infraconnected such that any T-filter $\mathcal{F}$ on E satisfies $\mathcal{P}(\mathcal{F}) = \emptyset$.*

Proof. If E is not infraconnected, it admits an empty annulus $\Gamma(a, r', r")$ and then by Theorem 7.3.14, $H(E)$ contains the characteristic function of $\mathcal{I}(\Gamma(a, r', r")) = d(a, r') \cap E$, hence clearly D is not analytic. Now let D have a T-filter $\mathcal{F}$ with a not empty beach.

We first suppose $\mathcal{F}$ increasing, of center a and diameter S. By Theorem 7.7.2, there exists $f \in H(E)$ strictly vanishing along $\mathcal{F}$ such that $f(x) = 0$ whenever $x \in \mathcal{P}(\mathcal{F})$. Now let $b \in D$ such that $|a - b| \geq S$ and let $s \in]0, S[$. We have $d(b, s) \cap D \subset \mathcal{P}(\mathcal{F})$ and therefore $f(x) = 0$ whenever $x \in d(b, s) \cap D$, while f is not identically zero in $d(a, S^-)$. Hence E is not analytic.

We now suppose $\mathcal{F}$ is decreasing and has diameter S. Since $\mathcal{P}(\mathcal{F}) \neq \emptyset$, $\mathcal{F}$ admits a center $a \in D$ and then by Theorem 7.7.2, there exists $f \in H(E)$ strictly vanishing along $\mathcal{F}$, such that $f(x) = 0$ whenever $x \in d(a, S) \cap D$.

Now reciprocally, we suppose E not to be analytic. There exist $f \in H(D)$, $a, b \in E$ and $r > 0$ such that $f(b) \neq 0$, and $f(x) = 0$ whenever $x \in d(a, r) \cap D$. Hence by Proposition 7.7.6, there exists a T-filter $\mathcal{F}$ such that $d(a, r) \subset \mathcal{P}(\mathcal{F})$. $\quad\square$

T-filters let us characterize the algebras $H(E)$ which are principal ideal rings.

Theorem 7.7.10. *Let E be a closed and bounded subset of $\mathbb{K}$. The following five properties (a), (b), (c), (d), (e) are equivalent.*

(a) *Every ideal of $H(E)$ is generated by a polynomial whose zeros belong to $\overset{\circ}{E}$.*

(b) *$H(E)$ is a principal ideal algebra with no idempotent different from 0 and 1.*

(c) *$H(E)$ is a Noetherian algebra with no idempotent different from 0 and 1.*

(d) *E is infraconnected, with no T-filter and E is open.*

(e) *Every element of $H(E)$ is quasi-invertible.*

Proof.   Obviously (a) implies (b) and (b) implies (c). Next, by Theorem 7.3.19, (e) implies (a). We check that (d) implies (e) by Theorem 7.7.5. Hence we only have to show that (c) implies (d). For this, we suppose that (d) is not satisfied.

First, if E is not infraconnected by Theorem 7.3.14, $H(E)$ admits idempotents other than 0 and 1 whence obviously (c) is not satisfied.

Henceforth, we assume E to be infraconnected. If E is not open, E admits a pierced Cauchy filter $\mathcal{F}$. Let a be its limit. Since E is closed, a belongs to E, hence $x - a$ is not quasi-invertible. Now we suppose that E admits a T-filter $\mathcal{F}$. Thus, E admits a pierced filter $\mathcal{F}$ with elements $f \in H(E)$ properly vanishing along $\mathcal{F}$. Let $\mathcal{H}$ be the ideal of the $f \in H(E)$ such that $\lim_{\mathcal{F}} f(x) = 0$. We will prove that $\mathcal{H}$ is not of finite type. Indeed, suppose that $\mathcal{H}$ is of finite type and let $g_1, \ldots, g_q$ be a system of generators. By Proposition 7.3.19 there exists $f \in \mathcal{H}$ and a sequence z_n in E thinner than $\mathcal{F}$ satisfying

(α) $g_1(z_n) \neq 0$ whenever $n \in \mathbb{N}$, and

(β) $\lim_{n \to \infty} \left(\frac{|f(z_n)|}{\max_{1 \leq i \leq q} |g_i(z_n)|} \right) = +\infty$.

For every $n \in \mathbb{N}$ let $t_n = \max_{1 \leq i \leq q} |g_i(z_n)|$. Since $\mathcal{H}$ admits $g_1, \ldots, g_q$ as a system of generators, there exist $h_1, \ldots, h_q \in H(E)$ such that $f = g_1 h_1 + \cdots + g_q h_q$, which is impossible since the h_i are bounded. Therefore we have proven that $\mathcal{H}$ is not of finite type. This finishes showing that (c) implies (d). This ends the proof of Theorem 7.7.10. $\square$

Theorem 7.7.11. *Let E be a closed and bounded subset of $\mathbb{K}$. The following statements are equivalent.*

(a) *Every ideal of $H(E)$ is principal.*

(b) *$H(E)$ is Noetherian*

(c) *E is open, has finitely many infraconnected components, and each one has no T-filter.*

Besides, when Statements (a), (b), (c) are satisfied, the algebra $H(E)$ is isomorphic to the direct product $H(E_1) \times \cdots \times H(E_q)$ with $E_1, \ldots, E_q$ the infraconnected components of E.

Proof. As (a) implies (b), we first suppose (b) to be satisfied and will show (c). Obviously $H(E)$ has finitely many idempotents, hence it has finitely many empty annuli. Therefore, E has finitely many infraconnected components, $E_1, \ldots, E_q$. Then by Theorem 7.3.14, for each $i = 1, \ldots, q$ the characteristic function u_i of E_i belongs to $H(E)$. But for each $i = 1, \ldots, q$, $H(E_i)$ is isometrically isomorphic to $u_i H(E)$. Besides by definition the $(u_i)_{1 \leq i \leq q}$ consists of a system of idempotents such that $u_i u_j = 0$ whenever $i \neq j$ and $\sum_{i=1}^{n} u_i = 1$. Hence $H(E)$ is isometrically isomorphic to the direct product $u_1 H(E) \times \cdots \times u_q H(E)$. Thus, finally, we see that $H(E)$ is isometrically isomorphic to $H(E_1) \times \cdots \times H(E_q)$. Obviously, each algebra $H(E_i)$ must be Noetherian. But then, since E_i is infraconnected, $H(E_i)$ has no idempotent other than 0 and 1, and then, by Theorem 7.7.10, E_i is open and E_i has no T-filter. Hence we have shown that E is open, that E has finitely many infraconnected components and that each one has no T-filter. Thus, we have shown (c) is satisfied. Further, we have seen that Statements (a), (b), (c), do imply that $H(E)$ is isomorphic to $H(E_1) \times \cdots \times H(E_q)$.

Now we suppose (c) satisfied and denote by $E_i, (1 \leq i \leq q)$, the infraconnected components of E. Hence every algebra $H(E_i)$ is a principal ideal algebra because E_i has no T-filter and E_i is open. Besides, by Theorem 7.3.14, for each $i = 1, \ldots, q$, the characteristic function u_i of E_i belongs to $H(E)$. As previously, the $(u_i)_{1 \leq i \leq q}$ consists of a system of idempotents such that $u_i u_j = 0$ whenever $i \neq j$ and $\sum_{i=1}^{n} u_i = 1$. Hence $H(E)$ is isometrically isomorphic to the direct product $H(E_1) \times \cdots \times H(E_q)$. Therefore, it is seen that every ideal of $H(E)$ is principal and therefore (a) is satisfied. This ends the proof of Theorem 7.7.11. $\qquad \square$

Now we will characterize the sets E such that $H(E)$ is an algebra with no divisors of zero.

Definition: Two monotonous filters $\mathcal{F}_1, \mathcal{F}_2$ will be said to be *complementary* if $\mathcal{P}(\mathcal{F}_1) \cup \mathcal{P}(\mathcal{F}_2) = D$.

Theorem 7.7.12. *Let E be a subset of $\mathbb{K}$. There exist f and $g \in H(E)$ not identically equal to zero such that fg is identically zero if and only if: either E is not infraconnected, or E has two complementary T-filters.*

Proof. If E is not infraconnected, it admits an empty annulus $\Gamma(a, r', r'')$ and then the characteristic functions u of $\mathcal{I}(\Gamma(a, r', r''))$ (resp., $w = 1 - u$ of $\mathcal{E}(\Gamma(a, r', r''s)))$ belong to $H(E)$ and satisfy $uw = 0$ though they both are not identically zero.

If E is infraconnected with two complementary T-filters $\mathcal{F}_1, \mathcal{F}_2$, there exists f_1 (resp., f_2) strictly vanishing along $\mathcal{F}_1$ (resp., $\mathcal{F}_2$) and identically equal to zero in $\mathcal{P}(\mathcal{F}_1)$ (resp., $\mathcal{P}(\mathcal{F}_2)$) hence we have $f_1(x)f_2(x) = 0$ whenever $x \in \mathcal{P}(\mathcal{F}_1) \cup \mathcal{P}(\mathcal{F}_2)$ hence whenever $x \in E$.

Now we suppose E infraconnected and we suppose that there exist f_1 and f_2 not identically equal to zero such that $f_1(x)f_2(x) = 0$ whenever $x \in D$. Let $a_1 \in E$ be such that $f_1(a_1) \neq 0$ and let $a_1 \in E$ be such that $f_2(a_2) \neq 0$. Let $r = |a_1 - a_2|$.

We first suppose that $_E\varphi_{a_1,r}(f_1) = 0$. Then by Corollary 7.4.14, there exists an increasing T-filter $\mathcal{F}_1$ of center a_1 and diameter $s \in]0, r]$ such that f_1 is strictly vanishing along $\varphi_{a_1,l}(f_1) \neq 0$, hence we have $_E\varphi_{a_1,l}(f_2) = 0$. If $_E\varphi_{a_2,r}(f_2) = 0$, then f_2 is strictly vanishing along an increasing T-filter $\mathcal{F}_2$ of center a_2 and diameter $t \in]0, r]$ and we have $\mathcal{P}(\mathcal{F}_1) \cup \mathcal{P}(\mathcal{F}_2) = D$. If $_E\varphi_{a_2,r}(f_2) \neq 0$, as $_E\varphi_{a_2,r}(f_2) = _E\varphi_{a_1,r}(f_2)$, at the same time we have $_E\varphi_{a_1,l}(f_2) = 0$ and $_E\varphi_{a_1,r}(f_2) \neq 0$. Let t be the supremum of the set of the $\xi \leq r$ such that $_E\varphi_{a_1,\xi}(f_2) = 0$. Then f_2 is strictly vanishing along a decreasing T-filter $\mathcal{F}_2$ of center a_1 and diameter t. Hence we have $\mathcal{P}(\mathcal{F}_1) \cup \mathcal{P}(\mathcal{F}_2) = E$.

We now suppose that $_E\varphi_{a_1,r}(f_1) \neq 0$, hence $_E\varphi_{a_2,r}(f_1) \neq 0$ and then we have $_E\varphi_{a_2,r}(f_2) = 0$. Since $f_2(a_2) \neq 0$ there exists $l \in]0, r[$ such that $_E\varphi_{a_2,l}(f_2) \neq 0$. Let s be the infimum of the set of the l such that $_E\varphi_{a,l}(f_2) = 0$. We see that f_2 is strictly vanishing along an increasing T-filter $\mathcal{F}_2$ of center a_2 and diameter s and then we have $_E\varphi_{a_2,l}(f_1) = 0$ whenever $l \in]0, s[$ hence f is strictly vanishing along a decreasing T-filter $\mathcal{F}_1$ of center a_2 and diameter $t \in [s, r]$. Finally we have $\mathcal{P}(\mathcal{F}_1) \cup \mathcal{P}(\mathcal{F}_2) = D$. This ends the proof. $\qquad\square$

Corollary 7.7.13. *Let E be a closed and bounded subset of $\mathbb{K}$. The algebra $H(E)$ has no divisor of zero if and only if E is infraconnected with no pair of complementary T-filters.*

Theorem 7.7.14. *Let $\mathcal{F}$ be a T-filter on D that admits no T-filter complementary to $\mathcal{F}$. Then $\mathcal{I}(\mathcal{F}) = \mathcal{I}_0(\mathcal{F})$.*

Proof. Of course we suppose $\mathcal{P}(\mathcal{F}) \neq \emptyset$. For instance, suppose $\mathcal{F}$ decreasing, and let $r = \mathrm{diam}(\mathcal{F})$. Suppose that there exists $f \in \mathcal{I}(\mathcal{F}) \backslash \mathcal{I}_0(\mathcal{F})$, and let $a \in \mathcal{P}(\mathcal{F})$ be such that $f(a) \neq 0$. Since we have $_D\varphi_{a,r}(f) = 0$, by Lemma 7.7.1, D admits an increasing T-filter of center a and diameter $s \in]0, r]$, which is obviously complementary to $\mathcal{F}$. In the same way, if $\mathcal{F}$ is increasing, we perform a symmetric proof. This ends the proof of Theorem 7.7.14. □

As consequences of Theorems 7.7.9, 7.7.10, and Corollary 7.7.13, we have Corollaries 7.7.15, and 7.7.16.

Corollary 7.7.15. *If $H(D)$ has no non-trivial idempotent and is Noetherian, then D is analytic. If D is analytic, then $H(D)$ has no divisors of zero.*

Corollary 7.7.16. *If $H(D)$ is Noetherian and has no divisors of zero, then it is a principal ideal ring.*

As a consequence of Theorem 7.5.14, we can give an answer to the question whether there exist $\mathbb{K}$-Banach algebras that are not multbijective.

Theorem 7.7.17. *Let $\mathbb{K}$ be weakly valued. By Theorem 7.5.14, there exists a closed bounded infraconnected subset D admitting an irregular distinguished circular filter $\mathcal{F}$ such that $D = \mathcal{Q}(\mathcal{F})$. Let $(\mathcal{F}_i)_{i \in J}$ be the T-family of $\mathcal{F}$ and for each $i \in J$, let b_i be a center of $\mathcal{F}_i$ and let r_i be its diameter. Then, for all $i, j \in J$ such that $|b_i - b_j| > \max(r_i, r_j)$, φ_{b_i,r_i} and φ_{b_j,r_j} are two distinct multiplicative semi-norms admitting $\mathcal{I}(\mathcal{F})$ for kernel.*

Proof. On the one hand, φ_{b_i,r_i} and φ_{b_j,r_j} obviously are two distinct multiplicative semi-norms admitting $\mathcal{I}(\mathcal{F})$ for kernel. On the other hand, $\mathcal{I}(\mathcal{F})$ is a maximal ideal of $H(D)$ because every $a \in D$ belongs to a set $\mathcal{B}(\mathcal{F}_i)$ and hence $\mathcal{I}(\mathcal{F})$ is not included in $\mathcal{I}(a)$. □

7.8. The p-adic Fourier transform

Notation: Here, we assume $\mathbb{K} = \mathbb{C}_p$. For every integer $s \in \mathbb{N}$, A_s will denote the multiplicative group of the p^s-th roots of 1, and $A = \bigcup_{s \in \mathbb{N}} A_s$ will denote the multiplicative group of all the p^sth roots of 1 for any $s \in \mathbb{N}$.

We define $r_s \in]0, 1[$ by

$$-\log r_s = \frac{1}{p^{s-1}(p-1)} \quad (s \in \mathbb{N}^*).$$

Following classical results in p-adic analysis [23], for every $s \in \mathbb{N}^*$ we have

(a) $A_s \setminus A_{s-1} \subset C(1, r_s)$,
(b) given any primitive p^sth root α of 1, then $A_s = \bigcup_{j=0}^{p-1} \alpha^j A_{s-1}$,
(c) given any $\alpha \in A_s$, the mapping θ defined in A_s by $\theta(x) = x - \alpha$ is an isometric bijection from A_s onto A_s.

Given n such that $1 < n < s$, we denote by $\phi_{s,n}$ the canonical surjection from A_s onto $\frac{A_s}{A_n}$.

Lemma 7.8.1 is easily deduced from Properties (a), (b), (c).

Lemma 7.8.1. *Let $n, s \in \mathbb{N}$ satisfy $0 < n < s$. The quotient group $\frac{A_s}{A_n}$ is equipped with a distance η defined by $\eta\phi_{s,n}(x), \phi_{s,n}(y)) = |x - y|$ whenever $x, y \in A_s$.*

Notations: In $A_s \setminus A_{s-1}$ there exists a subset $E_{s,n}$ isometric to $\frac{A_s}{A_n}$, satisfying $\operatorname{diam}(E_{s,n}) = r_{s-1}$, such that $E_{s,n}$ has p^{s-n} elements.

Given two topological groups $(A, +)$ and $(B, *)$ we denote by $\mathcal{H}om(A, B)$ the group of the continuous homomorphisms from A into B.

Lemma 7.8.2. *Let $\gamma \in A_s$. The group homomorphism ϕ_γ from $(\mathbb{Z}, +)$ into $(\mathbb{C}_p^*, \cdot)$ defined as $\phi_\gamma(n) = \gamma^n$ is continuous with respect to the p-adic absolute value on $\mathbb{Z}$ and on $\mathbb{C}_p^*$. Then ϕ_γ has continuation to a continuous group homomorphism from $(\mathbb{Z}_p, +)$ into $(\mathbb{C}_p^*, \cdot)$*

Proof.   Indeed we have $\phi_\gamma(n) = 1$ when $|n| < p^{-s}$, hence the group homomorphism ϕ_γ is continuous at 0 and therefore is continuous. As a consequence, by continuity ϕ_γ has continuation to a continuous group homomorphism from $(\mathbb{Z}_p, \cdot)$ into $(\mathbb{C}_p^*, \cdot)$. $\square$

Notation: Given $\gamma \in A$, we will denote by ψ_γ^* the unique continuous group homomorphism from $(\mathbb{Z}_p, +)$ into $(\mathbb{C}_p^*, \cdot)$ such that $\phi_\gamma^*(n) = \gamma^n$ whenever $n \in \mathbb{Z}$.

Lemma 7.8.3. *The mapping Y from A into $\mathcal{H}om((\mathbb{Z}_p, +), (\mathbb{C}_p^*, \cdot))$ defined as $Y(\gamma) = \phi_\gamma^*$ is a group isomorphism.*

Proof.   First, we check that Y is injective. Indeed, if $\gamma \in Ker(Y)$ then we have $\gamma^n = 1$ whenever $n \in \mathbb{Z}$ hence $\gamma = 1$. Second, we check that Y is surjective. Let $\theta \in \mathcal{H}om((\mathbb{Z}_p, +), (\mathbb{C}_p^*, \cdot))$ and let $\gamma = \theta(1)$. It is seen that $\theta(n) = \gamma^n$ whenever $n \in \mathbb{Z}$, therefore by continuity we have $\theta = \phi_\gamma^*$. $\square$

Notations: In order to simplify the notations, henceforth we put $\gamma^n = \phi_\gamma^*(n)$ for all $n \in \mathbb{Z}_p$. Let $\mathcal{R}$ denote the filter of the complementaries of the finite subsets of A. Let $L^1(A)$ denote the $\mathbb{C}_p$-Banach vector space of the functions f from A into $\mathbb{C}_p$ such that $\lim_{\mathcal{R}} f(x) = 0$, equipped with the norm $\| \cdot \|$ of uniform convergence on A.

Given $f, g \in L^1(A)$, the series $\sum_{\lambda \in A} (f(\lambda)\, g(\gamma\lambda^{-1}))$ is seen to converge. We denote by $f * g$ the mapping from A into $\mathbb{C}_p^*$ defined by $f * g(\gamma) = \sum_{\lambda \in A} f(\lambda)\, g(\gamma\lambda^{-1})$. Thus, $*$ is a convolution on $L^1(A)$.

Lemma 7.8.4. *$L^1(A)$ is equipped with a structure of $\mathbb{C}_p$-Banach algebra whose multiplication is $*$.*

Proof. First, we check that $f * g$ belongs to $L^1(A)$ whenever $f, g \in L^1(A)$. Indeed let $\epsilon \in]0, +\infty[$. Every finite subset of A is clearly included in a finite subgroup of A. Hence there exists a finite subgroup B of A such that $|f(\lambda)|\, \|G\| \leq \epsilon$ and $|g(\lambda)|\, \|f\| \leq \epsilon$ whenever $\lambda \in A \setminus B$. Hence when $\gamma \in A \setminus B$ we see that either $\lambda \in A \setminus B$ or $\gamma\,\lambda^{-1} \in A \setminus B$ and then we have $|f(\lambda)\, g(\gamma\lambda^{-1})| \leq \epsilon$ whenever $\lambda \in A$. This shows that $\lim_{\mathcal{R}} f * g(\gamma) = 0$. Thus, $*$ is an internal law in $L^1(A)$. Finally, formal calculations show that $L^1(A)$ is a $\mathbb{C}_p$-algebra with this law as a multiplication. $\qquad \square$

Notation: Let $\mathcal{B}(\mathbb{Z}_p, \mathbb{C}_p)$ be the Banach algebra of the continuous functions from $\mathbb{Z}_p$ into $\mathbb{C}_p$. For every $f \in L^1(A)$, let $\mathcal{F}(f)$ be the mapping from $\mathbb{Z}_p$ into $\mathbb{C}_p^*$ defined as

$$\mathcal{F}(f)(n) = \sum_{\gamma \in \Gamma} f(\gamma)\gamma^n.$$

Lemma 7.8.5. *For all $f \in L^1(A)$, $\mathcal{F}(f)$ belongs to $\mathcal{B}(\mathbb{Z}_p, \mathbb{C}_p)$.*

Proof. Let $u \in \mathbb{Z}_p$. Let $\epsilon \in]0, +\infty[$ and let A_s be a subgroup of A such that $|f(\gamma)| \leq \epsilon$ whenever $\gamma \in A \setminus A_s$. It is seen that $\gamma^{(pq)} = 1$ whenever $\gamma \in A_s$ and $q \geq s$. As a consequence we have $\gamma^n = \gamma^u$ for all $\gamma \in A_s$ when $|n - u| \leq \frac{1}{p^s}$ and therefore $|f(n) - f(u)| \leq \sup_{\gamma \in A \setminus A_s} |f(\gamma)| \leq \epsilon$. Thus, we have checked that $\mathcal{F}(f) \in \mathcal{B}(\mathbb{Z}_p, \mathbb{C}_p)$. $\qquad \square$

Definition: For all $f \in \mathcal{B}(\mathbb{Z}_p, \mathbb{C}_p)$, $\mathcal{F}(f)$ will be named *the Fourier Transform of f with respect to $\mathbb{Z}_p$*.

The problem on whether $\mathcal{F}$ is injective was asked by Bernard de Mathan and simultaneously got two different solutions in 1973 [2, 20]. Actually, Yvette Amice showed this problem to be equivalent to a problem of T-filter.

Notations: Let $\rho \in]0, p^{-\frac{1}{(p-1)}}[$ and let $D_\rho = \mathbb{C}_p \setminus (\bigcup_{\gamma \in A} d(\gamma, \rho))$. Let $\mathcal{G}$ be the increasing filter of center 1 and radius 1 on D_ρ. By Properties (a), (b), the set $A_s \setminus A_{s-1}$ consists of $(p-1)p^{s-1}$ points γ satisfying $\Psi(\gamma-1) = -\frac{1}{p^{s-1}(p-1)}$

and $-\Psi(\gamma-\lambda) \geq \frac{1}{(p-1)}$ whenever $\gamma, \lambda \in A$. Thus, it is seen that $\mathcal{G}$ is a pierced filter of piercing ρ.

Theorem 7.8.6 (Y. Amice). *If there exists an idempotent T-sequence associated to $\mathcal{G}$, then $\mathcal{F}$ is not injective.*

Proof. We assume that there exists an idempotent T-sequence associated to $\mathcal{G}$. We notice that $\mathcal{P}(\mathcal{G}) = \mathbb{C}_p \setminus d(1, 1^-)$. Since the $d(\gamma, r^-)$ are the only holes of D_r, by Theorem 7.4.15, there exists $g \in H(D_r)$, strictly vanishing along $\mathcal{G}$, equal to 0 in all of $\mathcal{P}(\mathcal{G})$, meromorphic on each hole $d(\gamma, r^-)$, admitting each $\gamma \in A$ as a pole of order 1 or 0 and having no other pole in $d(\gamma, r^-)$. Hence g is of the form $\sum_{\gamma \in A} \frac{a_\gamma}{1-\gamma x}$ with $\lim_{\mathcal{R}} a_\gamma = 0$, and certain $a_\lambda \neq 0$. So we have

(1) $\sum_{\gamma \in A} \frac{a_\gamma}{1-\gamma x} = \sum_{\gamma \in A} a_\gamma \sum_{n=0}^{\infty} (\gamma x)^n$.

Further, when $x \in d(0, 1^-)$, the series

$$\sum_{\gamma \in A} a_\gamma \sum_{n=0}^{\infty} (\gamma x)^n$$

is clearly equal to

$$\sum_{n=0}^{\infty} \Big(\sum_{\gamma \in A} a_\gamma \gamma^n \Big) x^n.$$

This is a power series that by (1) is identically equal to zero, whenever $x \in d(0, r^-)$. Hence we have

(2) $\sum_{\gamma \in A} a_\gamma \gamma^n = 0$ for all $n \in \mathbb{N}$.

Now, let $f \in \mathcal{B}(\mathbb{Z}_p, \mathbb{C}_p)$ be defined as $f(n) = \sum_{\gamma \in A} a_\gamma \gamma^n$. Since certain a_γ are different from zero, f is not identically zero. But by (2) we see that $F(f)(n) = 0$ for all $n \in \mathbb{N}$. Actually $\mathbb{N}$ is dense in $\mathbb{Z}_p$, and therefore we see that f is identically zero in $\mathbb{Z}_p$. This ends the proof of Theorem 7.8.6. $\qquad \square$

Theorem 7.8.7. *$\mathcal{G}$ admits an idempotent T-sequence.*

Proof. For each $m \in \mathbb{N}^*$, we put $S_m = C(1, r_m) \cap D$, we denote by u_m the integral part of $\log m$, and we put $q_m = p^{m-1-u_m}$. We know that $A_m \setminus A_{m-1} \subset S_m$ and then by Lemma 7.8.1, E_{m,u_m} has q_m elements. Let $E_{m,u_m} = (\alpha_{m,j})_{1 \leq j \leq q_m}$, let $\rho \in]0, \frac{1}{p}[$ and for every $j = 1, ..., q_m$ let $T_{m,j} = d(\alpha_{m,j}, \rho^-)$. We will prove the weighted sequence $(T_{m,j}, 1)_{1 \leq j \leq q_m}$, $m \in \mathbb{N}$ to be an idempotent T-sequence. For each $m \in \mathbb{N}^*$, let Q_m be the q_m-degree monic polynomial whose zeros are the $(\alpha_{m,j})_{1\leq j\leq q_m}$, each one being a simple zero, and let $\lambda_m = \|\frac{1}{Q_m}\|_{S_m}(r_m)^{q_m}$. We will prove the sequence (λ_m) to be bounded. By Lemma 7.8.3, there exists $\alpha \in E_{m,u_m}$ such that $\|\frac{1}{Q_m}\|_{S_m} = \varphi_{\alpha,\rho}\left(\frac{1}{Q}\right)$. Obviously we have $(r_m)^{q_m} = \varphi_{\alpha,r_m}(Q_m)$ hence $\|\frac{1}{Q_m}\|_{S_m}(r_m)^{q_m} = \frac{\varphi_{\alpha,r_m}(Q_m)}{\varphi_{\alpha,\rho}(Q_m)}$ and therefore $\lambda_m \leq \frac{\varphi_{\alpha,r_m}(Q_m)}{\varphi_{\alpha,\rho}(Q_m)}$. In terms of valuations we obtain

(1) $\log \lambda_m \leq \Psi_\alpha(Q_m, \log r_m) - \Psi_\alpha(Q_m, \log \rho)$.

We will compute $\Psi_\alpha(Q_m, \log r_m) - \Psi_\alpha(Q_m, \log \rho)$. We may put it in the form

(2) $\left(\sum_{h=u_m+1}^{m} \Psi_\alpha(Q_m, \log r_h) - \Psi_\alpha(Q_m, \log r_{h-1}) \right) + \Psi_\alpha(Q_m, \log r_{u_m}) - \Psi_\alpha(Q_m, \log \rho)$

Since E_{m,u_m} is isometric to the quotient group $\frac{A_m}{A_{u_m}}$, it is seen that Q_m admits exactly p^{h-u_m} zeros (taking multiplicities into account) inside $d(\alpha, r_h)$ and therefore we have

$$\Psi_\alpha(Q_m, \log r_h) - \Psi_\alpha(Q_m, \log r_{h-1}) = (\log r_h - \log r_{h-1})p^{h-1-u_m}.$$

But actually

$$\log r_h - \log r_{h-h} = \frac{1}{p-1}\left(\frac{1}{p^{h-1}} - \frac{1}{p^h} \right) = -\frac{1}{p^{h-1}}.$$

So we have

(3) $\Psi_\alpha(Q_m, \log r_h) - \Psi_\alpha(Q_m, \log r_{h-1}) = p^{-u_m-1}$.

Next, since α is the only zero of Q_m in $d(\alpha, r_{u_m}^-)$ we have

(4) $\Psi_\alpha(Q_m, \log r_{u_m}) - \Psi_\alpha(Q_m, \log \rho) = \log(r_{u_m}) - \log(\rho)$

$= -\log \rho - \frac{1}{(p-1)p^{u_m-1}}$.

Hence by (2), (3), (4) we obtain

(5) $\Psi_\alpha(Q_m, \log r_m) - \Psi_\alpha(Q_m, \log \rho)$

$= (m - u_m - 1)p^{-u_m-1} - \log \rho - \frac{1}{(p-1)p^{u_m-1}}$.

Actually by definition we have $p^{u_m-1} > m$ and then by (5) we obtain

(6) $\Psi_\alpha(Q_m, \log r_m) - \Psi_\alpha(Q_m, \log \rho) < \log \rho + 1$.

Thus, by (1) and (6), the sequence (λ_m) is clearly bounded. Then, bt Lemma 7.4.6, it is seen that the weighted sequence $(T_{m,j}, 1)_{1 \le j \le q_m, m \in \mathbb{N}}$ is a T-sequence if and only if

(7) $\lim_{m \to \infty} \left(\prod_{j=1}^{m} \left(\frac{r_j}{r_m} \right)^{q_j} \right) = 0.$

For convenience we consider

$$B_m = \sum_{j=1}^{m} q_j (\log r_m - \log r_j) = \sum_{j=1}^{m} p^{j-u_j} \left(\frac{1}{(p-1)(p^{j-1})} - \frac{1}{(p-1)p^{m-1}} \right)$$

$$= \frac{1}{p-1} \left(\sum_{j=1}^{m} p^{-u_j+1} - \frac{p^{j-u_j}}{p^{m-1}} \right).$$

By definition we have $p^j < j$ hence

(8) $\sum_{j=1}^{m} p^{-u_j+1} > p \sum_{j=1}^{m} \frac{1}{j}.$

Besides, it is seen that

(9) $\sum_{j=1}^{m} \frac{p^{j-u_j}}{p^{m-1}} < \sum_{j=1}^{m} p^{j-m+1} < \frac{1}{p-1}.$

Thus, by (8) and (9), we see that $\lim_{m \to \infty} B_m = +\infty$ and therefore (7) is true. This ends the proof of Theorem 7.8.7. $\qquad \square$

Corollary 7.8.8. *F is not injective.*

Chapter 8

Holomorphic Functional Calculus

8.1. Analytic elements on classic partitions

Classic partitions let us generalize the notion of holes for a subset D of $\mathbb{K}$, relatively to a disk containing D. So, we can generalize algebras of analytic elements, a notion introduced in [22, 26].

Definitions: Given a closed set $S \subset \mathbb{K}$, we call *a classic partition of S* a partition of S of the form $\left(d(b_j, r_j^-)\right)_{j \in J}$. The disks $d(b_j, r_j^-)$ are called *the holes of the partition.*

Let $\mathcal{O}$ be a classic partition of $d(a, r)$. A closed infraconnected set E included in $d(a, r)$, will be said to be a *sub-$\mathcal{O}$-set* if every hole of E is a hole of $\mathcal{O}$. Moreover, a sub-$\mathcal{O}$-set E will be called a $\mathcal{O}$-set if $\widetilde{E} = d(a, r)$.

A weighted sequence $(T_{m,i}, q_{m,i})_{1 \leq i \leq k_m, \, m \in \mathbb{N}}$ of any $\mathcal{O}$-set will be called *a weighted sequence of $\mathcal{O}$.* In particular, a T-sequence of any $\mathcal{O}$-set will be called *a T-sequence of $\mathcal{O}$*, and an idempotent weighted sequence of any $\mathcal{O}$-set will be called *an idempotent weighted sequence of $\mathcal{O}$.*

Let $a \in \mathbb{K}$ and $r > 0$, let S be a closed subset of $d(a, r)$, and let $\mathcal{O} = \left(d(b_j, r_j^-)\right)_{j \in I}$ be a classic partition of $d(a, r) \setminus S$. An annulus $\Gamma(b, r', r'')$ included in $d(a, r) \setminus S$ will be said to be *$\mathcal{O}$-minorated* if there exists $\lambda > 0$ such that $r_j \geq \lambda$ for every $j \in I$ such that $d(b_j, r_j^-) \subset \Gamma(b, r', r'')$.

Remarks. Given a set $D \subset \mathbb{K}$, the set $\widetilde{D} \setminus \overline{D}$ admits a unique classic partition $\left(d(b_j, r_j^-)\right)_{j \in J}$ such that $r_j = z(b_j, D) \, \forall j \in J$, and then the disks $d(b_j, r_j^-) \, (j \in J)$ are just the holes of D. This partition will be called the *natural partition of $\widetilde{D} \setminus \overline{D}$.*

Theorem 8.1.1 ([36]). *Let $\mathcal{O} = (d(\alpha_j, r_j^-))_{j \in J}$ be a classic partition of the annulus $\Gamma(a, r', r'')$. There exists $h \in J$ and a $\mathcal{O}$-minorated annulus of the form $\Gamma(\alpha_h, r_h, \rho)$ included in $\Gamma(a, r', r'')$.*

Proof. Suppose the claim is false. Without loss of generality we can assume $a = 0$. For every $i \in J$, we put $T_i = d(\alpha_i, r_i^-)$. Since our claim is false, for every $\epsilon > 0$, we can find an index $i \in J$ such that $r_i < \epsilon$. Let us fix $\epsilon > 0$, and let $h \in J$ be such that $r_h < \min(\epsilon, r')$ and Let $s_h \in]r_h, r_h + \epsilon[$. Then $\Gamma(\alpha_h, r_h, s_h)$ is included in $\Gamma(a, r', r'')$ and is not included in any hole T_i. Consequently, $\Gamma(\alpha_h, r_h, s_h)$ admits a classic partition of the form $\mathcal{O}_h = (T_i)_{i \in J_h}$, with $J_h \subset J$. Since our claim is false, we can find another index $k \in J_h$ such that $r_k < \frac{\epsilon}{2}$. In this way, by induction we can construct a sequence $(i_n)_{n\mathbb{N}}$ of J such that $T_{i_{n+1}} \subset \Gamma(a_{i_n}, r_{i_n}, r_{i_n} + \frac{\epsilon}{2^n})$ and $r_{i_n} < \frac{\epsilon}{2^n} \, \forall n \in \mathbb{N}$. The sequence (α_{i_n}) is a Cauchy sequence whose limit α obviously lies in $\Gamma(0, r', r'')$. Hence α belongs to a certain hole T_l. Therefore, when n is big enough then α_{i_n} lies in T_l, a contradiction to the definition of $\mathcal{O}$. $\square$

Proposition 8.1.2 is almost obvious.

Proposition 8.1.2. *Let $S \subset d(a, r)$ and let $\mathcal{O}$ be a classic partition of $d(a, r) \setminus S$ admitting a $\mathcal{O}$-minorated annulus $\Gamma(b, r'r'')$. For all $\rho \in]r', r''[$, there exists an increasing idempotent T-sequence of $\mathcal{O}$ of center b and diameter ρ, together with a decreasing idempotent T-sequence of $\mathcal{O}$ of same center and diameter.*

By a similar proof of Theorem 7.5.10 of Chapter 7, here we can state Proposition 7.1.3.

Proposition 8.1.3. *Let $\mathbb{K}$ be strongly valued. Let S be a bounded set included in a disk $d(a, r)$, and let $\mathcal{O}$ be a classic partition of $d(a, r) \setminus S$. We assume that for every $r \in]\delta(a, S), \mathrm{diam}(S)[\cap|\mathbb{K}|$, each class of $C(a, r)$, except maybe finitely many ones, contains at least one hole of $\mathcal{O}$. Then for each $r \in]\delta(a, S), \mathrm{diam}(S)]$ there exists in $\mathcal{O}$ an increasing idempotent T-sequence and a decreasing idempotent T-sequence of center a and diameter r.*

Corollary 8.1.4. *Let $\mathbb{K}$ be strongly valued. Let $\mathcal{O} = (d(b_j, r_j^-))_{j \in J}$ be a classic partition of the annulus $\Gamma(a, r', r'')$ and let $s \in]r', r''[$. Then $\mathcal{O}$ admits an increasing idempotent T-sequence and a decreasing idempotent T-sequence of center a and diameter s.*

Corollary 8.1.5. *Let $\mathbb{K}$ be strongly valued. Let S be a countable set, included in a disk $d(a, r)$ and let $\mathcal{O}$ be a classic partition of $d(a, r) \setminus S$. For every $l \in]0, r[$, $\mathcal{O}$ admits an increasing and a decreasing idempotent T-sequence of center a and diameter l.*

Notations: Let $\mathcal{D}$ be a compact subset of $Mult(\mathbb{K}[x])$ and let $D = \{a \in \mathbb{K} \mid \varphi_a \in \mathcal{D}\}$. The $\mathbb{K}$-algebra $R(D)$ is equipped with the semi-multiplicative norm $\| \cdot \|_{\mathcal{D}}$ defined as $\|h\|_{\mathcal{D}} = \sup\{\psi(h) \mid \psi \in \mathcal{D}\}$. Then we check that $\|h\|_{\mathcal{D}} \geq \|h\|_D \ \forall h \in R(D)$.

Notations: Let E be a closed subset of the disk $d(a,r)$ and let D be a closed subset of E. Let $\mathcal{O} = (d(b_i, r_i^-)_{i \in J})$ be a classic partition of $d(a,r) \backslash E$. Let $\mathcal{D} = \omega(W(D, \mathcal{O}))$. Then on $R(D)$ we put $\| \cdot \|_{D,\mathcal{O}} = \| \cdot \|_{\mathcal{D}}$.

We will denote by $H(D, \mathcal{O})$ the completion of $\mathbb{K}(x)$ for the norm $\| \cdot \|_{D,\mathcal{O}}$. In particular, if $D = \emptyset$, we denote by $H(\mathcal{O})$ the completion of $\mathbb{K}(x)$ for the norm $\| \cdot \|_{\mathcal{O}}$.

If $D = \emptyset$, we just put $\|h\|_{\mathcal{O}} = \sup\{\phi(h) \mid \phi \in \Phi(\mathcal{O})\}$. If $\mathcal{O}$ is the partition of holes of an infraconnected bounded set D, then by definition we have $\|f\|_{D,\mathcal{O}} = \|f\|_D \ \forall f \in R(D)$.

Each $\psi \in \mathrm{Mult}(R(D), \| \cdot \|_{D,\mathcal{O}})$ has continuation to $H(D, \mathcal{O})$ and will be denoted by ψ again.

We denote by ω the mapping from $W(D, \mathcal{O})$ into $\mathrm{Mult}(H(D, \mathcal{O}), \| \cdot \|_{D,\mathcal{O}})$, defined as $\omega(\mathcal{F}) = {}_{D,\mathcal{O}}\varphi_{\mathcal{F}}$,

By Lemma 2.3.4 of Chapter 2, we have Lemma 8.1.6.

Lemma 8.1.6. *Let $V = d(a,r)$, let E be a closed subset of V, let D be a closed subset of E and let $\mathcal{O}$ be a classic partition of $V \backslash E$. Then $\| \cdot \|_{D,\mathcal{O}}$ is a semi-multiplicative norm of $\mathbb{K}$-algebra on $R(D)$.*

Remark. Any $\mathbb{K}$-algebra $H(D, \mathcal{O})$ is a uniform Banach $\mathbb{K}$-algebra.

Lemma 8.1.7. *Let D be a bounded closed infraconnected subset of $\mathbb{K}$ and let $(T_i)_{i \in J}$ be the family of holes of D. Let $\mathcal{O} = (T_i)_{i \in J}$. Then $\| \cdot \|_{D,\mathcal{O}} = \| \cdot \|_D$.*

Proof. Let $T_i = d(b_i, r_i^-)$, $(i \in J)$ be a hole of $\mathcal{O}$, hence a hole of D. Since D is infraconnected, the circular filter $\mathcal{F}_i$ of center b_i and diameter r_i is secant with D, therefore $\varphi_{\mathcal{F}_i}(f) \leq \|f\|_D \ \forall f \in R(D)$, so the claim is obvious. $\qquad\qquad\square$

Lemma 8.1.8. *Let $\mathcal{O}$ be a classic partition of a disk $d(a,r)$, and let D be a sub-$\mathcal{O}$-set. Then we have $\|h\|_D \leq \|h\|_{\mathcal{O}}$ for every $h \in R(D)$. Moreover, if D is a $\mathcal{O}$-set, then $\|h\|_D = \|h\|_{\mathcal{O}}$ for every $h \in R(D)$.*

Proof. Let $h \in R(D)$. By Theorem 7.1.15 of Chapter 7, there exists a D-bordering filter $\mathcal{F}$ such that $\|h\|_D = \varphi_{\mathcal{F}}$. And $\varphi_{\mathcal{F}}$ is either $\varphi_{a,r}(h)$, or some $\varphi_{b,\rho}(h)$, where $d(b, \rho^-)$ is a hole of D. But since D is sub-$\mathcal{P}$-set, in all cases, $\varphi_{\mathcal{F}}$ belongs to $\mathrm{Mult}(R(\mathcal{O}), \| \cdot \|_{\mathcal{O}})$. Consequently, we have $\|h\|_D \leq \|h\|_{\mathcal{O}}$.

Now, suppose that D is a $\mathcal{O}$-set. Given a hole $d(b, \rho^-)$ of $\mathcal{O}$, this hole is a hole of D because D is a $\mathcal{O}$-set. And since D is infraconnected, this hole defines a circular filter secant with D. Therefore, we have $\varphi_{b,\rho}(h) \leq \|h\|_D$, and consequently $\|h\|_{\mathcal{O}} \leq \|h\|_D$. $\qquad\qquad\qquad\qquad\qquad\qquad\square$

Corollary 8.1.9. *Let $\mathcal{O}$ be a classic partition of a disk $d(a,r)$, and let E be a $\mathcal{O}$-set. Then $H(E)$ is isometrically isomorphic to a $\mathbb{K}$-subalgebra of $H(\mathcal{O})$.*

Henceforth, given a classic partition $\mathcal{O}$ of a disk $d(a,r)$, and a sub-$\mathcal{O}$-set E, we will consider $H(E)$ as a $\mathbb{K}$-subalgebra of $H(\mathcal{O})$.

8.2. Holomorphic properties on partitions

Throughout this chapter D is a closed bounded subset of the disk $d(a,r)$. Given a classic partition, we have a Mittag–Leffler theorem which generalizes the well known Mittag–Leffler theorem for analytic elements on an infraconnected set (7.1.13) of Chapter 7. Next, properties of T-filters also apply to analytic elements on partitions.

Theorem 8.2.1. *Let $\mathcal{O}$ be a classic partition of $d(a,r)\backslash D$. Let $f \in H(D,\mathcal{O})$. There exists a unique sequence of holes $(T_n)_{n\in\mathbb{N}^*}$ of $\mathcal{O}$ and a unique sequence $(f_n)_{n\in\mathbb{N}}$ in $H(D,\mathcal{O})$ such that $f_0 \in H(d(a,r))$, $f_n \in H_0(\mathbb{K}\setminus T_n)$ $(n > 0)$, $\lim_{n\to\infty} f_n = 0$ satisfying*

(i) $f = \sum_{n=0}^{\infty} f_n$ $\quad$ *and* $\quad \|f\|_{D,\mathcal{O}} = \sup_{n\in\mathbb{N}} \|f_n\|_{D,\mathcal{O}}.$

Moreover for every hole $T_n = d(a_n, r_n^-)$, we have

(ii) $\|f_n\|_{D,\mathcal{O}} = \|f_n\|_{\mathbb{K}\setminus T_n} = \varphi_{a_n,r_n}(f_n) \leq \varphi_{a_n,r_n}(f) \leq \|f\|_{D,\mathcal{O}}.$
(iii) $\|f_0\|_{D,\mathcal{O}} = \|f_0\|_{\tilde{D}} = \varphi_{a,r}(f_0) \leq \varphi_{a,r}(f) \leq \|f\|_{D,\mathcal{O}}.$

Let $D' = d(a,r) \setminus (\bigcup_{n=1}^{\infty} T_n)$. *Then f belongs to $H(D')$ and its decomposition in $H(D')$ is given again by (i) and then f satisfies $\|f\|_{D'} = \|f\|_{D,\mathcal{O}}$.*

Proof. Suppose first that f lies in $R(D)$. Without loss of generality, we may denote by $T_1, \ldots, T_q$ the holes of $\mathcal{O}$ that contain poles of f. Let $E = d(a,r)\setminus\bigcup_{j=1}^{q} T_j$. Then E is a $\mathcal{O}$-set. Since E is infraconnected, we can apply Theorem 7.1.13 of Chapter 7, that gives all statements above. Now, consider $f \in H(D,\mathcal{O})$ and let $(h_m)_{m\in\mathbb{N}}$ be a sequence in $R(D)$ converging to f in $H(D,\mathcal{O})$. The set of holes of $\mathcal{O}$ containing at least one pole of one term h_m of this sequence is obviously countable and may be defined as a sequence $(T_n)_{n\in\mathbb{N}^*}$. Thus, each h_m has a unique decomposition $h = \sum_{n=0}^{\infty} h_{m,n}$, with $h_{m,0} \in H(d(0,r))$, and $h_{m,n} \in H_0(\mathbb{K}\setminus T_n)$ $\forall n \in \mathbb{N}^*$. So, similarly to the classical proof of the Mittag–Leffler Theorem for infraconnected sets, for any m, $q \in \mathbb{N}$, we have $\|h_m - h_q\|_{D,\mathcal{O}} = \sup_{n \in \mathbb{N}}(\|h_{m,n} - h_{q,n}\|_{D,\mathcal{O}})$ and then, for each fixed $n \in \mathbb{N}^*$, (resp., $n = 0$) the sequence $(h_{m,n})_{m \in \mathbb{N}}$ converges in $H_0(\mathbb{K} \setminus T_n)$ (resp., in $H(d(a,r))$) to a limit f_n. We check that the series $\sum_{n=0}^{\infty} h_n$ converges to f in $H(D, \mathcal{O})$, and satisfies all statements above. $\square$

Corollary 8.2.2. *Let $(T_i)_{i \in I}$ be the family of holes of D. Let J be a subset of I and let $L = I \setminus J$. Let $E = D \bigcup (\bigcup_{i \in J} T_i)$ and let $F = \mathbb{K} \setminus \bigcup_{i \in L} T_i$. Then we have $H(D) = H(E) \oplus H_0(F)$, and for each $g \in H_0(E)$, $h \in H(F)$, we have $\|g + h\|_D = \max(\|g\|_E, \|h\|_F)$.*

Definitions and notation: Let $f \in H(D, \mathcal{O})$. We consider the series $\sum_{n=0}^{\infty} f_n$ obtained in Theorem 8.2.1, whose sum is equal to f in $H(D, \mathcal{O})$, with $f_0 \in H(\widetilde{D})$, $f_n \in H(\mathbb{K} \setminus T_n) \setminus \{0\}$ and with T_n holes of $\mathcal{O}$. Each T_n will be called a *f-hole* and f_n will be called the *Mittag–Leffler term of f associated to T_n*, whereas f_0 will be called the *principal term of f*. For each f-hole T of $\mathcal{O}$, the Mittag–Leffler term of f associated to T will be denoted by $\overline{f_T}$ whereas the principal term of f will be denoted by $\overline{f_0}$. The series $\sum_{n=0}^{\infty} f_n$ (with $f_n = \overline{f_{T_n}} \,\, \forall n \in \mathbb{N}^*$) will be called *the Mittag–Leffler series of f on* $(D, \mathcal{O})$.

Let E be a closed subset of the disk $d(a,r)$ and let D be a closed subset of E. Let $\mathcal{O} = (d(b_i, r_i^-)_{i \in J})$ be a classic partition of $d(a,r) \setminus E$. We will denote by $W(D, \mathcal{O})$ the set of circular filters $\mathcal{F}$ on $\mathbb{K}$ such that all elements $B \in \mathcal{F}$ contain points of D or holes of $\mathcal{O}$. In particular, if $D = \emptyset$, we just put $W(\mathcal{O}) = W(\emptyset, \mathcal{O})$.

Similarly to what was done with sets $\mathrm{Mult}(H(D), \| \cdot \|_D)$, here we have a Garandel–Guennebaud's theorem for algebras $H(D, \mathcal{O})$.

Theorem 8.2.3. *Let E be a closed set such that $D \subset E \subset d(a,r)$, and let $\mathcal{O}$ be a classic partition of $d(a,r) \setminus E$. For every $\mathcal{F} \in W(D, \mathcal{O})$, the multiplicative semi-norm $\varphi_{\mathcal{F}}$ defined on $R(D)$ extends by continuity to an element $_{D,\mathcal{O}}\varphi_{\mathcal{F}}$ of $\mathrm{Mult}(H(D, \mathcal{O}), \| \cdot \|_{D,\mathcal{O}})$ such that $_{D,\mathcal{O}}\varphi_{\mathcal{F}}(f) = \lim_{\mathcal{F}} |f(x)|$ whenever $f \in R(D)$. Moreover, the mapping ω from $W(D, \mathcal{O})$ into $\mathrm{Mult}(H(D, \mathcal{O}), \| \cdot \|_{D,\mathcal{O}})$, defined as $\omega(\mathcal{F}) = _{D,\mathcal{O}}\varphi_{\mathcal{F}}$, is a bijection.*

Notations: Let E be a closed set such that $D \subset E \subset d(a,r)$, and let $\mathcal{O}$ be classic partition of $d(a,r) \setminus E$. As we did in $\mathrm{Mult}(H(D), \| \cdot \|_D)$, in order to avoiding a too heavy notation, the extension $_{D,\mathcal{O}}\varphi_{\mathcal{F}}$ of $\varphi_{\mathcal{F}}$ to $\mathrm{Mult}(H(D, \mathcal{O}), \| \cdot \|_{D,\mathcal{O}})$ will just be denoted by $\varphi_{\mathcal{F}}$. Similarly, if $\mathcal{F}$ is the circular filter of center a and diameter ρ, then $\varphi_{a,\rho}$ will still denote the extension of $\varphi_{a,\rho}$ to $(H(D, \mathcal{O}), \| \cdot \|_{D,\mathcal{O}})$.

Let $\mathcal{O} = (d(b_i, r_i^-))_{i \in J}$ be a classic partition of $d(a, r) \setminus D$. We will denote by $R'(D, \mathcal{O}, (b_i)_{i \in J})$ (resp., $R''(D, \mathcal{O}, (b_i)_{i \in J})$) the $\mathbb{K}$-subvector space of $R(D)$ consisting of the $h \in R(D)$ of the form $\sum_{i \in I} \frac{\lambda_i}{x - b_i}$, with I a finite subset of J (resp., $\sum_{i \in L} \frac{\lambda_i}{x - b_i} + \frac{\mu_i}{(x - b_i)^2}$). The completion of $R'(D, \mathcal{O}, (b_i)_{i \in J})$ (resp., $R''(D, \mathcal{O}, (b_i)_{i \in J})$) is a Banach $\mathbb{K}$-subvector space of $H(D, \mathcal{O})$ that we will denote by $H'(D, \mathcal{O}, (b_i)_{i \in J})$ (resp., $H''(D, \mathcal{O}, (b_i)_{i \in J})$).

By Theorem 8.2.1, Proposition 8.2.4 is immediate.

Proposition 8.2.4. *Let $\mathcal{O} = (d(b_i, r_i^-)_{i \in J})$ be a classic partition of $d(a, r) \setminus D$. Then $H'(D, \mathcal{O}, (b_i)_{i \in J})$ (resp., $H''(D, \mathcal{O}, (b_i)_{i \in J})$) is equal to the set of $f \in H(D, \mathcal{O})$ of the form $\sum_{i \in I} \frac{\lambda_i}{x - b_i}$, (resp., $\sum_{i \in I} \frac{\lambda_i}{x - b_i} + \frac{\mu_i}{(x - b_i)^2}$) with I a countable subset of J, and $\lim_{\mathcal{H}} \frac{|\lambda_i|}{r_i} = 0$, whereas $\mathcal{H}$ is the filter of complements of finite subsets of I.*

Proposition 8.2.5 will be useful in the following chapters.

Proposition 8.2.5. *Let $\mathcal{O} = (d(b_i, r_i^-)_{i \in J})$ be a classic partition of $d(a, r) \setminus D$, and suppose that $\mathcal{O}$ admits an increasing (resp., a decreasing) idempotent T-sequence (T_n) of center b and diameter l. Let $s \in]0, l[$ (resp., $s \in]l, r[$) and let $\epsilon \in]0, 1[$. There exists $f \in H'(D, \mathcal{O}, (b_i)_{i \in J})$ satisfying*

(i) $|\psi(f - 1)|_\infty < \epsilon \; \forall \psi \in \mathrm{Mult}(H(D, \mathcal{O}), \| \cdot \|_{D, \mathcal{O}})$ *such that* $\psi(x - b) \leq s$ *and* $\psi(f) = 0 \; \forall \psi \in \mathrm{Mult}(H(D, \mathcal{O}), \| \cdot \|_{D, \mathcal{O}})$ *such that* $\psi(x - b) \geq l$,

(ii) $\psi(f) \neq 0 \; \forall \psi \in \mathrm{Mult}(H(D, \mathcal{O}), \| \cdot \|_{D, \mathcal{O}}) \setminus \mathrm{Mult}_a(H(D, \mathcal{O}), \| \cdot \|_{D, \mathcal{O}})$ *such that* $s < \psi(f) < l$.

(resp., (i) $\psi(f - 1) \leq \epsilon \; \forall \psi \in \mathrm{Mult}(H(D, \mathcal{O}), \| \cdot \|_{D, \mathcal{O}})$ such that $\psi(x - b) \geq s$ and $\psi(f) = 0 \; \forall \psi \in \mathrm{Mult}(H(D, \mathcal{O}), \| \cdot \|_{D, \mathcal{O}})$ such that $\psi(x - b) \leq l$, (ii) $\psi(f) \neq 0 \; \forall \psi \in \mathrm{Mult}(H(D, \mathcal{O}), \| \cdot \|_{D, \mathcal{O}}) \setminus \mathrm{Mult}_a(H(D, \mathcal{O}), \| \cdot \|_{D, \mathcal{O}})$ such that $l < \psi(f) < s$).

Proof. We set $E = \mathbb{K} \setminus (\bigcup_{n=0}^\infty T_n)$. Since the sequence $(T_n, 1)_{n \in \mathbb{N}}$ is an idempotent T-sequence of E, of center b and diameter l, by Theorem 7.4.13 of Chapter 7, there exists $f \in H(E)$, strictly vanishing along the increasing (resp., decreasing) T-filter of center b and diameter l, meromorphic on each hole T_n and admitting each b_n as a pole of order at most one, and having no other pole, satisfying further

$|f(x)| = 1$ for all $x \in D \cap d(b, s)$,
$f(x) = 0$ for all $x \in D \setminus d(b, l^-)$ (resp., $|f(x) - 1| \leq \epsilon$ for all $x \in D \setminus d(b, s^-)$),

$f(x) = 0$ for all $x \in D \cap d(b,l))$, and for every circular filter $\mathcal{G}$ of diameter $u < l$, secant with $d(b, l^-)$, $\varphi_\mathcal{G}(g) \neq 0$. Moreover, by Theorem 8.2.1 for each hole T_n, $\overline{f}_{T_n}$ is of the form $\frac{\lambda_n}{x - b_n}$, so all claims are shown. $\qquad\square$

Corollary 8.2.6. *Let $\mathcal{O}$ be a classic partition of $d(a,r) \setminus D$, and suppose that $\mathcal{O}$ admits a $\mathcal{O}$-minorated annulus $\Gamma(b, r', r'')$. Let l, $\rho', \rho'' \in]r', r''[$ satisfy $\rho' < l < \rho''$. There exist f, $g \in H(D, \mathcal{O})$ satisfying:*

on one hand $\psi(f) = 1$ $\forall \psi \in \mathrm{Mult}(H(D, \mathcal{O}), \| \cdot \|_{D, \mathcal{O}})$ such that $\psi(x - b) \leq \rho'$ and $\psi(f) = 0$ $\forall \psi \in \mathrm{Mult}(H(D, \mathcal{O}), \| \cdot \|_{D, \mathcal{O}})$ such that $\psi(x - b) \geq l$, on the other hand $\psi(g) = 1$ $\forall \psi \in \mathrm{Mult}(H(D, \mathcal{O}), \| \cdot \|_{D, \mathcal{O}})$ such that $\psi(x - b) \geq \rho''$ and $\psi(g) = 0$ $\forall \psi \in \mathrm{Mult}(H(D, \mathcal{O}), \| \cdot \|_{D, \mathcal{O}})$ such that $\psi(x - b) \leq l$).

Lemma 8.2.7. *Let $\mathbb{K}$ be strongly valued, let $\mathcal{O}$ be a classic partition of $d(a,r)$ and let $\mathcal{F}$, $\mathcal{G} \in W(\mathcal{O})$, $\mathcal{F} \neq \mathcal{G}$. There exists $f \in H(\mathcal{O})$ such that $\varphi_\mathcal{F}(f) = 0$, $\varphi_\mathcal{G}(f) \neq 0$.*

Proof. Suppose first that $\mathcal{F}$ and $\mathcal{G}$ are not comparable with respect to $\preceq$. Then, by Proposition 5.4.9 of Chapter 5, we can find disks $d(b, l) \in \mathcal{F}$ and $d(c, m) \in \mathcal{G}$ such that $d(b, l) \cap d(c, m) = \emptyset$. Let $t = |b - c|$. Then by Proposition 5.4.9 of Chapter 5 again, we have $t > \max(l, m)$, hence, denoting by $\mathcal{O}'$ the set of holes of $\mathcal{O}$ included in $\Gamma(b, l, t)$, this annulus admits $\mathcal{O}'$ as a classic partition.

Suppose now that $\mathcal{F}$ and $\mathcal{G}$ are comparable with respect to $\preceq$. So, we can suppose that $\mathcal{G}$ surrounds $\mathcal{F}$. Let $\rho = \mathrm{diam}(\mathcal{F})$, $t = \mathrm{diam}(\mathcal{G})$ and let $l \in]\rho, t[$. By Lemma 5.3.12 of Chapter 5, there exists a unique disk $d(b, l)$ which belongs to $\mathcal{F}$, and then denoting by $\mathcal{O}'$ the set of holes of $\mathcal{O}$ included in $\Gamma(b, l, t)$, this annulus admits $\mathcal{O}'$ as a classic partition.

Now, in both situations, we take $s \in]l, t[$. By Corollary 8.1.4, $\mathcal{O}$ admits an increasing idempotent T-sequence and a decreasing idempotent T-sequence, both of center b and diameter s. Therefore, the conclusion comes from Proposition 8.2.5. $\qquad\square$

Theorem 8.2.8. *Let $\mathbb{K}$ be strongly valued and let $\mathcal{O}$ be a classic partition of $d(a,r)$. Then $\mathrm{Mult}_m(H(\mathcal{O}), \| \cdot \|_\mathcal{O}) = \mathrm{Mult}(H(\mathcal{O}), \| \cdot \|_\mathcal{O})$.*

Proof. Let $\psi \in \mathrm{Mult}(H(\mathcal{O}, \| \cdot \|_\mathcal{O})$ and let $\mathcal{M}$ be a maximal ideal of $H(\mathcal{O})$ containing $\mathrm{Ker}(\psi)$. Then ψ is of the form $\varphi_\mathcal{F}$, with $\mathcal{F} \in W(\mathcal{O})$. On the other hand, by Theorem 2.5.13 of Chapter 2, there exists $\phi \in \mathrm{Mult}_m(H(\mathcal{O}), \| \cdot \|_\mathcal{O})$ such that $\mathrm{Ker}(\phi) = \mathcal{M}$, and ϕ is of the form $\varphi_\mathcal{G}$, with $\mathcal{G} \in W(\mathcal{O})$. Suppose $\mathcal{G} \neq \mathcal{F}$. Then by Lemma 8.2.7, there exists $g \in H(\mathcal{O})$ such that

$\varphi_{\mathcal{F}}(g) = 0$, and $\varphi_{\mathcal{G}}(g) \neq 0$, a contradiction to the hypothesis $\mathrm{Ker}(\psi) \subset \mathcal{M}$. Consequently, $\mathcal{M} = \mathrm{Ker}(\psi)$. $\hfill\square$

8.3. Shilov boundary for algebras $(H(D), \mathcal{O})$

In Section 2.3 of Chapter 2, we proved that, given a unital commutative ultrametric Banach $\mathbb{K}$-algebra A, a Shilov boundary does exist for $(A, \| \cdot \|_{\mathrm{sp}})$. However, this existence is abstract and doesn't let us determine its nature. When we consider a Banach $\mathbb{K}$-algebra $H(D)$, we are able to describe very precisely its Shilov boundary by using the characterization of multiplicative semi-norms by circular filters.

Definitions and notations: Throughout Section 8.3, D is a closed and bounded subset of the disk $d(a, r)$ and $\mathcal{O} = (d(b_i, r_i^-)_{i \in J})$ is a classic partition of $d(a, r) \setminus D$.

A circular filter $\mathcal{F}$ will be said to be *strictly $\mathcal{O}$-bordering* if it is peripheral to a hole of $\mathcal{O}$ or of $d(a, r)$. A circular filter $\mathcal{O}$ will be said to be *$\mathcal{O}$-bordering* if either it is strictly $\mathcal{O}$-bordering or all elements $B \in \mathcal{F}$ contain holes of $\mathcal{O}$.

We will denote by $\Sigma(\mathcal{O})$ the set of $\varphi_{\mathcal{F}} \in Mult(\mathbb{K}[x])$ such that $\mathcal{F}$ is $\mathcal{O}$-bordering and by $\Sigma_0(\mathcal{O})$ the set of $\varphi_{\mathcal{F}} \in Mult(\mathbb{K}[x])$ such that $\mathcal{F}$ is strictly $\mathcal{O}$-bordering.

Lemma 8.3.1. *Let $\mathcal{O}$ be a classic partition of $d(a, r) \setminus D$. A circular filter $\mathcal{F}$ on $\mathbb{K}$ which is $\mathcal{O}$-bordering but not strictly $\mathcal{O}$-bordering is secant with $d(a, r) \setminus D$ but not secant with any hole of $\mathcal{O}$. Moreover, $\Sigma_0(\mathcal{O})$ is dense in $\Sigma(\mathcal{O})$.*

Proof.   Let $\mathcal{O} = (d(b_i, r_i^-)_{i \in J})$ and let $\varphi_{\mathcal{F}} \in \Sigma(\mathcal{O}) \setminus \Sigma_0(\mathcal{O})$. Then every element of $\mathcal{F}$ contains holes of $\mathcal{O}$, hence is secant with $d(a, r) \setminus D$. Suppose that $\mathcal{F}$ is secant with a hole $T = d(b, \rho^-)$ of $\mathcal{O}$. Since $\mathcal{F}$ is not the peripheral of T it is secant with a disk $d(c, s) \subset d(b, \rho^-)$ with $s < \rho$, and of course it has elements B of diameter $l \in]s, \rho[$ that contain no hole of $\mathcal{O}$, a contradiction. So, $\mathcal{F}$ is not secant with any hole of $\mathcal{O}$. Now, let $h_1, \ldots, h_n \in R(D, \mathcal{O})$, let $\epsilon > 0$. Since $\mathcal{F}$ admits a basis of affinoids sets, we can find an affinoid set $E \in \mathcal{F}$ such that $| |h_j(x)| - \varphi_{\mathcal{F}}(h_j)|_\infty < \epsilon \ \forall x \in E, \forall j = 1, \ldots, n$. Then there exists a hole T of $\mathcal{O}$ included in E. Consequently, $|\varphi_T(h_j) - \varphi_{\mathcal{F}}(h_j)|_\infty < \epsilon \ \forall j = 1, \ldots, n$. This shows that $\Sigma_0(\mathcal{O})$ is dense in $\Sigma(\mathcal{O})$. $\hfill\square$

Theorem 8.3.2. *Let $\mathcal{O}$ be a classic partition of $d(a, r) \setminus D$. Then $\Sigma(\mathcal{O})$ is equal to the boundary of $Mult(H(D, \| \cdot \|_{D, \mathcal{O}})$ inside $Mult(\mathbb{K}[x])$.*

Proof. The proof roughly follows the same way as this of Theorem 6.3.1 of Chapter 6. Let $\mathcal{O} = (d(b_i, r_i^-)_{i \in J})$. Since $\Sigma_0(\mathcal{O})$ is dense inside $\Sigma(\mathcal{O})$, and since the boundary of $\mathrm{Mult}(H(D, \| \cdot \|_D,)$ inside $\mathrm{Mult}(\mathbb{K}[x])$ is obviously closed, it is sufficient to show that $\Sigma_0(\mathcal{O})$ is included in this boundary. Let $\varphi_{\mathcal{F}} \in \Sigma_0(\mathcal{O})$. If $\mathcal{F}$ is the peripheral of $d(a, r)$, it is obvious that $\varphi_{\mathcal{F}}$ belongs to the closure of the subset of $\mathrm{Mult}(\mathbb{K}[x]) \setminus \mathrm{Mult}(H(D, \mathcal{O}), \| \cdot \|_{D,\mathcal{O}})$ consisting of all $\varphi_{\mathcal{G}}$ such that $\mathcal{G}$ is secant with $\mathbb{K} \setminus d(a, s)$ for some $s > r$. Consequently, we are led to assume that $\mathcal{F}$ is the peripheral of certain hole $d(b_j, r_j^-)$ of $\mathcal{O}$, and then symmetrically, it is also obvious that $\varphi_{\mathcal{F}}$ belongs to the closure of $\mathrm{Mult}(\mathbb{K}[x]) \setminus \mathrm{Mult}(H(D, \mathcal{O}), \| \cdot \|_{D,\mathcal{O}})$ inside $\mathrm{Mult}(\mathbb{K}[x])$.

Conversely, let $\varphi_{\mathcal{F}} \in \mathrm{Mult}(H(D, \mathcal{O}), \| \cdot \|_{D,\mathcal{O}})$ belong to the closure of $\mathrm{Mult}(\mathbb{K}[x]) \setminus \mathrm{Mult}(H(D, \mathcal{O}), \| \cdot \|_{D,\mathcal{O}})$ in $\mathrm{Mult}(\mathbb{K}[x])$. If $\varphi_{\mathcal{F}}$ is the peripheral of $d(a, r)$, it obviously belongs to $\Sigma(\mathcal{O})$. So, we can suppose it is not the peripheral of $d(a, r)$. We can find an affinoid set $E \in \mathcal{F}$ such that $E \subset d(a, r)$. Then $\mathrm{Mult}(H(E), \| \cdot \|_E)$ is a neighborhood of $\varphi_{\mathcal{F}}$, hence it contains an open neighborhood W of $\varphi_{\mathcal{F}}$. Therefore, $W \cap (\mathrm{Mult}(\mathbb{K}[x]) \setminus \mathrm{Mult}(H(D, \mathcal{O}), \| \cdot \|_{D,\mathcal{O}})) \neq \emptyset$. Moreover, $W \cap (\mathrm{Mult}(\mathbb{K}[x]) \setminus \mathrm{Mult}(H(D, \mathcal{O}), \| \cdot \|_{D,\mathcal{O}}))$ is open in $\mathrm{Mult}(\mathbb{K}[x])$ because so are W and $\mathrm{Mult}(\mathbb{K}[x]) \setminus \mathrm{Mult}(H(D, \mathcal{O}), \| \cdot \|_{D,\mathcal{O}})$. Since $\mathrm{Mult}_a(\mathbb{K}[x])$ is dense inside $\mathrm{Mult}(\mathbb{K}[x])$, there exists

$$\varphi_a \in \left(\mathrm{Mult}_a(\mathbb{K}[x]) \setminus \mathrm{Mult}(H(D, \mathcal{O}), \| \cdot \|_{D,\mathcal{O}}) \right) \cap \mathrm{Mult}(H(E), \| \cdot \|_E).$$

Consequently, a lies in $E \setminus D$. But since $E \subset d(a, r)$, a must belong to Y, which proves that $\mathcal{F}$ is secant with Y. Therefore the boundary of $\mathrm{Mult}(H(D, \mathcal{O}), \| \cdot \|_{D,\mathcal{O}})$ inside $\mathrm{Mult}(\mathbb{K}[x])$ is included in $\Sigma(\mathcal{O})$, and finally these two sets are equal. $\qquad\square$

Theorem 8.3.3. *Let $\mathcal{O} = (d(b_i, r_i^-)_{i \in J})$ be a classic partition of $d(a, r) \setminus D$. Then $\Sigma(\mathcal{O})$ is the Shilov boundary for $(H(D, \mathcal{O}), \| \cdot \|_{D,\mathcal{O}})$.*

Proof. First, we notice that $\Sigma(\mathcal{O})$ is closed inside $\mathrm{Mult}(\mathbb{K}[x])$ because it is the boundary of $\mathrm{Mult}(H(D, \mathcal{O}), \| \cdot \|_{D,\mathcal{O}})$. By Theorem 8.2.1, it is easily seen that $\Sigma_0(\mathcal{O})$ is a boundary for $H(D, \mathcal{O})$.

Now, we will show that $\Sigma(\mathcal{O})$ is the smallest closed boundary. Suppose it is not the smallest. Then, there exists another closed boundary S which does not contain $\Sigma(\mathcal{O})$. Since S is closed, and since $\Sigma_0(\mathcal{O})$ is dense in $\Sigma(\mathcal{O})$ there exists $\psi \in \Sigma_0(\mathcal{O}) \setminus S$. Let $\mathcal{G}$ be the strictly $\mathcal{O}$-bordering filter such that $\psi = \varphi_{\mathcal{G}}$. Thus, ψ is of the form $\varphi_{b,s}$.

Since S is closed there exists an affinoid set E of the form $d(b, s'') \setminus \bigcup_{j=1}^q d(c_j, s')$, with $s' < s < s''$ such that $\mathrm{Mult}(H(E), \| \cdot \|_E)E \cap S = \emptyset$.

This means that for every $\varphi_{\mathcal{F}} \in S$, $\mathcal{F}$ is not secant with E. Suppose first that $s = r$. Consider $g(x) = \prod_{j=1}^{q}(x - c_j)$. Then $\varphi_{\mathcal{G}}(h) = r^q$, and for every $\varphi_{\mathcal{F}} \in S$ we can check that $\varphi_{\mathcal{F}}(g) \leq s'^q$, a contradiction to the hypothesis "S is a boundary".

Consequently we have $s < r$, and therefore, since $\varphi_{\mathcal{G}} \in \Sigma_0(D)$, the disk $d(b, s^-)$ is a hole of $\mathcal{O}$. Let $h(x) = \frac{\prod_{j=1}^{q}(x - c_j)}{(x-b)^{q+1}}$. Then $\varphi_{\mathcal{G}}(h) = \frac{1}{s}$ and for every $\varphi_{\mathcal{F}} \in S$ we can check that $\varphi_{\mathcal{F}}(h) = \max(\frac{s'}{s^2}, \frac{1}{s''}) < \frac{1}{s}$, a contradiction to the hypothesis "S is a boundary". Thus, we have proven that any closed boundary must contain $\Sigma(\mathcal{O})$, and this finishes proving that $\Sigma(\mathcal{O})$ is the Shilov boundary for $H(D, \mathcal{O})$. $\qquad \square$

Theorem 8.3.4 was given in [10].

Theorem 8.3.4 (K. Boussaf). $\Sigma(D)$ *is the Shilov boundary for* $(H(D), \| \cdot \|_D)$.

Proof. By Theorem 6.3.1 of Chapter 6, $\Sigma(D)$ is the boundary of $\mathrm{Mult}(R(D), \| \cdot \|_D)$ inside $\mathrm{Mult}(\mathbb{K}[x])$, hence similarly, is the boundary of $\mathrm{Mult}(H(D), \| \cdot \|_D)$ inside $\mathrm{Mult}(\mathbb{K}[x])$, and hence it is closed inside $\mathrm{Mult}(\mathbb{K}[x])$.

Suppose first that D is infraconnected. By considering the natural partition $\mathcal{O}$ of $\widetilde{D} \setminus D$, we have $\Sigma(\mathcal{O}) = \Sigma(D)$, so we can apply Theorem 8.3.3 showing that $\Sigma(D)$ is the Shilov boundary of $H(D)$. Now, we shall consider the general case.

First, we check that $\Sigma(D)$ is a boundary. Let $h \in R(D)$. If there exists an infraconnected component E of D such that $\|h\|_D = \|h\|_E$, then as we have seen, there exists $\varphi_{\mathcal{F}} \in \Sigma(E)$ such that $\varphi_{\mathcal{F}}(h) = \|h\|_D$, and since of course $\varphi_{\mathcal{F}} \in \Sigma(D)$, hence $\|h\|_D$ is reached in $\Sigma(D)$. Now, suppose that there exists no infraconnected component E of D such that $\|h\|_D = \|h\|_E$. Then there exists a sequence $(D_n)_{n \in \mathbb{N}}$ of infraconnected components of D such that $\|h\|_{D_n} > \|h\|_D - \frac{1}{n}$. Consequently, for each $n \in \mathbb{N}$, we can take a point $a_n \in D_n$ such that $|h(a_n)| > \|h\|_D - \frac{1}{n}$. By Theorem 5.2.1 of Chapter 5, from the sequence (a_n) we can extract a subsequence which is either a monotonous distances sequence, or an equal distances sequence, or a Cauchy sequence. Here, as D is closed, if we could extract a Cauchy sequence, there would exist $a \in D$ such that $|h(a)| = \|h\|_D$, which is excluded. Hence, we can extract a subsequence which is either a monotonous distances sequence, or an equal distances sequence. So, without loss of generality, we can assume that the sequence (a_n) is either a monotonous distances sequence, or an equal distances sequence, and therefore by Proposition 5.3.7 of Chapter 5, there exists a large circular filter $\mathcal{G}$ less thin than the sequence (a_n). Then $\mathcal{G}$ is obviously secant with D and satisfies $\varphi_{\mathcal{G}}(h) = \|h\|_D$. But by hypothesis, $\mathcal{G}$ is not secant with any infraconnected component because for each component E of D we have $\|h\|_D > \|h\|_E$. Consequently, $\mathcal{G}$ is secant with $\mathbb{K} \setminus D$, and therefore belongs to $\Sigma(D)$. This shows that $\Sigma(D)$ is a boundary.

Let $\Sigma_1(D)$ be the set of $\varphi_{\mathcal{F}} \in \Sigma(D)$ which are B-bordering for one of the infraconnected components B of D. We will show that $\Sigma_1(D)$ is dense in $\Sigma(D)$. Let $\varphi_{\mathcal{F}} \in \Sigma(D)$. If $\mathcal{F}$ is secant with one of the infraconnected components B of D, by definition $\mathcal{F}$ is D-bordering, hence $\varphi_{\mathcal{F}} \in \Sigma(D)$. Now, suppose that $\mathcal{F}$ is not secant with any infraconnected component of D. Consider an affinoid set $E \in \mathcal{F}$. Then $E \cap D \neq \emptyset$, hence there exists an infraconnected component B of D such that $E \cap B \neq \emptyset$. If $B \subset \tilde{E}$ then the peripheral of B is clearly secant with E, hence $\varphi_{\mathcal{F}} \in \Sigma(D)$. Now, suppose that B is not included in $\tilde{E}$. Then, $\tilde{E} \subset \tilde{B}$, but since $E \cap \mathbb{K}\setminus \neq \emptyset$, there exists a hole $T = d(b, r^-)$ of B included in $\tilde{E}$ and then, $\varphi_{b,r}$ belongs to $\Sigma_1(D)$, and is secant to E. This shows that $\mathrm{Mult}(H(E), \| \cdot \|_E) \cap \Sigma_1(D) \neq \emptyset$, and consequently $\Sigma_1(D)$ is dense inside $\Sigma(D)$.

Suppose now that $\Sigma(D)$ is not the smallest boundary. So, there exists a boundary S which does not contain $\Sigma(D)$. More precisely, since S is closed there exists $\varphi_{\mathcal{H}} \in \Sigma_1(D) \setminus S$. Then $\mathcal{H}$ has center $b \in B$ and diameter s. Since S is closed there exists an affinoid set E of the form $d(b, s'') \setminus \bigcup_{j=1}^q d(c_j, s')$, with $s' < s < s''$, $|c_i - c_j| = |c_i - b| = s \ \forall i \neq j$ and such that for every $\varphi_{\mathcal{F}} \in S$, $\mathcal{F}$ is not secant with E.

Suppose first that $\mathcal{H}$ is the peripheral of an infraconnected component B of D. There exists an annulus $\Gamma(b, z', z'')$ included in $\mathbb{K} \setminus D$ such that $s < z' < z'' < \sqrt[q]{s^q s''}$. Let $c \in \Gamma(b, z', z'')$, let $l = |c|$ and $h(x) = \frac{\prod_{j=1}^q (x - c_j)}{(x-c)^{q+1}}$. Then we have $\varphi_{\mathcal{H}}(h) = \frac{s^q}{l^{q+1}}$. Now, let $\varphi_{\mathcal{F}} \in S$ be such that $\mathcal{F}$ is secant with $d(b, s)$. Since $\mathcal{F}$ is not secant with E, we check that $\varphi_{\mathcal{F}}(h) \leq \frac{s'^q}{l^{q+1}} < \varphi_{\mathcal{H}}(h)$. And now, let $\varphi_{\mathcal{F}} \in S$ be such that $\mathcal{F}$ is not secant with $d(b, s)$. Since $\mathcal{F}$ is not secant with E, it can't be secant with $d(b, s'')$. Consequently, we have $\varphi_{\mathcal{F}}(h) \leq \frac{1}{s''} < \frac{s^q}{l^{q+1}} = \varphi_{\mathcal{H}}(h)$.

Thus, if $\mathcal{H}$ is the peripheral of an infraconnected component B of D, we are led to a contradiction, and this shows that $\mathcal{H}$ is just the peripheral of a hole $T = d(b, s^-)$ of an infraconnected component B of D. Since B is an infraconnected component of D, there exists an annulus $\Gamma(c, z', z'')$ included in $\mathbb{K} \setminus D$ such that $s' < z' < z'' < s$ with $c \in d(b, s^-)$. So, $T = d(c, s^-)$. Let $a \in \Gamma(c, z', z'')$, let $l = |a - c|$ and let $h(x) = \frac{\prod_{j=1}^q (x - c_j)}{(x-a)^{q+1}}$. Then we have $\varphi_{\mathcal{H}}(h) = \frac{1}{s}$. Now, let $\varphi_{\mathcal{F}} \in S$. If $\mathcal{F}$ is secant with $C(c, s)$, since $\mathcal{F}$ is not secant with E, we check that $\varphi_{\mathcal{F}}(h) \leq \frac{s'}{s^2} < \varphi_{\mathcal{H}}(h)$. And if $\mathcal{F}$ is secant with

$\mathbb{K} \setminus d(c, s)$, it must be secant with $\mathbb{K} \setminus d(c, s''^-)$, and then we check that $\varphi_{\mathcal{F}}(h) \leq \frac{1}{s''} < \frac{1}{s} = \varphi_{\mathcal{H}}(h)$. Finally, suppose that $\mathcal{F}$ is secant with $d(c, s^-)$. Then one of the c_j lies in $d(c, s)$ and we can assume $c = c_1$. Therefore, $\mathcal{F}$ must be secant with $d(c_1, s'^-)$, and we check that $\varphi_{\mathcal{F}}(h) \leq \frac{s' s^{q-1}}{l q+1} < \frac{1}{s} = \varphi_{\mathcal{H}}(h)$. This finishes showing that $\varphi_{\mathcal{F}}(h) < \varphi_{\mathcal{H}}(h)$ for every $\varphi_{\mathcal{F}} \in S$, and therefore we can conclude that any closed boundary must contain $\Sigma(D)$. Consequently, $\Sigma(D)$ is a Shilov boundary of $H(D)$. $\qquad\square$

Remarks.   A boundary is not necessarily closed. Let $(a_n)_{n \in \mathbb{N}^*}$ be sequence of $\mathbb{K}$ such that $|a_{n+1}| < |a_n| < 1$, $\lim_{n \to \infty} |a_n| = r > 0$, and let $D = \{x \in \mathbb{K} \mid |x| < 1, |x - a_n| \geq |a_n| \; \forall n \in \mathbb{N}^*\}$. For every $n \in \mathbb{N}^*$, let $\mathcal{F}_n$ be the circular filter of center a_n and diameter $|a_n|$, let $\phi_n = \varphi_{\mathcal{F}_n}$, let $\mathcal{G}$ be the circular filter of center 0 and diameter 1, and let $\mathcal{G}'$ be the circular filter of center 0 and diameter r, let $\psi = \varphi_{\mathcal{G}}$, and let $\psi' = \varphi_{\mathcal{G}'}$. Clearly, $\Sigma_0(D) = \{\phi_n \mid n \in \mathbb{N}^*\} \cup \{\psi\}$, and then by Theorem 7.1.15 of Chapter 7, this is a boundary for $(H(D), \| . \|_D)$. On the other hand by Theorem 8.3.4, the Shilov boundary for $(H(D), \| . \|_D)$ is $\Sigma(D) = \{\phi_n \mid n \in \mathbb{N}^*\} \cup \{\psi, \psi'\}$. But $\Sigma_0(D)$ is not closed because ψ' is the limit of the sequence (ϕ_n) inside $\mathrm{Mult}(H(D), \| . \|_D)$, with respect to the topology of simple convergence. One can ask whether $\Sigma_0(D)$ is equal to $\mathrm{Min}(H(D), \| . \|_D)$.

By Theorems 5.3.1, 5.3.2 of Chapter 5 and 8.3.4, we can state Theorem 8.3.5.

Theorem 8.3.5. *Let D be closed bounded and have finitely many infraconnected components. The three following sets are equal:*
    $\Sigma(D),$
    The boundary of $\mathrm{Mult}(H(D), \| . \|_D)$ inside $\mathrm{Mult}(\mathbb{K}[x])$ with respect to the topology of simple convergence,
    The Shilov boundary for $(H(D), \| . \|_D)$.

Proof.   Theorem 8.3.6 is just a corollary of Corollary 8.3.5 when D is infraconnected. Now, suppose that D has finitely many infraconnected components $D_1, \ldots, D_q$. Since by Corollary 7.3.16, $H(D) = H(D_1) \times \cdots \times H(D_q)$, and the theorem holds for each algebra $H(D_j)$, so we can easily see that the Shilov boundary of $H(D)$ is equal to the union of Shilov boundaries of all algebras $H(D_j)$, $1 \leq j \leq q$, hence the conclusion follows. $\qquad\square$

More precisely, we can state Corollary 8.3.6.

Corollary 8.3.6. *Let D be closed bounded having finitely many D-bordering filters. The four following sets are equal:*

$\Sigma(D)$,

The boundary of $\mathrm{Mult}(H(D), \| \, . \, \|_D)$ inside $\mathrm{Mult}(\mathbb{K}[x])$ with respect to the topology of simple convergence,,

The boundary of $\mathrm{Mult}(H(D), \| \, . \, \|_D)$ inside $\mathrm{Mult}(\mathbb{K}[x])$ with respect to the δ-topology,,

The Shilov boundary for $(H(D), \| \, . \, \|_D)$.

Proof. If D has finitely many D-bordering filters, then it has finitely many infraconnected components. □

Remark. In particular, Corollary 8.3.6 applies to $H(D)$ when D is an affinoid set.

8.4. Holomorphic functional calculus

Holomorphic functional calculus is well known and helpful in complex Banach algebra. We can also define it, from the spectrum of an element x, by considering first the rational functions without poles in $\mathrm{sp}(x)$. Now, considering an ultrametric Banach $\mathbb{K}$-algebra A, thanks to specific ultrametric properties given in Lemma 8.4.1, this calculus has continuation from a Banach algebra of the form $H(\mathrm{sp}(x), \mathcal{O})$ to A, where $\mathcal{O}$ is a natural partition defined by the norm of A. This calculus will let us solve several problems in spectral theory and about idempotents. The holomorphic functional calculus was first defined in [22, 26].

Notation: In all this chapter, A is a unital commutative Banach $\mathbb{K}$-algebra whose norm $\| \, . \, \|$ is ultrametric.

Lemma 8.4.1. Let $x \in A$ be invertible and let $b \in \mathbb{K}$ be such that $|b| < \frac{1}{\|x^{-1}\|}$. Then $x - b$ is invertible and satisfies $\|(x - b)^{-1}\| = \|x^{-1}\|$.

Proof. By hypothesis, we have $|b| < \frac{1}{\|x^{-1}\|_{\mathrm{sp}}}$, hence $\lim_{n \to \infty} \|(bx^{-1})^n\| = 0$. Consequently, the series $\sum_{n=0}^{\infty} \left(\frac{b}{x}\right)^n$ is converging in A. Then we check that $(x - b) \sum_{n=0}^{\infty} (b^n x^{-n-1}) = 1$. Now since $\|bx^{-1}\| < 1$, we have $\|x^{-1}\| > \|b^n x^{-n-1}\| \; \forall n \in \mathbb{N}^*$. But since $\| \, . \, \|$ is ultrametric, we have $\| \sum_{n=0}^{\infty} b^n x^{-n-1} \| = \|x^{-1}\|$, and therefore $\|(x - b)^{-1}\| = \|x^{-1}\|$. □

Corollary 8.4.2. Let $x \in A$ and for each $b \in \mathbb{K} \setminus \mathrm{sp}(x)$, let $\rho(b) = \frac{1}{\|(x-b)^{-1}\|}$. Then $\rho(a) = \rho(b)$ whenever $a, b \in \mathbb{K} \setminus \mathrm{sp}(x)$ such that $|a - b| < \frac{1}{\|(x-b)^{-1}\|}$. Let $r \geq \|x\|_{\mathrm{sp}}$. Then $d(0, r) \setminus \mathrm{sp}(x)$ admits a classic partition of the form $(d(b, \rho(b)^-))_{b \in d(0,r) \setminus \mathrm{sp}(x)}$, with $d(b, \rho(b)^-) = d(a, \rho(a)^-)$ whenever $a, b \in \mathbb{K} \setminus \mathrm{sp}(x)$ such that $|a - b| < \frac{1}{\|(x-b)^{-1}\|}$.

Definitions and notations: Let $x \in A$, let $a \in \mathbb{K}$ and let $r \in [\|x - a\|_{\mathrm{sp}}, \|x\|]$, with $r > \|x - a\|_{\mathrm{sp}}$ if A is not uniform. We will call *x-normal partition of center a and diameter r* a classic partition $\mathcal{O} = (d(b_i, r_i^-)_{i \in J})$ of $d(a, r) \setminus sp(x)$ satisfying $\frac{1}{\|(x - b_i)^{-1}\|} = r_i \ \forall i \in J$.

In the same way, we will call *x-spectral partition of center a and diameter r* a classic partition $\mathcal{O} = (d(b_i, r_i^-)_{i \in J})$ of $d(a, r) \setminus sp(x)$ satisfying $\frac{1}{\|(x - b_i)^{-1}\|_{\mathrm{sp}}} = r_i \ \forall i \in J$.

A *x-normal* (resp., *x-spectral*) partition will be said to be *centered* if it has center $a \in \widehat{sp(x)}$.

A *x-spectral partition of center a* will be said to be *wide* if it has diameter $r > \|x - a\|_{\mathrm{sp}}$. We will call *strict x-spectral partition* the unique centered *x-spectral partition* of diameter $\|x\|_{\mathrm{sp}}$. Then, if A is uniform, the strict *x-spectral partition* is a *x-normal partition* too.

Let $r = \|x - a\|$ and let $\mathcal{O} = (d(b_i, r_i^-)_{i \in J})$ be a centered *x-normal partition* of $d(a, r) \setminus sp(x)$. Then $x \in A$ will be said to have *bounded normal ratio* if the family $\left\{ \frac{\|(x - b_i)^{-1}\|}{\|(x - b_i)^{-1}\|_{\mathrm{sp}}} \mid i \in J \right\}$ is bounded. The $\mathbb{K}$-algebra A will be said *to have bounded normal ratio* if every element of A has bounded normal ratio.

Theorem 8.4.3. *Let $t \in A$. There exists a unique homomorphism $\mathcal{H}_t$ from $R(sp(t))$ into A such that $\mathcal{H}_t(P) = P(t)$ for all $P \in \mathbb{K}[X]$. Moreover, $\mathcal{H}_t$ is injective if and only if $S(t) \neq 0$ for every polynomial $S \in \mathbb{K}[X]$ different from 0. For every $h \in R(sp(t))$, we have $sp(h(t)) = h(sp(t))$ and $sa(h(t) = h(sa(t))$.*

Proof. Let $D = sp(t)$. We may obviously define $\mathcal{H}_t$ from $\mathbb{K}[x]$ to A as $\mathcal{H}_t(P) = P(t)$. Now let $Q \in \mathbb{K}[X]$ have its zeros in $\mathbb{K} \setminus D$. Then $Q(t)$ is invertible in A, so we may extend to $R(D)$ the definition of $\mathcal{H}_t$, as $\mathcal{H}_t(\frac{P}{Q}) = P(t) Q(t)^{-1}$, for all rational function $\frac{P}{Q} \in R(D)$ (with $(P, Q) = 1$). The uniqueness of $\mathcal{H}_t$ is then obvious. Next, $\mathrm{Ker}(\mathcal{H}_t)$ is an ideal of $R(D)$ which is obviously generated by a polynomial G. Then $G = 0$ if and only there is no polynomial $S \in \mathbb{K}[X]$ different from 0, such that $S(t) = 0$.

Now, let $h = \frac{P}{Q} \in R(sp(t))$, (with $(P, Q) = 1$). Let $\lambda \in sp(t)$), and let $\chi \in \mathcal{X}(A, \mathbb{K})$ be such that $\chi(h(t)) = \lambda$. Then $\chi(h(t)) = h(\lambda)$, hence $h(sp(t)) \subset sp(h(t))$. Conversely, let $\mu \in sp(h(t))$, let $\chi \in \mathcal{X}(A, \mathbb{K})$ be such that $\chi(h(t)) = \mu$, and let $\alpha = \chi(t)$. Then, we have $\chi(P(t)) - \mu\chi(Q(t)) = 0$, hence α is a zero of the polynomial $P(X) - \mu Q(X)$. Therefore α lies in $\mathbb{K}$ because $\mathbb{K}$ is algebraically closed. But then, as $t - \alpha$ belongs to the kernel of χ, α lies in $sp(t)$. Hence we have $sp(h(t)) = h(sp(t))$. Finally, $sa(h(t)) = \{\chi(h(t)) \mid \chi \in \mathcal{X}(A, \mathbb{K})\} = \{h(\chi(t)) \mid \chi \in \mathcal{X}(A, \mathbb{K})\} = h(sa(t))$. $\qquad\square$

Remarks. When the homomorphism $\mathcal{H}_t$ defined in Theorem 8.4.3 is injective, the $\mathbb{K}$-subalgebra $B = \mathcal{H}_t(R(sp(t)))$ is isomorphic to $R(sp(t))$, and in fact is the full subalgebra generated by t in A. So, in such a case, we will consider $R(sp(t))$ as a $\mathbb{K}$-subalgebra of A.

Definition and notation: Let $t \in A$. In this chapter and in the sequel we will denote by $\mathcal{H}_t$ the canonical homomorphism from $R(sp(t))$ into A defined in Theorem 8.4.3, and we will call it *the canonical homomorphism associated to t*. Meanwhile, when it is injective, we will currently confound a rational function $h \in R(sp(t))$ with its image $\mathcal{H}_t(h)$.

Given $\psi \in \mathrm{Mult}(A, \| \cdot \|)$ we will put $\psi_t = \psi \circ \mathcal{H}_t$. Thus, we have $\psi_t(P) = \psi(P(t)) \; \forall P \in \mathbb{K}[x]$.

We will put $\sigma_A(t) = \{\phi_t \mid \phi \in \mathrm{Mult}(A, \| \cdot \|)\}$, and when there is no risk of confusion on the algebra A, we will put $\sigma(t)$ instead of $\sigma_A(t)$.

Lemma 8.4.4. *Let $t \in A$. Then $\sigma_A(t)$ is a compact in $\mathrm{Mult}(\mathbb{K}[x])$. Let $a \in sp(t)$, let $s = \|t - a\|_{\mathrm{sp}}$, let $b \in d(a, s) \setminus sp(t)$ and let $r = \dfrac{1}{\left\|\frac{1}{t-b}\right\|_{\mathrm{sp}}}$. Let $\psi \in \mathrm{Mult}(A, \| \cdot \|)$ and let $\mathcal{F}$ be the circular filter such that $\psi_t = \varphi_{\mathcal{F}}$. Then $\mathcal{F}$ is secant with $d(a, s) \setminus d(b, r^-)$.*

Proof. The mapping which associates to each $\phi \in \mathrm{Mult}(A, \| \cdot \|)$ its image $\phi_t \in \mathrm{Mult}(\mathbb{K}[x])$ is clearly continuous, hence $\sigma_A(t)$ is compact. By definition of ψ_t, it is obvious that $\mathcal{F}$ is secant with $d(a, s)$. Next, it is clear that $\psi_t(t - b) \geq r$, hence $\mathcal{F}$ is secant with $d(a, s) \setminus d(b, r^-)$. $\qquad\square$

Theorem 8.4.5. *Let $t \in A$ and let $\mathcal{O}$ be a t-spectral partition. $\mathcal{H}_t$ satisfies $\|\mathcal{H}_t(f)\|_{\mathrm{sp}} = \|\mathcal{H}_t(f)\|_{\sigma(t)} \leq \|f\|_{\mathrm{sp}(t),\mathcal{O}}$ for all $f \in R(\mathrm{sp}(t), \mathcal{O})$.*

Proof. The equality $\|\mathcal{H}_t(f)\|_{\mathrm{sp}} = \|\mathcal{H}_t(f)\|_{\sigma(t)}$ is obvious and just comes from the definition of $\sigma(t)$. Let $D = sp(t)$, let $a \in sp(t)$, let $r_0 = \|f - a\|_{\mathrm{sp}}$. Let $\psi \in \mathrm{Mult}(A, \| \cdot \|_{\mathrm{sp}})$ and let $\mathcal{F}$ be the circular filter such that $\varphi_{\mathcal{F}} = \psi_t$. Let $(d(b_j, r_j))_{1 \leq j \leq n}$ be the family of holes of $\mathcal{O}$ containing at least one pole of f and for each $j = 1, \ldots, n$, let $\mathcal{F}_j$ be the circular filter of center b_j and diameter r_j. Finally, let $\mathcal{F}_0$ be the circular filter of center a and diameter r_0. According to Theorem 8.2.1 f is of the form $\sum_{j=0}^{n} f_j$, with $f_0 \in \mathbb{K}[x]$ and $f_j \in R_0(\mathbb{K} \setminus d(b_j, r_j^-))$, and we have $\|h\|_{\mathrm{sp}(t),\mathcal{O}} = \max_{0 \leq j \leq n} \|h_j\|_{D,\mathcal{O}}$. For each $j = 1, \ldots, n$, since by Lemma 8.4.4, $\mathcal{F}$ is secant with $\mathbb{K} \setminus d(b_j, r_j^-)$, we have $\varphi_{\mathcal{F}}(h_j) \leq \varphi_{b_j, r_j}(h_j)$. Similarly, as $\mathcal{F}$ is secant with $d(a, r_0)$, we have

$\varphi_{\mathcal{F}}(h_0) \leq \varphi_{a,r_0}(h_0)$. Consequently, we obtain $\psi(h(t)) \leq \|h\|_{D,\mathcal{O}}$. This is true for all $\psi \in \mathrm{Mult}(A, \| \cdot \|)$, and therefore by Theorem 2.5.17 of Chapter 2, we obtain $\|f(t)\|_{\mathrm{sp}} \leq \|f\|_{D,\mathcal{O}}$. $\square$

Remark.   Particularly, Theorem 8.4.5 is true when $\mathcal{O}$ is the strict t-spectral partition.

Theorem 8.4.6. *Let $t \in A$ and let $\mathcal{O}$ be a t-normal partition. $\mathcal{H}_t$ is continuous and has continuation to $H(sp(t), \mathcal{O})$. Moreover, if A is uniform, then $\mathcal{H}_t$ is continuous as a mapping from $(R(sp(t)), \| \cdot \|_{\sigma(t)})$ to A and has continuation to $H(\sigma(t))$ and to $H(sp(t), \mathcal{O})$. So, it satisfies $\|\mathcal{H}_t(f)\| = \|f\|_{\sigma(t)} \leq \|f\|_{sp(t),\mathcal{O}}$ for all $f \in H(sp(t), \mathcal{O})$.*

Proof.   The claim about $H(\sigma(t))$ is immediate and comes from Theorem 8.4.5. Now, let $\mathcal{O} = (d(b_i, r_i^-)_{i \in J})$ be a t-normal partition of diameter r. Let $D = sp(t)$. Let $T = d(a, \rho^-)$ be a hole of $\mathcal{O}$. It is easily checked that

$$\left\| \mathcal{H}_t\left(\frac{1}{(t-a)^k}\right) \right\| \leq \left\| \mathcal{H}_t\left(\frac{1}{t-a}\right) \right\|^k = \left(\frac{1}{\rho}\right)^k = \left(\left\|\frac{1}{t-a}\right\|_{D,\mathcal{O}}\right)^k$$

$$= \left\|\frac{1}{(t-a)^k}\right\|_{D,\mathcal{O}}.$$

Now, let $h \in R(D)$, let $T_1, \ldots, T_q$ be the h-holes of $\mathcal{O}$ and let $\sum_{n=0}^{q} g_n$ be the Mittag–Leffler series of h on $(D, \mathcal{O})$, with $g_n = \overline{h_{T_n}} \; \forall j = 1, \ldots, q$. Then we have $\|\mathcal{H}_t(g_j)\| \leq \|h\|_{D,\mathcal{O}}$ for each $j = 1, \ldots, q$.

Finally, we check that there exists a constant $L \geq 1$ such that $\|t^n\| \leq Lr^n$. Indeed, if A is uniform, we just take $L = 1$. And if A is not uniform, then we have $r > \|t\|_{\mathrm{sp}}$, hence $\lim_{n \to \infty} \frac{\|t^n\|}{r^n} = 0$. Thus, we have shown the existence of L in all cases. Consequently, in $H(D, \mathcal{O})$, we check that $\|\mathcal{H}_t(g_j)\| \leq L\|g_j\|_{D,\mathcal{O}}$ for each $j = 0, \ldots, q$. But by Theorem 8.2.1, we have $\|h\|_{D,\mathcal{O}} = \max_{0 \leq j \leq q} \|g_j\|_{D,\mathcal{O}}$. Therefore, we conclude $\|\mathcal{H}_t(h)\| \leq L\|h\|_{D,\mathcal{O}}$, which finishes proving that $\mathcal{H}_t$ is continuous. Consequently, it has continuation to $H(D, \mathcal{O})$. If A is uniform, we can take $L = 1$, hence $\|\mathcal{H}_t(h)\| \leq \|h\|_{D,\mathcal{O}}$. $\square$

Corollary 8.4.7. *Let $t \in A$, let $\mathcal{O}$ be a t-normal partition, and let $\psi \in \mathrm{Mult}(A, \| \cdot \|)$. Then ψ_t belongs to $\mathrm{Mult}(R(sp(t), \mathcal{O}), \| \cdot \|_{D,\mathcal{O}})$.*

Definitions and notations: Let $x \in A$, let $a \in sp(x)$, let $s = \|x - a\|_{\mathrm{sp}}$. An annulus $\Gamma(a, r', r'') \subset d(b, s)$ will be said to be x-*clear* if there exist ψ', $\psi'' \in \mathrm{Mult}(A, \| \cdot \|)$ such that $\psi'(x - a) = r'$, $\psi''(x - a) = r''$, and $\psi(x - a) \notin]r', r''[\; \forall \psi \in \mathrm{Mult}(A, \| \cdot \|)$.

Remarks. If $\Gamma(a, r', r'')$ is a x-clear annulus, in particular we have $\Gamma(a, r', r'') \cap sp(x) = \emptyset$. But the converse is not true.

Theorem 8.4.8. *Let $t \in A$. If there exists a t-clear annulus, then both* $\mathrm{Mult}(A, \| \cdot \|)$ *and* $\sigma(t)$ *are not connected. Moreover, there exists a t-clear annulus if and only if* $\sigma(t)$ *is not connected.*

Proof. Suppose first that there exists a t-clear annulus $\Gamma(a, r, s)$. Let $F = \{\phi \in \mathrm{Mult}(A, \| \cdot \|) \mid \phi(x - a) \leq r\}$ and let $G = \{\phi \in \mathrm{Mult}(A, \| \cdot \|) \mid \phi(x - a) \geq s\}$. Then, F and G are two closed subsets making a partition of $\mathrm{Mult}(A, \| \cdot \|)$, hence $\mathrm{Mult}(A, \| \cdot \|)$ is not connected. Moreover, the sets $F_t = \{\phi_t \mid \phi \in F\}$ and $G_t = \{\phi_t \mid \phi \in G\}$ are two closed subsets of $\sigma(t)$ making a partition of $\sigma(t)$, hence $\sigma(t)$ is not connected. Conversely, suppose now that $\sigma(t)$ is not connected. By Theorem 6.3.7 of Chapter 6, there exists an annulus $\Xi = \Gamma(a, r_0, s_0)$ together with filters $\mathcal{F}, \mathcal{G} \in T$ such that $\mathcal{F} \in \Omega^{-1}(\sigma(t))$ secant with $\mathcal{I}(\Xi)$, $\mathcal{G} \in \Omega^{-1}(\sigma(t))$ secant with $\mathcal{E}(\Xi)$, such that none of circular filters $\mathcal{H} \in \Omega^{-1}(\sigma(t))$ is secant with Ξ. Now, let $r = \sup\{\phi(t - a), \mid \phi \in \mathrm{Mult}(A, \| \cdot \|) \ \phi(t - a) \leq r_0\}$ and $s = \inf\{\phi(t - a), \ \phi \in \mathrm{Mult}(A, \| \cdot \|) \ \phi(t - a) \geq s_0\}$. Then, $\Gamma(a, r, s)$ is a t-clear annulus. $\qquad\square$

Lemma 8.4.9 is obvious.

Lemma 8.4.9. *Let $x \in A$, let $a \in sp(x)$, let $\Gamma(a, r', r'')$ be a x-clear annulus. For all $s', s'' \in]r', r''[$ such that $s' < s''$, for every $\psi \in \mathrm{Mult}(A, \| \cdot \|)$, $\Omega^{-1}(\psi_x)$ is not secant with $\Gamma(a, s', s'')$.*

Notation: Let $\phi \in \mathrm{Mult}(A, \| \cdot \|)$ and let $x \in A$. We will denote by ϕ_x the element of $\mathrm{Mult}(R(sp(x), \| \cdot \|_{\mathrm{sp(x)}})$ defined as $\phi_x(h) = \phi(h(x)) \ \forall h \in R(sp(x))$.

Theorem 8.4.10. *Let $x \in A$, let $a \in sp(x)$, let $s = \|x - a\|_{\mathrm{sp}}$, let $D = sp(x)$ and let $\mathcal{O}$ be the strict x-spectral partition. The equality $\|h(x)\|_{\mathrm{sp}} = \|h\|_{D,\mathcal{O}}$ holds for every $h \in R(sp(x))$ if and only if there exists no x-clear annulus.*

Proof. Suppose first that there exists a x-clear annulus $\Gamma(a, r', r'')$ and let $\Gamma(a, s', s'') \subset \Gamma(a, r', r'')$ be such that $\Omega^{-1}(\phi_x)$ is not secant with $\Gamma(a, s', s'')$, whenever $\phi \in \mathrm{Mult}(A, \| \cdot \|)$, while there exists $\phi', \phi'' \in \mathrm{Mult}(A, \| \cdot \|)$ such that $\Omega^{-1}(\phi'_x)$ is secant with $\mathbb{K} \setminus d(a, r''^-)$ and $\Omega^{-1}(\phi''_x)$ is secant with $d(a, r')$. Let $r \in]s', s''[\cap |\mathbb{K}|$, let $b \in C(a, r)$ and let $f(x) = \frac{x - a}{(x - b)^2}$. We can easily check that $\|f\|_{\mathrm{sp}} \leq \max(\frac{s'}{r^2}, \frac{1}{s''}) < \frac{1}{r}$. Now, by definition of $\mathcal{O}$, it is clear that $d(b, r^-)$ belongs to $\mathcal{O}$, therefore $\|f\|_{D,\mathcal{O}} \geq \varphi_{b,r}(f) = \frac{1}{r}$: this shows that the two norms are not equal when there exists a x-clear annulus.

Conversely, we now suppose that there exists no x-clear annulus. Let $h \in R(D)$. By Theorem 8.4.5, we have

$$\|h(x)\|_{\mathrm{sp}} \leq \|h\|_{D,\mathcal{O}} \tag{8.4.1}$$

Let $T = d(b, r^-)$ be a hole of $\mathcal{O}$ containing a pole of h and let $\mathcal{F}$ be the circular filter of center b and diameter r. We will show that there exists $\psi \in \mathrm{Mult}(A, \| \cdot \|)$ such that $\psi_x = \varphi_{\mathcal{F}}$. Indeed, suppose that it is not true. By definition of $\mathcal{O}$, we have $r = \inf\{\xi(x - b) \mid \xi \in \mathrm{Mult}(A, \| \cdot \|)\}$. Consequently, either there exists $\zeta \in \mathrm{Mult}(A, \| \cdot \|)$ such that $\zeta(x - b) = r$, or there exists a sequence $(\zeta_n)_{n \in \mathbb{N}}$ such that $\zeta_n(x - b) > r$ and $\lim_{n \to \infty} \zeta_n(x - b) = r$. If there exists a sequence $(\zeta_n)_{n \in \mathbb{N}}$ such that $\zeta_n(x - b) > r$ and $\lim_{n \to \infty} \zeta_n(x - b) = r$, then the sequence $(\zeta_n)_{n \in \mathbb{N}}$ admits a cluster ψ in $\mathrm{Mult}(A, \| \cdot \|)$ such that $\psi(x - b) = r$ and then, since $\zeta_n(x - b) > r \; \forall n \in \mathbb{N}$, we can see that $\psi_x = \varphi_{\mathcal{F}}$.

Now, we just suppose that there exists $\zeta \in \mathrm{Mult}(A, \| \cdot \|)$ such that $\zeta(x - b) = r$. Let $\mathcal{G}$ be the circular filter such that $\zeta_x = \varphi_{\mathcal{G}}$. Since we suppose that $\mathcal{G} \neq \mathcal{F}$, then $\mathcal{G}$ is secant with a class $d(c, r^-)$ of $C(b, r)$ and has a diameter $\rho < r$. Then by Lemma 5.3.12 of Chapter 5, we can find a disk $d(c, l)$ which belongs to $\mathcal{G}$, with $\rho < l < r$. Since there exists no x-clear annulus, for every $n \in \mathbb{N}^*$ there exists $\zeta_n \in \mathrm{Mult}(A, \| \cdot \|)$ such that $r - \frac{1}{n} < \zeta_n(x - c) < r$. Therefore, the sequence $(\zeta_n)_{n \in \mathbb{N}}$ admits a cluster ψ in $\mathrm{Mult}(A, \| \cdot \|)$ such that ψ_x is the limit of the sequence $((\zeta_n)_x)_{n \in \mathbb{N}}$, and thereby we have again $\psi_x = \varphi_{\mathcal{F}}$.

So, we have proven in all cases that there exists $\psi \in \mathrm{Mult}(A, \| \cdot \|)$ such that $\psi_x = \varphi_{\mathcal{F}}$. Let $\mathcal{F}_0$ be the circular filter of center a and diameter s. Let $(d(b_j, r_j))_{1 \leq j \leq n}$ be the family of holes of $\mathcal{O}$ containing at least one hole of h and for each $j = 1, \ldots, n$, let $\mathcal{F}_j$ be the circular filter of center b_j and diameter r_j. As we just saw, for every $j = 1, \ldots, n$, there exists $\psi_j \in \mathrm{Mult}(A, \| \cdot \|)$ such that $(\psi_j)_x = \varphi_{\mathcal{F}_j}$.

In the same way, we can easily show that there exists $\psi_0 \in \mathrm{Mult}(A, \| \cdot \|)$ such that $(\psi_0)_x = \varphi_{\mathcal{F}_0}$. Consequently, we have $\|h\|_{\mathrm{sp}} \geq \max_{0 \leq j \leq n} \varphi_{\mathcal{F}_j}(h)$. On the other hand by Theorem 8.2.1, we know that $\max_{0 \leq j \leq n} \varphi_{\mathcal{F}_j}(h) = \|h\|_{D,\mathcal{O}}$, hence by (8.4.1) we obtain $\|h(x)\|_{\mathrm{sp}} = \|h\|_{D,\mathcal{O}}$. $\qquad\square$

Theorem 8.4.11. *Let $t \in A$ have bounded normal ratio and let $\mathcal{O}$ be a wide t-spectral partition of diameter r. Let $(d(b_i, r_i^-))_{i \in J}$ be the family of holes of $\mathcal{O}$. The restriction of $\mathcal{H}_t$ to $(R'(sp(t), (b_i)_{i \in J}), \| \cdot \|_{sp(t), \mathcal{O}})$ is a continuous $\mathbb{K}$-linear mapping and has continuation to $H'(s(t), (b_i)_{i \in J}), \mathcal{O})$.*

Proof. Let $D = sp(t)$. Let $f \in (R'(sp(t), \mathcal{O}, (b_i)_{i \in J})$. Let $\overline{f}_0 + \sum_{n=1}^{\infty} \overline{f}_{T_n}$ be the Mittag–Leffler series of f on $\mathcal{O}$. For each $n \in \mathbb{N}^*$, let $b_n \in T_n$. Then $\overline{f}_{T_n}$ is of the form $\frac{\lambda_n}{x-b_n}$. Since t has bounded normal ratio, there exists a constant B such that $\|\overline{f}_{T_n}\|_{D,\mathcal{O}} \leq B\|\mathcal{H}_t(\overline{f}_{T_n})\|$ $\forall n \in \mathbb{N}^*$. Finally, let $\overline{f}_0 = \sum_{m=0}^q a_m x^m$. By hypothesis, we have $\lim_{m\to\infty} |a_m| r^m = 0$. If $\|t\|_{\mathrm{sp}} = \|t\|$, then $r = \|x\|$, hence $\|\mathcal{H}_t(\overline{f}_0)\| \leq \|\overline{f}_0\|_{D,\mathcal{O}}$. Now, if $\|t\|_{\mathrm{sp}} < \|t\|$, then $r > \|t\|_{\mathrm{sp}}$, hence there exists a constant C such that $\|t^m\| < Cr^m$ $\forall m \in \mathbb{N}$, and therefore, $\|\mathcal{H}_t(\overline{f}_0)\| \leq C\|\overline{f}_0\|_{D,\mathcal{O}}$. Thus, putting $Q = \max(B,C)$, we have $\|\mathcal{H}_t(f)\| \leq Q\|f\|_{D,\mathcal{O}}$ $\forall f \in (R'(sp(x),(b_i)_{i\in J})$. $\qquad\square$

By Lemma 5.3.12 of Chapter 5, we have Lemma 8.4.12.

Lemma 8.4.12. *Let $t \in A$ be such that $S(t) \neq 0$ for every polynomial $S \in K[X]$ different from 0. Let $a \in \mathbb{K}$, let $\phi \in \mathrm{Mult}(A, \| \cdot \|)$ and let $r = \phi(t-a)$. Then $\Omega^{-1}(\phi_t)$ is (a,r)-approaching.*

Proposition 8.4.13. *Let $t \in A$ be such that the mapping $\mathcal{H}_t$ is injective. Let $a \in \mathbb{K} \setminus sp(t)$, and let $r = \|(t-a)^{-1}\|_{\mathrm{sp}}^{-1}$. There exists $\theta \in \mathrm{Mult}(A, \| \cdot \|)$ whose restriction to $R(sp(t))$ has a circular filter (a,r)-approaching.*

Proof. We consider $R(sp(t))$ as a $\mathbb{K}$-subalgebra of A. Let $\phi_t \in \mathrm{Mult}(A, \| \cdot \|)$. If $\Omega^{-1}(\phi_t)$ is secant with a disk $d(a,\rho)$ for some $\rho \in]0,r[$, then clearly we have $\phi_t(t-a) \leq \rho$ hence $\phi_t((t-a)^{-1}) > \frac{1}{r}$ and therefore $\|(t-a)^{-1}\|_{\mathrm{sp}} > \frac{1}{r}$ which contradicts the hypothesis. So $\Omega^{-1}(\phi_t)$ is secant with $\mathbb{K} \setminus d(a,r^-)$.

Suppose that there exists $\rho > r$ such that, for every $\phi \in \mathrm{Mult}(A, \| \cdot \|)$, $\Omega^{-1}(\phi)$ is not secant with $d(a,\rho)$. Clearly we have $\phi(t-a) \geq \rho$ for all $\phi \in \mathrm{Mult}(A, \| \cdot \|)$ and therefore $\|(t-a)^{-1}\|_{\mathrm{sp}} < \frac{1}{r}$, a contradiction. Consequently, for each $n \in \mathbb{N}^*$ we can find $\phi_n \in \mathrm{Mult}(A, \| \cdot \|)$ such that $\Omega^{-1}((\phi_n)_t)$ is secant with $d(a, r + \frac{1}{n})$. And since it is also secant with $\mathbb{K} \setminus d(a,r^-)$, finally, it is secant with $\Gamma(a, r, r + \frac{1}{n})$. Since $\mathrm{Mult}(A, \| \cdot \|)$ is compact, we can take a point of adherence θ of the filter associated to the sequence $(\phi_m)_{m\in\mathbb{N}}$. So, for every $m \in \mathbb{N}$, there exists $q_m \in \mathbb{N}$ such that $|\phi_{q_m}(t) - \theta(t)|_\infty \leq \frac{1}{m}$. Since this is true for all $m \in \mathbb{N}$, and since $\lim_{n\to\infty} \phi_n(t-a) = r$, we have $\theta(t-a) = r$, thereby $\Omega^{-1}(\theta(t))$ is (a,r)-approaching. $\qquad\square$

Chapter 9

Spectral Properties in Uniform Algebras

9.1. Spectral properties

The chapter is first devoted to present several semi-multiplicative semi-norms and compare them to the spectral semi-norm. We introduce properties (o), (p), (q), (r), (s) which will keep the same meaning throughout the book. Most of results in this chapter were stated in [26]. Next, we consider two kinds of Gelfand transform in an ultrametric $\mathbb{K}$-algebra: one is similar to the definition in complex analysis, but does not show strong properties. The other is more sophisticated and shows certain specific properties linked to circular filters.

Notation: Let us recall that $\mathbb{K}$ denotes an algebraically closed field complete for a non-trivial ultrametric absolute value. Throughout the chapter, $(A, \| \ \|)$ will be a unital commutative normed $\mathbb{K}$-algebra.

Let us recall that by Corollary 2.5.9 of Chapter 2, we have $\|x\|_{\mathrm{sp}} = \sup\{\varphi(x) \mid \varphi \in \mathrm{Mult}(A, \| \, . \, \|)\}$.

We can define several other semi-multiplicative semi-norms:

$\|x\|_{\mathrm{sa}} = \sup\{\varphi(x) \mid \varphi \in \mathrm{Mult}_a(A, \| \, . \, \|)\},$
$\|x\|_{\mathrm{sm}} = \sup\{\varphi(x) \mid \varphi \in \mathrm{Mult}_m(A, \| \, . \, \|)\},$

Moreover, we put $\tau(x) = \sup\{|\lambda| \mid \lambda \in sp(x) \cup \{0\}\}$.

Remark 1. In certain cases, $sp(x)$ might be empty (for instance in a field extension of $\mathbb{K}$). This is why $\tau(x)$ involves $sp(x) \cup \{0\}$.

Proposition 9.1.1. $\tau(x) \leq \|x\|_{\mathrm{sm}} \forall x \in A$. If all maximal ideals of A are of codimension 1, then τ equals $\| \, . \, \|_{\mathrm{sm}}$.

Proof. Let $\lambda \in sp(x)$. There exists a homomorphism χ from A to an extension of $\mathbb{K}$ such that $\chi(x) = \lambda$, hence $|\lambda| \leq \|x\|_{\mathrm{sm}}$. This is true for every λ such that $|\lambda| \leq \tau(x)$, therefore $\tau(x) \leq \|x\|_{\mathrm{sm}}$.

Suppose now that all maximal ideals of A are of codimension 1. Take $x \in A$. For every $\lambda \in sp(x)$, there exists a maximal ideal M such that $x - \lambda \in M$, M is the kernel of a homomorphism χ from A to $\mathbb{K}$ such that $\chi(x) = \lambda$, therefore $|\lambda| = |\chi(x)| \leq \|x\|_{\mathrm{sm}}$. $\qquad \square$

Remark 1. If A has maximal ideals of infinite codimension, τ is not necessarily a semi-norm. Suppose that A admits a unique maximal ideal of infinite codimension and no other maximal ideal. Let χ be a homomorphism from A onto an extension of $\mathbb{K}$, let $x \in A$ be such that $\chi(x) \in \mathbb{K}$ and let $y \in A$ be such that $\chi(y) \notin \mathbb{K}$. Then $\tau(x) = |\chi(x)|$ but $\tau(x + y) = \tau(x - y) = 0$. If τ were a semi-norm, we should have $\tau(2x) = 0$. But if $\mathbb{K}$ has characteristic different from 2, we have $\tau(2x) \neq 0$.

Proposition 9.1.2 is immediate from Theorems 2.5.11, 2.5.15 and Corollary 2.5.8 of Chapter 2.

Proposition 9.1.2. *Assume* $\mathrm{Max}_1(A) \neq \emptyset$. *Then* $\|x\|_{\mathrm{sa}} = \sup\{|\chi(x)| \mid \chi \in \mathcal{X}(A, \mathbb{K})\}$ $\forall x \in A$.

Notation: In A we denote by (o), (p), (q), (r), (s) these properties:

(o) $\|x\|_{\mathrm{sa}} = \tau(x)$ $\forall x \in A$,
(p) $\|x\|_{\mathrm{sa}} = \|x\|_{\mathrm{sp}}$ $\forall x \in A$,
(q) $\tau(x) = \|x\|_{\mathrm{sp}}$ $\forall x \in A$,
(r) $\|x\|_{\mathrm{sa}} = \|x\|_{\mathrm{sm}}$ $\forall x \in A$,
(s) $\|x\|_{\mathrm{sm}} = \|x\|_{\mathrm{sp}}$ $\forall x \in A$.

Thus, by definition, Property (p) implies Properties (o), (q), (r). Next, (s) and (r) implies (o).

Proposition 9.1.3. *Assume that A satisfies Property p). Then A is semi-simple if and only if $\| \, . \, \|_{\mathrm{sp}}$ is a norm.*

Theorem 9.1.4. *Let A be complete, let $\| \, . \, \|_{\mathrm{sp}}^A$ be its spectral semi-norm and let B be a unital commutative normed $\mathbb{K}$-algebra whose spectral semi-norm $\| \, . \, \|_{\mathrm{sp}}^B$ satisfies Property (q). Then every $\mathbb{K}$-algebra homomorphism ϕ from A to B satisfies $\|\phi(x)\|_{\mathrm{sp}}^B \leq \|x\|_{\mathrm{sp}}^A$ $\forall x \in A$.*

Proof. Suppose that for certain $x \in A$ we have $\|\phi(x)\|_{\mathrm{sp}}^B > \|x\|_{\mathrm{sp}}^A$ and let $y = \phi(x)$. Then there exists $\lambda \in sp_B(\phi(x))$ such that $|\lambda| > \|x\|_{\mathrm{sp}}^A$. Then by Theorem 2.5.11 of Chapter 2, $\lambda - x$ is invertible in A, hence $\phi(\lambda - x)$

is invertible in B. But $\phi(\lambda - x) = \lambda - y$, a contradiction to the hypothesis $\lambda \in sp_B(\phi(x))$. □

Corollary 9.1.5. *Assume A is complete, let B be a uniform unital commutative normed $\mathbb{K}$-algebra satisfying Property (q) and let ϕ be a $\mathbb{K}$-algebra homomorphism from A to B. Then ϕ is continuous.*

By Theorems 2.5.11 of Chapter 2 and 9.1.4 we have Corollary 9.1.6.

Corollary 9.1.6. *Let A be complete. Let F be a set and let B be a $\mathbb{K}$-subalgebra with unity of the algebra of bounded functions from F to $\mathbb{K}$ equipped with the norm $\| \cdot \|_F$. Then every $\mathbb{K}$-algebra homomorphism ϕ from A to B is continuous and satisfies $\|\phi(f)\|_F \leq \|f\|_{sp} \; \forall f \in A$.*

Notations: In A we have defined two kinds of "Gelfand transform" $\mathbf{G}_A$ and $\mathbf{GM}_A$ (see Section 6.4 of Chapter 6). The first one is similar to that in complex analysis and was already used in C.5, consisting of associating to each element f of A the mapping f^0 from $\mathcal{X}(A, \mathbb{K})$ to $\mathbb{K}$ defined as $f^0(\chi) = \chi(f)$, $(\chi \in \mathcal{X}(A, \mathbb{K}))$.

The second one, denoted by $\mathbf{GM}_A$ consists of associating to each element f of A the mapping f^* from $\mathrm{Mult}(A, \| \cdot \|)$ to $\mathrm{Mult}(\mathbb{K}[x])$ defined by $f^*(\phi)(P) = \phi(P \circ f)$, $(P \in \mathbb{K}[x])$.

As in complex analysis, Propositions 9.1.7–9.1.9 are immediate.

Proposition 9.1.7. $\mathbf{G}_A$ *is injective if and only if the intersection of all maximal ideals of codimension 1 is null.*

Proposition 9.1.8. $\mathcal{X}(A, \mathbb{K})$ *being equipped with the topology of pointwise convergence, for every $f \in A$, f^0 belongs to $\mathcal{C}(\mathcal{X}((A, \mathbb{K}), \mathbb{K})$.*

Proposition 9.1.9. $\mathcal{C}(\mathcal{X}((A, \mathbb{K}), \mathbb{K})$ *being equipped with the norm of uniform convergence, $\mathbf{G}_A$ satisfies $\|f^0\| = \|f\|_{sa}$ for every $f \in A$.*

Theorem 9.1.10. *Assume that A satisfies Property p). Then $\mathcal{C}(\mathcal{X}(A, \mathbb{K}), \mathbb{K})$ being equipped with the norm of uniform convergence, then $\mathbf{G}_A$ is an isometry if and only if $\|x^2\| = \|x\|^2 \; \forall x \in A$.*

Proof. Suppose $\|x^2\| = \|x\|^2$ is true for all $x \in A$, let $x \in A$ and let $\rho = \frac{\|x\|_{sp}}{\|x\|}$. Then we check that $\frac{\|x^{2^n}\|_{sp}}{\|x^{2^n}\|} = \rho^{(2^n)}$ hence $\|x\|_{sp} = \rho \lim_{n \to \infty} \|x^n\|^{\frac{1}{n}}$, and therefore $\rho = 1$. So, $\mathbf{G}_A$ is an isometry. The converse is trivial. □

Theorem 9.1.11. *Assume that A satisfies Property (p). The following two properties are equivalent on A:*

A is semi-simple and $\mathbf{G}_A(A)$ is closed in $\mathcal{C}(\mathcal{X}(A, \mathbb{K}), \mathbb{K})$,
A is uniform.

Proof. First, suppose that A is uniform. Let x belong to the Jacobson radical of A. In particular, x belongs to the intersection of maximal ideals of codimension 1 of A. So, we have $\|x\|_{\text{sa}} = 0$, so, by Property (p), $\|x\|_{\text{sp}} = 0$, and therefore $x = 0$, hence A is semi-simple. Let h belong to the closure of $\mathbf{G}_A(A)$ in $\mathcal{C}(\mathcal{X}(A, \mathbb{K}), \mathbb{K})$. Since $\mathcal{C}(\mathcal{X}(A, \mathbb{K}), \mathbb{K})$ is equipped with the norm of uniform convergence, we just have to consider a sequence f_n^0 of $\mathbf{G}_A$ converging to h. Thus, $\|h - f_n^0\| = \sup\{|\chi(f_n) - h(\chi)| \,|\, \chi \in \mathcal{X}(A, \mathbb{K})\}$. The sequence $(f_n)_{n \in \mathbb{N}}$ is a Cauchy sequence in A, with respect to $\|x\|_{\text{sp}} = 0$, and therefore with respect to the norm of A because A is uniform. Let $f = \lim_{n \to \infty} f_n$. Then we can check that $f^0 = \lim_{n \to \infty} f_n^0$ in $\mathcal{C}(\mathcal{X}(A, \mathbb{K}), \mathbb{K})$ and therefore $h = f^0$.

Conversely, suppose that A is semi-simple and that $\mathbf{G}_A(A)$ is closed in $\mathcal{C}(\mathcal{X}(A, \mathbb{K}), \mathbb{K})$. Since A is semi-simple and satisfies Property (p), $\| \cdot \|_{\text{sa}}$ is a norm equal to $\|x\|_{\text{sp}}$. Consider a Cauchy sequence (f_n) with respect to the norm $\| \cdot \|_{\text{sp}}$. Then the sequence (f_n^0) is a Cauchy sequence in $\mathcal{C}(\mathcal{X}(A, \mathbb{K}), \mathbb{K})$, and has a limit h which actually lies in $\mathbf{G}_A(A)$ because $\mathbf{G}_A(A)$ is closed in $\mathcal{C}(\mathcal{X}(A, \mathbb{K}), \mathbb{K})$. Hence, there exists $f \in A$ such that $f^0 = h$. Consequently, f is the limit of the sequence (f_n) with respect to the norm $\| \cdot \|_{\text{sp}}$. Thus, A is a $\mathbb{K}$-algebra complete for both $\| \cdot \|_{\text{sp}}$ and $\| \cdot \|$. And since $\|x\|_{\text{sp}} \leq \|x\| \; \forall x \in A$, by Hahn–Banach's Theorem the two norms are equivalent. $\qquad \square$

Now we will examine the mapping $\mathbf{GM}_A$.

Theorem 9.1.12. *Given $f \in A$, the mapping f^* from $\text{Mult}(A, \| \cdot \|)$ to $\text{Mult}(\mathbb{K}[x])$ is continuous with respect to the topology of pointwise convergence.*

Proof. Let $\mathcal{G} = f^*(\phi)$. By Theorem 6.2.10 of Chapter 6, the family of sets $\text{Mult}(H(E), \| \cdot \|_E)$, where E is an infraconnected affinoid subset of $\mathbb{K}$ lying in $\mathcal{G}$, makes a basis of neighborhoods of $\varphi_\mathcal{G}$. So, we take an arbitrary $\mathcal{G}$-affinoid B and will show that there exists a neighborhood W of ϕ inside $\text{Mult}(A, \| \cdot \|)$ whose image by f^* is included in $\text{Mult}(H(B), \| \cdot \|_B)$. If $\mathcal{G}$ has center b and diameter $r > 0$, we can take an arbitrary $\mathcal{G}$-affinoid B of the form $d(b, r + \epsilon) \setminus \left(\bigcup_{j=1}^q d(b_j, (r - \epsilon^-)) \right)$, and if $\mathcal{G}$ has no center and but has diameter r, or is a Cauchy filter, we can take an arbitrary $\mathcal{G}$-affinoid B of the form $d(b, r + \epsilon)$. Putting $b = b_0$, we consider now the neighborhood of ϕ

$W = \{\psi \in \mathrm{Mult}(A, \| \cdot \|) \mid |\psi(f - b_j) - \phi(f - b_j)| \leq \epsilon, (0 \leq j \leq q)\}$. Then we can check that $f^*(W) \subset \mathrm{Mult}(H(B), \| \cdot \|_B)$, which proves the claim. $\square$

Lemma 9.1.13 is immediate.

Lemma 9.1.13. *Let $\chi \in \mathcal{X}(A, \mathbb{K})$ and let $f \in A$. Then $f^*(|\chi|) = \varphi_{\chi(f)}$.*

Proof. Let $\mathcal{F}$ be the circular filter such that $\varphi_{\mathcal{F}} = f^*(|\chi|)$. Then $\varphi_{\mathcal{F}}$ clearly belongs to $\mathrm{Mult}_m(\mathbb{K}[x])$ and $\mathcal{F}$ is the Cauchy filter of neighborhoods of $\chi(f)$ because $\varphi_{\mathcal{F}}(P) = |P(\chi(f))| \ \forall P \in \mathbb{K}[x]$. $\square$

Theorem 9.1.14. *Assume that A is uniform and that the intersection of maximal ideals of codimension 1 is null. Then $\mathbf{GM}_A$ is injective.*

Proof. Suppose $f^* = g^*$ and $f \neq g$. let $r = \|f - g\|$. Since the intersection of maximal ideals of codimension 1 is null, there exists ($\chi \in \mathcal{X}(A, \mathbb{K})$ such that $\chi(g - f) \neq 0$. By Lemma 9.1.13, we have $f^*(|\chi|) = \varphi_{\chi(f)}$ and $g^*(|\chi|) = \varphi_{\chi(g)}$. But since $f^* = g^*$, we have $f^*(|\chi|) = g^*(|\chi|)$. Let $\mathcal{F}$ be the circular filter such that $\varphi_{\mathcal{F}} = f^*(|\chi|) = g^*(|\chi|)$. Then $\varphi_{\mathcal{F}}$ clearly belongs to $\mathrm{Mult}_m(\mathbb{K}[x])$ and $\mathcal{F}$ is the Cauchy filter of neighborhoods of $\chi(f)$ because $\varphi_{\mathcal{F}}(P) = |P(\chi(f))|$. Similarly, it is the Cauchy filter of neighborhoods of $\chi(g)$. Consequently, $\chi(f) = \chi(g)$, hence $\chi(f - g) = 0$, a contradiction to the hypothesis. $\square$

Theorem 9.1.15. *Let A be uniform. Let $\phi \in \mathrm{Mult}(A, \| \cdot \|)$, let $f, g \in A$ be such that $f^*(\phi) \neq g^*(\phi)$. Let $\varphi_{\mathcal{F}} = f^*(\phi)$, $\varphi_{\mathcal{G}} = g^*(\phi)$. Then $\mathrm{diam}(\sup(\mathcal{F}, \mathcal{G})) = \phi(f - g)$.*

Proof. Suppose that $f^*(\phi) \neq g^*(\phi)$. Let $r = \mathrm{diam}(\mathcal{F})$, $s = \mathrm{diam}(\mathcal{G})$. Suppose first that $\mathcal{F} \prec \mathcal{G}$, hence $r < s$. Let us take $l \in]r, s[$. By Lemma 5.3.12 of Chapter 5, there exists a unique disk $d(a, l)$ which belongs to $\mathcal{F}$ Then a is a center of $\mathcal{G}$. Thus, we have $\phi(f - a) \leq l$, $\phi(g - a) = s$, hence, by Lemma 2.3.3 of Chapter 2, $\phi(g - f) = s = \mathrm{diam}(\sup(\mathcal{F}, \mathcal{G}))$.

Now, suppose that $\mathcal{F}$ and $\mathcal{G}$ are uncomparable. Let $\Sigma = \sup(\mathcal{F}, \mathcal{G})$. By Proposition 2.4.9 of Chapter 5, we can find disks $F = d(a, r') \in \mathcal{F}$, $G = d(b, s') \in \mathcal{G}$ such that $\delta(F, G) = \lambda(\mathcal{F}, \mathcal{G}) = |a - b| = \mathrm{diam}(\sup(\mathcal{F}, \mathcal{G}))$. Of course we have $|b - a| > r'$ and $|b - a| > s'$. Then, $\phi(g - a) = |b - a|$ and $\phi(f - a) < |b - a|$, hence $\phi(g - f) = |b - a|$, and therefore $\phi(g - f) = \mathrm{diam}(\sup(\mathcal{F}, \mathcal{G}))$. $\square$

Corollary 9.1.16. *Let A be uniform. Let $\phi \in \mathrm{Mult}(A, \| \cdot \|)$, let $f, g \in A$. Let $\varphi_{\mathcal{F}} = f^*(\phi)$, $\varphi_{\mathcal{G}} = g^*(\phi)$. Then $\delta(f^*(\phi), g^*(\phi))) \leq \phi(f - g) \leq \|f - g\|$.*

Notation: Given $\phi \in \mathrm{Mult}(A, \| \, . \, \|)$, we will denote by Z_ϕ the mapping from A into $\mathrm{Mult}(\mathbb{K}[x])$ defined as $Z_\phi(f) = f^*(\phi)$.

Corollary 9.1.17. *Let A be uniform. The family of functions Z_ϕ, $\phi \in \mathrm{Mult}(A, \| \, . \, \|)$ is uniformly equicontinuous with respect to the δ-topology on $\mathrm{Mult}(\mathbb{K}[x])$.*

Corollary 9.1.18. *Let A be uniform. Each function Z_ϕ, $\phi \in \mathrm{Mult}(A, \| \, . \, \|)$ is continuous with respect to the topology of pointwise convergence on $\mathrm{Mult}(\mathbb{K}[x])$.*

Lemma 9.1.19. *Let $x \in A$ be such that $\tau(x) < \|x\|$, let $r = \tau(x)$ and let $s = \|x\|$. There exists $\phi \in \mathrm{Mult}(A, \| \, . \, \|)$ and a numbers $\mathcal{S} > 0$ and an affinoid F which belongs to the filter $\mathcal{F} = \Omega^{-1}(\phi_x)$, such that $\mathrm{diam}(\Omega^{-1}(\psi_x)) \geq \sigma$, for all $\psi \in \mathrm{Mult}(A, \| \, . \, \|)$ such that $\Omega^{-1}(\psi_x) \in \Phi(F)$.*

Proof. We assume lemma is not true. Let $\mathcal{O}$ be the strict x-spectral partition, let $m \in \,]r, s[$ and let $\phi_0 \in \mathrm{Mult}(A, \| \, . \, \|)$ be such that $\Omega^{-1}((\phi_0)_x)$ is secant with $\Gamma(0, m, s)$. Let $\mathcal{F}_0 = \Omega^{-1}((\phi_0)_x)$ and let $s_0 = \mathrm{diam}(\mathcal{F}_0)$.

Suppose we have already constructed a finite sequence $(\phi_n)_{n \in \mathbb{N}}$ for $n = 0$, $\ldots, q$, satisfying $\mathrm{diam}(\Omega^{-1}((\phi_n)_x)) < \frac{s}{n}$ and $\delta(\Omega^{-1}((\phi_n)_x), \Omega^{-1}((\phi_{n+1})_x)) < \frac{2s}{n+1}$. Let $\mathcal{F}_n = \Omega^{-1}((\phi_n)_x)$. Since the lemma is supposed to be false, there exists no affinoid $F \in \mathcal{F}$ such that $\mathcal{F}_n$ is the only circular filter of the form $\Omega^{-1}(\psi_x)$, $\psi \in \mathrm{Mult}(A, \| \, . \, \|)$, which is secant with F.

Suppose that we can't find $\phi_{n+1} \in \mathrm{Mult}(A, \| \, . \, \|)$ such that the filter $\mathcal{F}_{n+1} = \Omega^{-1}((\phi_{n+1})_x))$ satisfies $\mathrm{diam}(\mathcal{F}_{n+1}) < \frac{s}{n+1}$, and $\delta(\mathcal{F}_{n+1}, \mathcal{F}_n) < \frac{2s}{n+1}$. This means that for every $\psi \in \mathrm{Mult}(A, \| \, . \, \|)$ such that $\delta(\Omega^{-1}(\psi_x), \mathcal{F}_n) < \frac{2s}{n+1}$, we have $\mathrm{diam}(\Omega^{-1}(\psi_x)) \geq \frac{1}{n}$. Let F be the unique disk of diameter $\frac{2s}{n+1}$ such that $\mathcal{F}_n \in \Phi(F)$. Then for all $\mathcal{G} \in \phi(F)$ we have $\delta(F, \mathcal{G}) \leq \frac{2s}{n+1}$, hence hence $\mathrm{diam}(\Omega^{-1}(\psi_x)) \geq \frac{s}{n+1}$ for all $\psi \in \mathrm{Mult}(A, \| \, . \, \|)$ such that $\Omega^{-1}(\psi_x) \in \Phi(F)$, which proves the conclusion. Thus, since we have supposed that the conclusion is false, we can find $\phi_{n+1} \in \mathrm{Mult}(A, \| \, . \, \|)$ such that the filter $\mathcal{F}_{n+1} = \Omega^{-1}((\phi_{n+1})_x))$ satisfies $\mathrm{diam}(\mathcal{F}_{n+1}) < \frac{s}{n+1}$, and $\delta(\mathcal{F}_{n+1}, \mathcal{F}_n) < \frac{2s}{n+1}$. And this is true for all $n \in \mathbb{N}$, hence the sequence $((\phi_n)_x)_{n \in \mathbb{N}}$ is a Cauchy sequence with respect to the δ-topology. Consequently, its limit in $\mathrm{Mult}(\mathbb{K}[x])$ is of the form φ_α, with $\alpha \in \Gamma(0, r, s)$. On the other hand, since $\mathrm{Mult}(A, \| \, . \, \|)$ is compact for the topology of pointwise convergence, the sequence $(\phi_n)_{n \in \mathbb{N}}$ admits a point of adherence θ with respect to the topology of pointwise convergence. Consequently, θ_x is a point of adherence of the sequence $((\psi_n)_x)_{n \in \mathbb{N}}$. Next, since $\mathrm{Mult}(\mathbb{K}[x])$ is sequentially compact, we can extract from the sequence $(\psi_n)_x$ a subsequence converging to θ_x with respect to the topology of pointwise convergence. Thus, without loss of generality, we can assume that the sequence $(\psi_n)_x$ converges to θ_x with respect to the topology of pointwise convergence in $\mathrm{Mult}(\mathbb{K}[x])$. But since the sequence $(\psi_n)_x$ converges to φ_α for the δ-topology, so much the more does it converge to φ_α for the topology of pointwise convergence, hence $\varphi_\alpha = \theta_x$. Thus, θ_α is punctual, a contradiction to the fact that $\alpha \notin sp(x)$. Consequently, our hypothesis "the conclusion is false" is wrong, and this finishes proving Lemma 9.1.19.

$\square$

9.2. Uniform $\mathbb{K}$-Banach algebras and properties (s) and (q)

Notations: Throughout this chapter, A is a unital commutative ultrametric Banach $\mathbb{K}$-algebra, whose norm is $\| \, . \, \|$. As in Section 8.4 of Chapter 8, given $t \in A$ we denote by $\mathcal{H}_t$ the canonical morphism associated to t.

The question whether semi-norms $\| \, . \, \|_{\mathrm{sp}}$ and $\| \, . \, \|_{\mathrm{sm}}$ are equal is an old one. Actually, when $\mathbb{K}$ is weakly valued, there exist unital commutative ultrametric Banach $\mathbb{K}$-algebras where the semi-norm $\| \, . \, \|_{\mathrm{sp}}$ is strictly superior to the semi-norm $\| \, . \, \|_{\mathrm{sm}}$ for certain elements. But if we assume that $\| \, . \, \|_{\mathrm{sp}}$ is a norm and that A is complete for this norm, then we can prove the equality (see [24, 36]). In Corollary 9.2.13, we find again a theorem due to B. Guennebaud stating that the completion of a field with respect to a semi-multiplicative norm, admitting at least two continuous absolute values, has divisors of zero. In particular, this shows why Corollary 4.3.4 in [3] is wrong.

Lemma 9.2.1. *Let $t \in A$ have bounded normal ratio and let $\mathcal{O} = (d(a_i, r_i))_{i \in J}$ be a wide t-spectral partition. The restriction of $\mathcal{H}_t$ to $R'(sp(t), \mathcal{O}, (a_i)_{i \in J})$ and $R''(sp(t), \mathcal{O}, (a_i)_{i \in J})$ are continuous.*

Proof. The restriction of $\mathcal{H}_t$ to $R'(sp(t), \mathcal{O}, (a_i)_{i \in J})$ is obviously continuous. Next, we check that the family of $\{ \dfrac{\|(x - b_j)^{-2}\|}{\|(x - b_j)^{-2}\|_{\mathrm{sp}}} \, | i \in J \}$ is bounded because on one hand $\|(x - b_j)^{-2}\|_{\mathrm{sp}} = \left(\|(x - b_j)^{-1}\|_{\mathrm{sp}} \right)^2$ and on the other hand $\|(x - b_j)^{-2}\| \leq \left(\|(x - b_j)^{-1}\| \right)^2$.

$\square$

Lemma 9.2.2. *Let $t \in A$. Assume that $\mathrm{Ker}(\mathcal{H}_t) \neq \{0\}$. Then $\|t\|_{\mathrm{sp}} = \|t\|_{\mathrm{sm}}$. Moreover, if $A \neq \mathbb{K}$, then A has divisors of zero.*

Proof. Let $D = sp(t)$ and let $B = \mathcal{H}_t(R(D))$. Then $\mathrm{Ker}(\mathcal{H}_t)$ is an ideal of $R(D)$ generated by a monic polynomial $G(x) = \prod_{i=1}^{q}(x - a_i)$. Since $G(t) = 0$, for every $\psi \in \mathrm{Mult}(A, \| \cdot \|)$, we have $\psi(G(t)) = 0$, hence there exists an integer $l(\psi) \in \{1, \ldots, q\}$ such that $\psi(t - a_{l(\psi)}) = 0$, hence $\psi(t) = |a_{l(\psi)}|$. Then, $t - a_{l(\psi)}$ lies in $\mathrm{Ker}(\psi)$ and therefore belongs to a maximal ideal $\mathcal{M}$ of A. But by Theorem 2.5.13 of Chapter 2, there exists $\theta_\psi \in \mathrm{Mult}_m(A, \| \cdot \|)$ such that $\mathrm{Ker}(\theta_\psi) = \mathcal{M}$. Hence we have $\theta_\psi(t) = |a_{l(\psi)}| = \psi(t)$. Thus, we have shown that $\psi(t) \leq \|t\|_{\mathrm{sm}}$. But this is true for every $\psi \in \mathrm{Mult}(A, \| \cdot \|)$. Consequently by Theorem 2.5.17 of Chapter 2, we have $\|t\|_{\mathrm{sp}} = \|t\|_{\mathrm{sm}}$, which proves the first statement.

Next, we notice that $\mathrm{Ker}(\mathcal{H}_t)$ admits a generator $G(x) \in \mathbb{K}[x]$ whose zeros lie in D. If $\deg(G) = 1$, then t lies in $\mathbb{K}$ (considered as a $\mathbb{K}$-subalgebra of A), and obviously we have $\psi(t) = |t| \ \forall \psi \in \mathrm{Mult}(A, \| \cdot \|)$, and therefore, $\sup\{|\lambda| \mid \lambda \in D\} = \|t\|_{\mathrm{sp}}$. Now, if $\deg(G) > 1$, then $\mathrm{Ker}(\mathcal{H}_t)$ is not prime, hence A contains divisors of zero, so the second statement is trivial. $\qquad\square$

Theorem 9.2.3 ([24]). *Let $t \in A$ have bounded normal ratio. Then $\|t\|_{\mathrm{sp}} = \|t\|_{\mathrm{sm}}$.*

Proof. Let $D = sp(t)$. We put $r = \|t\|_{\mathrm{sp}}$, $r' = \|t\|_{\mathrm{sm}}$ and we suppose $r' < r$. If $\|t\| = \|t\|_{\mathrm{sp}}$ we put $u = r$, and if $\|t\| > \|t\|_{\mathrm{sp}}$ we take $u \in]r, \|t\|[$. Let $B = \mathcal{H}_t(R(D))$. By Lemma 9.2.1, we can assume that $\mathrm{Ker}(\mathcal{H}_t) = \{0\}$. Hence B is isomorphic to $R(D)$. Let $\mathcal{O}$ be a wide t-spectral partition of diameter u, and for convenience, we put $\| \cdot \|_t = \| \cdot \|_{D,\mathcal{O}}$. Let $(d(a_i, r_i))$, $(i \in J)$ be the family of holes of $\mathcal{O}$ and let $B = \mathcal{H}_t(R'(D, \mathcal{O}, (a_i)_{i \in J}))$.

By Theorem 8.4.6 of Chapter 8, the restriction V'_t of $\mathcal{H}_t$ to $R'(D, (a_i)_{i \in J})$ is continuous once $R'(D, \mathcal{O}, (a_i)_{i \in J})$ is equipped with the norm $\| \cdot \|_t$. Therefore, V'_t has continuation to a continuous $\mathbb{K}$-vector space homomorphism from $H'(D, \mathcal{O}, (a_i)_{i \in J})$ into the closure $\overline{B}$ of B in A. We will still denote it by V'_t.

Let $s' \in]r', r[$. The annulus $\Gamma(0, s', r)$, admits a partition by a subfamily $\mathcal{S}$ of holes of $\mathcal{O}$. Hence by Theorem 8.1.1 of Chapter 8, $\Gamma(0, s', r)$, contains a $\mathcal{O}$-minorated annulus $\Gamma(b, \rho, v)$. Of course, we may choose v as close to ρ as we want. Particularly, if $|b| > \rho$, we take $v \in]\rho, |b|[$. Next, we take $\lambda \in]\rho, v[$. Clearly b does not lie in D. Let $r_b = \frac{1}{\|(x-b)^{-1}\|_{\mathrm{sp}}}$. By Lemma 8.4.12 of Chapter 8, there exists $\theta \in \mathrm{Mult}(A, \| \cdot \|)$ such that $\Omega^{-1}(\theta)$ is (b, r_b)-approaching. In fact, by definition, we have $r_b \leq \rho$, hence $C(b, r_b)$ is included in $d(b, \rho)$, hence $\Omega^{-1}(\theta)$ is secant with $d(b, \rho)$. On the other hand, there certainly exists $\phi \in \mathrm{Mult}(A, \| \cdot \|)$ such that $\Omega^{-1}(\phi)$ is (b, r)-approaching. Then by Propositions 8.2.5 of Chapter 8, there exist $f, g \in H'(D, \mathcal{O}, (a_i)_{i \in J})$ satisfying:

$\psi(f) = 1$ for all $\psi \in \mathrm{Mult}(H(D, \mathcal{O}), \| \cdot \|_{D,\mathcal{O}})$ such that $\psi(t - b) > v$, and
$\psi(f) = 0$ for all $\psi \in \mathrm{Mult}(H(D, \mathcal{O}), \| \cdot \|_{D,\mathcal{O}})$ such that $\psi(t - b) \leq \lambda$,
$\psi(g) = 1$ for all $\psi \in \mathrm{Mult}(H(D, \mathcal{O}), \| \cdot \|_{D,\mathcal{O}})$ such that $\psi(t - b) < \rho$, and
$\psi(g) = 0$ for all $\psi \in \mathrm{Mult}(H(D, \mathcal{O}), \| \cdot \|_{D,\mathcal{O}})$ such that $\psi(t - b) \geq \lambda$.

We put $F = \mathcal{H}_t(f)$, and $G = V'_t(g)$. Though $\mathcal{H}_t$ is not supposed to be continuous on all $H(D, \mathcal{O}, (a_i)_{i \in J})$, it is continuous on $H'(D, \mathcal{O}, (a_i)_{i \in J})$. Given $\phi \in \mathrm{Mult}(A, \| \cdot \|)$, we will denote by ϕ_t the $\mathbb{K}$-vector space semi-norm defined on $R(D) + H'(D, \mathcal{O}), \| \cdot \|_{D,\mathcal{O}}$.

Suppose first $|b| \leq \rho$. It is seen that for every $\psi \in \mathrm{Mult}_m(A, \| \cdot \|)$, we have $\psi(t) \leq r'$, hence $\psi_t(x) \leq r'$. Consequently, $\psi_t(g) = 1$, and therefore $\psi(G) = 1$, hence G is invertible in A. On the other hand, $\Omega^{-1}(\phi_t)$ is $(0, r)$-approaching, hence $\phi_t(g) = 0$, and therefore $\phi(G) = 0$, a contradiction to the conclusion "G is invertible in A".

Finally, suppose $|b| > \rho$. in the same way we have $d(0, r') \subset \mathbb{K} \setminus d(b, \lambda)$, and therefore, $\psi(F) = 1$ for all $\psi \in \mathrm{Mult}_m(A, \| \cdot \|)$, and $\psi(F) = 0$ for all ψ such that $\Omega^{-1}(\psi_t)$ is secant with $d(b, \lambda)$. So, F is invertible. But $\Omega^{-1}(\theta_t)$ is secant with $d(b, \lambda)$ and therefore, $\theta(F) = 0$, a contradiction to the property "F invertible". $\qquad \square$

Corollary 9.2.4. *If A has bounded normal ratio, then A satisfies Property (s).*

Corollary 9.2.5. *If A is uniform, then A satisfies Property (s).*

Corollary 9.2.6. *If A has bounded normal ratio and has no maximal ideals of infinite codimension, then A satisfies Property (p).*

Corollary 9.2.7. *If A is uniform and has no maximal ideals of infinite codimension, then A satisfies Property (p).*

Corollary 9.2.8. *If A is uniform then its Jacobson radical is null.*

Proposition 9.2.9. *Let $t \in A$ have bounded normal ratio and assume that there exist $\psi', \psi'' \in \mathrm{Mult}(A, \| \cdot \|)$ and r, r', r'' such that $\psi'(t) \leq r' < r < r'' \leq \psi''(t)$ and $\Gamma(0, r', r'') \cap sp(t) = \emptyset$. Then A has divisors of zero.*

Proof. Let $D = sp(t)$, and let $B = \mathcal{H}_t(R(D))$. Let $\mathcal{O} = (d(a_i, r_i^-))_{i \in J}$ be a wide t-spectral partition. The annulus $\Gamma(0, r, r')$ admits a partition by a subfamily $\mathcal{S}$ of holes of $\mathcal{O}$. Hence by Theorem 8.1.1 of Chapter 8, $\Gamma(0, r, r')$ contains a $\mathcal{O}$-minorated annulus $\Gamma(b, \rho, v)$. Moreover, we notice that there exists at least one element ϕ of $\mathrm{Mult}(A, \| \cdot \|)$ such that $\Omega^{-1}(\phi_t)$ is secant with $d(b, \rho)$. Indeed, if $t - b$ is not invertible, this is obvious.

Now, suppose that $t - b$ is invertible. Since $\Gamma(b, \rho, \upsilon)$ admits a partition by holes of the t-spectral partition, according to Corollary 8.4.2 of Chapter 8, we have $\|(t - b)^{-1}\|^{-1} \leq \rho$ and by and Lemma 8.4.12 of Chapter 8, there exists $\phi \in \text{Mult}(A, \| \cdot \|)$ such that $\Omega^{-1}(\phi_t)$ is secant with $d(b, \rho)$. Then by Corollary 8.2.6 of Chapter 8, there exist $f, g \in H'(D, \mathcal{O}, (a_i)_{i \in J})$ satisfying $fg = 0$ and $\psi'_t(f) = \phi_t(g) = 1$. By Theorem 8.4.6 of Chapter 8, we know that $\mathcal{H}_t$ is continuous on $R''(D, \mathcal{O}, (a_i)_{i \in J})$. And by Proposition 8.2.4 of Chapter 8, fg belongs to $H''(\mathcal{O}, (a_i)_{i \in J})$. Consequently, $\mathcal{H}_t(f)\mathcal{H}_t(g) = 0$. But $\psi'(\mathcal{H}_t(f)) = \phi(\mathcal{H}_t(f)) = 1$ hence $\mathcal{H}_t(f)$, $\mathcal{H}_t(g)$ are divisors of zero in A. $\qquad\square$

Theorem 9.2.10. *Let A have no divisors of zero and let $t \in A$ have bounded normal ratio and be such that $sp(t) \neq \emptyset$. Then $\tau(t) = \|t\|_{\text{sp}}$.*

Proof. Suppose $\tau(t) < \|t\|_{\text{sp}}$. Since $sp(t) \neq \emptyset$ there obviously exist $\psi_1 \in \text{Mult}(A, \| \cdot \|)$ such that $\psi_1(t) = \tau(t)$ and by Corollary 2.5.8 of Chapter 2, there exists $\psi_2 \in \text{Mult}(A, \| \cdot \|)$ such that $\psi_2(t) = \|t\|_{\text{sp}}$. Putting $r_1 = \psi_1(t)$ and $r_2 = \psi_2(t)$, we have $\Gamma(0, r_1, r_2) \cap sp(t) = \emptyset$. Hence by Proposition 9.2.8, A has divisors of zero, a contradiction. $\qquad\square$

Corollary 9.2.11. *Let A be uniform and have no divisors of zero, and be such that $sp(t) \neq \emptyset \; \forall t \in A$. Then A satisfies Property (q).*

Corollary 9.2.12. *Let B be a unital commutative uniform ultrametric Banach $\mathbb{K}$-algebra without divisors of zero, such that $sp_B(t) \neq \emptyset \; \forall t \in B$ and let ϕ be a $\mathbb{K}$-algebra homomorphism from A to B. Let $\| \cdot \|_{\text{sp}}^A$ (resp., $\| \cdot \|_{\text{sp}}^B$) be the spectral norm of A (resp., B). Then ϕ satisfies $\|\phi(x)\|_{\text{sp}}^B \leq \|\phi(x)\|_{\text{sp}}^A \; \forall x \in A$.*

Thanks to Proposition 9.2.9, Guennebaud's theorem [32] on the completion of a field which is a normed $\mathbb{K}$-algebra having more than one continuous absolute value, appears as a simple corollary.

Theorem 9.2.13. *Let F be a field extension of $\mathbb{K}$ equipped with a semi-multiplicative norm and admitting two different continuous absolute values. Then the completion of F has divisors of zero.*

Proof. Let A be the completion of F, with respect to its norm. It is a uniform Banach $\mathbb{K}$-algebra such that $\text{Mult}'(A, \| \cdot \|)$ contains at least the expansions ψ_1, ψ_2 of two different continuous absolute values defined on F. Let $t \in F$ be such that $\psi_1(t) < \psi_2(t)$. As an element of F, the spectrum of t is empty, hence it is also empty as an element of A. Putting $r_1 = \psi_1(t)$, $r_2 = \psi_2(t)$, we can apply Proposition 9.2.9, therefore A has divisors of zero. $\qquad\square$

Remarks. In [3, proof of Corollary 4.3.4], the completion of a field for a spectral norm (denoted by $\mathcal{H}(\Sigma)$) is assumed to be a field, which is not true.

However, Theorem 9.2.14 shows that, even in a uniform $\mathbb{K}$-algebra, Properties (s) and (q) are not equivalent.

Theorem 9.2.14. *There exists a uniform unital commutative Banach $\mathbb{K}$-algebra satisfying Property (s) but not Property (q).*

Proof. Let $l \in]0,1[$, let $D = d(0,l^-)$ and let $\mathcal{O}$ be the classic partition of $d(0,1)$ that consists of $d(0,l^-)$ and of the disks $d(a,|a|^-)$ for $a \in d(0,1)$ and $l \leq |a| \leq 1$. For every $h \in R(D)$, we have $\|h\|_{\mathcal{O}} = \sup\{\varphi_{0,r}(h), | \ l \leq r \leq 1\}$. Let A be the completion of $R(D)$ for the norm $\| \cdot \|_{\mathcal{O}}$. By construction, A is complete for its norm $\| \cdot \|_{\mathrm{sp}}$. Consequently, by Corollary 9.2.5, it satisfies Property s). On the other hand, for every $\lambda \in \mathbb{K} \setminus d(0,l)$, $x - \lambda$ is invertible in $R(D)$, hence in A. Consequently, we have $\tau(x) < \|x\|_{\mathrm{sp}}$. So, A does not satisfy Property (q). □

Theorem 9.2.15. *Let A be a Banach $\mathbb{K}$-algebra having a bounded normal ratio, such that $sp(x)$ is infraconnected for all $x \in A$. Then A satisfies Property (q).*

Proof. Since $sp(x)$ is infraconnected for all $x \in A$, we can see that $\mathrm{Ker}(\mathcal{H}_x) = \{0\} \ \forall x \notin \mathbb{K}$. Suppose there exists $t \in A$ such that $\tau(t) < \|t\|$. Let $r' = \tau(t)$, let $r'' = \|t\|$ and let $\mathcal{O} = (d(a_i, r_i^-))_{i \in J}$ be a wide t-spectral partition. By Theorem 8.1.1 of Chapter 8, there exists a $\mathcal{O}$-minorated annulus $\Gamma(a, s', s'')$ included in $\Gamma(0, r', r'')$, and by Corollary 8.4.2 of Chapter 8, and Lemma 8.4.12 of Chapter 8, there exists $\phi \in \mathrm{Mult}(A, \| \cdot \|)$ such that $\Omega^{-1}(\phi_t)$ is secant with $d(a, s')$. Let $r \in]s', s''[$. Suppose first $\Gamma(a, s', s'')$ admits 0 as a center. Since $\Gamma(0, s', s'')$ is a $\mathcal{O}$- minorated annulus, we can find an increasing distances holes sequence $(T'_n)_{n \in \mathbb{N}}$ of $\mathcal{O}$ of center 0 and diameter r such that the weighted sequence $(T_n, 1)_{n \in \mathbb{N}}$ makes a T-sequence. The T-filter of center 0 and diameter r will be denoted by $\mathcal{F}$. Let $D = d(0, r'') \setminus \bigcup_{n \in \mathbb{N}} T_n$. Since D is clearly a $\mathcal{O}$-set, $H(D)$ is a $\mathbb{K}$-subalgebra of $H(sp(t), \mathcal{O}, (a_i)_{i \in J})$. Now, we denote by D' the set $D \cup (\mathbb{K} \setminus d(0, r)$. Hence $H_b(D')$ is a Banach $\mathbb{K}$-subalgebra of $H(D)$. Let $R'(D') = R_b(D') \cap R'((sp(t), \mathcal{O}, (a_i)_{i \in J})$ and $H'(D') = H_b(D') \cap H'((sp(t), \mathcal{O}, (a_i)_{i \in J})$. Particularly, by Theorem 8.4.11 of Chapter 8, the canonical homomorphism $\mathcal{H}_t$ is defined and continuous on $H'(D')$. Then, by Theorem 7.4.15 of Chapter 7, there exists $f \in H(D)$, meromorphic in each hole T_n, admitting a_n in T_n as a unique pole, having order at most 1, such that $|f(c)| = 1 \ \forall c \in d(0, s')$, $f(c) = 0 \ \forall c \in D \setminus d(0, r-)$. Thus, we see that f belongs to $H'(D')$. Let $F = \mathcal{H}_t(f)$. Since $r'' > r$, there exists $\psi \in \text{Mult}(A, \| \cdot \|)$ such that $\psi(t) > r$, hence $\psi(\mathcal{H}_t(f)) = 0$, hence F belongs to a maximal ideal containing $\text{Ker}(\psi)$, and therefore 0 belongs to $sp(F)$. On the other hand for every $c \in sp(x)$, we have $|f(c)| = 1$, hence $sp(F) \cap C(0,1) \neq \emptyset$. Actually, we will show that $sp(F) \subset C(0,1) \cup \{0\}$, i.e., $F - l$ is invertible for all $l \in \mathbb{K}$ such that $0 < |l| < 1$.

Let $l \in \mathbb{K}$ be such that $0 < |l| < 1$ and suppose that $F - l$ is not invertible in A. Let $\zeta \in \mathcal{X}(A)$ be such that $\zeta(F - l) = 0$. Let $\mathcal{M} = \text{Ker}(\zeta)$, and let $\Omega = \frac{A}{\mathcal{M}}$. Let $\phi \in \text{Mult}(A, \| \cdot \|)$ be such that $\text{Ker}(\phi) = \mathcal{M}$. Then ϕ defines an element ϕ_t of $\text{Mult}(R(D))$ which of course is of the form $\varphi_{\mathcal{G}}$, with $\mathcal{G}$ a circular filter on $\mathbb{K}$. Moreover, in $R_b(D')$, the norm $\| \cdot \|_{D'}$ obviously satisfies $\|h\|_{D'} \geq \|h\|_{sp(t),\mathcal{O}}$, hence the restriction of $\varphi_{\mathcal{G}}$ to $R'(D')$ is clearly continuous with respect to the norm $\| \cdot \|_{D'}$ and has extension to $H'(D')$. So, in particular, we have $\phi_t(f - l) = 0$, hence $\lim_{\mathcal{G}} f(x) - l = 0$, and therefore $\mathcal{G}$ is not secant with $d(0, s') \cup (\mathbb{K} \setminus d(0, r))$. In particular, $\mathcal{G} \neq \mathcal{F}$. But since $\mathcal{F}$ is the only T-filter on D', this implies that $\mathcal{G}$ is a Cauchy filter on D', of limit a, hence we have $\zeta(t) = a$, a contradiction to the hypothesis $sp(t) \subset d(0, r')$. Consequently, $F - l$ is invertible in A for all $l \in \mathbb{K}$ such that $0 < |l| < 1$, and therefore $sp(F)$ is not infraconnected.

We have a symmetric proof when 0 is not a center of $\Gamma(a, s', s'')$. Then $\Gamma(a, s', s'')$ is included in a class of $C(0, |a|)$, therefore $s'' < |a|$. Indeed, consider $u = \frac{1}{t-a}$. We can check that $sp(u) \subset C(0, \frac{1}{|a|})$, and since there exists $\psi \in \text{Mult}(A, \| \cdot \|)$ such that $\Omega^{-1}(\psi_t)$ is secant with $d(a, r')$, we have $\psi(t-a) \leq s''$, hence $\psi(u) \geq \frac{1}{s''} > \frac{1}{|a|}$. Consequently, we check that $\tau(u) \leq \frac{1}{|a|}$, and $\|u\| > \frac{1}{s''}$, so we are led to the same situation with u instead of t. $\square$

9.3.   Properties (o) and (q) in uniform Banach $\mathbb{K}$-algebras

We shall show that in a uniform Banach $\mathbb{K}$-algebra, Property (o) implies Property (q). Conversely, we shall see that Property (q) doesn't imply Property (o).

Notation: Throughout this chapter, A is a unital commutative ultrametric Banach $\mathbb{K}$-algebra. Several results come from [26].

Lemma 9.3.1. *Let $x \in A$ be such that $\tau(x) < \|x\|$. There exists $\phi \in \text{Mult}(A, \| \cdot \|)$, a number $v > 0$ and an affinoid set F which belongs to the filter $\mathcal{F} = \Omega^{-1}(\phi_x)$, such that $\text{diam}(\Omega^{-1}(\psi_x)) \geq v$, for all $\psi \in \text{Mult}(A, \| \cdot \|)$ such that $\Omega^{-1}(\psi_x) \in \Phi(F)$.*

Proof.   We assume that Lemma 9.3.1 is not true. Let $r = \tau(x)$ and let $s = \|x\|$, let $m \in]r, s[$, let $\mathcal{O}$ be the strict x-spectral partition, and let

$\phi_0 \in \text{Mult}(A, \| \cdot \|)$ be such that $\Omega^{-1}((\phi_0)_x)$ is secant with $\Gamma(0, m, s)$. Let $\mathcal{F}_0 = \Omega^{-1}((\phi_0)_x)$ and let $s_0 = \text{diam}(\mathcal{F}_0)$.

Suppose we have already constructed a finite sequence $(\phi_n)_{n \in \mathbb{N}}$ for $n = 0, \ldots, q$, satisfying $\text{diam}(\Omega^{-1}((\phi_n)_x)) < \frac{s}{n+1}$ and $\delta(\Omega^{-1}((\phi_n)_x), \Omega^{-1}((\phi_{n+1})_x)) < \frac{2s}{n+1}$. Let $\mathcal{F}_n = \Omega^{-1}((\phi_n)_x)$. Since the Lemma is supposed to be false, there exists no affinoid $F \in \mathcal{F}_n$ such that $\mathcal{F}_n$ is the only circular filter of the form $\Omega^{-1}(\psi_x)$, $\psi \in \text{Mult}(A, \| \cdot \|)$, which be secant with F.

Suppose that we can't find $\phi_{n+1} \in \text{Mult}(A, \| \cdot \|)$ such that the filter $\mathcal{F}_{n+1} = \Omega^{-1}((\phi_{n+1})_x))$ satisfies $\text{diam}(\mathcal{F}_{n+1}) < \frac{s}{n+2}$, and $\delta(\mathcal{F}_{n+1}, \mathcal{F}_n) < \frac{2s}{n+1}$. This means that for every $\psi \in \text{Mult}(A, \| \cdot \|)$ such that $\delta(\Omega^{-1}(\psi_x), \mathcal{F}_n) < \frac{2s}{n+1}$, we have $\text{diam}(\Omega^{-1}(\psi_x)) \geq \frac{1}{n+1}$. Let F be the unique disk of diameter $\frac{2s}{n+1}$ such that $\mathcal{F}_n \in \Phi(F)$. Then for all $\mathcal{G} \in \Phi(F)$ we have $\delta(\mathcal{F}, \mathcal{G}) \leq \frac{2s}{n+1}$, hence $\text{diam}(\Omega^{-1}(\psi_x)) \geq \frac{s}{n+1}$ for all $\psi \in \text{Mult}(A, \| \cdot \|)$ such that $\Omega^{-1}(\psi_x) \in \Phi(F)$, which proves the conclusion of the lemma. Thus, since we have supposed that the conclusion is false, we can find $\phi_{n+1} \in \text{Mult}(A, \| \cdot \|)$ such that the filter $\mathcal{F}_{n+1} = \Omega^{-1}((\phi_{n+1})_x))$ satisfies $\text{diam}(\mathcal{F}_{n+1}) < \frac{s}{n+2}$, and $\delta(\mathcal{F}_{n+1}, \mathcal{F}_n) < \frac{2s}{n+1}$. And this is true for all $n \in \mathbb{N}$, hence the sequence $((\phi_n)_x)_{n \in \mathbb{N}}$ is a Cauchy sequence with respect to the δ-topology. Consequently, its limit in $\text{Mult}(\mathbb{K}[x])$ is of the form φ_α, with $\alpha \in \Gamma(0, r, s)$. On the other hand, since $\text{Mult}(A, \| \cdot \|)$ is compact for the topology of pointwise convergence, the sequence $(\phi_n)_{n \in \mathbb{N}}$ admits a a cluster θ with respect to the topology of pointwise convergence. Consequently, θ_x is a cluster of the sequence $((\psi_n)_x)_{n \in \mathbb{N}}$. Next, since $\text{Mult}(\mathbb{K}[x])$ is sequentially compact, we can extract from the sequence $(\psi_n)_x$ a subsequence converging to θ_x with respect to the topology of pointwise convergence. Thus, without loss of generality, we can assume that the sequence $(\psi_n)_x$ converges to θ_x with respect to the topology of pointwise convergence in $\text{Mult}(\mathbb{K}[x])$. But since the sequence $(\psi_n)_x$ converges to φ_α with respect to the δ-topology, by Theorem 6.3.3 of Chapter 6, so much the more does it converges to φ_α for the topology of pointwise convergence, hence $\varphi_\alpha = \theta_x$. Thus, θ_α is punctual, a contradiction to the fact that $\alpha \notin sp(x)$. Consequently, our hypothesis "the conclusion is false" is wrong. $\qquad\square$

Theorem 9.3.2. *If A does not satisfy Property (q), then GM_A is not injective.*

Proof. Suppose that A does not satisfy Property (q) and let $x \in A$ be such that $\tau(x) < \|x\|$. Consider now the strict x-spectral partition $\mathcal{O}$ of $d(0, s)(x)$. Let $r = \tau(x)$ and let $s = \|x\|$. By Lemma 9.3.1, there exists

$\phi \in \text{Mult}(A, \| \cdot \|)$, a number $v > 0$ and an affinoid set F which belongs to the filter $\mathcal{F} = \Omega^{-1}(\phi_x)$, such that $\text{diam}(\Omega^{-1}(\psi_x)) \geq v$, for all $\psi \in \text{Mult}(A, \| \cdot \|)$ such that $\Omega^{-1}(\psi_x) \in \Phi(F)$. Whether or not $\mathcal{F}$ has a center, we are going to construct an element $G \in A$ such that, given $\psi \in \text{Mult}(A, \| \cdot \|)$, either $\phi(G) = 0$, or $\text{diam}(\Omega^{-1}(\psi_x)) \geq v$.

Suppose first that $\mathcal{F}$ has no center. By Lemma 5.3.12 of Chapter 5, there exists disks $d(a, t) \in \mathcal{F}$, $d(a, l) \in \mathcal{F}$ included in F, with $l < t$. By Proposition 8.2.5 of Chapter 8, there exists $g \in H(sp(x), \mathcal{O})$ such that $\varphi_{\mathcal{G}}(g) = 1 \ \forall \mathcal{G} \in \Phi(d(a, l))$ and $\varphi_{\mathcal{G}}(g) = 0 \ \forall \mathcal{G} \in \Phi(d(0, s) \setminus d(a, t^-))$.

Suppose now that $\mathcal{F}$ has center a and diameter m. Without loss of generality we can assume that F is of the form $d(a, m + \epsilon) \setminus \bigcup_{j=1}^{q} d(a_j, (m - \epsilon)^-)$, with $m + \epsilon < s$ when $m < s$. For each $j = 1, \ldots, q$, by Proposition 8.2.5 of Chapter 8, there exists $g_j \in H(sp(x), \mathcal{O})$ such that $\varphi_{\mathcal{G}}(g_j) = 1 \ \forall \mathcal{G}$ secant with $d(0, s) \setminus d(a, m)^-$ and $\varphi_{\mathcal{G}}(g_j) = 0 \ \forall \mathcal{G}$ secant with $d(a_j, m - \epsilon)$. Moreover, if $m = s$ we put $g_0 = 1$, and if $m < s$, by Proposition 8.2.4 of Chapter 8, there exists $g_0 \in H(sp(x), \mathcal{O})$ such that $\varphi_{\mathcal{G}}(g) = 1 \ \forall \mathcal{G}$ secant with $d(a, m)$ and $\varphi_{\mathcal{G}}(g) = 0 \ \forall \mathcal{G}$ secant with $d(0, s) \setminus d(a, (m + \epsilon)^-)$. Then we put $g = \prod_{j=0}^{q} g_j$.

In both cases, we now put $G = \mathcal{H}_x(g)$ and we can check that given $\psi \in \text{Mult}(A, \| \cdot \|)$, either $\phi(G) = 0$, or $\text{diam}(\Omega^{-1}(\psi_x)) \geq v$. Let $\xi \in \mathbb{K}$ satisfy $\|\xi G\| < v$ and let $y = x + \xi G$. Then $\|y - x\| < v$. Consider $\psi \in \text{Mult}(A, \| \cdot \|)$, let $\varphi_{\mathcal{R}} = x^*(\psi)$, $\varphi_{\mathcal{T}} = y^*(\psi)$. If $\psi(G) = 0$, then $\psi(y - x) = 0$, hence by Theorem 9.1.15 we have $x^*(\psi) = y^*(\psi)$. And now, if $\psi(G) \neq 0$, then we can see that $\psi(y - x) < v \leq \text{diam}(\mathcal{R}) \leq \text{diam}(sup(\mathcal{R}, \mathcal{T}))$, therefore $x^*(\psi) = y^*(\psi)$ again. Consequently, $x^*(\psi) = y^*(\psi) \ \forall \psi \in Mult(A, \| \cdot \|)$. However, $x \neq y$ because $\phi(G) \neq 0$. $\qquad \square$

Corollary 9.3.3. *Let A be uniform, having at least one maximal ideal of codimension 1 and such that the intersection of all maximal ideals of codimension 1 is equal to $\{0\}$. Then A satisfies Property (q).*

Proof. In Theorem 9.1.14 we saw that GM_A is injective, hence by Theorem 9.3.2, A satisfies Property (q). $\qquad \square$

Theorem 9.3.4. *Let A be uniform and satisfy Property (o) and $\text{Max}_1(A) \neq \emptyset$. Then A satisfies Property (p).*

Proof. Suppose A does not satisfy Property (p), hence it does not satisfy Property (q). Let $t \in A$ be such that $\tau(t) < \|t\|$. Since $\text{Max}_1(A) \neq \emptyset$ we have $sp(t) \neq \emptyset$. Taking $\alpha \in sp(t)$, we notice that $\tau(t - \alpha) = \tau(t)$, and

$\|t - \alpha\| = \|t\|$. So, without loss of generality, we can assume $0 \in sp(t)$. Let $s = \|t\|$, let $r = \tau(t)$, and let $\mathcal{O}$ be the strict t-spectral partition. There exists a $\mathcal{O}$-minorated annulus $\Gamma(b, r', r'')$ included in $\Gamma(0, r, s)$.

Suppose 0 is (resp., is not) a center of $\Gamma(b, r', r'')$. We can find in $\Gamma(0, r', r'')$ (resp., in $\Gamma(b, r', r'')$) an increasing (resp., a decreasing) idempotent T-sequence $(T_n)_{n \in \mathbb{N}}$ of center 0 (resp., of center b) and diameter $\rho \in]r', r''[$ whose holes are holes of $\mathcal{O}$. Consider the $\mathcal{O}$-set D admitting each T_n as a hole, and no other holes. We notice that D admits a unique T-filter $\mathcal{F}$ run by the T-sequence $(T_n)_{n \in \mathbb{N}}$. By Theorem 7.4.13 of Chapter 7, there exists $f \in H(D)$ such that $\varphi_{0,\rho}(f) = 0$, $|f(a)| = 1 \forall a \in d(0, r')$ (resp., $f(a) = 1 \; \forall a \in d(0, s) \setminus d(0, r'')$), $f(a) = 0 \; \forall a \in d(0, s) \setminus d(0, \rho^-)$, (resp., $f(a) = 0 \; \forall a \in d(b, \rho))$, and such that f does not vanish along any large circular filter secant with $d(0, r)$ (resp., with $d(0, s) \setminus d(b, \rho)$). Then, since the T_n are the only holes of D, D admits no T-filter different from $\mathcal{F}$. Therefore, for every $\lambda \in \mathbb{K}^*$, $f - \lambda$ is quasi-invertible.

On the other hand, since D is a $\mathcal{O}$-set, $H(D)$ is a $\mathbb{K}$-subalgebra of $H(sp(t), \mathcal{O})$. And since A is uniform, $\mathcal{H}_t$ is defined and continuous on $H(D)$. For every $\psi \in \text{Mult}(A, \| \cdot \|)$ we put $\psi_t = \mathcal{H}_t \circ \psi$. Since $s > \rho$ we can find $\psi \in \text{Mult}(A, \| \cdot \|)$ such that ψ_t has a filter secant with $d(0, s) \setminus d(0, \rho^-)$ (resp., since $\Gamma(b, r', r'')$ is not included in any hole of $\mathcal{O}$, we can find $\psi \in \text{Mult}(A, \| \cdot \|)$ such that ψ_t has a filter secant with $d(0, r')$). Consequently, we have $\psi(\mathcal{H}_t(f)) = 0$. Let $F = \mathcal{H}_t(f)$: we see that F is not invertible in A.

Now, let $\chi \in \mathcal{X}(A, \mathbb{K})$. Since A is uniform, $\chi \circ \mathcal{H}_t$ defines a character $\chi_t \in \mathcal{X}(H(D), \mathbb{K})$ which, by hypothesis, satisfies $\chi_t(x) \leq r$, hence $sa(\mathcal{H}_t(f))$ is included in $C(0, 1)$. Let $\alpha \in \mathbb{K}$ be such that $< |\alpha| < 1$. Since D has no T-filter except $\mathcal{F}$, we notice that $f - \alpha$ is not vanishing along any T-filter of D and therefore by Corollary 7.7.3 of Chapter 7, $f - \alpha$ is quasi-invertible, hence it is of the form $P(x)h$, with P a polynomial and h an invertible element of $H(D)$. Let $g = Ph$ and let $G = \mathcal{H}_t(g)$. Suppose that G is not invertible in A. Let $\theta \in \mathcal{X}(A, \mathbb{K})$ be such that $\theta(G) = 0$. Then $\theta(P(t)) = 0$, hence $\theta(t)$ is one of the zeros of P, and therefore lies in $sp(t)$. Consequently, we have $\theta(t) \leq r$, hence $|\theta_t(f)| = 1$, hence $|\theta(F)| = 1$ and therefore we check that $|\theta(G)| = 1$ a contradiction. Thus, we see that G is invertible in A. Then, $sa(G) \subset C(0, 1)$, but α belongs to $sp(G)$ because F is not invertible in A. And since G is invertible in A, $sa(\frac{1}{G})$ is also included in $C(0, 1)$, but $\frac{1}{\alpha}$ belongs to $sp(\frac{1}{G})$. Thus, we have $\left\|\frac{1}{G}\right\|_{sa} = 1$, but $\tau(\frac{1}{G}) \geq |\frac{1}{\alpha}| > 1$, a contradiction to Property (o). $\square$

On the other hand, we will prove that there exist uniform Banach ultrametric unital commutative $\mathbb{K}$-algebras satisfying Property (q), but not Property (o).

According to Theorems 9.1.4 and 9.3.4 we have Corollary 9.3.5.

Corollary 9.3.5. *Let B be a uniform normed unital commutative $\mathbb{K}$-algebra satisfying Property (o) and let ϕ be a $\mathbb{K}$-algebra homomorphism from A to B. Let $\| \cdot \|_{\mathrm{sp}}^B$ be the norm of B and let $\| \cdot \|_{\mathrm{sp}}^A$ be the spectral semi-norm of A. Then ϕ is continuous and satisfies $\|\phi(x)\|_{\mathrm{sp}}^B \leq \|x\|_{\mathrm{sp}}^A \ \forall x \in A$.*

Definitions and notations: Let E be a unital commutative ultrametric normed $\mathbb{K}$-algebra. Let $n \in \mathbb{N}$ and let $\| \cdot \|_0$ be the norm of E. The $\mathbb{E}$-algebra of polynomials in n variables $E[X_1, \ldots, X_n]$ is equipped with the *Gauss norm* $||| \cdot |||$ defined as $||| \sum_{i_1,\ldots,i_n} a_{i_1,\ldots,i_n} X_1^{i_1}, \ldots, X_n^{i_n} ||| = \sup_{i_1,\ldots,i_n} \|a_{i_1,\ldots,i_n}\|_0$. By Proposition 2.5.16 of Chapter 2, the norm is a $\mathbb{K}$-algebra norm and if the norm of E is multiplicative, this norm $||| \cdot |||$ is known to be a multiplicative norm of $\mathbb{K}$-algebra.

Further, by Proposition 2.5.16 of Chapter 2, if E is complete, the completion of $E[X_1, \ldots, X_n]$ with respect to this norm, denoted by $E\{X_1, \ldots, X_n\}$, consists of the set of power series in n variables $\sum_{i_1,\ldots,i_n} a_{i_1,\ldots,i_n} X_1^{i_1} \cdots X_n^{i_n}$ such that $\lim_{i_1+\cdots+i_n \to \infty} a_{i_1,\ldots,i_n} = 0$. The elements of such an algebra are called *the restricted power series in n variables.*

Henceforth, in this chapter we denote by M the disk $d(0, 1^-)$ and by A the $\mathbb{K}$-algebra $H(M)\{Y\}$. We denote by x the identical mapping on M, we fix $r \in]0, 1[$, we put $t = 1 - xY$ and we denote by S the multiplicative subset generated in A by the $t - a$, $a \in d(0, r^-)$.

We denote by B the $\mathbb{K}$-algebra $S^{-1}A$. Finally, we denote by $\mathcal{T}(r)$ the set of $\psi \in \mathrm{Mult}(A, \| \cdot \|)$ such that $\psi(X) \geq r$.

Lemma 9.3.6. *Let $P \in E[Y]$ and let $Q(Y) = P(1 - Y)$. Then $|||P||| = |||Q|||$.*

Proof.   Let $P(Y) = \sum_{j=0}^n a_j Y^j$ and let $Q(Y) = \sum_{j=0}^n b_j Y^j$. Then each b_j is of the form $(-1)^j \sum_{k=j}^n \binom{i}{j} a_k$. Hence $|b_j| \leq |||P|||$ and thereby $|||Q||| \leq |||P|||$. But on the other hand $P(Y) = Q(1 - Y)$, hence $|||P||| \leq |||Q|||$, consequently the equality holds. $\square$

Lemma 9.3.7. *Let $\psi \in \mathcal{T}(r)$ and let $h = \frac{f}{g} \in B$, with $f \in A$, $g \in S$. Then $0 \notin \psi(S)$. The mapping ψ^* defined in B by $\psi^*\left(\frac{f}{g}\right) = \frac{\psi(f)}{\psi(g)}$ belongs* to $\text{Mult}(B)$. *Further, the mapping ϕ from B to $[0, +\infty]$ defined as $\phi(h) = \sup_{\psi \in \mathcal{T}(r)} \psi^*(h)$ belongs to $SM(B)$.*

Proof. First, we will show that $\phi(h) < +\infty$. We can write g in the form $\prod_{j=1}^{n}(t - a_j)^{q_j}$, $a_j \in d(0, r^-)$, $q_j \in \mathbb{N}^*$. We set $q = \sum_{j=1}^{n} q_j$. Since $\psi(t) \geq r$, clearly $\psi(t - a_j) = \psi(t)$, and hence we have $\psi(g) \geq r^q$. Consequently, $\psi^*\left(\frac{f}{g}\right) \leq \frac{\psi(f)}{r^q} \leq \frac{\|f\|}{r^q}$, and therefore $\phi(h) \leq \frac{\|f\|}{r^q}$. Thus, $\phi \in SM(B)$. On the other hand, since the norm $\| \, . \, \|_M$ of $H(M)$ is an absolute value, we can find in $\mathcal{T}(r)$ a multiplicative norm which obviously lies in $\mathcal{T}(r)$. Consequently ϕ is a norm. $\qquad\square$

Notations: Henceforth, $\| \, . \, \|_r$ will denote the norm defined on B in Lemma 9.3.7 and $\widehat{B}$ will denote the completion of B with respect to this norm. Then by definition, $\widehat{B}$ is a unital commutative uniform ultrametric Banach $\mathbb{K}$-algebra.

Given a normed $\mathbb{K}$-algebra E, we will denote by $E\{\{Y\}\}_r$ the set of Laurent series with coefficients in E such that $\lim_{n \to +\infty} a_n = 0$, and $\lim_{n \to -\infty} \|a_n\| r^n = 0$.

In $\mathbb{K}[Y]$ (resp., in $\mathbb{K}\{Y\}$) we will denote by $\mathbb{K}[Y]_0$ (resp., $\mathbb{K}\{Y\}_0$) the ideal of polynomials (resp., of series) f such that $f(0) = 0$. And we put $W = H(M) + \mathbb{K}\{Y\}_0$.

Lemmas 9.3.8 and 9.3.9 are immediate.

Lemma 9.3.8. *W is the direct sum of the $\mathbb{K}$-vector spaces $H(M)$ and $\mathbb{K}\{Y\}_0$, and is equipped with the product norm $\| \, . \, \|$ defined as $\|g + h\| = \max(\|g\|_M, \|h\|)$, with $g \in H(M)$ and $h \in \mathbb{K}\{Y\}_0$.*

Lemma 9.3.9. *$B \subset H(M)\{t\}\{\frac{1}{t}\}_r \subset \widehat{B}$.*

Lemma 9.3.10. *For every $a \in C(0,1)$, $(a - xY)A$ is a maximal ideal of infinite codimension of A. Moreover, we have $sa(xY) = M$, $sp(xY) = U$.*

Proof. Let $N = sa(xY)$ and let $D = sp(xY)$. Since $\|\|xY\|\| = 1$, $sp(xY)$ is obviously included in U. On the other hand D is clearly included in M because given $a \in M$, the homomorphism $\chi_{a,1}$ from A to $\mathbb{K}$ defined as $\chi_{a,1}(\sum_{j=0}^{\infty} f_j Y^j) = \sum_{j=0}^{\infty} f_j(a)$ satisfies $\chi_{a,1}(xY) = a$. Consequently, we have

$$M \subset N \subset D \subset U. \tag{9.3.1}$$

Let $a \in U \setminus M$ and consider the ideal I generated by $a - xY$. Clearly $I \cap H(M) = \{0\}$. Consequently the canonical surjection θ from A onto the algebra $A' = \frac{A}{I}$ induces an injection from $H(M)$ to A'.

We will show that $\frac{A}{I}$ is just the completion of the field of fractions F of $H(M)$. Since the norm of A is multiplicative, I is obviously closed in A, hence A' is equipped with the quotient norm, which makes it a Banach $\mathbb{K}$-algebra. Let $a \in M$. In A', we have $\theta(x) = \left(\theta\left(\frac{Y}{a}\right)^{-1}\right)$, and $\left\|\theta\left(\frac{Y}{a}\right)\right\| \le \left\|\left\|\frac{Y}{a}\right\|\right\| = 1$, and similarly $\|\theta(x)\| \le 1$, hence $\|\theta(x)\| = \left\|\theta\left(\frac{Y}{a}\right)\right\| = 1$. Consequently, the series $\sum_{j=0}^{\infty} \frac{b^n}{(\theta(x))^{n+1}}$ converges in $\mathbb{K}$ for every $a \in M$, which proves that $\theta(b - x)$ is invertible in A' and admits for inverse $\sum_{j=0}^{\infty} \frac{b^n}{(\theta(x))^{n+1}}$. Thus, A' contains a field isomorphic to F and we will confound it with F, due to the fact that the restriction of θ to $H(M)$ may be considered as the identity.

In the field F, it is clearly seen that the relation $\|(\theta(x - b))^{-1}\| = \|(\theta(x - b))\|^{-1}$ holds for all $b \notin M$ because $x - b$ is invertible in $H(M)$, and that $\|\| \cdot \|\|$ is an absolute value. Consider now $b \in M$. Indeed, $(\theta(b - x)^{-1}) = \sum_{j=0}^{\infty} \frac{b^n}{(\theta(x))^{n+1}} = \sum_{j=0}^{\infty} b^n \left(\theta\left(\frac{Y}{a}\right)\right)^{n+1}$, which proves that $\|(\theta(b - x)^{-1})\| \le 1$. So, actually, $\|(\theta(b - x)^{-1})\| = 1$, and of course $\|\theta(b - x)\| = 1$ because $\|\|b - x\|\| = 1$. Thus, the norm of A' induces on F the continuation of the absolute value of $H(M)$. Since $\theta(H(M)[Y]) \subset F$, then F is dense in A', hence A' is the completion of F with respect to its absolute value. Therefore, I is a maximal ideal of infinite codimension. Thus, given $a \in U \setminus M$, then a belongs to $D \setminus N$, hence $U \setminus M = D \setminus N$, and consequently, by (1), $L = M$, and $D = U$. $\qquad\square$

Theorem 9.3.11 ([24]). *$\widehat{B}$ is a unital commutative ultrametric uniform Banach $\mathbb{K}$-algebra without divisors of zero, satisfying Property (q) but not Property (o). Moreover, it is equal to the set of Laurent series $\sum_{-\infty}^{+\infty} a_n Y^n$, with $a_n \in W$, $\lim_{n \to +\infty} a_n = 0$, $\lim_{n \to -\infty} \|a_n\| r^n = 0$, and we have $\|\sum_{-\infty}^{+\infty} a_n Y^n\| = \max(\sup_{n \ge 0} \|a_n\|, \sup_{n < 0} \|a_n\| r^n)$.*

Proof. First, we will show that $\widehat{B}$ does not satisfy Property (o). Let $t = 1 - xY$. By Lemma 9.3.10, we have $sp_A(xY) = U$, $sa_A(xY) = M$. Consequently, we can see that $sa_A(t) = d(1, 1^-)$. Let $\chi \in \mathcal{X}(A, \mathbb{K})$. For all $g \in S$, we notice that $\chi(g) \ne 0$. Thus, χ has continuation to a $\mathbb{K}$-algebra homomorphism $\tilde{\chi}$ from B to $\mathbb{K}$ defined as $\tilde{\chi}\left(\frac{f}{g}\right) = \frac{\chi(f)}{\chi(g)}$ $\forall f \in A$, $g \in S$. Since $|\chi|$ clearly lies in $\mathcal{T}(r)$, $\tilde{\chi}$ is continuous with respect to the norm $\| \cdot \|$ of B, therefore $\widehat{\chi}$ has continuation to a $\mathbb{K}$-algebra homomorphism $\widehat{\chi}$ from $\widehat{B}$ to $\mathbb{K}$. Consequently, $sa_A(xY) \subset sa_{\widehat{B}}(xY)$. But of course, the inverse inequality is true because $A \subset \widehat{B}$, so we have $sa_{\widehat{B}}(xY) = M$, thereby

$$\left\|\frac{1}{t}\right\|_{sa}^{\widehat{B}} = 1. \tag{9.3.2}$$

Now, let $\chi \in \mathcal{X}(A, \mathbb{K})$ satisfy

$$|\chi(t)| \geq r \tag{9.3.3}$$

By (9.3.3) we have $\chi(g) \neq 0 \ \forall g \in S$, therefore it has continuation to a $\mathbb{K}$-algebra homomorphism from $\widehat{B}$ to $\chi(A) = \mathbb{K}$. Thus, we have proven that $sp_{\widehat{B}}(t) \supset \Lambda 0, r, 1)$. But since B is a $\mathbb{K}$-subalgebra of $\widehat{B}$, we have the inverse inclusion, and consequently, the equality holds. Hence $sp_{\widehat{B}}\left(\frac{1}{t}\right) = \Lambda(0, 1, \frac{1}{r})$ thereby $\sup\left\{0, \ |\lambda| \ \lambda \in sp_{\widehat{B}}\left(\frac{1}{t}\right)\right\} = \frac{1}{r}$. Thus, by (9.3.2) we can see that $\widehat{B}$ does not satisfy Property (o).

Now, we will show that $\widehat{B}$ has no divisors of zero, and first we will prove that B is the $\mathbb{K}$-vector space of the power series f of the form $\sum_{n=0}^{\infty} a_n t^n$, with $a_n \in W$, $\lim_{n \to +\infty} a_n = 0$, $\|f\| = \sup_{n \geq 0} \|a_n\|$.

Let $h \in H(M)$ and let $j \in \mathbb{N}^*$. Considering the power series of h at 0, by Theorem 8.1.9 of Chapter 8, we can write $h = \sum_{j=0}^{n-1} a_j x^j + x^n g$, with $g \in H(M)$, and then we have $\|h\|_M = (|a_0|, \ldots |a_{n-1}|, \|g\|_M)$. Consequently,

$$h(x)Y^k = \sum_{j=0}^{k-1} a_j x^j Y^k + (xY)^k g = \sum_{j=0}^{k-1} (a_j)(xY)^j (Y^{k-j}) + (xY)^k g.$$

So, we can write

$$hY^k = \sum_{j=0}^{k-1} a_j (1-t)^j Y^{k-j} + (1-t)^k y. \tag{9.3.4}$$

Now consider the polynomial $P(Z) = \sum_{j=0}^{k-1} a_j Y^{k-j} Z^j \in \mathbb{K}[Y]_0[Z]$ and let $Q(Z) = P(1-Z)$. By Lemma 9.3.6, we have $|||Q||| = |||P||| \sup_{0 \leq j \leq k-1} |a_j|$. Putting $Q(Z) = \sum_{j=0}^{k-1} b_j(Y) Z^j$, we get

$$|||Q(Z)||| = \sup_{0 \leq j \leq k-1} \|b_j(Y)\| = \sup_{0 \leq j \leq k-1} |a_j|. \tag{9.3.5}$$

Consequently, we obtain

$$\left\| \sum_{j=0}^{k-1} a_j (1-t)^j Y^{k-j} \right\| = \left\| \sum_{j=0}^{k-1} b_j(Y) Z^j \right\| = |||Q||| = |||P||| = \sup_{0 \leq j \leq k-1} |a_j|.$$

And finally,

$$\|Q(Y)\| = \max(|a_0|, \ldots, |a_{k-1}|). \tag{9.3.6}$$

On the other hand, we have $|||g(1 - Z)^j||| = |||gY^j||| = ||g||$. But since $||1 - t|| = 1$, in A we obtain

$$||g(1 - t)^j|| = ||g||. \qquad (9.3.7)$$

By (9.3.4) we deduce that $hY^j = Q(t) + (1 - t)^j g$, $g \in H(M)$, $Q(Z) \in \mathbb{K}[Y]_0[Z]$. And by (9.3.5)–(9.3.7) we obtain

$$||h|| = \max(||Q(t)||, ||(1 - t)^j g||). \qquad (9.3.8)$$

Now, consider $f = \sum_{j=0}^{+\infty} f_j Y^j \in A$. For each $j \in \mathbb{N}$, $f_j Y^j$ is of the form $Q_j(t) + g_j S_j(t)$, with $Q_j(Z) \in \mathbb{K}[Y]_0[Z]$, $g_j \in H(M)$, $S_j(Z) \in \mathbb{K}[Z]$, satisfying further

$$|||Q_j||| \leq ||f_j||. \qquad (9.3.9)$$

and

$$|||g_j||| \leq ||f_j||. \qquad (9.3.10)$$

Consequently, we have

$$|||Q_j(Z) + g_j S_j||| \leq ||f_j||. \qquad (9.3.11)$$

By (9.3.9), the series $\sum_{j=0}^{\infty} Q_j(Z)$ is clearly converging in $\mathbb{K}\{Y\}_0\{Z\}$ to a limit $\sum_{j=0}^{\infty} \lambda_j(Y)Z^j$. In the same way, by (9.3.10) the series $\sum_{j=0}^{\infty} g_j S_j(t)$ converges in $H(M)\{Z\}$ to a limit $\sum_{j=0}^{\infty} \mu_j Z^j$. Moreover, by (9.3.9) and (9.3.10), we can see that $\sup_{j \in \mathbb{N}} ||\lambda_j|| \leq ||f||$ and $\sup_{j \in \mathbb{N}} ||\mu_j|| \leq ||f||$. On the other hand we have $||f|| \leq \max(\sup_{j \geq 0} ||\lambda_j||, \sup_{j \geq 0} ||\mu_j||)$, thereby $||f|| = \max(\sup_{j \geq 0} ||\lambda_j||, \sup_{j \geq 0} ||\mu_j||)$. Putting $a_j = \lambda_j + \mu_j$, we have $||a_j|| = \max(||\lambda_j||, ||\mu_j||)$. This shows A to be the set of series $G = \sum_0^{+\infty} a_n t^n$ with coefficients $a_j \in \Lambda$, satisfying $\lim_{n \to +\infty} a_n = 0$, $||G|| = \sup_{n \geq 0} ||a_n||$.

Now, we will show that $\widehat{B}$ is the set $\mathcal{T}$ of series $G = \sum_{-\infty}^{+\infty} a_n t^n$, with coefficients $a_j \in W$, satisfying $\lim_{n \to +\infty} a_n = 0$, $\lim_{n \to -\infty} ||a_n|| r^n = 0$, and $||G|| = \sup_{n \geq 0} ||a_n||$. Since $\left|\left|\frac{1}{t}\right|\right| = \frac{1}{r}$, it is seen that $\mathcal{T} \subset \widehat{B}$. Given $a \in d(0, r^-)$, we notice that $\frac{1}{1-a} = \sum_{n=0}^{+\infty} \frac{a^n}{t^{n+1}}$ belongs to $\mathcal{T}$. So, by Lemma 9.3.9, B is included in $\mathcal{T}$, and therefore $\mathcal{T}$ is dense in $\widehat{B}$.

Since the norm of $H(M)$ is an absolute value, on $H(M)[Z]$ we can consider the two absolute values ψ_0, ψ_1 defined by

$$\psi_0\left(\sum_{j=0}^{q} a_j Z^j\right) = \sup_{0 \le j \le q} \|a_j\| \psi_1\left(\sum_{j=0}^{q} a_j Z^j\right) = \sup_{0 \le j \le q} \|a_j\| r^j$$

$(a_j \in H(M), 0 \le j \le q)$. These two absolute values have continuation to $H(M)[Z, \frac{1}{Z}]$, so this apply to $H(M)[t, \frac{1}{t}]$. But $Y = \frac{1-t}{x}$ does belong to $H(M)[t, \frac{1}{t}]$. Therefore, we can consider the restrictions ψ_0', ψ_1' of ψ_0, ψ_1 to $H(M)[Y]$. Thus, by hypothesis we have $\psi_i'(Y) = \frac{\psi_i'(1-t)}{\psi' i(x)} = \psi'(1-t) = 1$ $(i = 0, 1)$. Then given $\sum_{k=0}^{q} b_k Y^k$ $H(M)[Y]$, we have $\psi_i'(\sum_{k=0}^{q} b_k Y^k) \le \sup_{0 \le k \le q} \|b_k\|$ $(i = 0, 1)$. Consequently, $\psi'(f) \le \|f\|$ $\forall f \in H(M)[Y]$, hence each ψ_i' has continuation to an absolute value ψ_i'' defined on A, and actually belongs to $\mathrm{Mult}(A, \| \cdot \|)$ $(i = 0, 1)$. On the other hand, we notice that each ψ_i'' belongs to $\mathcal{T}(r)$, and therefore has extension ψ_i''' to B and $\widehat{B}$ $(i = 0, 1)$. In particular we have

$$\psi_i'''(f) \le \|G\| \quad \forall G \in \widehat{B} \ (i = 0, 1). \tag{9.3.12}$$

Now, let $l = \sum_{-\infty}^{+\infty} a_n t^n \in \mathcal{T}$. By (9.3.12), we can check that $\|G\| \ge \psi_0(G) = \sup_{n \ge 0} \|a_n\|$ and $\|G\| \ge \psi_1(G) = \sup_{n > 0} \|a_n\| r^n$. Thus, we have $\|G\| = \max(\sup_{n \ge 0} \|a_n\|, \sup_{n < 0} \|a_n\| r^n)$. Consequently, $\mathcal{T}$ is complete for the norm $\| \cdot \|$ and therefore is equal to $\widehat{B}$.

Let F be the field of fractions of $H(M)$ and let f^0 be its completion with respect to the absolute value $\| \cdot \|_M$ of $H(M)$ extended to F. Then, F^0 is equipped with an absolute value that we still denote by $\| \cdot \|_M$. We can consider $F^0 + \mathbb{K}\{Y\}_0$ as a $\mathbb{K}$-vector space equipped with the norm $\| \cdot \|$ defined as it follows: given $g \in \widehat{F}$, $h \in \mathbb{K}\{Y\}_0$, we set $\|g + h\| = \max(\|g\|_M, \|h\|)$. Let J be the $\mathbb{K}$-vector space consisting of all series $\sum_{-\infty}^{+\infty} a_n t^n$ $a_n \in F^0 + \mathbb{K}\{Y\}_0)\{\{t\}\}_r$. It is equipped with the norm defined by

$$\left\|\sum_{-\infty}^{+\infty} a_n t^n\right\| = \max(\sup_{k \ge 0} \|a_k\|, \sup_{k < 0} \|a_k\| r^k).$$

Thus, $\widehat{B}$ is a $\mathbb{K}$-subvector space of J which actually is included in $F^0\{\{t\}\}_r$ because so is $\mathbb{K}\{Y\}$, due to the fact that $Y = \frac{1-t}{x}$ and $\left\|\frac{1}{x}\right\| = 1$. Since both the multiplication of $\widehat{B}$ and its norm are induced by those of $F^0\{\{t\}\}_r$, $\widehat{B}$ is a $\mathbb{K}$-subalgebra of $F^0\{\{t\}\}_r$, and therefore $\widehat{B}$ is an integral domain and hence it has no divisor of zero. But the norm of $\widehat{B}$ is semi-multiplicative, and therefore $\widehat{B}$ is uniform, hence by Corollary 9.2.11, $\widehat{B}$ satisfies Property (q). $\qquad\square$

9.4. Properties (o) and (q) and strongly valued fields

In Section 9.3, we saw that in a commutative uniform Banach $\mathbb{K}$-algebra, Property (o) implies Property (q). Here we examine whether this remains true when the algebra is not supposed to be uniform. Actually, this is depending on the ground field [24]: if it is strongly valued, then the property is always true. If it is weakly valued, counter-examples show that the property sometimes doesn't hold.

Notation: Throughout the chapter, A is a commutative ultrametric Banach $\mathbb{K}$-algebra with unity.

By Theorem 7.5.14 of Chapter 7, we have Theorem 9.4.1.

Theorem 9.4.1. *If $\mathbb{K}$ is weakly valued, there exist unital commutative ultrametric Banach $\mathbb{K}$-algebra, without divisors of zero, all maximal ideals of which have codimension 1, satisfying property (o) but neither Property (q) nor Property (s).*

Theorem 9.4.2. *Let $\mathbb{K}$ be strongly valued. Then A satisfies Property (s).*

Proof. Suppose Property (s) is not satisfied. So, there exists $t \in A$ such that $\|t\|_{\mathrm{sp}} > \|t\|_{\mathrm{sm}}$, and therefore, there exists $\psi_0 \in \mathrm{Mult}(A, \| \cdot \|)$ such that $\psi_0(t) > \|t\|_{\mathrm{sm}}$. Let $r = \|t\|_{\mathrm{sm}}$ and $u = \psi_0(t)$. Thus, $sp(t)$ is included in $d(0, r)$. Let $\mathcal{O}$ be a t-normal partition. By Corollary 8.1.5 of Chapter 8, there exists an increasing idempotent T-sequence $(T_n)_{n \in \mathbb{N}}$ of center 0 and diameter u with all T_n included in the annulus $\Gamma(0, r, u)$. Therefore, by Proposition 8.2.5 of Chapter 8, there exists $f \in H(D, \mathcal{O})$ such that $\psi(f) = 1 \; \forall \psi \in \mathrm{Mult}(H(D, \mathcal{O}), \| \cdot \|_{D,\mathcal{O}})$ such that $\psi(t) \leq r$, and $\psi(f) = 0 \; \forall \psi \in \mathrm{Mult}(H(D, \mathcal{O}), \| \cdot \|_{D,\mathcal{O}})$ such that $\psi(t) \geq u$. Let $\mathcal{M}$ be a maximal ideal of A containing $\mathrm{Ker}(\psi_0)$, and let $\phi \in \mathrm{Mult}(A, \| \cdot \|)$ such that $\mathrm{Ker}(\phi) = \mathcal{M}$. Since $\mathcal{O}$ is a t-normal partition, by Theorem 8.3.6 of Chapter 8, we have a canonical continuous morphism $\mathcal{H}_t$ from $H(D, \mathcal{O})$ into A. Let $F = \mathcal{H}_t(f)$. Then, $\psi_0(F) = 0$, and therefore $\phi(F) = 0$. But since $\phi \in \mathrm{Mult}_m(A, \| \cdot \|)$, we have $\phi(t) \leq r$. Let $\tilde{\phi} = \phi \circ \mathcal{H}_t$. Then $\tilde{\phi}(F) = 0$. Now, $\tilde{\phi}$ is of the form $\varphi_{\mathcal{F}}$ and $\mathcal{F}$ is secant with $d(0, r)$, so we have $\tilde{\phi}(f) = 1$, and therefore $\tilde{\phi}(F) = 1$, a contradiction. $\qquad\square$

Proposition 9.4.3. *Assume* $\mathbb{K}$ *is strongly valued. Suppose there exists* $t \in A$, ψ_1, $\psi_2 \in \mathrm{Mult}(A, \| \cdot \|)$ *such that* $\psi_1(t) < \psi_2(t)$ *and* $\Gamma(0, \psi_1(t)$, $\psi_2(t)) \cap sp(t) = \emptyset$. *Then* A *admits divisors of zero and* A *contains elements whose spectrum is not infraconnected. Moreover, if* $\psi_1(t) = \tau(t)$, *then* A *does not satisfy Property* (o) *and the intersection of all maximal ideals of codimension 1 is not null.*

Proof. Let $D = sp_A(t)$. We put $r_1 = \psi_1(t)$, $r_2 = \psi_2(t)$. Let $s \in$ $]\|t\|_{\mathrm{sp}}, \|t\|[$. Let $\mathcal{O}$ be a t-normal partition of diameter s. Let $r \in]r_1, r_2[$. By Corollary H.1.5, there exist an increasing T-sequence and a decreasing T-sequence of center 0 and diameter r, and therefore by Propositions 8.2.5 of Chapter 8, there exist $f, g \in H(D, \mathcal{O})$ such that:

$$\psi_t(f) = 1 \ \forall \psi \in \mathrm{Mult}(H(D, \mathcal{O}), \| \cdot \|_{D, \mathcal{O}}) \ \text{satisfying} \ \psi_t(t) \leq r_1, \quad \text{and}$$

$$\psi_t(f) = 0 \ \forall \psi \in \mathrm{Mult}(H(D, \mathcal{O}), \| \cdot \|_{D, \mathcal{O}}) \ \text{satisfying} \ \psi_t(t) \geq r, \quad (9.4.1)$$

$$\psi_t(f) \neq 0 \ \forall \psi \in \mathrm{Mult}(A, \| \cdot \|) \setminus \mathrm{Mult}_a(A, \| \cdot \|) \ \text{such that}$$

$$r_1 < \psi_t(f) < r, \quad (9.4.2)$$

$$\psi_t(g) = 1 \ \forall \psi \in \mathrm{Mult}(H(D, \mathcal{O}), \| \cdot \|_{D, \mathcal{O}}) \ \text{such that} \ \psi(t) \geq r_2 \ \text{and}$$

$$\psi_t(g) = 0 \ \forall \psi \in \mathrm{Mult}(H(D, \mathcal{O}), \| \cdot \|_{D, \mathcal{O}}) \ \text{such that} \ \psi(t) \leq r. \quad (9.4.3)$$

Thus, $fg = 0$. Since $\mathcal{O}$ is a t-normal partition, by Theorem 8.4.6 of Chapter 8, we have the canonical continuous morphism $\mathcal{H}_t$ from $H(D, \mathcal{O})$ into A. Now, $\mathcal{H}_t$ is defined on $H(D, \mathcal{O})$. Let $F = \mathcal{H}_t(f)$, $G = \mathcal{H}_t(g)$. Let $\mathcal{F}_j$ be the circular filter of $(\psi_j)_t$ ($j = 1, 2$). Then $\mathcal{F}_1$ is secant with $d(0, r_1)$, hence $(\psi_1)_t(f) = 1$, and therefore $\psi_1(F) = 1$. In the same way, $\mathcal{F}_2$ is secant with $d(0, s) \setminus d(0, r^-)$, hence $(\psi_2)_t(g) = 1$, and therefore $\psi_2(G) = 1$. Thus, both F, G are different from 0 though $FG = 0$ and therefore A contains divisors of 0.

Now, we will show that $sp_A(F)$ is not infraconnected. Consider $l \in$ $d(0, 1^-) \setminus \{0\}$, and suppose that $F - l$ is not invertible in A. Let $\zeta \in \mathcal{X}(A)$ be such that $\zeta(F-l) = 0$. Let $\mathcal{M} = Ker(\zeta)$. Let $\phi \in \mathrm{Mult}(A, \| \cdot \|)$ be such that $Ker(\phi) = \mathcal{M}$. Then ϕ defines an element ϕ_t of $\mathrm{Mult}(R(D, \| \cdot \|_D)$ which of course is of the form $\varphi_{\mathcal{G}}$, with $\mathcal{G}$ a circular filter on $\mathbb{K}$. But since $0 < |l| < 1$, by (9.4.1) we can see that $\mathcal{G}$ is not secant with $d(0, s_1)$ and is not secant with $\mathbb{K} \setminus d(0, r^-)$ either. Consequently, $\mathcal{G}$ approaches a circle $C(0, u)$, with $s_1 < u < r$. But by (9.4.2) we can see that $\varphi_{\mathcal{G}} \in \mathrm{Mult}_a(A, \| \cdot \|)$, hence there exists $\chi \in \mathcal{X}(A, \mathbb{K})$ such that $\varphi_{\mathcal{G}} = |\chi|$, therefore $|\chi(t)| \leq s_1$, a contradiction.

Consequently, $F - l$ is invertible in A for all $l \in \mathbb{K}$ such that $0 < |l| < 1$, and therefore $sp(F)$ is not infraconnected.

Now, suppose that $\tau(t) = \psi_1(t)$ and let $\chi \in \mathcal{X}(A, \mathbb{K})$. Then $|\chi(t)| \leq \tau(t) = s_1$, hence $\chi(G) = 0$, therefore G belongs to the intersection of all maximal ideals of codimension 1. So, this intersection is not null.

Since $(\psi_2)_t(f) = 0$, F is not invertible in A, hence $0 \in sp_A(F)$. Consider $l \in d(0, 1^-) \backslash \{0\}$, and let $y = \frac{1}{F-l}$. Thus, we have $sp_A(F-l) \subset C(0,1) \cup \{-l\}$ and $sp_A(y) \subset C(0,1) \cup \left\{ \frac{1}{-l} \right\}$. More precisely, since F is not invertible in A, $-l$ does belong to $sp_A(F - l)$, and consequently, $\frac{1}{-l}$ does belong to $sp_A(y)$. Hence we have $\tau(y) = \frac{1}{|l|}$. However, for every $\chi \in \mathcal{X}(A, \mathbb{K})$, $\chi(t)$ lies in D, hence $\chi(x)$ belongs to $d(0, s_1)$, and therefore $|\chi(F)| = |f(\chi(t))| = 1$, thereby $|\chi(y)| = 1$. Consequently, we have $\|y\|_{sa} = 1$, which proves that Property (o) is not satisfied. $\square$

Corollary 9.4.4. *Assume* $\mathbb{K}$ *is strongly valued. Suppose there exists* $t \in A$ *such that* $sp(t)$ *is not infraconnected. Then* A *admits divisors of zero.*

Proof.   Let $\Gamma(a, r_1, r_2)$ be an empty annulus of $sp_A(t)$. By Theorem 5.1.15 of Chapter 5, there exists $\psi_1, \psi_2 \in \text{Mult}(A, \| \, . \, \|)$ such that $r_1 = \psi_1(t), r_1 = \psi_2(t)$. So, we can apply Theorem 9.4.3. $\square$

Theorem 9.4.5. *Assume* $\mathbb{K}$ *is strongly valued. There exists* $t \in A$, $\psi_1, \psi_2 \in$ $\text{Mult}(A, \| \, . \, \|)$ *such that* $\psi_1(t) < \psi_2(t)$ *and* $\Gamma(0, \psi_1(t), \psi_2(t)) \cap sp(t) = \emptyset$ *if and only if* A *contains elements whose spectrum is not infraconnected.*

Proof.   Suppose that there exists $t \in A$, $\psi_1, \psi_2 \in \text{Mult}(A, \| \, . \, \|)$ such that $\psi_1(t) < \psi_2(t)$ and $\Gamma(0, \psi_1(t), \psi_2(t)) \cap sp(t) = \emptyset$. Then by Proposition 9.4.3, A admits an element u such that $sp(u)$ is not infraconnected. Conversely, if A admits an element u such that $sp(u)$ is not infraconnected, $sp(u)$ admits an empty annulus $\Gamma(a, r_1, r_2)$ and then by Lemma 5.1.15 of Chapter 5, there exist $\psi_1, \psi_2 \in \text{Mult}(A, \| \, . \, \|)$ such that $\psi_1(t) = r_1$, $\psi_2(t) = r_2$. $\square$

Theorem 9.4.6. *Assume* $\mathbb{K}$ *is strongly valued and let* A *be such that* $\text{Max}_1(A) \neq \emptyset$. *If* A *satisfies one of the following four conditions bellow, then* A *satisfies Property (q).*

(i)   *Property (o),*
(ii) A *has no divisors of zero,*
(iii) $sp(x)$ *is infraconnected for every* $x \in A$,
(iv)   *the intersection of all maximal ideals of codimension 1 is null.*

Proof.   We assume that A does not satisfies Property (q), and will show that A does not satisfy the four properties at the bottom of Theorem 9.4.6.

Let $t \in A$ be such that $\tau(t) < \|t\|_{\text{sp}}$. Let $s_1 = \tau(t)$, let $s_2 = \|t\|_{\text{sp}}$. Both s_1, s_2 lie in the closure of $\{\phi(t) \mid \phi \in \text{Mult}(A, \| \cdot \|) \}$, hence there exists $\psi_1 \in \text{Mult}(A, \| \cdot \|)$ such that $\psi_1(t) = s_1$, and by Theorem 2.5.17 of Chapter 2, there exists $\psi_2 \in \text{Mult}(A, \| \cdot \|)$ such that $\psi_2(t) = s_2$. By hypothesis we have $\Gamma(0, s_1, s_2) \cap sp(t) = \emptyset$ hence by Proposition 9.4.3, A has divisors of zero, the intersection of all maximal ideals of codimension 1 is not null, $sp_A(t)$ is not infraconnected. Thus, there exists an element $f \neq 0$ which belongs to the intersection of all maximal ideals of codimension 1, hence $\|f\|_{\text{sa}} = 0$, but $\tau(f) > 0$ which ends the proof. $\square$

9.5. Multbijective Banach $\mathbb{K}$-algebras

Definition and notation: Throughout this chapter, A is a unital commutative ultrametric Banach $\mathbb{K}$-algebra.

In Theorem 2.5.13 of Chapter 2, we saw that the mapping from $\text{Mult}_m(A, \|. \|)$ into $\text{Max}(A)$ associating to each $\psi \in \text{Mult}_m(A, \|. \|)$ its kernel, is surjective. The natural question coming next, is whether the mapping is also injective. The answer was given in [21, 23].

Remarks. In $\mathbb{C}$, it is well known that every maximal ideal of a $\mathbb{C}$-Banach algebra is the kernel of one and only one multiplicative semi-norm, and that every multiplicative semi-norm has a kernel which is a maximal ideal.

When $\mathbb{K}$ is strongly valued, we have a different answer. First we can establish this theorem.

Theorem 9.5.2. *Let $\mathbb{K}$ be strongly valued, and let F be a field extension of $\mathbb{K}$ which is an ultrametric normed $\mathbb{K}$-algebra admitting two different absolute values continuous for this norm of F. Then the completion of F is a Banach $\mathbb{K}$-algebra which admits divisors of zero.*

Proof. Let $\widehat{F}$ be the completion of F. Let ψ_1, ψ_2 be two different continuous absolute values. They have continuation to $\widehat{F}$: $\widehat{\psi}$, $\widehat{\phi}$. Now, we can find $x \in F$ such that $\psi_1(x) < \psi_2(x)$. Since ψ and ϕ are continuous absolute value on F, on $\mathbb{K}$ they coincide with the absolute value of $\mathbb{K}$, hence $x \notin \mathbb{K}$. We put $r_1 = \psi_1(x)$, $r_2 = \psi_2(x)$. Since x does not belong to $\mathbb{K}$, we have $sp(x) = \emptyset$. Consequently, by Theorem 9.4.3 $\widehat{F}$ has divisors of zero. $\square$

As a consequence, we obtain Theorem 9.5.3.

Theorem 9.5.3. *If $\mathbb{K}$ is strongly valued, then A is multbijective.*

Proof. Suppose $\mathbb{K}$ is strongly valued. Let $\mathcal{M}$ be a maximal ideal of A, and suppose there exist ψ_1, $\psi_2 \in \text{Mult}_m(A, \| \cdot \|)$ such that $\text{Ker}(\psi_1) = \text{Ker}(\psi_2) = \mathcal{M}$. Let $F = \frac{A}{\mathcal{M}}$. Thus, the field F admits two different absolute values $| \cdot |_j$

quotients of ψ_j, respectively, $(j = 1, 2)$ which by hypothesis, are continuous for the norm of F quotient of the norm of A. Since F is a Banach $\mathbb{K}$-algebra for its quotient norm, by Theorem 9.5.2 we are led to a contradiction. $\square$

By Theorem 7.5.14 we have Corollary 9.5.4.

Corollary 9.5.4. *There exist non-multbijective unital commutative ultrametric Banach $\mathbb{K}$-algebras if and only if $\mathbb{K}$ is weakly valued.*

Remarks. Counter-examples of non-multbijective Banach $\mathbb{K}$-algebras are very hard to construct, and seem to be a very strange case. It would be interesting to obtain sufficient conditions to prevent this kind of problem. We notice that our counter-example is an algebra of analytic elements $H(D)$. Hence in particular, $\text{Mult}_m(H(D), \| \cdot \|_D)$ is dense in $\text{Mult}(H(D), \| \cdot \|_D)$. Consequently, density does not imply multbijectivity. However, we are not able to construct a Noetherian non-multbijective ultrametric Banach $\mathbb{K}$-algebra. So, we can ask whether Noetherian ultrametric Banach $\mathbb{K}$-algebras are multbijective, no mater what the field.

Theorem 9.5.5. *Let $\mathbb{K}$ be strongly valued, let $\psi \in \text{Mult}(A, \| \cdot \|)$ and suppose that the set of maximal ideals containing $\text{Ker}(\psi)$ is countable. Then ψ belongs to $\text{Mult}_m(A, \| \cdot \|)$.*

Proof. Let $\mathcal{J} = \text{Ker}(\psi)$, let $B = \frac{A}{\mathcal{J}}$ and let θ be the canonical surjection from A onto B. Then $\text{Max}(B) = \theta(\text{Max}(A))$ is countable. Suppose that $\mathcal{J} \notin \text{Max}(A)$. Let $\mathcal{M}$ be a maximal ideal of A containing $\mathcal{J}$ and let $\mathcal{M}' = \theta(\mathcal{M})$. Then B admits an absolute value ψ' such that $\psi = \psi' \circ \theta$. Moreover, there exists (a unique) $\phi \in \text{Mult}_m(A, \| \cdot \|)$ such that $\text{Ker}(\phi) = \mathcal{M}$. Let $\phi' \in \text{Mult}_m(B, \| \cdot \|')$ be such that $\text{Ker}(\phi') = \mathcal{M}'$. Since $\mathcal{M} \neq \mathcal{J}$, $\mathcal{M}'$ is not null. Let $t \in \mathcal{M}' \setminus \{0\}$: we have $\phi'(t) = 0$, $\psi'(t) \neq 0$. Let $D = sp(t)$, and let $l = \psi'(t)$ and let $\mathcal{S}$ be a t-normal partition. Since $\text{Max}(B)$ is countable, so is D. We notice that 0 lies in D and that none of the classes of the circles $C(0, r)$, with $0 < r < \|t\|$ can meet D, except at most a countable subfamily. Consequently, by Theorem 7.5.10 of Chapter 7, $\mathcal{S}$ admits an increasing idempotent T-sequence of center 0 and diameter l. Therefore, by Proposition 8.2.5 of Chapter 8, there exists $f \in H(D, \mathcal{S})$ such that $f(0) = 1$ and $\varphi(f) = 0 \; \forall \varphi \in \text{Mult}(H(D, \mathcal{S}), \| \cdot \|_{D, \mathcal{S}})$ such that $\phi(t) \geq l$. Since $\mathcal{O}$ is a t-normal partition, by Theorem 2.4.6 of Chapter 8, we have a canonical continuous morphism $\mathcal{H}_t$ from $H(D, \mathcal{S})$ into A. Let $F = \mathcal{H}_t(f)$, let $L = \frac{B}{\mathcal{M}'}$ and let Ω be the canonical surjection from B onto L. Then Ω is continuous with respect to the quotient topologies of B and L. And since $t \in \mathcal{M}'$, we have $\Omega(\mathcal{H}_t(x)) = 0$ and therefore, $\Omega(\mathcal{H}_t(f)) = f(0) = 1$. Consequently,

$F \neq 0$. Next, as $\psi'(t) = l$, ψ' defines on $H(sp(t), \mathcal{S})$ an element ψ'_t such that $\psi'_t(x) = l$, hence $\psi'_t(f) = 0$ and therefore $\psi'(F) = 0$, a contradiction to the fact that ψ' is an absolute value on B. □

9.6. Polnorm on algebras and algebraic extensions

Throughout this section is designed to help us prove that a reduced affinoid algebra is complete for its spectral norm. In particular, this will be useful in characteristic $p \neq 0$.

Notations: Given a field $\mathbb{E}$ of characteristic $p \neq 0$, we denote by $\mathbb{E}^{\frac{1}{p}}$ the extension of $\mathbb{E}$ containing all p^{th}-roots of elements of $\mathbb{E}$. Let $\mathbb{L}$ be a subfield of $\mathbb{K}$.

Let B be a unital commutative $\mathbb{L}$-algebras without divisors of zero and let A be a $\mathbb{L}$-subalgebra of B containing the unity of B, equipped with a $\mathbb{L}$-algebra semi-multiplicative semi-norm ψ.

Let $P(X) = \sum_{j=0}^{q} a_j X^j \in A[X]$ be monic. We put $\mathbb{S}(P, \psi) = \max_{0 \leq j < q}(\psi(a_j))^{\frac{1}{q-j}}$. When there is no risk of confusion on the $\mathbb{L}$-algebra semi-multiplicative semi-norm ψ of A, we will only write $\mathbb{S}(P)$ instead of $\mathbb{S}(P, \psi)$. In particular, when the monic polynomial P belongs to $\mathbb{L}[X]$, then $\mathbb{S}(P)$ will denote $\max_{0 \leq j < q}(|a_j|^{(\frac{1}{q-j})})$.

Now, suppose that B is integral over A. The function $\| \, . \, \|_{pol}^{A}$ defined on B as $\|y\|_{pol}^{A} = \mathbb{S}(irr(y, A), \psi)$ will be called *the A-polnorm of B*.

Theorem 9.6.1 is well known in fields theory.

Theorem 9.6.1. *Every field of characteristic 0 is perfect. A field $\mathbb{E}$ of characteristic $p \neq 0$ is perfect if and only if $\mathbb{E} = \mathbb{E}^{\frac{1}{p}}$.*

Lemma 9.6.2. *Let B be equipped with a $\mathbb{L}$-algebra semi-multiplicative semi-norm ψ. Let $P(X) \in A[X]$ and let $b \in B$ satisfy $P(b) = 0$. Then $\mathbb{S}(P, \psi) \geq \psi(b)$.*

Proof. Let $P(X) = \sum_{j=0}^{q} a_j X^j$ and suppose $\psi(b) > \mathbb{S}(P)$. Thus, we have $\psi(a_j) < \psi(b)^{q-j}$, hence $\psi(a_j b^j) \leq \psi(a_j)\psi(b)^j < \psi(b)^j \psi(b)^{q-j} = \psi(b)^q$ whenever $j = 0, \ldots, q - 1$. Consequently, $\mathbb{S}(\sum_{j=0}^{q-1} a_j X^j) < \psi(b^q)$. But by Lemma 2.3.3 of Chapter 2, ψ is ultrametric. But since $P(b) = 0$ we have $\psi(\sum_{j=0}^{q-1} a_j b^j) = \psi(b^q)$, a contradiction. □

Lemma 9.6.3 is almost classical and comes from the behavior of polynomials in an algebraically closed complete ultrametric field recalled in Corollary 6.2.6 and Theorem 6.2.7 of Chapter 6.

Lemma 9.6.3. *Let $P(X) = \sum_{j=0}^{q} a_j X^j \in \mathbb{K}[X]$ be monic and let $\alpha_1, \ldots, \alpha_q$ be its zeros in K. Then $\max_{1 \leq j \leq q} |\alpha_j| = \mathbb{S}(P)$.*

Proof. Let $\{|\alpha_1|, \ldots, |\alpha_q|\} = \{r_1, \ldots, r_h\}$, with $r_j < r_{j+1}$ ($1 \leq j < h$). For each $j = 1, \ldots, h$, let m_j be the number of zeros of P inside $C(0, r_j)$, let $t_j = N^+(P, -\log(r_j))$, and $s_j = N^-(P, -\log(r_j))$. On one hand, by Corollary 7.3.2 of Chapter 7, we have $m_j = t_j - s_j$ ($1 \leq j < h$) and $\log(r_j) = \frac{1}{t_j - s_j}(\log(|a_{s_j}| - \log(|a_{t_j}|))$. In particular $\log(r_h) = \frac{1}{m_h} \log(|a_{s_h}|) = \max_{1 \leq j < h} \log(|\alpha_j|)$. But since $N^-(P, -\log(r_h)) = s_h$ and $N^+(P, -\log(r_h)) = q$, by definition we have $r_h^q = |a_{s_h}|(r_h)^{s-h} \geq |a_j|(r_h)^j$ $1 \leq j < q$, hence $\frac{1}{q-j} \log(|a_j|) \leq \log(|a_{s_h}|) \leq r_h = \frac{1}{q-s_h} \log(|a_{s_h}|)$ $\forall j = 0, \ldots, q - 1$. Consequently $\max_{1 \leq j < q} \log(|\alpha_j|) = \frac{1}{q-s_h} \log(|a_{s_h}|) \geq \frac{1}{q-j} \log(|a_j|)$ $\forall j = 0, \ldots, q - 1$, therefore $\mathbb{S}(P) = \max_{0 \leq j \leq q} |\alpha_j|$. $\square$

Proposition 9.6.4. *Let A be integrally closed such that B is finite over A. Let $\mathcal{T}$ be a set of $\mathbb{L}$-algebra homomorphisms from A into $\mathbb{K}$ such that, for each $t \in A$, there exists $\xi_t \in \mathcal{T}$ satisfying $|\xi_t(t)| = \sup\{|\chi(t)| \mid \chi \in \mathcal{T}\}$. Let $\mathcal{H}$ be the set of $\mathbb{L}$-algebra homomorphisms from B to $\mathbb{K}$ whose restrictions to A belong to $\mathcal{T}$. On B we put $\|x\| = \sup\{|\chi(x)| \mid \chi \in \mathcal{H}\}$. Let $y \in B$ and let $P = irr(y, A)$. Then $\mathbb{S}(P, \| . \|) = \|y\|$. Moreover there exists $\zeta \in \mathcal{H}$ such that $|\zeta(y)| = \|y\|$.*

Proof. By Lemma 2.3.4 of Chapter 2, the function $\| . \|$ defined on B is a $\mathbb{L}$-algebra semi-multiplicative semi-norm. Therefore, by Lemma 9.6.1, we have $\mathbb{S}(P, \| . \|) \geq \|y\|$.

We will prove the opposite inequality. Let $P(X) = \sum_{j=0}^{q} a_j X^j \in A[X]$ and let $h \in \{0, \ldots, q - 1\}$ be such that $\mathbb{S}(P, \| . \|_{sp}^A) = \sqrt[q-h]{\|a_h\|^A}$. By hypothesis there exists $\xi \in \mathcal{T}$ such that $|\xi(a_h)| = \|a_h\|$. Let $f(X) = \sum_{j=0}^{q} \xi(a_j) X^j \in \mathbb{L}[X]$. Then, of course, $\mathbb{S}(f, | . |) = \mathbb{S}(P, \| . \|)$ because $|\xi(a_j)| \leq \|a_j\|$ $\forall j = 0, \ldots, q$. Let $\alpha_1, \ldots, \alpha_q$ be the zeros of f in $\mathbb{K}$. By Lemma 9.6.3, we have $\max_{1 \leq j \leq q} |\alpha_j| = \mathbb{S}(f, | . |)$. Let β be a zero of f in $\mathbb{K}$ such that $|\beta| = \mathbb{S}(f, | . |)$. Let ϕ be the homomorphism from $A[X]$ into $\mathbb{K}$ defined as $\phi(\sum_{j=0}^{q} c_j X^j) = \sum_{j=0}^{q} \xi(c_j)\beta^j$ and let θ be the canonical surjection from $A[X]$ onto $A[y]$. Since $P(X)A[X] \subset Ker(\phi)$, ϕ factorizes in the form $\xi \circ \theta$, and then ξ is a $\mathbb{L}$-algebra homomorphism from $A[y]$ into $\mathbb{K}$ such that $\xi(y) = \beta$. Then by Lemma 1.1.17 of Chapter 1, ξ has continuation to a $\mathbb{L}$-algebra homomorphism $\check{\xi}$ from B to $\mathbb{K}$. Thus, $\check{\xi}$ satisfies $\mathbb{S}(P, \| . \|) = |\check{\xi}(y)|$, and of course $|\check{\xi}(y)| \leq \|y\|$, hence $\mathbb{S}(P, \| . \|) = |\check{\xi}(y)|$, which finishes proving our claim. $\square$

Corollary 9.6.5. *Let B be finite over A and let A be integrally closed and equipped with a semi-multiplicative norm $\| \cdot \|$ such that for each $t \in A$ there exists a $\mathbb{L}$-algebra homomorphism χ from A into $\mathbb{K}$ satisfying $\|t\| = |\chi(t)|$. Then the A-polnorm of B is a semi-multiplicative norm.*

Theorem 9.6.6. *Let F be an algebraic extension of $\mathbb{L}$. The $\mathbb{L}$-polnorm of F is a $\mathbb{L}$-algebra semi-multiplicative norm.*

Proof. We can consider that $\mathbb{K}$ is the complete algebraic closure of $\mathbb{L}$. Now let $x \in F$, let $P = irr(x, \mathbb{L})$ and let $\alpha_1, \ldots, \alpha_q$ be the zeros of P in $\mathbb{K}$. By Lemma 9.6.3, we have $\max_{1 \leq j \leq q} |\alpha_j| = \mathbb{S}(P, | \cdot |)$. Let G be the Galois group of F over $\mathbb{L}$. Thus, considering F as a subfield of $\mathbb{K}$, we have $\mathbb{S}(P, | \cdot |) = \max_{g \in G} |g(x)|$ and then we can easily check that this is a semi-multiplicative $\mathbb{L}$-algebra norm. $\qquad \square$

Theorem 9.6.7. *Let $\mathbb{L}$ be perfect and let $\mathbb{K}$ be the complete algebraic closure of $\mathbb{L}$. Then $\mathbb{K}$ is $\mathbb{L}$-productal.*

Proof. Let $\mathbb{E}$ be a finite field extension of $\mathbb{L}$ and let Tr be the trace function of $\mathbb{E}$ over $\mathbb{L}$. Then Tr is obviously continuous with respect to the $\mathbb{L}$-polnorm on $\mathbb{K}$. Since $\mathbb{L}$ is perfect, $\mathbb{E}$ is a separable extension of $\mathbb{L}$, hence for every $a \in \mathbb{E}$, $a \neq 0$, there exists $b \in \mathbb{E}$ such that $Tr(ab) \neq 0$. Now consider the $\mathbb{L}$-linear mapping ϕ from $\mathbb{E}$ into $\mathbb{L}$ defined by $\phi(x) = Tr(xb)$. We have $\phi(a) \neq 0$, hence by Corollary 2.3.1 of Chapter 2, $\mathbb{K}$ is $\mathbb{L}$-productal. $\qquad \square$

Proposition 9.6.8. *Let $\mathbb{E}$ be an algebraic extension of $\mathbb{L}$ and let $\mathbb{F}$ be an algebraic extension of $\mathbb{E}$. If the $\mathbb{L}$-polnorm on F is multiplicative, $\mathbb{E}$ being normed with this absolute value, then the $\mathbb{L}$-polnorm and the $\mathbb{E}$-polnorm coincide on $\mathbb{F}$.*

Proof. Let $\| \cdot \|_{pol}^{\mathbb{L}}$ (resp., $\| \cdot \|_{pol}^{\mathbb{E}}$) be the $\mathbb{L}$-polnorm (resp., $\mathbb{E}$-polnorm) on $\mathbb{F}$. Since $\| \cdot \|_{pol}^{\mathbb{E}}$ is obviously a semi-multiplicative $\mathbb{L}$-algebra norm, by Lemma 9.6.2, we have $\|x\|_{pol}^{\mathbb{L}} \geq \|x\|_{pol}^{\mathbb{E}} \; \forall x \in \mathbb{F}$. Next, $\| \cdot \|_{pol}^{\mathbb{L}}$ induces on $\mathbb{E}$ the $\mathbb{L}$-polnorm of $\mathbb{E}$ because given $x \in \mathbb{E}$, its minimal polynomial over $\mathbb{L}$ is the same, whether we consider x as an element of $\mathbb{E}$ or as an element of $\mathbb{F}$. Consider now the $\mathbb{L}$-polnorm of $\mathbb{F}$: since it is a $\mathbb{L}$-algebra norm, it satisfies $\|xy\|_{pol}^{\mathbb{L}} \leq \|x\|_{pol}^{\mathbb{L}} \|y\|_{pol}^{\mathbb{L}} \forall \, \lambda \in \mathbb{L}, \; x \in \mathbb{F}$, hence $\| \cdot \|_{pol}^{\mathbb{L}}$ is a mapping ψ from $\mathbb{F}$ to $\mathbb{R}_+$ satisfying $\psi(x + y) \leq \max(\psi(x), \psi(y)), \quad \psi(xy) \leq \psi(x)\psi(y)$ and $\psi(x^n) = \psi(x)^n \; \forall n \in \mathbb{N}^*, \; \forall x, y \in \mathbb{F}$. Consequently, by Lemma 9.6.2, we have $\|x\|_{pol}^{\mathbb{E}} \geq \|x\|_{pol}^{\mathbb{L}}$. $\qquad \square$

Proposition 9.6.9. *Let* $\mathbb{L}$ *have characteristic* $p > 0$. *For every* $n \in \mathbb{N}$, *the* $\mathbb{L}$-*polnorm of* $\mathbb{L}^{\frac{1}{p^n}}$ *is an absolute value extending this of* $\mathbb{L}$.

Proof. Since the $\mathbb{L}$-polnorm is a $\mathbb{L}$-algebra norm, we just have to check that it is multiplicative. On the other hand, by induction, it is sufficient to show the claim when $n = 1$. Let x, $y \in \mathbb{L}^{\frac{1}{p}}$. By hypothesis, $irr(x, \mathbb{L})$ is of the form $X^p - a$, $irr(y, \mathbb{L})$ is of the form $X^p - b$, with a, $b \in \mathbb{L} \setminus \mathbb{L}^{\frac{1}{p^n}}$. Thus, we have $\|x\|_{\mathrm{pol}}^{\mathbb{L}} = \sqrt[p]{|a|}$, $\|y\|_{\mathrm{pol}}^{\mathbb{L}} = \sqrt[p]{|b|}$. Now, $(xy)^p = ab$, hence $irr(xy, \mathbb{L})$ divides $X^p - ab$, and is of the form $X^p - c$. Hence $c = ab$, and thereby, $\|xy\|_{\mathrm{pol}}^{\mathbb{L}} = \sqrt[p]{|ab|} = \|x\|_{\mathrm{pol}}^{\mathbb{L}} \|y\|_{\mathrm{pol}}^{\mathbb{L}}$. $\square$

Theorem 9.6.10. *Let* $\mathbb{L}$ *have characteristic* $p > 0$. *If* $\mathbb{L}^{\frac{1}{p}}$ *is* $\mathbb{L}$-*productal with respect to the* $\mathbb{L}$-*polnorm, then so is the algebraic closure of* $\mathbb{L}$.

Proof. Let $\mathbb{L}^{\infty} = \bigcup_{n=1}^{\infty} \mathbb{L}^{\frac{1}{p^n}}$, and let Ω be an algebraic closure of $\mathbb{L}$ containing $\mathbb{L}^{\infty}$. Then by Theorem 9.6.1 $\mathbb{L}^{\infty}$ is perfect, and by Proposition 9.6.9, the $\mathbb{L}$-polnorm is an absolute value that continues this of $\mathbb{L}$. By Theorem 9.6.7, since $\mathbb{L}^{\infty}$ is perfect, $\mathbb{K}$ is $\mathbb{L}^{\infty}$- productal with respect to the $\mathbb{L}^{\infty}$-polnorm. But by Proposition 9.6.8, the $\mathbb{L}^{\infty}$-polnorm is equal to the $\mathbb{L}$-polnorm on $\mathbb{K}$. Consequently $\mathbb{K}$ is $\mathbb{L}^{\infty}$-productal with respect to the $\mathbb{L}$-polnorm which is an absolute value on $\mathbb{L}^{\infty}$. Hence by Lemma 2.3.3 of Chapter 2, $\mathbb{K}$ is $\mathbb{L}$-productal with respect to the $\mathbb{L}$-polnorm. $\square$

Chapter 10

Algebras Topologically of Finite Type

10.1. Hensel lemma

Hensel lemma is indispensable in many works on polynomials and restricted series on ultrametric fields.

Definition and notations: Recall that $\mathbb{L}$ denotes a complete field equipped with a non-trivial ultrametric absolute value $| \, . \, |$.

Let E be a unital commutative ultrametric normed $\mathbb{L}$-algebra. Let $n \in \mathbb{N}$ and let $\| \, . \, \|_0$ be the norm of E. The $\mathbb{E}$-algebra of polynomials in n variables $E[X_1, \ldots, X_n]$ is equipped with the *Gauss norm* $\| \, . \, \|$ defined as $\| \sum_{i_1, \ldots, i_n} a_{i_1, \ldots, i_n} X_1^{i_1} \cdots X_n^{i_n} \| = \sup_{i_1, \ldots, i_n} \| a_{i_1, \ldots, i_n} \|_0$. By Proposition 2.5.16 of Chapter 2, this norm is a $\mathbb{L}$-algebra norm. In particular, if the norm of E is multiplicative, so is the norm $\| \, . \, \|$.

We denote by $E\{X_1, \ldots, X_n\}$ the set of power series in n variables $\sum_{i_1, \ldots, i_n} a_{i_1, \ldots, i_n} X_1^{i_1} \cdots X_n^{i_n}$ such that $\lim_{i_1 + \cdots + i_n \to \infty} a_{i_1, \ldots, i_n} = 0$. The elements of such an algebra are called *the restricted power series in n variables, with coefficients in E*. Hence by definition $E[X_1, \ldots, X_n]$ is dense in $E\{X_1, \ldots, X_n\}$. By Proposition 2.5.16 of Chapter 2, if E is complete, $E\{X_1, \ldots, X_n\}$ is just the completion of $E[X_1, \ldots, X_n]$. Particularly, when $E = \mathbb{L}$, $\mathbb{L}\{X_1, \ldots, X_n\}$ is denoted T_n.

The Gauss norm $\| \, . \, \|$ defined on $E[X_1, \ldots, X_n]$ obviously has continuation to $E\{X_1, \ldots, X_n\}$. For convenience, given $F(X_1, \ldots, X_n) \in E\{X_1, \ldots, X_n\}$ we put $\Psi\Big(F(X_1, \ldots, X_n)\Big) = \log\Big(\|F(X_1, \ldots, X_n)\|\Big)$.

Throughout the chapter, A will denote a unital commutative $\mathbb{L}$-algebra equipped with an absolute value $| \, . \, |$ which extends that of $\mathbb{L}$ and satisfies $\{|t| \mid x \in A\} = |\mathbb{L}|$ and we put again $\Psi(t) = \log(|t|)$. We then denote by A_0 the subring $\{t \in A \mid |x| \leq 1\}$ of A and by $\mathcal{M}_A$ the prime ideal $\{t \in A \mid |t| < 1\}$

of A_0. Moreover, we denote by U the subring of $\mathbb{L}$ $\{x \in \mathbb{L} \ |x| \leq 1\}$ and by M its maximal ideal $\{x \in \mathbb{L} \ |x| < 1\}$.

Let $\mathcal{V} = \{f \in A\{X_1, \ldots, X_n\} \ | \ \|f\| \leq 1\}$ and let $\mathcal{N} = \{f \in A\{X\} \ | \ \|f\| < 1\}$. Given $f \in \mathcal{V}$ we will denote by $\overline{a}$ the residue class of a in $\frac{\mathcal{V}}{\mathcal{N}}$.

Given $f = \sum_{n=0}^{\infty} a_n x^n \in A\{x\}$, here we will denote by $J(f)$ the biggest of the integers m such that $|a_m| = \|f\|$.

Lemmas 10.1.1 and 10.1.2 are immediate.

Lemma 10.1.1. $\mathcal{M} = MU$, $\mathcal{N} = MW$, $\frac{W}{N} = \frac{U}{M}[\overline{x}]$.

Lemma 10.1.2. *Let* $f \in A\{x\}$ *be such that* $\|f\| = 1$. *Then* $J(f) = \deg(\overline{f})$.

Lemma 10.1.3. *Let* $f, g \in A\{x\}$. *Then* $J(fg) = J(f) + J(g)$.

Proof. Suppose first that $\Psi(f) = \Psi(g) = 0$. Let $q = J(f)$, $t = J(g)$. By Lemma 10.1.1, in $\frac{U}{M}[\overline{x}]$, $\overline{f}$ is a polynomial of degree q, $\overline{g}$ is a polynomial of degree t, hence $\overline{fg}$ is a polynomial of degree $q+t$, which shows the statement. Now consider the general case. Since $\{|t| \ | \ t \in A\} = |\mathbb{L}|$ we can find $\lambda \in \mathbb{L}$ such that $\Psi(\lambda f) = \Psi(\lambda g) = 0$, so we are led to the same conclusion. $\quad\square$

Lemma 10.1.4. *Let* $P \in A[x]$ *and let* $D \in U[x]$ *be monic. Let* $Q, R \in A[x]$ *satisfy* $P = DQ + R$ *and* $\deg(R) < \deg(D)$. *Then we have* $\Psi(Q) \leq \Psi(P)$ *and* $\Psi(R) \leq \Psi(P)$.

Proof. We can clearly assume $P \neq 0$. Then, by multiplying P by a suitable constant $\lambda \in \mathbb{L}$, we can also assume $\Psi(P) = 0$. Since D is monic, the Euclidean division of P by D is clearly possible in $U[x]$, and therefore Q is the quotient, R is the rest of this division, due to the fact that $\deg(R) < \deg(D)$. So we have $\Psi(Q) \leq 0$, $\Psi(R) \leq 0$ because both Q, R belong to $U[x]$. $\quad\square$

Theorem 10.1.5. *Let* A *be complete with respect to* $| \ . \ |$. *Let* $f \in A\{x\}$ *and let* $D \in U[x]$ *be monic. There exists a unique pair* $(g, R) \in A\{x\} \times A[x]$ *such that* $f = Dg + R$, *and* $\deg(R) < \deg(D)$. *Moreover, we have* $\Psi(g) \leq \Psi(f)$ *and* $\Psi(R) \leq \Psi(f)$.

Proof. Since A is complete, then so is $A\{x\}$ with respect to the norm $\| \ . \ \|$. Consider the mapping ϕ, defined on $A[x]$, which associates to each $P \in A[x]$ the quotient Q in the division by D. By Lemma 10.1.4, ϕ is continuous with respect to the absolute value $\| \ . \ \|$. Since $A\{x\}$ is complete and since $A[x]$ is dense in $A\{x\}$, then ϕ has continuation to a mapping $\widehat{\phi}$ from $A\{x\}$ to $A\{x\}$ such that $\Psi(\widehat{\phi}(f)) \leq \Psi(f)$. In the same way, consider the mappings θ defined on $A[x]$ which associates to each $P \in A[x]$ the rest R in the division of P by D. By Lemma 10.1.4, θ is continuous, therefore it can be continuously extended to a mapping $\widehat{\theta}$ from $A\{x\}$ to $A[x]$ such that $\Psi(\widehat{\theta}(f)) \leq \Psi(f)$ and $\deg(\widehat{\theta}(f)) < \deg(D)$. Thus, by putting $g = \widehat{\phi}(f)$ and $R = \widehat{\theta}(f)$ we have proven the existence of the pair (g, R).

Now, suppose we have another pair $(h, S) \in A\{x\} \times A[x]$ such that $f = Dh + S$, and $\deg(S) < \deg(D)$. Then $D(h - g) = R - S$. Since D is monic, by Lemma 10.1.3 we have $J(D(h - g)) \geq \deg(D)$, while $J(R - S) \leq \deg(D) - 1$, a contradiction. Hence $h = g$, $S = R$. $\qquad\square$

Corollary 10.1.6. *Let A be complete with respect to $| \, . \, |$. Let $f \in A\{x\}$ and let $D \in U[x]$ be monic. If D divides f in $A\{x\}$ then it divides f in $A[x]$.*

Notation: Given a ring B and $g, h \in B[x]$, $\mathcal{I}(g, h)$ will denote the ideal generated by g and h in $B[x]$.

Lemma 10.1.7. *Let g, $h \in U[x]$ be monic, such that $\mathcal{I}(\overline{g}, \overline{h}) = \frac{U}{\mathcal{M}}[\overline{x}]$ and let $q \in \mathbb{N}$ be such that $q < \deg(g) + \deg(h)$. There exist $S, W \in \mathbb{L}[x]$ satisfying $\Psi(Sg + Wh - x^q) < 0$, $\Psi(S) \leq 0$, $\Psi(W) \leq 0$, $\deg(S) < \deg(h)$, $\deg(W) < \deg(g)$.*

Proof. Since $\mathcal{I}(\overline{g}, \overline{h}) = \frac{U}{\mathcal{M}}[\overline{x}]$, there exists φ and $\phi \in \mathcal{L}[x]$ such that $\varphi\overline{g} + \phi\overline{h} = \overline{1}$, $\deg(\varphi) < \deg(\overline{h})$, $\deg(\phi) < \deg(\overline{g})$. Let $B, E \in U[x]$ satisfy $\overline{B} = \varphi$, $\overline{E} = \phi$, $\deg(B) = \deg(\varphi)$, $\deg(E) = \deg(\phi)$. Thus we have $\overline{Bg + Eh - 1} = \overline{0}$, i.e.,

$$\Psi(Bg + Eg - 1) < 0. \tag{10.1.1}$$

Moreover, $\Psi(B) = 0$. We now consider the Euclidean division of Bx^q by h and Ex^q by g, respectively. We obtain $Bx^q = B_0 h + S$ and $Ex^q = E_0 g + W$. By Lemma 10.1.4, we have $\max(\Psi(S), \Psi(B_0)) \leq \Psi(Bx^q) = 0$. Next, by hypothesis we have

$$\deg(S) < \deg(h), \tag{10.1.2}$$

and

$$\deg(W) < \deg(g). \tag{10.1.3}$$

Let $F = Bg + Eh - 1$. Then we have $Bx^q = (B_0 + E_0)gh + Sg + Wh - x^q$. Since $q < \deg(g) + \deg(h)$, by (10.1.2) and (10.1.3) we see that $\deg(Sg + Wh - x^q) < \deg(g) + \deg(h)$ and therefore $Sg + Wh - x^q$ is just the remainder of the Euclidean division of Fx^q by gh. But then, by Lemma 10.1.4, we have $\Psi(Sg + Wh - x^q) \leq \Psi(Fx^q) = \Psi(F)$. And by (10.1.1) and by definition of F it is seen that $\Psi(F) < 0$. This finishes proving that $\Psi(Sg + Wh - x^q) < \Psi(x^q)$. $\qquad\square$

Notations: Let $g, h \in U[x]$ be monic and satisfy $\mathcal{I}(\overline{g}, \overline{h}) = \frac{U}{\mathcal{M}}[\overline{x}]$. We will denote by $E(g, h)$ the set of constants $\lambda \in \mathbb{R}_+$ such that, for every polynomial $Q \in \mathbb{L}[x]$ satisfying $\deg(Q) < \deg(g) + \deg(h)$, there exist $F, G \in \mathbb{L}[x]$ satisfying $\Psi(Fg + Gh - Q) \leq \Psi(Q) + \lambda$, $\Psi(F) \leq \Psi(Q)$, $\Psi(G) \leq \Psi(Q)$, $\deg(F) < \deg(h)$, $\deg(G) < \deg(g)$.

Lemma 10.1.8. *Let $g, h \in U[x]$, be monic and satisfy $\mathcal{I}(\overline{g}, \overline{h}) = \frac{U}{\mathcal{M}}[\overline{x}]$, and let $d = \deg(g) + \deg(h)$. Then $E(g, h)$ is a non-empty interval whose lower bound is 0. Moreover, given $\lambda \in E(g, h)$ and monic polynomials ϕ, $\theta \in U[x]$ such that $\Psi(g - \phi) \leq \lambda$, $\Psi(h - \theta) \leq \lambda$, then $E(\phi, \theta) = E(g, h)$.*

Proof. We can apply Lemma 10.1.7 to each polynomial $Q_n = x^n$ for every $n = 0, \ldots, d - 1$. Thus, we have polynomials S_n, W_n satisfying $\Psi(S_n g + W_n h - x^n) < 0$, $\Psi(S_n) \leq 0$, $\Psi(W_n) \leq 0$, $\deg(S_n) < \deg(h)$, $\deg(T_n) < \deg(g)$. We put $\lambda_n = \Psi(S_n g + W_n h - x^n)$, $(0 \leq n \leq d - 1)$. Now let $Q = \sum_{n=0}^{d-1} a_n x^n$, let $S = \sum_{n=0}^{d-1} a_n S_n$, $W = \sum_{n=0}^{d-1} a_n W_n$ and let $\lambda = \min_{0 \leq n \leq d-1} \lambda_n$. Clearly we have

$$\Psi(Sg + Wh - Q) \leq \max_{0 \leq n \leq d-1} (\Psi(a_n) + \lambda_n) \leq \max_{0 \leq n \leq d-1} \Psi(a_n)$$
$$+ \max_{0 \leq n \leq d-1} \lambda_n = \Psi(Q) + \lambda.$$

But trivially: $\Psi(S) \leq \max_{0 \leq n \leq d-1} \Psi(a_n), \Psi(S) \leq \Psi(Q),$

$$\deg(W) \leq \max_{0 \leq n \leq d-1} (\deg(S_n)) < \deg(h),$$

$$\deg(W) \leq \max_{0 \leq n \leq d-1} (\deg(W_n)) < \deg(h).$$

So, λ lies in $E(g, h)$. Then it is obviously seen that $E(g, h)$ is a non-empty interval, and that its lower bound is 0.

Now, let $y \in E(g, h)$ and let ϕ, $\theta \in U[x]$ be monic and satisfy $\Psi(g - \phi) < y$, $\Psi(h - \theta) < y$. Since $\Psi(S) \leq \Psi(Q)$, $\Psi(W) \leq \Psi(Q)$, it is easy to see that $\Psi(S(g - \phi) + W(h - \theta)) < y + \Psi(Q)$, and therefore $\Psi(S\phi + W\theta - Q) < y + \Psi(Q)$. This shows that $y \in E(\phi, \theta)$, and therefore $E(g, h) \subset E(\phi, \theta)$. But similarly we have $E(\phi, \theta) \subset E(g, h)$. $\square$

Lemma 10.1.9. *Let A be complete and let $f(x) = \sum_{j=0}^{\infty} a_j x^j \in A\{x\}$ be such that $\|f\| = |a_q| = 1$, with $J(f) = q$. Let $g, h \in U[x]$ be monic and satisfy $\mathcal{I}(\overline{g}, \overline{h}) = \frac{U}{\mathcal{M}}[\overline{x}]$. Let $\lambda \in E(g, h)$. There exist monic polynomials S, $W \in \mathbb{L}[x]$ satisfying $\Psi(Sg + Wh - f) \leq \lambda + \Psi(f), \deg(S) < \deg(g),$*

$$\deg(W) \leq \max(\deg(h), q - \deg(g)), \Psi(S) \leq \Psi(f), \quad \Psi(W) \leq \Psi(f).$$

Proof. By Theorem 10.1.5, we consider the Euclidean division of f by $gh : f = \ell gh + Q_1$. Hence $\deg(Q_1) < \deg(g) + \deg(h)$. By Theorem 10.1.5, we have

$$\Psi(Q_1) \leq \Psi(f), \qquad (10.1.4)$$

$$\Psi(\ell) \leq \Psi(f). \qquad (10.1.5)$$

By Lemma 10.1.8, there exist $S_1, W_1 \in \mathbb{L}[x]$ satisfying

$$\Psi(S_1 g + W_1 h - Q_1) \leq \Psi(Q_1) + \lambda(g, h), \qquad (10.1.6)$$

$$\Psi(S_1) \leq \Psi(Q_1), \qquad (10.1.7)$$

$$\Psi(W_1) \leq \Psi(Q_1), \qquad (10.1.8)$$

$$\deg(S_1) < \deg(h), \qquad (10.1.9)$$

$$\deg(W_1) < \deg(g). \qquad (10.1.10)$$

Now we put $S = S_1 + \ell h$, $W = W_1$. So we have $Sg + Wh - f = S_1 + \ell gh + W_1 h - \ell gh - Q_1$ and therefore by (10.1.6) we obtain $\Psi(Sg + Wh - f) \leq \Psi(Q_1) + \lambda$. Hence by (10.1.4) we obtain

$$\Psi(Sg + Wh - f) \leq \Psi(f) + \lambda. \qquad (10.1.11)$$

By (10.1.4), (10.1.8) it is seen that

$$\Psi(W) \leq \Psi(f). \qquad (10.1.12)$$

Next, we have $\Psi(h) = 0$, hence by (10.1.5) we see that

$$\Psi(\ell h) \leq \Psi(f). \qquad (10.1.13)$$

But by (10.1.7) we have $\Psi(S_1) \leq \Psi(f)$ and therefore by (10.1.13) we obtain

$$\Psi(S) \leq \Psi(f). \qquad (10.1.14)$$

Finally, by definition we have $\deg(\ell) = \deg(f) - \deg(gh)$ and therefore

$$\deg(W) \leq \max\left(\deg(S_1), \deg(\ell h)\right) \leq \max\left(\deg(h), q - \deg(g)\right). \qquad (10.1.15)$$

Thanks to (10.1.10), (10.1.11), (10.1.12), (10.1.14), (10.1.15). $\square$

Theorem 10.1.10 (Hensel lemma). *A is supposed to be complete. Let $f \in A\{x\}$ be such that $\|f\| = 1$ and such that $\overline{f}$ splits in $\frac{U}{M}[\overline{x}]$ in the form $\gamma\eta$ with $\mathcal{I}(\gamma, \eta) = \frac{U}{M}[\overline{x}]$. There exists a unique pair $(g, h) \in A[x] \times A\{x\}$ such that g is monic and satisfies $f = gh$, $\overline{g} = \gamma$, $\overline{h} = \eta$, $\deg(g) = \deg(\gamma)$.*

Proof.   We can obviously take monic polynomials $g_0, h_0 \in U[x]$ such that $\overline{g_0} = \gamma$, $\overline{h_0} = \eta$. We put $\nu = \Psi(f - g_0 h_0)$, and take $\tau \in E(g_0, h_0)$ satisfying $\tau \geq \nu$. We will construct sequences $(g_n)_{n \in \mathbb{N}}, (h_n)_{n \in \mathbb{N}}$ in $\mathbb{L}[x]$ satisfying for all $n \geq 0$:

(i_n) $\Psi(f - g_n h_n) \leq (n+1)\tau$,
(ii_n) $\Psi(g_n - g_{n-1}) \leq n\tau$, $\Psi(h_n - h_{n-1}) \leq n\tau$,
(iii_n) $\deg(h_n) \leq \deg(f) - \deg(g_0)$, $\deg(g_n) = \deg(g_0)$,
(iv_n) $\overline{g_n} = \gamma$, $\overline{h_n} = \eta$,
(v_n) $\tau \in E(g_n, h_n)$,
(vi_n) g_n is monic.

First we put $f_1 = f - g_0 h_0$. We notice that $\deg(f_1) = \deg(f)$. We now apply Lemma 10.1.9, to the case when $(Q, g, h) = (f_1, g_0, h_0)$: there exist $S_1, W_1 \in A[x]$ satisfying

$$\deg(W_1) < \deg(g_0),$$

$$\deg(S_1) < \deg(f) - \deg(g_0), \tag{10.1.16}$$

$$\Psi(S_1) \leq \tau, \tag{10.1.17}$$

$$\Psi(W_1) \leq \tau, \tag{10.1.18}$$

$$\Psi(S_1 g_0 + W_1 h_0 - P_1) \leq \tau + \Psi(P_1). \tag{10.1.19}$$

Next we put $g_1 = g_0 + W_1$, $h_1 = h_0 + S_1$. We check that $(i_1), (ii_1), (iii_1), (iv_1)$ are satisfied. Moreover, by (10.1.17) and (10.1.18), and by Lemma 10.1.8, τ lies in $E(g_1, h_1)$, hence v_1 is satisfied.

Now we suppose we have already constructed pairs (g_m, h_m) satisfying $(i_m), (ii_m), (iii_m), (iv_m), (v_m)$ for every $m = 0, \dots, n$. Then we put $f_{n+1} = f - g_n h_n$. We can apply Lemma 10.1.9 to the case when (Q, g, h) is equal to (f_{n+1}, g_n, h_n). So, we can obtain $S_{n+1}, W_{n+1} \in A[x]$ satisfying

$$\Psi(S_{n+1} g_n + W_{n+1} h_n - f_{n+1}) \leq \tau + \Psi(f_{n+1}), \tag{10.1.20}$$

$$\deg(W_{n+1}) < \deg(g_n), \tag{10.1.21}$$

$$\deg(S_{n+1}) \leq \max(\deg(h_n), \deg(f_{n+1}) - \deg(g_n)), \tag{10.1.22}$$

$$\Psi(S_{n+1}) \leq \Psi(f_{n+1}), \ \Psi(W_{n+1}) \leq \Psi(f_{n+1}). \tag{10.1.23}$$

By (10.1.20), and by v_n) we obtain

$$\Psi(S_{n+1}g_n + W_{n+1}h_n - f_{n+1}) \leq (n+2)\tau. \qquad (10.1.24)$$

Now we put $g_{n+1} = g_n + W_{n+1}$, $h_{n+1} = h_n + S_{n+1}$. We check that

$$f - g_{n+1}h_{n+1} = (f_{n+1} - h_n W_{n+1} - g_n S_{n+1}) - S_{n+1}W_{n+1}$$
$$= f_{n+1} - hW_{n+1} - g_{n+1}S_{n+1} + (h_{n+1} - h_n)W_{n+1}$$
$$+ (g_{n+1} - g_n)S_{n+1} + S_{n+1}W_{n+1}.$$

By (10.1.21) we notice that G_{m+1} is monic, hence vi_{n+1}) is satisfied.
By ii_m) true for $m \leq n$, we notice that

$$\Psi(g_n - g_{n+1}) \leq (n+1)\tau, \quad \Psi(h_n - h_{n+1}) \leq (n+1)\tau \qquad (10.1.25)$$

which gives (ii_{n+0}), and by (10.1.23) and (10.1.24), we obtain $(i_{n+1})v(f - g_{n+1}h_{n+1}) \geq (n+2)\tau$. Relations (iii_{n+1}), (iv_{n+1}) are easily checked. By (10.1.25) and by Lemma 10.1.8, Relation (v_{n+1}) is also clear. Therefore the sequences $(g_n)_{n \in \mathbb{N}}$, $(h_n)_{n \in \mathbb{N}}$ satisfying (i_n), (ii_n), (iii_n), (iv_n), (v_n) are now constructed. Since A is complete, the A-module $F_q[x]$ of the polynomials of degree $m \leq q$ is obviously complete with respect to the Gauss norm. Then by Relations ii_n) both sequences $(g_n)_{n \in \mathbb{N}}$, $(h_n)_{n \in \mathbb{N}}$ converge in $F_q[x]$. We put $g = \lim_{n \to \infty} g_n$, $h = \lim_{n \to \infty} h_n$. Clearly by (iii_n) we have $\deg(g) = \deg(g_0) = \deg(\gamma)$. By iv_n) we have $\overline{g} = \gamma$, $\overline{h} = \eta$, and finally by i_n) we have $\Psi(f - gh) = -\infty$ hence $f = gh$.

Thus, we have shown the existence of the pair (g, h). Now, since g is a monic polynomial, the equality $f = gh$ appears to be the Euclidean division of f by g. Consequently, such a pair is unique. $\qquad \square$

Corollary 10.1.11. *A is supposed to be complete. Let $f(x) = \sum_{j=0}^{\infty} a_j x^j \in A\{x\}$ and let $q = J(f)$. If a_q is invertible in A then f is associated to a monic polynomial $P \in U[x]$ of degree q, in $A\{x\}$.*

10.2. Definitions of affinoid algebras

Affinoid algebras were introduced by John Tate in [44] who called them *algebras topologically of finite type*. As this last name suggests, such an algebra is the completion of an algebra of finite type for a certain norm. We will recall definitions about algebras topologically of finite type, now mainly called affinoid algebras [3–5, 44].

Definitions and notations: Recall that throughout the book, $\mathbb{L}$ is a complete ultrametric field and here A, B are unital commutative ultrametric Banach $\mathbb{L}$-algebras.

Recall that $U = d(0,1)$, $M = d(0,1^-)$ and the residue class field of $\mathbb{L}$ is $\mathcal{L}$.

Let $(E, \| \cdot \|)$ be a normed $\mathbb{L}$-vector space which also is a A-module. The norm of E will be called a *A-quasi-algebra-norm* if it satisfies $\|af\| \leq \|a\|\|f\| \, \forall a \in A, \, \forall f \in E$. If E is equipped with such a A-quasi-algebra-norm, $(E, \| \cdot \|)$ will be called a *A-quasi-algebra-normed A-module*.

A U-submodule E_1 of A will be said *to define the topology of E* if it is a bounded neighborhood of zero, i.e., if there exist r, $s \in]0, +\infty[$ with $r < s$ such that $\{x \in E| \, \|x\| \leq r\} \subset E_1 \subset \{x \in E| \, \|x\| \leq s\}$.

We will denote by E_0 (resp., A_0) the U-submodule of the $x \in E$ (resp., $x \in A$) such that $\|x\| \leq 1$.

Given $n \in \mathbb{N}$, the algebra $\mathbb{L}\{X_1, \dots, X_n\}$ is called *a topologically pure extension of $\mathbb{L}$ of dimension n* and is denoted by T_n. By definition, an algebra T_m is included in T_n for all $n > m$ and the Gauss norm of T_m is induced by this of T_n. Such a Gauss norm on T_n will be just denoted by $\| \cdot \|$.

Let B be a unital commutative Banach A-algebra. B is said to be a *A-affinoid algebra* if it is isomorphic to a quotient of any algebra of the form $A\{X_1, \dots, X_n\}$ by one of its ideals.

Let $t_1, \dots, t_n \in B$ such that $\|t_j\| \leq 1$ and let ϕ be the $\mathbb{L}$-algebra homomorphism from $A\{X_1, \dots, X_n\}$ into B defined as $\phi(F(X_1, \dots, X_n)) = F(t_1, \dots, t_n)$. We will denote by $A\{t_1, \dots, t_n\}$ its image i.e., the set of sums of series in $t_1, \dots, t_n$ whose coefficients tend to 0 along the filter of complements of finite subsets of $\mathbb{N}^n$.

Remarks. (1) The quotient of a topologically pure extension T_n is a $\mathbb{L}$-affinoid algebra.

(2) If A is a $\mathbb{L}$-affinoid algebra and if B is a quotient of A, then B is a $\mathbb{L}$-affinoid algebra.

(3) By definition, given $f = \sum_{i_1, \dots, i_n} a_{i_1, \dots, i_n} X_1^{i_1} \cdots X_n^{i_n} \in T_n$, there are finitely many coefficients $a_{i_1, \dots, i_n}$ such that $|a_{i_1, \dots, i_n}| = \|f\|$. Consequently, $\|f\|$ lies in $|\mathbb{L}|$ for every $f \in T_n$.

(4) Let $f(X) = \sum_{i=0}^{\infty} a_i X^i \in U\{X\} = (T_1)_0$ and let $q = N^+(f, 0)$. Consider the residue class $\mathcal{L}$-algebra $\overline{A} = \frac{(T_1)_0}{M(T_1)_0}$. In $\overline{A}$ we denote by $\overline{f}$ the residue class of $f(X)$. Then $\overline{f}$ is the polynomial $\sum_{i=0}^{q} \overline{a}_i X^i$.

(5) Since the Gauss norm on T_n is multiplicative, it is obviously equal to the associated semi-norm $\| \cdot \|_{\text{sp}}$.

Lemma 10.2.1. *Let* $\mathbb{L} = \mathbb{K}$, *let* $f_1, \ldots, f_m \in T_1$. *There exists* $\alpha \in U_\mathbb{K}$ *such that* $|f_j(\alpha)| = \|f_j\| \; \forall j = 1, \ldots, m$.

Proof. Indeed, for each $j = 1, \ldots, m$ the equality $|f_j(x)| = \|f_j\|$ holds in all classes of $U_\mathbb{K}$ except in finitely many ones. Consequently, all equalities $|f_j(x)| = \|f_j\|$ hold in all classes of $U_\mathbb{K}$ except in finitely many. $\qquad\square$

Proposition 10.2.2. *Let* A, B *be* $\mathbb{L}$-*affinoid algebras such that* B *is* A-*affinoid. Suppose that* $\frac{B_0}{MB_0}$ *is finite over* $\frac{A_0}{MA_0}$. *Then* B *is finite over* A.

Proof. By hypothesis, A is the quotient of certain $\mathbb{L}$-algebra $A\{T_1, \ldots, T_r\}$, and therefore is of the form $A\{t_1, \ldots, t_r\}$. Let $\lambda \in M$, $\lambda \neq 0$. For each $t \in B_0$, here we denote by $\tilde{t}$ the residue class of t in $\frac{B_0}{\lambda B_0}$. Thus, we have $\frac{B_0}{\lambda B_0} = \frac{A_0}{\lambda A_0}[\tilde{t}_1, \ldots, \tilde{t}_r]$. Now, by hypothesis for each $i = 1, \ldots, r$, there exists a polynomial $P_i(T) = T^{q_i} + f_{q_i-1}T^{q_i-1} + \cdots + f_0$, with $f_i \in A_0$, and $P_i(t_i) \in MB_0$. Let $P_i(t_i) = x_i g_i$, with $x_i \in M$, and $g_i \in B_0$. Now, we can choose $l_i \in \mathbb{N}$ such that $x_i^{l_i} \in \lambda A_0$, and thus $P_i(t_i)^{l_i}$ lies in λB_0, hence $\overline{P}_i(\tilde{t}_i)^{l_i} = 0$. Therefore, $\frac{B_0}{\lambda B_0}$ is finite over $\frac{A_0}{\lambda A_0}$, and finally, by Proposition 2.6.10 of Chapter 2, B is finite over A. $\qquad\square$

Lemma 10.2.3. *Let* A *be a* $\mathbb{L}$-*affinoid algebra which is a quotient of a topologically pure extension* $\mathbb{L}\{X_1, \ldots, X_n\}$ *of* $\mathbb{L}$ *by an ideal. Let* θ *be the canonical surjection, and for each* $i = 1, \ldots, n$, *we put* $t_i = \phi(X_i)$. *Then* $U\{t_1, \ldots, t_n\}$ *defines the topology of* A.

Proof. By definition, $U\{t_1, \ldots, t_n\}$ is included in B_0. So we just have to check that $U\{t_1, \ldots, t_n\}$ is a neighborhood of 0 in D. Let $\lambda \in]0, 1[$ and let $f \in B$ satisfy $\|f\| \leq \lambda$. Since $\lambda < 1$, by definition of the quotient norm of B, there exist $F \in U\{X_1, \ldots, X_n\}$ such that $\phi(F) = f$. Hence $f = F(t_1, \ldots, t_n)$ lies in $U\{t_1, \ldots, t_n\}$, and therefore $U\{t_1, \ldots, t_n\}$ is a neighborhood of 0 in B. $\qquad\square$

Theorem 10.2.4. *Suppose the topology of* A *is defined by a* U-*algebra* A_1. *Let* $y_1, \ldots, y_s \in A_1$ *be such that their residue classes in* $\frac{A_1}{MA_1}$ *be algebraically independent over* $\mathcal{L}$. *Then the canonical homomorphism* ϕ *from* $\mathbb{L}\{Y_1, \ldots, Y_s\}$ *into a* A *defined as* $\phi(Y_j) = y_j$, $(1 \leq j \leq s)$ *is an isomorphism from* $\mathbb{L}\{Y_1, \ldots, Y_s\}$ *onto a closed subalgebra of* A.

Proof. Let $F(Y_1, \ldots, Y_s) \in L\{Y_1, \ldots, Y_s\}$ be such that $F(y_1, \ldots, y_s) \in A_1$. Suppose that $F(Y_1, \ldots, Y_s) \notin U\{Y_1, \ldots, Y_s\}$. Since $\|F\|$ lies in $|\mathbb{L}|$, there exists $\lambda \in M$ such that $\|\lambda F\| = 1$. Let $P = \lambda F$. Then the residue class $\overline{P}$ of P in $\mathcal{L}[X_1, \ldots, X_s]$ is different from 0, but satisfies $\overline{P}(\overline{y}_1, \ldots, \overline{y}_s) = 0$ in $\frac{A_1}{MA_1}$, a contradiction. Consequently, $F(Y_1, \ldots, Y_s)$ belongs to $U\{Y_1, \ldots, Y_s\}$, and therefore ϕ is open. Hence, ϕ is a bicontinuous isomorphism from $\mathbb{L}\{Y_1, \ldots, Y_s\}$ onto a closed $\mathbb{L}$-subalgebra of A. □

Proposition 10.2.5. *A is a $\mathbb{L}$-affinoid algebra if and only if it is a finite extension of a topologically pure extension of $\mathbb{L}$ of the form $T[y_1, \ldots, y_q]$ where T is a topologically pure extension and $y_1, \ldots, y_q$ are elements of A such that $\max_{1 \leq j \leq q} \|y_j\| \leq 1$, so that A is isomorphic to a quotient of $\mathbb{L}\{X_1, \ldots, X_n, Y_1, \ldots, Y_q\}$.*

Proof. First suppose that A is a finite extension of a topologically pure extension of $\mathbb{L}$, say $A = \mathbb{L}\{X_1, \ldots, X_n\}[y_1, \ldots, y_q]$. Without loss of generality, we may clearly assume that $\max(|y_1|, \ldots, |y_q|) \leq 1$. Then the canonical homomorphism ϕ from $\mathbb{L}[X_1, \ldots, X_n, Y_1, \ldots, Y_q]$ into A defined by $\phi(X_i) = X_i \; \forall i = 1, \ldots, n$, and $\phi(Y_j) = y_j \; \forall j = 1, \ldots, q$, is clearly continuous with respect to the Gauss norm on $\mathbb{L}[X_1, \ldots, X_n, Y_1, \ldots, Y_q]$ and extends by continuity to a homomorphism $\hat{\phi}$ from $\mathbb{L}\{X_1, \ldots, X_n, Y_1, \ldots, Y_q\}$ into A. Then $\hat{\phi}$ is surjective by construction, so A is a quotient of $\mathbb{L}\{X_1, \ldots, X_n, Y_1, \ldots, Y_q\}$.

Conversely, we now assume that A is a $\mathbb{L}$-affinoid algebra, and suppose that A is a quotient of a topologically pure extension $T = \mathbb{L}\{X_1, \ldots, X_n\}$ of $\mathbb{L}$. Let θ be the canonical surjection, and for each $i = 1, \ldots, n$, let $t_i = \phi(X_i)$. Then by Lemma 10.2.3, the U-algebra $U\{t_1, \ldots, t_n\}$ defines the topology of T. Now, consider the residue algebra $\frac{T_1}{MT_1}$, and for each $i = 1, \ldots, n$, let τ_i be the residue class of t_i in $\frac{T_1}{MT_1}$. Then $\frac{T_1}{MT_1} = \mathcal{L}[\tau_1, \ldots, \tau_n]$. By Theorem 1.1.16 of Chapter 1, we can choose $\xi_1, \ldots \xi_s \in \frac{T_1}{MT_1}$ such that $\frac{T_1}{MT_1}$ be finite over $\mathcal{L}[\xi_1, \ldots, \xi_s]$. For each $i = 1, \ldots, s$, we can choose $y_i \in T_1$ whose residue class is ξ_i. Let $E = \mathbb{L}\{y_1, \ldots, y_s\}$. By Proposition 10.2.4, E is a topologically pure extension of $\mathbb{L}$. Now, putting $E_0 = U\{y_1, \ldots, y_s\}$, we can see that $\frac{T_1}{MT_1}$ is finite over $\frac{E_0}{ME_0}$, and therefore, by Proposition 10.2.2, A is finite over E. □

Proposition 10.2.6. *Let A be a $\mathbb{L}$-affinoid algebra without divisors of zero of the form $\mathbb{L}\{Y\}[x]$, with x integral over $\mathbb{L}\{Y\}$ and $\|x\| \leq 1$ and let $F(x) = irr(x, \mathbb{L}\{Y\})$. Then A is isomorphic to both $\frac{\mathbb{L}\{Y\}[X]}{F(X)\mathbb{L}\{Y\}[X]}$ and to $\frac{\mathbb{L}\{Y,X\}}{F(X)\mathbb{L}\{Y,X\}}$.*

Proof. By Theorem 10.2.5, A is isomorphic to a quotient of $\mathbb{L}\{Y, X\}$ by a prime ideal $\mathcal{I}$ that contains F. Hence A is isomorphic to a quotient of the $\mathbb{L}$-algebra $\frac{\mathbb{L}\{Y,X\}}{F(X)\mathbb{L}\{Y,X\}}$, hence there exists a surjective homomorphism ϕ from $\frac{\mathbb{L}\{Y,X\}}{F(X)\mathbb{L}\{Y,X\}}$ onto A. On the other hand, there exists an obvious injective homomorphism θ from the $\mathbb{L}$-algebra $D = \frac{\mathbb{L}\{Y\}[X]}{F(X)\mathbb{L}\{Y\}[X]}$, which is isomorphic to A, into B inducing the identity on $\mathbb{L}\{Y\}$. □

Theorem 10.2.7. *Suppose that* $\mathbb{L} = \mathbb{K}$. *Let* $f \in T_n$. *Then* f *is invertible in* T_n *if and only if* $|f((0))| > \|f - f((0))\|$.

Proof. If $|f((0))| > \|f - f((0))\|$, f is invertible by Theorem 2.5.11. Now suppose $|f((0))| \le \|f - f((0))\|$. Since $sp(f) = d(f(0), \|f - f(0)\|)$, it is seen that 0 belongs to $sp(f)$, hence f is not invertible. $\qquad \square$

10.3. Salmon's theorem

The chapter is designed to show that topologically pure extensions are factorial, a theorem due to Paolo Salmon [42].

Definitions and notations: Recall that $U = d(0,1)$, $M = d(0,1^-)$ and the residue class field of $\mathbb{L}$ is $\mathcal{L}$. Here, T_n denotes the topologically pure extension $\mathbb{L}\{X_1, \ldots, X_n\}$ and we put $V_n = \{x \in T_n \mid \|x\| \le 1\}$.

Let $f(X_1, \ldots, X_n) \in T_n$. We denote by $\overline{f}$ the residue class of f in $\frac{V_n}{M_{\mathbb{L}} V_n}$. Then f is said to be k-regular if it is of the form cg with $c \in \mathbb{L}$ and $g \in V_n$ such that $\overline{g}$ is monic as a polynomial in $\overline{X}_k$.

Lemma 10.3.1. *Let* $P, Q \in V_n[Y]$ *be monic polynomials associated to each other in* $T_n\{Y\}$. *Then* $P = Q$.

Proof. By hypothesis there exists u invertible in $T_n\{Y\}$ such that $Q = uP$. Without loss of generality, we can assume that P and Q belong to $\mathbb{K}[X]$. So, by Theorem 10.2.7, u is of the form $l + h$ with $l \in \mathbb{L}$, $h \in T_n\{Y\}$ and $\|h\| < |l|$. Since the Gauss norm is multiplicative, and since both P, Q are monic, we have $\|P\| = \|Q\| = 1$ and then $\|u\| = |l| = 1$. Suppose $P \ne Q$. Since both $\overline{P}$, $\overline{Q}$ are monic and associated in $\frac{V_n}{M_{\mathbb{L}} V_n}[\overline{X}_n]$, they have same degree, and therefore we check that $\overline{l} = \overline{1} = \overline{u}$. Consequently, u is of the form $1 + f$, with $f \in T_n\{Y\}$ and $\|f\| < 1$. Therefore, $Pf = Q - P$. Let $\lambda \in \mathbb{L}$ be such that $\|\lambda(P - Q)\| = 1$. Then the degree in Y of $\overline{\lambda(P - Q)}$ is at most $q - 1$. But since P is monic, and since $\|\lambda f P\| = 1$, the degree in Y of $\overline{\lambda P f}$ is at least q, a contradiction. Hence $P = Q$. $\qquad \square$

Lemma 10.3.2. *Let* $f \in T_n$ *be* n-*regular. There exists one and only one monic polynomial* $P \in U_{n-1}[X_n]$ *associated to* f *in* T_n.

Proof. Since f is n-regular of order q, it is of the form $\sum_{j=0}^{\infty} a_j X_n^j$, with $a_j \in U_{n-1}$ and a_q invertible in U_{n-1}. Hence by Corollary 10.1.11 f is associated in T_n to a monic polynomial in X_n. Suppose we can find two monic polynomials P and Q associated to f in T_n. In particular, Q is associated to P in T_n. Without loss of generality we can clearly assume that both P, Q belong to V_n. Then by Lemma 10.3.1, they are equal. $\qquad \square$

Definition: Let $f \in T_n$ be k-regular. The unique monic polynomial $P \in U\{X_1, \ldots, X_{k-1}, X_{k+1}, \ldots, X_n\}[X_k]$, associated to f in T_n will be called *the k-canonical associate to f*.

Let $f \in T_n$ be k-regular and let $f = cg$ with g the k-canonical associate to f. By definition, the degree q of $\bar{g}$ in $\overline{X}_k$ is such that $|a_{0,\ldots,0,q,0,\ldots,0} - 1| < 1$, $|a_{j_1,\ldots,j_{k-1},q,j_{k+1},\ldots,j_n}| < 1 \ \forall (j_1, \ldots, j_{n-1}) \neq (0, \ldots, 0)$, and $|a_{j_1,\ldots,j_n}| < 1$ whenever $j_n > q$. The integer q will be called *the k-order of f*.

Theorem 10.3.3. *Let $f \in T_n$ be k-regular of k-order q. There exists a unique monic polynomial $P(X_k) \in \mathbb{L}\{X_1, \ldots, X_{k-1}, X_{k+1}, \ldots, X_n\}[X_k]$ which is associated to f in T_n. Moreover, the degree of P in X_k is q.*

Proof. Without loss of generality, we can obviously assume that $k = n$, what we do for convenience. We put $T_{n-1} = E$ and $Y = X_n$. Without loss of generality, we can also assume that the coefficient $a_{0,\ldots,0,q}$ of Y in f is 1. Let q be the n-order of f. In $E\{Y\}$ f is of the form $\sum_{j=0}^{\infty} f_j Y^j$ with $\lim_{j \to \infty} f_j = 0$, $\|f_j\| \leq 1 \ \forall j \in \mathbb{N}$, $f_q = 1$ and $\|f_j\| < 1 \ \forall j > q$. Then $\bar{f}$ is a monic polynomial Π of degree q, hence by Corollary 10.1.11, there exists $Q \in E[Y]$ such that $\bar{Q} = \bar{f}$ and $h \in E\{Y\}$ such that $\bar{h} = 1$ and $f = Qh$. Since $\bar{Q} = \bar{f}$, the coefficient l of Y^q in Q is of the form $1 + \epsilon$ with $\|\epsilon\| < 1$, therefore by Theorem 2.5.11 of Chapter 2, Q is invertible in the Banach $\mathbb{L}$-algebra T_n. In the same way, since $\bar{h} = 1$, h is invertible in T_n. Consequently, the q-degree monic polynomial $P = l^{-1}Q$ is associated to f in T_n.

Now, we will show such a factorization is unique. Let wS be another factorization of f, with w invertible in T_n and S a monic polynomial in $E[Y]$. Thus in $E[Y]$ P and S are two monic polynomials associated to each other. Consequently, by Lemma 10.3.1, they are equal and this shows the uniqueness of the polynomial P. $\square$

Theorem 10.3.4. *Let $f \in T_n$ be k-regular of k-order q and factorize in T_n in the form $f_1 f_2$. Then, each f_j is k-regular of order q_j $(j = 1, 2)$, such that $q = q_1 + q_2$.*

Proof. Without loss of generality, we can assume that f, f_1, f_2 lie in V_n and that $\bar{f}$ is monic of degree q. Then $\bar{f} = \bar{f}_1 \bar{f}_2$. Putting $q_j = \deg(\bar{f}_j(X_k))$, $(j = 1, 2)$, we have $q = q_1 + q_2$. Since $\bar{f}$ is monic, each $\bar{f}_j$ has a coefficient of degree q_j in $\overline{X}_k$ which is constant, hence each f_j is k-regular, of k-order q_j. $\square$

Theorem 10.3.5. *Let $f \in U_{n-1}[X_n]$ be monic and irreducible in $T_{n-1}[X_n]$. Then it is irreducible in $T_{n-1}\{X_n\}$.*

Proof. Suppose f is not irreducible in T_n. By Theorem 10.3.4, there exists n-regular elements f_1, $f_2 \in T_n$ such that $f = f_1 f_2$, and therefore by Theorem 10.3.3, there exists a unique monic polynomial $g_1(X_n) \in T_{n-1}[X_n]$ associated to f_1 in T_n and a unique monic polynomial $g_2(X_n) \in T_{n-1}[X_n]$ associated to f_2 in T_n. Then $g_1 g_2$ is a monic polynomial in X_n and is associated to f hence it is equal to f, a contradiction to the hypothesis: f irreducible in $T_{n-1}[X_n]$. $\square$

Notations: In $\mathbb{N}^n$ we will denote by $\mathcal{O}_n$ the order relation defined as it follows: given $(i_1, \dots, i_n)$, $(j_1, \dots, j_n) \in \mathbb{N}^n$, we put $(i_1, \dots, i_n) < (j_1, \dots, j_n)$ if for certain $t \leq n$ we have $i_{t-1} < j_{t-1}$ and $i_m = j_m \ \forall m = t, \dots, n$, and we put $(i_1, \dots, i_n) \leq (j_1, \dots, j_n)$ if either $(i_1, \dots, i_n) < (j_1, \dots, j_n)$ or $(i_1, \dots, i_n) = (j_1, \dots, j_n)$. Then this relation is a total order relation in $\mathbb{N}^n$.

Proposition 10.3.6. *Let $h \in V_n$ be such that in $\overline{h}$, the coefficient of maximal index, with respect to $\mathcal{O}_n$ is equal to 1. There exists a $\mathbb{L}$-automorphism θ of T_n preserving V_n, such that $\overline{\theta(h)}$ is a monic polynomial in $\overline{X}_n$.*

Proof. Let $(k_1, \dots, k_n)$ be the maximum index (with respect to $\mathcal{O}_n$) whose coefficient in $\overline{h}$ is not 0. We can clearly find $t_2, \dots, t_n \in \mathbb{N}$ such that $h(X_1, X_2 + X_1^{t_2}, \dots, X_n + X_1^{t_n})$ is a monic polynomial in X_1 of degree $k_1 + k_2 t_2 + \cdots + k_n t_n$. Let ϕ be the $\mathbb{L}$-algebra endomorphism defined in $\mathbb{L}[X_1, \dots X_n]$ as $\phi(f) = f(X_1, X_2 + X_1^{t_2}, \dots, X_n + X_1^{t_n})$. Then we can check that ϕ is a $\mathbb{L}$-algebra automorphism of $\mathbb{L}[X_1, \dots X_n]$. Moreover, it also induces on $\mathcal{L}[\overline{X}_1, \dots, \overline{X}_n]$ a $\mathcal{L}$-algebra automorphism, so it satisfies $\|\psi(f)\| = \|f\|$. Therefore ϕ has continuation to T_n and preserves V_n. Now, let ω be the $\mathbb{L}$-algebra automorphism of T_n defined as $\omega(f(X_1, \dots, X_n)) = f(X_n, X_2, \dots, X_{n-1}, X_1)$. Then $\omega \circ \phi$ is a continuous $\mathbb{L}$-algebra automorphism θ of T_n such that $\overline{\theta(h)}$ is a monic polynomial in $\overline{X}_n$. $\square$

Theorem 10.3.7 ([42]). *T_n is factorial for every $n \in \mathbb{N}$.*

Proof. We assume the theorem true for every $n < m$, and will prove it when $n = m$. For convenience we put $E = T_{n-1}$. Let g be an irreducible element of T_n. We will prove that $g T_n$ is a prime ideal of T_n. Without loss of generality we can obviously assume that g belongs to V_n and that in $\overline{g}$ the coefficient of maximal index, with respect to $\mathcal{O}_n$, is equal to 1. Consequently, by Proposition 10.3.6, there exists a $\mathbb{L}$-algebra automorphism θ of T_n preserving V_n, such that $\overline{\theta(g)}$ is a monic polynomial in $\overline{X}_n$. Thus, without loss of generality, we can also assume that $\overline{g}$ is a monic polynomial in $\overline{X}_n$. Then, by Theorem 10.3.2, g is associated to a monic polynomial

$P(X_n) \in U_{n-1}[X_n]$. Let $q = \deg(P)$. Suppose that P admits a factorization in $U_{n-1}[X_n]$ in the form $Q_1 Q_2$, with $Q_i(X_n) = \sum_{j=0}^{q_i} a_{j,i} X_n^j$ $(i = 1, 2)$. Then $q_1 + q_2 = q$ and $a_{q_1} a_{q_2} = 1$. Since g is irreducible in T_n, so is P, hence one of the Q_i is invertible in T_n. Both polynomials $\frac{Q_i}{a_{q_i}}$ are monic and $\frac{Q_1 Q_2}{a_{q_1} a_{q_2}}$ divides P in T_n. Now, we consider T_n as $E\{X_n\}$. By Theorem 10.3.3, there exists only one monic polynomial in X_n associated to a n-regular series such as P. Hence one of the polynomials $\frac{Q_i}{a_{q_i}}$ is equal to 1, and therefore P is irreducible in $E[X_n]$.

Now, suppose that P divides a product $f_1 f_2$ in $E\{X_n\}$ with f_1, $f_2 \in E\{X_n\}$. And according to Theorem 10.1.5, let R_i be the rest of the Euclidean division of f_i by P $(i = 1, 2)$ in $E\{X_n\}$. Then P divides $R_1 R_2$ in B. But by Theorem 10.1.5, the Euclidean division in $E[X_n]$ is induced by the one in $E\{X_n\}$. Consequently, P divides $R_1 R_2$ in $E[X_n]$. And since, by induction hypothesis, E is factorial, so is $E[X_n]$. Consequently P must divide one of the R_i, say R_1, in $E[X_n]$. But since $\deg(R_1) < q$, of course $R_1 = 0$, hence P divides f_1, and therefore generates a prime ideal in $E[X_n]$ and so does g in $E\{X_n\}$. Thus every irreducible element of T_n generates a prime ideal, and consequently, by Lemma 1.1.2 of Chapter 1, we know that $E\{X_n\}$ is factorial, hence the theorem holds when $m = n$. $\qquad \square$

Corollary 10.3.8. *T_n is Noetherian and every ideal of T_n is closed.*

As consequence of the division, preparation theorems and Noether normalization, T_n is that an analog of Hilbert's Nullstellensatz is valid: the radical of an ideal is equal to the intersection of all maximal ideals containing this ideal [5].

Theorem 10.3.9 (Nullstellensatz for T_n). *Let $\mathcal{I}$ be an ideal of T_n. The radical of $\mathcal{I}$ is equal to the intersection of all maximal ideals containing $\mathcal{I}$.*

Corollary 10.3.10. *Let $\mathcal{I}$ be a prime ideal ideal of T_n. Then $\mathcal{I}$ is equal to the intersection of all maximal ideals containing $\mathcal{I}$.*

10.4. Algebraic properties of affinoid algebras

We shall recall the proofs of many classical algebraic properties of $\mathbb{L}$-affinoid algebras such as Noetherianness, finite codimension of maximal ideals and other easy spectral properties [44].

Notations: As previously, we just denote bu U the unit ball of $\mathbb{L}$ and by M the disk $d(0, 1^-)$. Moreover, we denote by U_n the unit ball of $\mathbb{L}^n$. Throughout the chapter, A is a $\mathbb{L}$-affinoid algebra.

Let us first notice Lemma 10.4.1 that is immediate.

Lemma 10.4.1. *If B a A-affinoid algebra, then B is a $\mathbb{L}$-affinoid algebra.*

Theorem 10.4.2. *A is Noetherian, every ideal is closed. Furthermore, each maximal ideal of A is of finite codimension.*

Proof. By Corollary 10.3.8, every topologically pure extension T_n is Noetherian, hence so is any quotient of T_n, i.e., any $\mathbb{L}$-affinoid algebra. Consequently, by Corollary 2.6.13 of Chapter 2, every ideal is closed.

Finally, consider a maximal ideal $\mathcal{M}$ of B. Then, $\frac{B}{\mathcal{M}}$ is a field F and a $\mathbb{L}$-affinoid algebra. Thus, F is finite over a topologically pure extension $\mathbb{L}\{X_1, \ldots, X_q\}$, and then $\mathbb{L}\{X_1, \ldots, X_q\}$ is a field. But such a ring is never a field except when $q = 0$. Hence, F is a finite field extension of $\mathbb{L}$ and hence $\mathcal{M}$ is of finite codimension. $\square$

Corollary 10.4.3. *Let $B = T_n[y_1, \ldots, y_q]$ be finite over a topologically pure extension T_n. Then B is isomorphic to a $\mathbb{L}$-affinoid algebra of the form $\frac{T_n\{Y_1, \ldots, Y_q\}}{I}$ where I is an ideal of $T_n\{Y_1, \ldots, Y_q\}$ such that $T_n \cap I = \{0\}$.*

Corollary 10.4.4. *In A there are finitely many minimal prime ideals.*

Corollary 10.4.5. *Let A admit an idempotent u. Then uA is a $\mathbb{L}$-affinoid algebra admitting u for unity.*

Theorem 10.4.6. *A is multbijective.*

Proof. Let $\mathcal{M}$ be maximal ideal of A. Then the field $M = \frac{A}{\mathcal{M}}$ is a finite extension of $\mathbb{L}$, Hence it admits a unique absolute value $|\,.\,|$ expansion of the absolute value of $\mathbb{L}$. Let χ be the canonical morphism from M onto A. Then $|\chi|$ is the unique continuous multiplicative semi-norm on A admitting $\mathcal{M}$ for kernel, which ends the proof. $\square$

Theorem 10.4.7. *For all $\alpha \in U_n$ and $f \in T_n$, we put $\chi_\alpha(f) = f(\alpha)$. The mapping ϕ from U_n into $\mathcal{X}(T_n)$ defined as $\phi(\alpha) = \chi_\alpha$ is a bijection.*

Proof. Indeed ϕ is an injection from U_n into $\mathcal{X}(T_n)$. Now, let $\gamma \in \mathcal{X}(T_n)$, and let $T_n = \mathbb{L}\{X_1, \ldots, X_n\}$. By Theorem 10.4.2, γ takes values in $\mathbb{L}$. For every $j = 1, \ldots, n$, let $\alpha_j = \varphi(X_j)$, and let $\alpha = (\alpha_1, \ldots, \alpha_n)$. Then we see that $\gamma = \chi_\alpha$. $\square$

Corollary 10.4.8. *Let $\chi \in \mathcal{X}(T_{n-1})$ and let $\alpha \in U$. There exists $\hat{\chi} \in \mathcal{X}(T_n)$ such that $\hat{\chi}(h) = \chi(h)$ $\forall h \in T_{n-1}$, and $\hat{\chi}(X_n) = \alpha$.*

Proposition 10.4.9. *The following three properties are equivalent:*

(i) $|\chi(f)| < 1 \ \forall \chi \in \mathcal{X}(A)$,

(ii) *In $A\{Y\}$, $1 - fY$ is invertible,*

(iii) $\|f\|_{\mathrm{sp}} < 1$.

Proof. Let us show that (i) implies (ii). By Lemma 10.4.1, $A\{Y\}$ is a $\mathbb{L}$-affinoid algebra. Next, by definition, we have $\|Y\| = 1$. Let $\chi \in \mathcal{X}(A\{Y\})$. Then $|\chi(fY)| < 1$, hence $\chi(1 - fY) \neq 0$. Consequently, by Theorem 10.4.2, $1 - fY$ does not belong to any maximal ideal of $A\{Y\}$, which proves (ii).

Let us show that (ii) implies (iii). Since $1 - fY$ is invertible, its inverse is a series $\sum_{n=0}^{\infty} g_n Y^n$, with $g_n \in A$ and $\lim_{n \to \infty} g_n = 0$. Since $(1 - fY)g = 1$, we check that $g_0 = 1$, and $g_{n+1} = fg_n$, $\forall n \in \mathbb{N}$. Consequently, $g_n = f^n$, and $\lim_{n \to \infty} f^n = 0$, hence by Theorem 2.5.7 of Chapter 2, $\|f\|_{\mathrm{sp}} < 1$.

Finally, we check that that (iii) implies (i): by Theorem 2.5.17, of Chapter 2, we have $|\chi(f)| \leq \|f\|_{\mathrm{sp}}$. $\qquad \square$

Theorem 10.4.10. *Every $\mathbb{L}$-affinoid algebra owns Property (p).*

Proof. Suppose a $\mathbb{L}$-affinoid algebra A does not own Property p), and let $f \in A$ be such that $\|f\|_{\mathrm{sa}} < \|f\|_{\mathrm{sp}}$. We can find power f^t of f such that the interval $]\|f^t\|_{\mathrm{sa}}, \|f\|_{\mathrm{sp}}[$ contains some $|\lambda| \in |\mathbb{L}|$. So we can assume that $\|f\|_{\mathrm{sa}} < |\lambda| < \|f\|_{\mathrm{sp}}$ and we have $\left\|\frac{f}{\lambda}\right\|_{\mathrm{sa}} < 1 < \left\|\frac{f}{\lambda}\right\|_{\mathrm{sp}}$, hence Condition (i) is satisfied but Condition (iii) is not, a contradiction. $\qquad \square$

Lemma 10.4.11. *Let $\mathcal{I}$ be an ideal of T_n, and let $A = \frac{T_n}{\mathcal{I}}$. Let θ be the canonical surjection from T_n onto A. Let $\phi \in \mathrm{Mult}(T_n, \| \cdot \|)$. There exists $\varphi \in \mathrm{Mult}(A, \| \cdot \|)$ such that $\phi = \varphi \circ \theta$ if and only if $\mathcal{I} \subset \mathrm{Ker}(\theta)$.*

Proof. Let $\| \cdot \|$ be the Gauss norm on T_n, and let $\| \cdot \|_q$ be the quotient $\mathbb{L}$-algebra norm of A. Let $f \in T_n$. Suppose that $\mathcal{I} \subset \mathrm{Ker}(\theta)$. For every $t \in \mathcal{I}$ we have $\theta(f) = \theta(f + t)$ hence $\phi(f) = \phi(f + t)$. Therefore, we can put $\varphi(\theta(f)) = \phi(f)$.

Conversely, if φ is of the form $\phi \circ \theta$, with $\phi \in \mathrm{Mult}(A, \| \cdot \|)$, then $\mathrm{Ker}(\varphi)$ obviously contains $\mathcal{I}$. $\qquad \square$

Lemma 10.4.12. *Let A be a $\mathbb{L}$-affinoid algebra, let $\mathcal{I}$ be an ideal of A, let $B = \frac{A}{\mathcal{I}}$ and let $\| \cdot \|_q$ be the quotient norm of B. Then $\mathrm{Mult}(B, \| \cdot \|_q)$ is homeomorphic to the subset of $\mathrm{Mult}(A, \| \cdot \|)$ consisting of the ϕ such that $\mathcal{I} \subset \mathrm{Ker}(\phi)$.*

Proof. Let θ be canonical surjection from A onto B and let ψ be the mapping from $\mathrm{Mult}(B, \| \cdot \|_q)$ into $\mathrm{Mult}(A, \| \cdot \|)$ defined by $\psi(\varphi) = \varphi \circ \theta$. Then ψ is an injection because given φ_1 and $\varphi_2 \in \mathrm{Mult}(B, \| \cdot \|_q)$ such that $\varphi_1 \circ \theta = \varphi_2 \circ \theta$ we have $\varphi_1(\theta(x)) = \varphi_2(\theta(x)$ $\forall x \in A$, hence $\varphi_1(u)) = \varphi_2(u)$ $\forall u \in B$. Now, let $\phi \in \mathrm{Mult}(A, \| \cdot \|)$ be such that $\mathcal{I} \subset \mathrm{Ker}(\phi)$ and let $J = \mathrm{Ker}(\phi)$. Let $x \in A$. Then $\theta(x)$ lies in B. Given x, $y \in A$ such that $\theta(x) = \theta(y)$ we have $x - y \in \mathcal{I}$, hence $x - y \in J$, therefore $\phi(x) = \phi(y)$. Consequently, we can put $\varphi(\theta(x)) = \phi(x)$, i.e., $\varphi(u) = \phi(x)$ whenever $\theta(x) = u$. Consequently, ψ is a surjection, and hence an bijection from $\mathrm{Mult}(B, \| \cdot \|_q)$ onto the subset of $\mathrm{Mult}(A, \| \cdot \|)$ consisting of the ϕ such that $\mathcal{I} \subset \mathrm{Ker}(\phi)$. The homeomorphism is then immediate. $\qquad\square$

Theorem 10.4.13. *Suppose that* $\mathbb{L} = \mathbb{K}$. *For all* $x \in T_n$ *there exists* $\chi \in \mathcal{X}(T_n)$ *such that* $|\chi(x)| = \|x\|_{\mathrm{sp}}$.

Proof. Let U be the disk $d(0,1)$ in $\mathbb{K}$. Since the Gauss norm on T_n satisfies $\| \cdot \| = \| \cdot \|_{\mathrm{sp}}$, we are reduced to show that for every $f \in T_n$ there exists $\chi \in \mathcal{X}(T_n)$ such that $|\chi(f)| = \|f\|$. We will proceed by induction. When $n = 1$, by Theorem 7.3.5 of Chapter 7, there exists $\alpha \in U$ such that $|f(\alpha)| = \|f\|$. Consequently, the mapping $\chi_\alpha \in \mathcal{X}(T_1)$ defined as $\chi_\alpha(g) = g(\alpha)$ $(g \in T_1)$ satisfies $|\chi_\alpha(f)| = \|f\|_{\mathrm{sp}}$. Now, suppose we have already proven the following property.

$(\mathcal{Q}_n)$: For every $n \le q$, given any $f_1, \ldots, f_m \in T_n$, there exists $\chi \in \mathcal{X}(T_n)$ such that $|\chi(f_j)| = \|f_j\|$ $\forall j = 1, \ldots, m$. Consider $f_1, \ldots, f_m \in T_{q+1}$. We can write each f_h in the form $\sum_{k=0}^{\infty} g_{k,h} X_{q+1}^k$, with $g_{k,h} \in T_q$. There obviously exists $N \in \mathbb{N}$ such that $\|g_{k,h}\| < \|f_h\|$ whenever $k \ge N$, whenever $h = 1, \ldots, m$. Now by the induction hypothesis we can find $\chi \in \mathcal{X}(T_q)$ such that $|\chi(\prod_{0 \le k \le N,\ 1 \le h \le m} g_{k,h})| = \|\prod_{0 \le k \le N,\ 1 \le h \le m} g_{k,h}\|$. And since both χ, $\| \cdot \|$ are multiplicative, and obviously satisfy $|\chi(g_{k,h})| \le \|g_{k,h}\|$ $\forall k = 0, \ldots, N$, $\forall h = 1, \ldots, m$, it is clear that $|\chi(g_{k,h})| = \|g_{k,h}\|$ $\forall k = 0, \ldots, N$, $\forall h = 1, \ldots, m$. For each $h = 1, \ldots, m$, consider $\phi_h(X) = \sum_{k=0}^{\infty} \chi(g_{k,h}) X^k$ which obviously lies in T_1. Since the property holds for $n = q$, there exists $\chi \in \mathcal{X}(T_1)$ such that $\chi(\phi_h) = \|\phi_h\|$. Putting $\alpha = \chi(X)$, we have $|\phi_h(\alpha)| = \|\phi_h\|$ $\forall h = 1, \ldots, m$. Therefore, we can clearly give χ an extension to an element $\hat\chi \in \mathcal{X}(T_{q+1})$ defined as $\hat\chi(\sum_{k=0}^{\infty}(g_k) X_{q+1}^k) = \sum_{k=0}^{\infty} \chi(g_k) \alpha^k$, and this homomorphism $\hat\chi$ satisfies $\|f_h\| = |\hat\chi(f_h)|$ $\forall h = 1, \ldots, m$. Thus, we have proven the property when $n = q + 1$, and therefore our claim holds for all $\mathbb{K}$-algebra T_n. $\qquad\square$

Notations: In this chapter, and in the following, given $f(X_1, \ldots, X_n) \in T_n$, we will write f in the form $\sum_{(i) \in \mathbb{N}^n} a_{(i)} X^{(i)}$, where $(i) = (i_1, \ldots, i_n)$, and where $X^{(i)}$ means $X_1^{i_1} \cdots X_n^{i_n}$. The set of indices $\mathbb{N}^n$ is equipped with this external law from $\mathbb{N} \times \mathbb{N}^n$ to $\mathbb{N}^n$: given $q \in \mathbb{N}$ and $(i) = (i_1, \ldots, i_n) \in \mathbb{N}^n$, $q.(i)$ will mean $(qi_1, \ldots, qi_n)$. We will denote by (u) the unit index $(i_1, \ldots, i_n)$ when $i_1 = \cdots = i_n = 1$, and by (0) the zero of $\mathbb{N}^n$.

Lemma 10.4.14. *Suppose that* $\mathbb{L} = \mathbb{K}$. *Let* $f \in T_n$. *Then* $sp(f)$ *is the disk* $d(f((0)), \|f - f((0))\|)$.

Proof. Without loss of generality, we can assume that $f((0)) = 0$. Let $r = \|f\|$. Of course $sp(f) \subset d(0, r)$. Then f is of the form $\sum_{(j) \in \mathbb{N}^n} b_{(j)} X^{(j)}$, and there exists $(k) \in \mathbb{N}^n$ such that $\|b_{(k)}\| = \|f\|$. Since $f((0)) = 0$, at least one of the coordinates of (k) is not 0. For convenience we can assume that (k) is of the form $(k_1, \ldots, k_n)$, with $k_n > 0$. Now, we can write T_n in the form $T_{n-1}\{Y\}$, and put $f = \sum_{n=0}^{\infty} a_n Y^n$. Let $l = k_n$. So, we have $\|f\| = \|a_l\|$. Let $\alpha \in d(0, r)$. Since $l > 0$, we have $\|f - \alpha\| = \|f\| = \|a_l\|$. By Theorem 10.4.13, there exists $\chi \in \mathcal{X}(T_{n-1})$ such that $|\chi(a_l)| = \|a_l\|$. Let $P(Y) = \sum_{n=0}^{\infty} \chi(a_n) Y^n - \alpha \in \mathbb{K}\{Y\}$. Clearly $\|P\| = |\chi(a_l)| = r$, hence by Theorem 10.3.5 of Chapter 7, P has its zeros in $d(0, r)$. Now, by Corollary 10.4.8 there exists $\hat{\chi} \in \mathcal{X}(T_n)$ such that $\hat{\chi}(h) = \chi(h) \ \forall h \in T_{n-1}$, and $\hat{\chi}(Y) = \alpha$. Then $\hat{\chi}(f) = \alpha$. □

We will now show that a $\mathbb{L}$-affinoid algebra is a finite extension of a topologically pure extension.

Proposition 10.4.15. *Let* E *be a normed* $\mathbb{L}$-*vector space which also is a* A-*module. Let* $t \in M_{\mathbb{L}}$. *Let* E_1 *be a* A_0-*submodule defining the topology of* E. *Assume that there exists* $e_1, \ldots, e_q \in E_1$ *such that* $E_1 = \sum_{i=1}^{q} A_0 e_i + t E_1$. *Let* ϕ *be the mapping from* A^q *into* E *defined as* $\phi(t_1, \ldots, t_q) = \sum_{i=1}^{q} t_i e_i$. *Then* ϕ *is surjective and* E *is finite over* A.

Proof. Let $y \in E_1$. We can write y in the form $\sum_{i=1}^{q} f_{1,i} e_i + t y_1$, with $f_{1,i} \in A_0$ and $y_1 \in E_1$, and in the same way, we can write $y_1 = \sum_{i=1}^{q} f_{2,i} e_i + t y_2$, with $f_{2,i} \in A_0$, and $y_2 \in E_1$. By induction, it clearly appears that for every $m \in \mathbb{N}^*$, we obtain $y = \sum_{i=1}^{q} (\sum_{j=1}^{m} f_{m,i} t^{m-1}) e_i + t^m y_m$. And since all terms $f_{j,i}$ lie in A_0, and all y_j lie in E_1, each series $\sum_{j=1}^{m} f_{m,i} t^{m-1}$ converges in the Banach algebra A to an element f_i. Then, since each set $t^m E_1$ is closed, $y - \sum_{i=1}^{q} f_i e_i$ lies in $t^m E_1$ for all $m \in \mathbb{N}^*$. Therefore, $y = \sum_{i=1}^{q} f_i e_i$. Consequently $\phi(A^q)$ contains E_1 and therefore ϕ is surjective on E. Moreover, E is finite over A. □

10.5. Jacobson radical of affinoid algebras

Proposition 10.5.1. *Let $A = T_n[x_1, \ldots, x_q]$ be a $\mathbb{L}$-affinoid algebra without divisors of zero. Let $\phi \in Mult'(T_n, \| . \|)$. There exists $\psi \in Mult'(A, \| . \|)$ whose restriction to T_n is ϕ.*

Proof. Let F be the completion of the field of fractions of T_n with respect to ϕ. Then $x_1, \ldots, x_q$ are algebraic over F. Let $A' = F[x_1, \ldots, x_q]$. Then A is isomorphic to a subring of A' through an isomorphism which induces the identity on T_n. By Theorem 1.3.8 of Chapter 1, there exists a unique absolute value ψ extending that of F to A' and then, considering A as included in A', ψ defines on A an absolute value ψ extending ϕ. Let $m = \max_{1 \le i \le q} \psi(x_i)$, let $\lambda \in \mathbb{L}$ be such that $|\lambda| \le \frac{1}{m}$, and let $y_i = \lambda x_i$ ($1 \le i \le q$). Let θ be the canonical surjection from $T_n\{X_1, \ldots, X_q\}$ onto A inducing the identity on T_n and such that $\theta(X_i) = y_i$ ($1 \le i \le q$). Since $\psi(y_i) \le 1$, and since ϕ belongs to $Mult(T_n, \| . \|)$, $\psi \circ \theta$ clearly satisfies

$$\psi \circ \theta(Y) \le \|Y\| \ \forall Y \in T_n\{X_1, \ldots, X_q\}. \tag{10.5.1}$$

Now, let $t \in A$. By definition $\|t\| = \inf\{\|Y\| \mid \theta(Y) = t\}$. So, given $Y \in T_n\{X_1, \ldots, X_q\}$ such that $\theta(Y) = t$, we have $\psi(t) = \psi \circ \theta(Y)$, which obviously does not depend on the Y such that $\theta(Y) = t$. So, by (10.5.1), considering such a Y we obtain $\psi(t) \le \inf\{\|Y\| \mid \theta(Y) = t\} = \|t\|$, and therefore ψ is continuous and belongs to $Mult(A, \| . \|)$. $\qquad\square$

Remarks. Such a continuous absolute value extending ϕ to A is not unique, due to the fact that elements which are conjugate over the field of fractions E of A are not always conjugate over F. For example suppose that $A = T_n[x, y]$ and that x and y are conjugate over E but not over F. Thus, we can have several isomorphism from A onto subrings of A', so we can have $\psi(x) \ne \psi(y)$ and then ϕ may admit two different expansions to A: ϕ_1, ϕ_2, satisfying $\phi_2(x) = \phi_1(y)$. Such a situation will be illustrated later by an example of a Krasner–Tate algebra.

Corollary 10.5.2. *A $\mathbb{L}$-affinoid algebra without divisors of zero, admits continuous absolute values.*

Corollary 10.5.3. *Every prime ideal of a $\mathbb{L}$-affinoid algebra A is the kernel of at least one continuous multiplicative semi-norm of A.*

Remarks. As previously noticed, there exist Krasner algebras $H(D)$ without divisors of zero, admitting no continuous absolute values [9, 23] and therefore in certain unital Banach ultrametric commutative $\mathbb{L}$-algebras, certain closed prime ideals are not the kernel of any continuous multiplicative semi-norm. Nevertheless, as long as Krasner algebras $H(D)$ are concerned, it is uneasy to construct a counter-example of an algebra $H(D)$ without divisors of zero admitting no continuous absolute value. Here, we see that such counter-examples don't exist among affinoid $\mathbb{L}$-algebras.

Remarks. Let A be a $\mathbb{L}$-affinoid algebra. Since each maximal ideal M is of finite codimension, the field $E = \frac{A}{M}$ admits a unique absolute value $| \, . \, |$ expanding this of $\mathbb{L}$ and each character χ from A onto E defines a unique element ϕ of Mult$(A, \text{Vert} \, . \, \|)$ as $\phi(x) = |\chi(x)|$.

Theorem 10.5.4 (Guennebaud). *Let A be a $\mathbb{L}$-affinoid algebra. Then* Mult$_m(A, \| \, . \, \|)$ *is dense inside* Mult$(A, \| \, . \, \|)$. *Let B be a $\mathbb{L}$-algebra of finite type. Then* Mult(B) *is locally compact and* Mult$_m(B)$ *is dense inside* Mult(B).

Proof. Let us fix $\psi \in \text{Mult}(B)$ and let $F = [x_1, \ldots, x_n]$ be a finite generating subset of B over $\mathbb{L}$ such that $\psi(x_j) < 1 \; \forall j = 1, \ldots, n$. Let θ be the canonical surjection from $\mathbb{L}[X_1, \ldots, X_n]$ onto B such that $\theta(X_j) = x_j$, $(1 \leq j \leq n)$. Then θ has continuation to a continuous morphism $\widehat{\theta}$ from $\mathbb{L}\{X_1, \ldots, X_n\}$ onto a $\mathbb{L}$-affinoid algebra $\widehat{B}$ quotient of $\mathbb{L}\{X_1, \ldots, X_n\}$ and $\widehat{B}$ is a Banach $\mathbb{L}$-algebra with respect to the quotient norm that we denote by $\| \, . \, \|_F$. Let $\mathcal{T}(\psi)$ be the set of finite subsets of generators F of B such that $\psi(x) \leq 1 \; \forall x \in F$.

Until the end of the proof, given a subset S of Mult$(B, \| \, . \, \|_F)$, $\overline{S}$ will denote the closure of S in Mult$(B, \| \, . \, \|_F)$.

We shortly check that the family of compact subsets $\{\overline{\text{Mult}_m}(B, \| \, . \, \|_F)$, $F \in \mathcal{T}(\psi)\}$ is a basis of a filter because given F, $G \in \mathcal{T}(\psi)$, $(\overline{\text{Mult}_m}(B, \| \, . \, \|_F)) \cap (\overline{\text{Mult}_m}(B, \| \, . \, \|_G))$ contains $\overline{\text{Mult}_m}(B, \| \, . \, \|_{F \cup G})$. Let $Y = \bigcap_{F \in \mathcal{T}(\psi)} \text{Mult}_m(B, \| \, . \, \|_F)$. Then Y is not empty. For every $x \in \widehat{B}$, we put $\nu(x) = \sup\{\varphi(x)| \; \varphi \in Y\}$. Then we have $\nu(x) \leq \inf\{\|x\|_F \mid F \in \mathcal{T}(\psi)\} \; \forall x \in \widehat{B}$. But $\inf\{\|x\|_F \mid F \in \mathcal{T}(\psi)\} = \psi(x) \; \forall x \in \widehat{B}$. Consequently:

$$\nu(x) \leq \psi(x) \; \forall x \in \widehat{B}. \tag{10.5.2}$$

Now, we fix $x \in \widehat{B}$ such that $\nu(x) < 1$. Since Y is compact, we can easily construct in Mult(B) an open neighborhood W of Y such that $\varphi(x) < 1 \; \forall \varphi \in W$. Suppose that for every $F \in \mathcal{T}(\psi)$, $\overline{\text{Mult}_m}(B, \| \, . \, \|_F)$ is not included in W. Then the family of compact sets $\overline{\text{Mult}_m}(B, \| \, . \, \|_F) \setminus W$ is a basis of a filter because $(\overline{\text{Mult}_m}(B, \| \, . \, \|_F) \setminus W) \cap (\overline{\text{Mult}_m}(B, \| \, . \, \|_G) \setminus W)$ contains $\overline{\text{Mult}_m}(B, \| \, . \, \|_{F \cup G}) \setminus W$. But since W is open, each set $\overline{\text{Mult}_m}(B, \| \, . \, \|_F)$ $(F \in \mathcal{T}(\psi))$ is a compact and therefore the intersection S is not empty. And since W is open, we have $S \cap W = \emptyset$, a contradiction.

Hence, there exists $J \in \mathcal{T}(\psi)$ such that $\overline{\text{Mult}_m}(B, \| \cdot \|_J) \subset W$. Consequently, $\|x\|_J < 1$, and therefore, $\psi(x) < 1$. Thus $\nu(x) < 1$ implies $\psi(x) < 1$. Consequently, since, by Lemma 10.5.1 of Chapter 2, ψ and ν are two semi-multiplicative semi-norms, we can conclude that $\psi(x) \leq \nu(x)$. So, by (10.5.2) the two semi-norms are equal. As a consequence, ψ lies in Y. Then, for every $F \in \mathcal{T}(\psi)$, ψ lies in $\overline{\text{Mult}_m}(B, \| \cdot \|_F)$. Since $\text{Mult}_m(B, \| \cdot \|_F)$ is a subset of $\text{Mult}_m(B)$, this finishes proving that $\text{Mult}_m(B)$ is dense in $\text{Mult}(B)$.

Now, since A is a quotient of a topologically pure extension $T_n = \mathbb{L}\{X_1, \ldots, X_n\}$ by an ideal J, we have a canonical homomorphism θ from T_n onto A and then, putting $x_n = \theta(X_n)$, we can consider the algebra of finite type $B = \mathbb{L}[x_1, \ldots, x_n]$ which is dense in A. Putting $F = \{x_1, \ldots, x_n\}$, the norm of A is equivalent to $\| \cdot \|_F$. We now take $\phi \in \text{Mult}(A, \| \cdot \|)$. As we just saw above, since ϕ is an element of $\text{Mult}(B, \| \cdot \|_F)$, it lies in $\overline{\text{Mult}_m}(B, \| \cdot \|_F)$. But by Theorem 2.5.6 of Chapter 2, $\text{Mult}(B, \| \cdot \|_F)$ and $\text{Mult}(A, \| \cdot \|_F)$ are homeomorphic. Therefore, $\text{Mult}_m(B, \| \cdot \|_F)$ and $\text{Mult}_m(A, \| \cdot \|_F)$ also are homeomorphic. Thus, all elements of $\overline{\text{Mult}_m}(B, \| \cdot \|_F)$ have continuation to elements of $\overline{\text{Mult}_m}(A, \| \cdot \|_F)$, and finally ϕ lies in $\overline{\text{Mult}_m}(A, \| \cdot \|)$. □

Theorem 10.5.5. *The nilradical of a $\mathbb{L}$-affinoid algebra A is equal to its Jacobson radical.*

Proof. Let $\mathcal{R}$ be the Jacobson radical and let $x \in \mathcal{R}$. Let $\mathcal{J}$ be a prime ideal of A, let $B = \frac{A}{\mathcal{R}}$ and let θ be the canonical surjection from A onto B. Since B has no divisors of zero different from 0, by Corollary 10.5.2 it admits a continuous absolute value ϕ, hence $\phi \circ \theta$ lies in $\text{Mult}(A, \| \cdot \|)$. Let $\psi = \phi \circ \theta$. Since $\text{Mult}_m(A, \| \cdot \|)$ is dense in $\text{Mult}(A, \| \cdot \|)$, for all $\epsilon > 0$ we can find $\zeta_\epsilon \in \text{Mult}_m(A, \| \cdot \|)$ such that $|\zeta_\epsilon(x) - \psi(x)|_\infty \leq \epsilon$. But since x belongs to $\mathcal{R}$, by hypothesis we have $\zeta_\epsilon(x) = 0$. Consequently $\psi(x) = 0$. But since ϕ is an absolute value of B, $\text{Ker}(\psi)$ is equal to $\mathcal{J}$. Thus, x belongs to $\mathcal{J}$, and therefore $\mathcal{R}$ is equal to the nilradical of A. □

Corollary 10.5.6. *The semi-norm $\| \cdot \|_{\text{sp}}$ is a norm if and only if A is reduced.*

Theorem 10.5.7. *A $\mathbb{L}$-affinoid algebra A is a Jacobson ring.*

Proof. Let $\mathcal{J}$ be a prime ideal of A and let θ be the canonical homomorphism from A onto the quotient algebra $B = \frac{A}{\mathcal{J}}$. Since B has no divisors of zero, its nilradical is $\{0\}$, and since it is a $\mathbb{L}$-affinoid algebra, the intersection of all its maximal ideals also equals $\{0\}$. But the maximal ideals of B are the images by θ of the maximal ideals of A containing $\mathcal{J}$. Consequently, we have

$$J = \theta^{-1}(0) = \theta^{-1}\left(\bigcap_{\mathcal{H} \in \mathrm{Max}(B)} \mathcal{H}\right) = \bigcap_{\mathcal{H} \in \mathrm{Max}(B)} \theta^{-1}(\mathcal{H}) = \bigcap_{J \subset \mathcal{N}, \mathcal{N} \in \mathrm{Max}(A)} \mathcal{N}.$$

Thus, J is equal to the intersection of all maximal ideals that contain it. $\square$

Theorem 10.5.8. *Suppose that* $\mathbb{L} = \mathbb{K}$. *Let A be a $\mathbb{L}$-affinoid algebra and let $f \in A$. If the interior of $\mathcal{Z}(F)$ is not empty, then either $f = 0$ or f is a divisor of zero.*

Proof. Suppose that the interior of $\mathcal{Z}(f)$ is not empty and that f is not identically zero. There exists a ball $\mathbb{B}_{T_n}(\phi, r) \subset \mathcal{Z}(f)$. Since $\mathrm{Mult}_m(A, \| \cdot \|)$ is dense in $\mathrm{Mult}(A, \| \cdot \|)$, there exists χ $\mathcal{X}(A)$ and a ball $B(|\chi|, r) \subset \mathcal{Z}(f)$.

Suppose first that A is T_n. Now, each character is characterized by a point $(x_1, \ldots, x_n)$ of U_n and hence χ is characterized by a point $(\alpha_1, \ldots, \alpha_n)$. Consequently, there exists a ball $d((\alpha_1, \ldots, \alpha_n), r) = \{(\lambda_1, \ldots, \lambda_n) \in \mathbb{K} \mid |\lambda_j - \alpha_j| \leq r\} \subset U_n$ such that $f(X) = 0 \; \forall X \in d((\alpha_1, \ldots, \alpha_n), r)$. Now, we can conclude that f is identically zero. Indeed, this is obvious if $n = 1$ and then we can generalize by induction on n. This is a contradiction and hence ends the proof when $f \in T_n$.

Consider now the general case and let A be a finite ring extension of T_n. Let $\mathcal{O}$ be the interior of $\mathcal{Z}(f)$. For each $\chi \in \mathcal{X}(A)$, we denote by $\widehat{\chi}$ its restriction to T_n. By Theorem 10.5.4, there exists $\chi \in \mathcal{X}(A)$ such that $|\chi|$ lies in $\mathcal{O}$. Given $\zeta \in \mathcal{X}(T_n)$ and $\rho > 0$, we denote by $B_{T_n}(\zeta, \rho)$ the ball of the $\eta \in \mathcal{X}(T_n)$ such that $|\eta(X) - \zeta(X)| \leq \rho \; \forall X \in U_n$. Then, there exists $r \in]0, 1[$ such that $\mathbb{B}_{T_n}(|\widehat{\chi}|, r) \subset \{|\widehat{\sigma}| \mid \sigma \in \mathcal{X}(A), |\sigma| \in \mathcal{O}\}$ and we can fix $\rho > 0$ such that $|\chi| \in \mathcal{O} \; \forall \widehat{\chi} \in B_{T_n}(\zeta, \rho)$. Therefore, $\sigma(f) = 0$ for all $\widehat{\sigma} \in \mathcal{X}(T_n)$ such that $\sigma \in B_A(\chi, r)$. For each $j = 1, \ldots, n$, we put $\chi(X_j) = \alpha_j$. Now, for each $(\xi_1, \ldots, \xi_n) \in d((\alpha_1, \ldots, \alpha_n), \rho)$, we can find $\gamma \in \mathcal{X}(T_n)$ such that $\gamma(X_j) = \xi_j \; \forall j = 1, \ldots, n$ and of course γ belongs to $B_{T_n}(\widehat{\chi}, \rho)$.

On the other hand, we denote by $P(X) = X^d + a_{d-1}X^{d-1} + \cdots + a_0 \in T_n[X]$ the minimal polynomial of f over T_n. Since $P(f) = 0$, for every $\phi \in \mathcal{O}$ we have $\phi(P(f)) = 0$ together with $\phi(f) = 0 \; \forall \phi \in \mathbb{B}_A(|\chi|, r)$. Particularly, we have $\sigma(a_0) = 0$ for all $\sigma \in \mathbb{B}_A(|\chi|, r)$. But since a_0 lies in T_n, actually we obtain $\widehat{\sigma}(a_0) = 0 \; \forall \sigma \in B_{T_n}(\chi, \rho)$, hence $a_0(\xi_1, \ldots, \xi_n) = 0 \; \forall (\xi_1, \ldots, \xi_n) \in d((\alpha_1, \ldots, \alpha_n), \rho)$ and consequently, $a_0 = 0$. Thus, we have $f(f^{d-1} + a_{d-1}f^{d-2} + \cdots + a_1) = 0$. But $f^{d-1} + a_{d-1}f^{d-2} + \cdots + a_1$ may not be identically zero because P is the minimal polynomial of f. Consequently, both f and $f^{d-1} + a_{d-1}f^{d-2} + \cdots + a_1$ are divisors of zero. $\square$

10.6. Spectral norm of affinoid algebras

In Section 10.5, we saw that if A is reduced, its spectral semi-norm is a norm. Then it is a natural question to ask whether A is uniform. The answer is quite easy and was soon given in [44] when the characteristic p of $\mathbb{L}$ is 0. It is much more difficult when $p \neq 0$.

Notations: Throughout the chapter and in the next one we denote by A a $\mathbb{L}$-affinoid algebra and by p the characteristic of $\mathbb{L}$. Next, considering $\mathbb{Z}^n$ as a $\mathbb{Z}$-module, for all $(i) = (i_1, \ldots, i_n) \in \mathbb{Z}$, and $q \in \mathbb{Z}$ we put $q(i) = (qi_1, \ldots, qi_n)$. Let $f(Y_1, \ldots, Y_n) = a_{i_1, \ldots, i_n} Y_1^{i_1} \cdots Y_n^{i_n} \in T_n$. We put $(Y) = (Y_1, \ldots, Y_n)$, $(Y)^{(i)} = (Y_1^{i_1}, \ldots, Y_n^{i_n})$ and $f(Y_1, \ldots, Y_n) = \sum_{(i) \in \mathbb{N}^n} \in a_{(i)}(Y)^{(i)}$.

For each $n \in \mathbb{N}^*$ we denote by L_n the field of fractions of T_n and A is a $\mathbb{L}$-affinoid algebra. By Corollary 10.4.4 A has finitely many minimal prime ideals $\mathcal{S}_1, \ldots, \mathcal{S}_l$. For each $j = 1, \ldots, l$, we denote by A_j the $\mathbb{L}$-affinoid algebra $\frac{A}{\mathcal{S}_j}$, by θ_j the canonical surjection from A onto A_j, and by $\| \cdot \|_{\mathrm{sp},j}$ the spectral norm of A_j.

Theorem 10.6.1. *Let A be reduced. For every $f \in A$, we have $\|f\|_{\mathrm{sp}} = \max(\|\theta_j(f)\|_{\mathrm{sp},j})$.*

Proof. By Theorem 10.4.10, we have $\|f\|_{\mathrm{sp}} = \sup_{\lambda \in \mathrm{sp}_A(f)} |\lambda|$, and $\|\theta_j(f)\|_{\mathrm{sp},j} = \sup_{\lambda \in \mathrm{sp}_{A_j}(\theta_j(f))} |\lambda|$, $(j = 1, \ldots, l)$. But by Lemma 1.1.15 of Chapter 1, $sp_A(f) = \bigcup_{j=1}^l sp_{A_j}(\theta_j(f))$. So, our claim is obvious. $\square$

Theorem 10.6.2. *For all $x \in A$ there exists $\chi \in \mathcal{X}(A)$ such that $|\chi(x)| = \|x\|_{\mathrm{sp}}$. Moreover, if A is an integral domain finite over T_n, then for all $f \in A$, we have $S(irr(f, T_n)) = \|f\|_{\mathrm{sp}}$.*

Proof. Let $f \in A$. First, we will show that there exists $\chi \in \mathcal{X}(A)$ such that $|\chi(x)| = \|x\|_{\mathrm{sp}}$. When A is a topologically pure extension, the statement is proven in Theorem 10.4.13. $\square$

Here we first consider a $\mathbb{L}$-affinoid algebra A without divisors of zero. There exists a topologically pure extension T_n such that A is a finite over T_n. By Lemma 1.1.18 of Chapter 1, every $\chi \in \mathcal{X}(T_n)$ admits at least one extension to an element of $\mathcal{X}(A)$. By Theorem 10.4.10, the spectral norm $\| \cdot \|_{\mathrm{sp}}$ of A satisfies $\|f\|_{\mathrm{sp}} = \sup\{|\chi(f)| \mid \chi \in \mathcal{X}(A)\}$. On the other hand, by Proposition 9.6.4 of Chapter 9, we have $\sup\{|\chi(f)| \mid \chi \in \mathcal{X}(A)\} = S(irr(f, T_n))$ and there exists $\chi \in \mathcal{X}(A)$ such that $|\chi(f)| = \sup\{|\chi(f)| \mid \chi \in \mathcal{X}(A)\}$. Consequently, there exists $\chi \in \mathcal{X}(A)$ such that $|\chi(f)| = \|f\|_{\mathrm{sp}}$. Thus, our claim is now proven when A is an integral domain.

We now consider the case when A is just reduced. By Theorem 10.6.1, we have $\max_{1 \leq j \leq l}(\|\theta_j(f)\|_{\mathrm{sp},j}) = \|f\|_{\mathrm{sp}}$. Thus, $\|f\|_{\mathrm{sp}}$ lies in $|\mathbb{L}|$. Let $\lambda \in \mathbb{L}$ be such that $|\lambda| = \|f\|_{\mathrm{sp}}$ and let $g = \frac{f}{\lambda}$. So, $\|g\|_{\mathrm{sp}} = 1$. By Theorem 10.4.7, if $|\chi(g)| < 1 \ \forall \chi \in \mathcal{X}(A)$ then $\|g\|_{\mathrm{sp}} < 1$, hence there exists $\chi \in \mathcal{X}(A)$ such that $|\chi(g)| = 1$, and therefore $|\chi(f)| = \|f\|_{\mathrm{sp}}$.

We can immediately generalize this property when A is not reduced. Indeed, by Theorem 10.4.10, we have $\sup\{|\chi(f)| \mid \chi \in \mathcal{X}(A)\} = \|f\|_{\mathrm{sp}} \ \forall f \in A$. Now, consider the nilradical $\mathcal{R}$ of A, let $B = \frac{A}{\mathcal{R}}$, and let ϕ be the canonical surjection of A onto B. Then, we have $\sup\{|\chi(f)| \mid \chi \in \mathcal{X}(A)\} = \sup\{|\chi \circ \phi(f)| \mid \chi \in \mathcal{X}(B)\}$. Since B is reduced, the property holds for B, and therefore also holds for A.

Theorem 10.6.3. *The algebraic closure of $\mathbb{L}_n$ is $\mathbb{L}_n$-productal with respect to $\| \cdot \|_{\mathrm{pol}}^{\mathbb{L}_n}$.*

Proof. If $p = 0$, the claim comes from Theorems 9.6.1 and 9.6.7 of Chapter 9. So, we suppose $p \neq 0$. By Theorem 9.6.10 of Chapter 9, it is sufficient to prove that $\mathbb{L}_n^{\frac{1}{p}}$ is $\mathbb{L}_n$-productal. For each $k = 1, \dots, n$, we put $Y_k = \sqrt[p]{X_k}$ and $Y = (Y_1, \dots, Y_n)$. So, we have $Y^{p.(u)} = X$, and more generally $Y^{jp.(u)} = X^{j.(u)}$. Let $g \in \mathbb{L}_n^{\frac{1}{p}}$. Then g is of the form $\sum_{j=0}^{p-1} \frac{a_j}{b_j} Y^{j.(u)}$, with $a_j, b_j \in T_n \ \forall j = 1, \dots, n$. And $g^p = \sum_{j=0}^{p-1} \frac{(a_j)^p Y^{jp.(u)}}{(b_j)^p} = \sum_{j=0}^{p-1} \frac{(a_j)^p X^{j.(u)}}{(b_j)^p}$. Let $b = \prod_{j=0}^{p-1} b_j$ and for each $j = 0, \dots, p-1$ we set $c_j = \frac{b}{b_j}$, which belongs to T_n. Then $(bg)^p = \sum_{j=1}^{p-1} (a_j c_j)^p X^{j.(u)}$. Now, for each $j = 0, \dots, p-1$, $a_j c_j$ is of the form $\sum_{(i) \in \mathbb{N}^n} \lambda_{(i),j} X^i$, with $\lambda_{(i),j} \in \mathbb{L}$, hence $(a_j c_j)^p = \sum_{(i) \in \mathbb{N}^n} \lambda_{(i),j} X^i$ and thereby $b^p g^p = \sum_{(i) \in \mathbb{N}^n, \ 0 \leq j \leq p-1} \lambda_{(i),j}^p X^{p.(i)+j.(u)}$.

Next, it is clear that the mapping ω from $\mathbb{N}^n \times \{0, \dots, p-1\}$ into $\mathbb{N}^n$ defined as $\omega((i), j) = p.(i) + j.(u)$ is an injection. Consequently, we have $\|b^p g^p\| = \max\{|\lambda_{(i),j}^p| \mid (i) \in \mathbb{N}^n, \ 0 \leq j \leq p-1\}$. On the other hand,

$$\max\{|\lambda_{(i),j}^p| \mid (i) \in \mathbb{N}^n, \ 0 \leq j \leq p-1\} = \max_{0 \leq j \leq p-1}(\max\{|\lambda_{(i),j}^p| \mid (i) \in \mathbb{N}^n\})$$

$$= \max_{0 \leq j \leq p-1} \|a_j c_j\|^p$$

hence $\|g^p\| = \max_{0 \leq j \leq p-1}\left(\frac{\|a_j c_j\|^p}{\|b\|^p}\right)$. Since $\| \cdot \|$ is an absolute value on T_n, finally we have $\|g^p\| = \max_{0 \leq j \leq p-1}\left(\left\|\frac{a_j}{b_j}\right\|\right)^p$. But $\|g\|_{\mathrm{pol}}^{\mathbb{L}_n} = \sqrt[p]{\|g^p\|}$, hence $\|g\|_{\mathrm{pol}}^{\mathbb{L}_n} = \max_{0 \leq j \leq p-1}\left(\left\|\frac{a_j}{b_j}\right\|\right)$. Thus, $\mathbb{L}_n^{\frac{1}{p}}$ is $\mathbb{L}_n$-productal with respect to $\| \cdot \|_{\mathrm{pol}}^{\mathbb{L}_n}$. $\square$

Theorem 10.6.4. *Let A be reduced. Then the norm of Banach $\mathbb{L}$-algebra of A is equivalent to the norm $\| \cdot \|_{sp}$.*

Proof. First we suppose that A is an integral domain. Let T_n be a topologically pure extension of $\mathbb{L}$ such that A is finite over T_n. As a finite T_n-module, A admits a basis $\{e_1, \ldots, e_q\}$. Putting $M = \max_{1 \leq j \leq q} \|e_j\|$, an element $f = \sum_{j=1}^q a_j e_j \in A$, with $a_j \in T_n \; \forall j = 1, \ldots, q$, satisfies

$$\|f\| \leq M \max_{1 \leq j \leq q} \|a_j\|. \tag{10.6.1}$$

Let L_n be the field of fractions of T_n. Of course $\{e_1, \ldots, e_q\}$ is also a basis of $L_n[x_1, \ldots, x_q]$ as a L_n-vector space. By Theorem 10.6.3, the algebraic closure of L_n is L_n-productal with respect to $\| \cdot \|_{pol}^{L_n}$. So we have a constant $V > 0$ such that

$$\|f\|_{pol}^{L_n} \geq V \max_{1 \leq j \leq q} \|a_j\|. \tag{10.6.2}$$

Thus, by (10.6.1), (10.6.2) and by Theorems 10.6.2, we obtain $V \max_{1 \leq j \leq q} \|a_j\| \leq \|f\|_{pol}^{L_n} = \|f\| = \|f\|_{sp} \leq M \max_{1 \leq j \leq q} \|a_j\|$. Thereby, $\| \cdot \|_{sp}$ is clearly equivalent to $\| \cdot \|$ on A.

We now consider the general case, when A is reduced but not necessarily an integral domain. Let $B = \prod_{j=1}^m \frac{A}{S_j}$. Let ϕ be the $\mathbb{L}$-algebra homomorphism from A into B defined as $\phi(x) = (\theta_1(x), \ldots, \theta_m(x))$. Since A is reduced, the intersection of all prime ideals is null, and therefore, $\bigcap_{j=1}^m S_j = \emptyset$. Thus, ϕ is injective and consequently, B is equipped with a structure of finite A-module whose external law is defined as follows: given $g \in A$, $(f_1, \ldots, f_m) \in B$, then $g(f_1, \ldots, f_m)$ is defined as $(\theta_1(g)f_1, \ldots, \theta_m(g)f_m)$. Then, as a A-submodule, A itself is isomorphic to the A-submodule of B consisting of the $g(f_1, \ldots, f_m)$ where $f_j = 1 \; \forall j = 1, \ldots, m$. Each $\mathbb{L}$-algebra A_j is a $\mathbb{L}$-affinoid algebra without divisors of zero, hence is complete for its own spectral norm $\| \cdot \|_{sp,j}$, and of course, B is obviously complete for the product $\mathbb{L}$-algebra norm defined as $\|(f, \ldots, f_m)\|_B = \max_{1 \leq j \leq m} \|f_j\|_{sp,j}$.

On the other hand, the norm $\| \cdot \|^B$, B is equipped with, is a A-quasi-algebra-norm because for each $j = 1, \ldots, m$, we have $\|\theta_j(g)f_j\|_{sp,j} \leq \|\theta_j(g)\|_{sp,j}\|f_j\|_{sp,j} \leq \|g\|_{sp}\|f_j\|_{sp,j} \leq \|g\|\|f_j\|_{sp,j}$. Consequently, since B is finite over A, by Corollary 10.6.13 of Chapter 2, every A-submodule of B is closed with respect to its norm $\| \cdot \|^B$. In particular, as a A-submodule, A is closed in B with respect to $\| \cdot \|^B$.

Now, given $f \in A$, we have $\|f\|_{\text{sp}} \geq \max_{1 \leq j \leq m} \|\theta_j(f)\|_{\text{sp},j}$. The inequality actually is an equality. Indeed, according to Theorem 10.6.2, we can find $\chi \in \mathcal{X}(A)$ such that $|\chi(f)| = \|f\|_{\text{sp}}$, and there does exist an index h such that $\mathcal{S}_h \subset \text{Ker}(\chi)$, so χ factorizes in the form $\chi_h \circ \theta_h$, hence $\|\theta(f)\|_{\text{sp},h} \geq \|f\|_{\text{sp}}$, so we get $\|f\|_{\text{sp}} = \max_{1 \leq j \leq m} \|\theta_j(f)\|_{\text{sp},j}$.

Finally, it is easily seen that the spectral norm $\| \, . \, \|_{\text{sp}}^B$ of B is just the product norm B has been equipped with, because $\mathcal{X}(B) = \prod_{j=1}^m \mathcal{X}(A_j)$. Thus, we have $\max_{1 \leq j \leq m} \|\theta_j(f)\|_{\text{sp},j} = \|(\theta_1(f), \ldots, \theta_m(f))\|_{\text{sp}}^B$, and therefore, the spectral norm $\| \, . \, \|_{\text{sp}}$ of A is induced by that of B. Consequently, A is complete for both norm $\| \, . \, \|_{\text{sp}}$ and $\| \, . \, \|$, and therefore, by Theorem 1.3.4 of Chapter 1, the two norms are equivalent. $\qquad \square$

Corollary 10.6.5. *A reduced affinoid algebra is uniform.*

Theorem 10.6.6. *Let A be reduced and let $f \in A$ be such that $\widetilde{sp(f)} = U$. Then the closure of $\mathbb{L}[f]$ in A is equal to $\mathbb{L}\{f\}$. Let $g \in A$ be such that $\|f - g\|_{\text{sp}} < 1$. Then $\widetilde{sp(g)} = U$. Moreover, if $g \in \mathbb{L}\{f\}$, then $\mathbb{L}\{f\} = \mathbb{L}\{g\}$.*

Proof. Suppose first that A is reduced, and therefore admits $\| \, . \, \|_{\text{sp}}$ as its norm. The restriction of the norm to $\mathbb{L}[f]$ is the Gauss norm. Indeed, let $P(X) \in \mathbb{L}[X]$, let $\| \, . \, \|$ be the Gauss norm on $\mathbb{L}[X]$ and let $\mathcal{F}$ be the circular filter of center 0 and diameter 1. Then by Lemma 4.1.1 of Chapter 4, we have $\|P\| = \lim_{\mathcal{F}} |P(x)|$. On the other hand $\|P(f)\|_{\text{sp}} = \sup_{\lambda \in \text{sp}(f)} |\lambda|$. But since $\widetilde{sp(f)} = U$, $\mathcal{F}$ is secant with $\widetilde{sp(f)}$, therefore $\lim_{\mathcal{F}} |P(x)| = \sup_{\lambda \in \text{sp}(f)} |\lambda|$, thereby we get the equality. Consequently, the closure of $\mathbb{L}[f]$ in A is equal to $\mathbb{L}\{f\}$. Now, consider $g \in A$ such that $\|f - g\|_{\text{sp}} < 1$. Then $\text{sp}(g)$ is included in U, as $\text{sp}(f)$. And since $\text{diam}(\text{sp}(f)) = 1$, we have $\text{diam}(\text{sp}(g)) = 1$, hence of course $\widetilde{sp(g)} = U$.

Now, suppose $g \in \mathbb{L}\{f\}$, hence g is of the form $\sum_{m=0}^\infty b_m f^m$, with $|b_0 - 1| < 1$, $|b_m| < 1 \; \forall m \in \mathbb{N}$. Without loss of generality, we can clearly assume that $b_0 = 1$. Thus, g is of the form $f + h(f)$, with $h(Y) \in \mathbb{L}\{Y\}$, and $\|h\| < 1$. Let $l(Y) = Y + h(Y)$. Since $\mathbb{L}\{Y\}$ is identical to $H(U)$, and since $\|h\| < 1$, it clearly satisfies the hypothesis of Theorem 7.3.11 of Chapter 7, hence $l(Y)$ is a strictly injective analytic element in U, making a bijection from U onto U. Consequently we can apply Theorem 7.3.13 of Chapter 7 showing that, if we put $Z = l(Y)$, then Y is a strictly injective function in Z, hence of the form $Z + t(Z)$, with $t \in \mathbb{L}\{Z\}$ and $\|t\| < 1$. Consequently, applying this to f, we see that f belongs to $\mathbb{L}\{g\}$. $\qquad \square$

10.7. Spectrum of an element of an affinoid algebra

Notations: Throughout Section 10.7 we assume that the groundfield is $\mathbb{K}$. Keeping notations introduced in Section 10.5, here we put $\|(i_1, \ldots, i_n)\|_\infty = \max(|i_1|_\infty, \ldots, |i_n|_\infty)$. In $\mathbb{K}^n$, (0) denotes $(0, \ldots, 0)$.

Proposition 10.7.1. *Let* $F(X) = \sum_{i=0}^q f_i X^i \in T_n[X]$ *be monic and irreducible in* $T_n[X]$. *The set of* $\lambda \in \mathbb{K}$ *such that* $F(\lambda)$ *is not invertible is an affinoid subset of* $\mathbb{K}$.

Proof. For each $i = 0, \ldots, q$, we put $f_i = \sum_{(j) \in \mathbb{N}^n} a_{(j),i} T^{(j)}$. So, $F(X)$ is of the form $\sum_{(j) \in \mathbb{N}^n} b_{(j)}(X) T^{(j)}$, whereas each $b_{(j)}$ lies in $\mathbb{K}[X]$ and satisfies $\deg(b_{(j)}) < q \,\, \forall (j) \neq (0)$, and $\deg(b_{(0)}) = q$ because F is monic. Moreover, since F is irreducible, we notice that there exist no $\alpha \in \mathbb{K}$ such that $b_{(j)}(\alpha) = 0 \,\, \forall (j) \in \mathbb{N}^n$. Further, since $\lim_{\|(j)\|_\infty \to +\infty} a_{(j),i} = 0 \,\, \forall i = 0, \ldots, q$, we have $\lim_{\|(j)\|_\infty \to +\infty} \|b_{(j)}\| = 0$. For every $\lambda \in \mathbb{K}$ we put $B(\lambda) = \sup_{(j) \in \mathbb{N}^n} |b_{(j)}(\lambda)|$. By Theorem 10.4.15, $F(\lambda)$ is not invertible in T_n if and only if $|b_{(0)}(\lambda)| \leq \sup_{(j) \neq (0)} |b_{(j)}|$. Let $D = \{\lambda \in \mathbb{K} \mid |b_{(0)}(\lambda) \leq \sup_{(j) \neq (0)} |b_{(j)}(\lambda)|\}$ and let $s = \sup_{(j) \neq (0)} \|b_{(j)}\|$. We first notice that D is bounded. Indeed, let $d(0, \rho)$ be a disk containing all zeros of $b_{(0)}$, and let $r = \max(\rho, s, 1)$. Let $\lambda \in \mathbb{K}$ be such that $|\lambda| > r$. Since all zeros of $b_{(0)}$ lie in $d(0, r)$ and since $b_{(0)}$ is monic, by Corollary 6.1.8 of Chapter 6, we have $|b_{(0)}(\lambda)| = |\lambda|^q$. On the other hand, for each $(j) \neq (0)$, if $t = \deg b_{(j)}$, we have the basic inequality $|b_{(j)}(\lambda)| \leq \|b_{(j)}\|.|\lambda|^t \leq \|b_{(j)}\|.|\lambda|^{q-1}$ because $|\lambda| > 1$. Consequently, $|b_{(j)}(\lambda)| \leq s|\lambda|^{q-1} < |\lambda|^q$, hence $\lambda \notin D$. This shows that $D \subset d(0, r)$.

Now, we will show that there exists a finite subset S of $\mathbb{N}^n \setminus (0)$ such that $\sup_{(j) \neq (0)} |b_{(j)}(\lambda)| = \sup_{(j) \in S} |b_{(j)}(\lambda)| \,\, \forall \, \lambda \in D$. Indeed, suppose this is not true. There exist injective sequences $(\lambda_m)_{m \in \mathbb{N}}$ in D and $(k_m)_{m \in \mathbb{N}}$ in $\mathbb{N}^n$ such that $\sup_{(j) \neq (0)} |b_{(j)}(\lambda_m)| = |b_{(k)_m}(\lambda_m)| \,\, \forall \, m \in \mathbb{N}$. Since D is bounded, by Proposition 5.3.7 of Chapter 5, we can extract from the sequence (λ_m) a subsequence thinner than a circular filter $\mathcal{F}$. Thus, without loss of generality, we can directly assume for convenience that the sequence (λ_m) itself is thinner than $\mathcal{F}$. Since the sequence $((k)_m)$ is injective, by definition of T_n we have $\lim_{m \to \infty} \|b_{(k)_m}\| = 0$, therefore $\varphi_{\mathcal{F}}(b_{(0)}) = 0$, hence $\varphi_{\mathcal{F}}$ is not an absolute value on $\mathbb{K}[X]$. Consequently, by Theorem 6.2.1 of Chapter 6, $\mathcal{F}$ is the filter of neighborhoods of a point $a \in D$, hence $b_{(0)}(a) = 0$. But as we noticed above there exists $(h) \in \mathbb{N}^n$ such that $b_{(h)}(a) \neq 0$, hence the inequality $|b_{(h)}(x)| > |b_{(0)}(x)|$ holds in a neighborhood of a, a contradiction to the hypothesis: $\sup_{(j)\neq(0)} |b_{(j)}(\lambda_m)| = |b_{(k)_m}(\lambda_m)| \; \forall \, m \in \mathbb{N}$. Thus we have proven this existence of S.

Now, for each $(j) \in S$, we set $D_{(j)} = \{\lambda \in \mathbb{K} \mid |b_{(0)}(\lambda)| \leq |b_{(j)}(\lambda)|\}$. Then $D = \bigcup_{(j) \in S} D_{(j)}$. Since $\deg(b_{(0)}) > \deg(b_{(j)}) \; \forall (j) \neq (0)$, by Theorem 6.1.10 of Chapter 6, each set $D_{(j)}$ is affinoid and therefore by Lemma 5.1.14 of Chapter 5 so is D. □

Theorem 10.7.2 is proven in [37] in another way.

Theorem 10.7.2. *For all $x \in A$, $sp(x)$ is an affinoid subset of $\mathbb{K}$.*

Proof. Assume that A is finite over T_n and let $x \in A$. First, suppose that A is an integral domain and set $F(X) = \sum_{i=0}^{q} f_i X^i = irr(x, T_n)$. For each $i = 0, \ldots, q$, we put $f_i = \sum_{(j) \in \mathbb{N}^n} a_{(j),i} Y^{(j)}$, $P(X) = \sum_{i=0}^{q} a_{(0),i} X^i$ and $Q(X) = F(X) - P(X)$. Now let $\lambda \in \mathbb{K}$ and $G_\lambda(X) = irr(x - \lambda, T_n)$. One checks that $G_\lambda(0) = F(\lambda)$, so by Proposition 1.1.7 of Chapter 1, $x - \lambda$ is invertible in A if and only if $F(\lambda)$ is invertible in T_n. Thus, $sp(x)$ is equal to the set of $\lambda \in \mathbb{K}$ such that $F(\lambda)$ is not invertible, and then by Proposition 10.7.1, this set is affinoid.

We now suppose that A is not necessarily an integral domain. By Corollary 10.4.4 A admits finitely many minimal prime ideals, $\mathcal{P}_1, \ldots, \mathcal{P}_k$. For each $j = 1, \ldots, k$, let $A_j = \frac{A}{\mathcal{P}_j}$ and let θ_j be the canonical surjection from A to A_j. Each algebra A_j is a $\mathbb{K}$-affinoid algebra without divisors of zero, hence $sp(\theta_j(x))$ is an affinoid set. Then by Lemma 1.1.15 of Chapter 1, we have $sp(x) = \bigcup_{j=1}^{k} sp(\theta_j(x))$, and therefore by Lemma 5.1.14 of Chapter 5, $sp(x)$ is an affinoid set. So our claim is proven in the general case. □

10.8. Topologically separable fields

When the field $\mathbb{K}$ is separable (in the topological meaning) we can prove that, given a $\mathbb{K}$-affinoid algebra A, the topology of pointwise convergence on $\mathrm{Mult}(A, \| \, . \, \|)$ is metrizable and we construct an equivalent metric, a study first made by Mainetti [36]. Of course, such a metric defines a topology weaker than this defined by δ in the case of one variable, as it was seen in Section 6.2 of Chapter 6.

Remarks. By construction, $\mathbb{C}_p$ is obviously separable. A field which is separable is weakly valued. But a weakly valued field is not necessarily separable. For instance, it is known that the spherical completion of a $\mathbb{C}_p$ is not separable, but is weakly valued as $\mathbb{C}_p$.

Notation: In this chapter, the algebraically closed field $\mathbb{K}$ is supposed to be separable, i.e., to have a dense countable subset that we will denote by

S, and we will denote by $S[Z_1, \ldots, Z_q]$ the set of polynomials in q variables with coefficients in S such that $S[x_1, \ldots, x_q]$ is dense in A (the x_j are not supposed to be algebraically independent).

As in previous chapters, we denote by $\| \cdot \|$ the Gauss norm on an algebra of polynomials $\mathbb{K}[Y_1, \ldots, Y_q]$ and on $\mathbb{K}\{Y_1, \ldots, Y_q\}$. In order to avoiding any confusion, $\| \cdot \|_a$ will denote the norm of $\mathbb{K}$-affinoid algebra of A.

Lemma 10.8.1. *A admits a dense countable subset of the form $S[x_1, \ldots, x_q]$, with $\|x_j\|_a \leq 1 \ \forall \ j = 1, \ldots, q$.*

Proof. Since A is a $\mathbb{K}$-affinoid algebra, it is of the form $T_m[y_1, \ldots, y_t]$ with $T_m = \mathbb{K}\{Y_1, \ldots, Y_m\}$. Consequently, the $\mathbb{K}$-algebra of finite type $\mathbb{K}[Y_1, \ldots, Y_m, y_1, \ldots, y_t]$ is dense in A, and of course $S[X_1, \ldots, X_m, y_1, \ldots, y_t]$ is dense in $\mathbb{K}[Y_1, \ldots, Y_m, y_1, \ldots, y_t]$, hence in A. By hypothesis the Y_j satisfy $\|Y_j\|_a = 1 \ \forall \ j = 1, \ldots, m$ and of course we can take the y_j such that $\|y_j\|_a = 1 \ \forall \ j = 1, \ldots, t$. $\qquad\square$

Lemma 10.8.2 is now immediate.

Lemma 10.8.2. *There exists a sequence $(P_n)_{n \in \mathbb{N}^*}$ in $S[Z_1, \ldots, Z_q]$ satisfying $P_n = Z_n, \ \forall n = 1, \ldots, t$, such that, given $Q \in \mathbb{K}[x_1, \ldots, x_q]$ and $\epsilon > 0$, there exists $m \in \mathbb{N}$ such that $\|Q - P_m\| < \epsilon$ and $\deg_h(Q) = \deg_h(P_m) \ \forall h = 1, \ldots, q$.*

Definitions and notations: Henceforth, we consider $x_1, \ldots, x_q \in A$ and a sequence $(P_n(Z_1, \ldots, Z_q))_{n \in \mathbb{N}^*}$ in $S[Z_1, \ldots, Z_q]$ satisfying $P_n = Z_n \ \forall n < q$, such that, given $Q \in \mathbb{K}[Z_1, \ldots, Z_q]$ and $\epsilon > 0$, there exists $m \in \mathbb{N}$ such that $\|Q(Z_1, \ldots, Z_q) - P_m(Z_1, \ldots, Z_q)\| < \epsilon$ (with respect to the Gauss norm in $\mathbb{K}[Z_1, \ldots, Z_q]$) and $\deg_h(Q(Z_1, \ldots, Z_q)) = \deg_h(P_m(Z_1, \ldots, Z_q)) \ \forall h = 1, \ldots, q$. Such a sequence will be called a *a S-appropriate sequence of* $\mathbb{K}[Z_1, \ldots, Z_q]$.

Now, we fix $b \in \mathbb{K}$ such that $|b| < 1$ and given $\phi, \psi \in \mathrm{Mult}(A, \| \cdot \|)$ we put $\mathcal{D}(\phi, \psi) = \sum_{n=1}^{\infty} |b|^n \min(|\phi(P_n(x_1, \ldots, x_q)) - \psi(P_n(x_1, \ldots, x_q))|_{\infty}, 1)$.

$\mathcal{D}$ obviously is a distance on $\mathrm{Mult}(A, \| \cdot \|_a)$. We will call $\mathcal{D}$ *the metric associated to the sequence* $(P_n)_{n \in \mathbb{N}^*}$ *and to* $x_1, \ldots, x_q$.

In $\mathrm{Mult}(A, \| \cdot \|_a)$ an open ball of center ϕ and diameter r will be denoted by $B(\phi, r)$.

Theorem 10.8.3. *Let $(P_n)_{n \in \mathbb{N}^*}$ be a S-appropriate sequence of $\mathbb{K}[Z_1, \ldots, Z_q]$. Let $\mathcal{D}$ be the metric associated to the S-appropriate sequence $(P_n)_{n \in \mathbb{N}^*}$. Then $\mathcal{D}$ is a distance on $\mathrm{Mult}(A, \| \cdot \|_a)$ which defines the topology of pointwise convergence.*

Proof. Let $\phi \in Mult(A, \| \cdot \|_a)$ and let $\epsilon \in]0, 1[$. In order to show the statement, we will first show that $B(\phi, \epsilon)$ contains a certain neighborhood $\mathcal{W}(\phi, Q_1(x_1, \ldots, x_q), \ldots, Q_s(x_1, \ldots, x_q), \eta)$. Let $s \in \mathbb{N}$ be such that $\sum_{n=s+1}^{\infty} |b|^n < \frac{\epsilon}{2}$ and let $\eta = \frac{\epsilon(1-|b|)}{2|b|^{s+1}}$. Let $\psi \in \mathcal{V}(\phi, P_1(x_1, \ldots, x_q), \ldots, P_s(x_1, \ldots, x_q), \eta)$. Clearly, on one hand, we have

$$\sum_{n=s+1}^{\infty} |b|^n \min |(\phi(P_n(x_1, \ldots, x_q)) - \psi(P_n(x_1, \ldots, x_q))|_{\infty}, 1)$$

$$\leq \sum_{n=s+1}^{\infty} |b|^n = \frac{|b|^{s+1}}{1 - |b|} < \frac{\epsilon}{2}$$

and on the other hand

$$\sum_{n=1}^{s} |b|^n \min |(\phi(P_n(x_1, \ldots, x_q)) - \psi(P_n(x_1, \ldots, x_q))|_{\infty}, 1) \leq \eta \sum_{n=1}^{s} |b|^n = \frac{\epsilon}{2}.$$

Consequently, $\mathcal{D}(\phi, \psi) < \epsilon$, therefore $\mathcal{W}(\phi, P_1(x_1, \ldots, x_q), \ldots, P_s(x_1, \ldots, x_q), \eta)$ is included in $B(\phi, \epsilon)$. Thus, a ball of center ϕ, with respect to $\mathcal{D}$, is a neighborhood of ϕ with respect to the topology of pointwise convergence, and therefore is an open set with respect to the topology of pointwise convergence, so the topology defined by the semi-distance $\mathcal{D}$ is weaker (in the large sens) than the topology of pointwise convergence.

Conversely, we now consider a neighborhood $\mathcal{W}(\phi, Q_1(x_1, \ldots, x_q), \ldots, Q_t(x_1, \ldots, x_q), \epsilon)$, with $Q_j \in A$ and $\epsilon \in]0, 1[$ and we will look for a ball $B(\phi, \eta)$ included in $\mathcal{W}(\phi, Q_1(x_1, \ldots, x_q), \ldots, Q_t(x_1, \ldots, x_q), \epsilon)$.

By hypothesis, the sequence $(P_n(Z_1, \ldots, Z_q))_{n \in \mathbb{N}}$ is dense in $\mathbb{K}[Z_1, \ldots, Z_q]$, with respect to the Gauss norm, hence the sequence $P_n(x_1, \ldots, x_q)_{n \in \mathbb{N}}$ is dense in $\mathbb{K}[x_1, \ldots, x_q]$ and in A with respect to the norm $\| \cdot \|_a$ of A. Consequently, for each $j = 1, \ldots, t$ there exists $n_j \in \mathbb{N}^*$ such that

$$\|P_{n_j}(x_1, \ldots, x_q) - Q_j(x_1, \ldots, x_q)\|_a < \frac{\epsilon}{3}. \tag{10.8.1}$$

Let $m = \max(n_1, \ldots, n_t)$ and let $\eta = \frac{\epsilon|b|^m}{3}$. Let $\psi \in B(\phi, \eta)$. Then we have $\sum_{n=1}^{m} |b|^n |\phi(P_n(x_1, \ldots, x_q)) - \psi(P_n(x_1, \ldots, x_q))|_{\infty} < \eta$. In particular for each $n = 1, \ldots, m$ we obtain $|b|^n |\phi(P_n(x_1, \ldots, x_q)) -$

$\psi(P_n(x_1, \ldots, x_q))|_\infty < \eta$, and therefore

$$|\phi(P_n(x_1, \ldots, x_q)) - \psi(P_n(x_1, \ldots, x_q))|_\infty < \frac{\eta}{|b|^n} \le \frac{\epsilon}{3} \; \forall n = 1, \ldots, m.$$

(10.8.2)

Now, by (10.8.1) we know that $|\phi(P_{n_j}(x_1, \ldots, x_q)) - \phi(Q_j(x_1, \ldots, x_q))|_\infty \le \frac{\epsilon}{3}$ and $|\psi(P_{n_j}(x_1, \ldots, x_q)) - \psi(Q_j(x_1, \ldots, x_q))|_\infty \le \frac{\epsilon}{3} \; \forall j = 1, \ldots, t$ and therefore by (10.8.2) and by Lemma 10.8.2, we can get $|\psi(Q_j(x_1, \ldots, x_q)) - \phi(Q_j(x_1, \ldots, x_q))|_\infty < \epsilon \; \forall j = 1, \ldots, t$.

Thus, $B(\phi, \eta)$ is included in $\mathcal{W}(\phi, Q_1, \ldots, Q_t, \epsilon)$ and hence the two topologies are equal on $\mathrm{Mult}(A, \| \cdot \|_a)$. $\qquad\square$

Notation: Given tow subsets X, Y of $\mathrm{Mult}(A, \| \cdot \|_a)$, as usual, we put $\mathcal{D}(X, Y) = \inf\{\mathcal{D}(\phi, \psi) \, | \phi \in X, \; \psi \in Y\}$.

Corollary 10.8.4. $\mathrm{Mult}(A, \| \cdot \|_a)$ *is metrizable and sequentially compact. If X, Y are two disjoint compact subsets of* $\mathrm{Mult}(A, \| \cdot \|_a)$*, then $\mathcal{D}(X, Y) > 0$.*

Then, being compact with respect to a metric, $\mathrm{Mult}(A, \| \cdot \|_a)$ is complete.

Corollary 10.8.5. $\mathrm{Mult}(A, \| \cdot \|_a)$ *is complete with respect to the distance $\mathcal{D}$.*

10.9.   Krasner–Tate algebras

In this section, we consider a Banach $\mathbb{K}$-algebra which is isomorphic to both a $\mathbb{K}$-algebra of analytic elements $H(D)$ and a $\mathbb{K}$-affinoid algebra. Such a Banach $\mathbb{K}$-algebra is called *a Krasner–Tate algebra* [18]. The issue is to characterize Krasner–Tate algebras among algebras $H(D)$ on one hand and among affinoid algebras on the other hand.

Notation: Henceforth, for each ideal $\mathcal{J}$ of T_n, we denote by $Z(\mathcal{J})$ the set of maximal ideals of T_n containing $\mathcal{J}$ and similarly, for every $h \in T_n$, we put $Z(h) = Z(hT_n)$.

Theorem 10.9.1. *Let P, $Q \in \mathbb{K}[X]$ be relatively prime polynomials. Then the ideal $(P(X) - YQ(X))\mathbb{K}\{Y, X\}$ of $\mathbb{K}\{Y, X\}$ is equal to its radical.*

Proof. By Theorem 10.3.7, $\mathbb{K}\{Y, X\}$ is factorial. Suppose its factorization into irreducible factors contains a factor g with a power $q > 1$. Then g divides $\frac{\partial(P(X) - YQ(X))}{\partial Y} = -Q(X)$. Consequently, g divides both P and Q in $\mathbb{K}\{Y, X\}$. Let $\mathcal{M}$ be a maximal ideal of $\mathbb{K}\{Y, X\}$ such that $g \in \mathcal{M}$ and let $\chi \in \mathcal{X}(\mathbb{K}\{Y, X\})$ be such that $Ker(\chi) = \mathcal{M}$. Let $\chi(X) = \alpha$. Then we have $\chi(P(X)) = P(\alpha) = 0$, $\chi(Q(X)) = Q(\alpha) = 0$, a contradiction to the hypothesis "P and Q relatively prime". This shows that $P(X) - YQ(X)$ has no multiple irreducible factors in its decomposition. But since, by Theorem 10.3.7, $\mathbb{K}\{Y, X\}$ is factorial, then the ideal $(P(X) - YQ(X))\mathbb{K}\{Y, X\}$ of $\mathbb{K}\{Y, X\}$ is equal to its radical. $\qquad\square$

Theorem 10.9.2. *Let D be affinoid. Let P, $Q \in \mathbb{K}[X]$ be relatively prime polynomials such that $\deg(P) > \deg(Q)$, $\frac{P}{Q} \in R(D)$, $\frac{P}{Q}(D) = U$, $\left(\frac{P}{Q}\right)^{-1}(U) = D$. Then the $\mathbb{K}$-algebra $H(D)$ is isomorphic to the $\mathbb{K}$-affinoid algebra $\frac{\mathbb{K}\{Y,X\}}{(P(X)-YQ(X))\mathbb{K}\{Y,X\}}$ which is also isomorphic to $\mathbb{K}\{h\}[x]$, with $h = \frac{P(x)}{Q(x)}$.*

Proof. Since $\deg(P) > \deg(Q)$, in $H(D)$ x is integral over $\mathbb{K}[h]$ and satisfies $P(x) - hQ(x) = 0$. Now, since $sp_{H(D)}(h) = h(D) = U$, the norm $\| \cdot \|_D$ induces the Gauss norm on $\mathbb{K}[h]$. So, by Theorem 10.6.6, the closure of $\mathbb{K}[h]$ in $H(D)$ is isometrically isomorphic to T_1 that we can denote here by $\mathbb{K}\{h\}$. Without loss of generality we can obviously assume that $\|x\|_D \le 1$. Thus, in $H(D)$, x is integral over $\mathbb{K}\{h\}$ and satisfies $P(x) - hQ(x) = 0$. Let $B = \mathbb{K}\{h\}[x]$. By Corollary 10.4.3, B is a $\mathbb{K}$-affinoid algebra isomorphic to a quotient of $\mathbb{K}\{Y, X\}$ by an ideal $\mathcal{I}$ that contains $P(X) - YQ(X)$ and the canonical surjection θ from $\mathbb{K}\{Y, X\}$ onto B satisfies $\theta(Y) = h$, $\theta(X) = x$. On the other hand, by definition B is a $\mathbb{K}$-subalgebra of $H(D)$. For every $a \in \mathbb{K} \setminus D$ since $|h(a)| > \|h\|_D$, by Theorem 10.2.7, $h(a) - h$ is invertible in B, hence $R(D) \subset B \subset H(D)$. In particular $sp_B(x) = sp_{H(D)}(x) = D$. We notice that $H(D)$ is reduced, hence so is B. Consequently, $\mathcal{I}$ is equal to its radical and therefore, by Theorem 10.6.4, B is complete for its spectral norm. But since $sp_B(x) = sp_{H(D)}(x)$ we can easily deduce that the spectral norm of B is induced by the norm $\| \cdot \|_D$. Indeed, given $\chi \in \mathcal{X}(B)$, since $\chi(x) \in D$, χ has continuation by continuity to $H(D)$, and therefore belongs to $\mathcal{X}(H(D))$, hence for all $f \in B$, the spectral norm $\| \cdot \|_{sp}^B$ of B satisfies $\|f\|_{sp}^B = \|f\|_D$. Therefore, since $R(D) \subset B$, we have $B = H(D)$.

Since $P(X) - YQ(X) \in \mathcal{I}$, of course $Z(\mathcal{I}) \subset Z(P(X) - YQ(X))$. Now, consider a maximal ideal $\mathcal{M}$ which contains $\mathcal{I}$ and let $\chi \in \mathcal{X}(\mathbb{K}\{Y, X\})$ be such that $Ker(\chi) = \mathcal{M}$. Let $\chi(X) = \alpha$, $\chi(Y) = \beta$. Since $x - a$ is invertible whenever $a \notin D$, we see that α lies in D, and then $\chi(Y) = h(\alpha)$, thereby $P(\alpha) - h(\alpha)Q(\alpha) = 0$, hence $P(X) - YQ(X)$ belongs to $\mathcal{M}$. Consequently, we have $Z(\mathcal{I}) = Z(P(X - YQ(X))$. Therefore, the ideals $(P(X) - YQ(X))\mathbb{K}\{Y, X\}$ and $\mathcal{I}$ have the same radical. Thus, on one hand, by Proposition 10.9.1, $(P(X) - YQ(X))\mathbb{K}\{Y, X\}$ is equal to its radical, on the other hand, since $H(D)$ is reduced, $\mathcal{I}$ also is equal to its radical, hence $(P(X) - YQ(X))\mathbb{K}\{Y, X\} = \mathcal{I}$. □

Theorem 10.9.3. *A $\mathbb{K}$-algebra $H(D)$ is a Krasner–Tate algebra if and only if D is affinoid.*

Proof. Indeed, if $H(D)$ is a Krasner-Tate algebra, then by Theorem 10.7.2, $\mathrm{sp}(x)$, i.e., D, is affinoid. Conversely, suppose that D is affinoid. By Theorem 6.1.11 of Chapter 6 there exists $h \in R(D)$ such that $h(D) = U$, $D = h^{-1}(U)$, with $h = \frac{P}{Q}$, P, Q relatively prime and such that $\deg(P) > \deg(Q)$. Consequently, Theorem 10.9.2 shows that $H(D)$ is a Krasner–Tate algebra isomorphic to $\frac{\mathbb{K}\{Y,X\}}{(P-YQ)\mathbb{K}\{Y,X\}}$. □

Theorem 10.9.4. *Let D be affinoid and let $f \in H(D)$. Then $f(D)$ is equal to $\mathrm{sp}(f)$ and is affinoid.*

Proof. Indeed by Theorem 10.9.3, $H(D)$ is a $\mathbb{K}$-affinoid algebra, hence by Theorem 10.7.2 $\mathrm{sp}(f)$ is an affinoid subset of $\mathbb{K}$. Next, since by Theorem 10.4.2, all maximal ideals are of finite codimension, hence here are of codimension 1, we have $\mathrm{sp}(f) = f(D)$.

Another way to describe Krasner–Tate algebras $H(D)$ when the set D is not infraconnected consists of focusing on the infraconnected components of D, in order to show a product of Krasner–Tate algebras. □

Theorem 10.9.5. *Let D be affinoid and let $D_1, \ldots, D_q$ be its infraconnected components. For every $j = 1, \ldots, q$, let P_j, $Q_j \in \mathbb{K}[X]$ be relatively prime and satisfy:*

$$\frac{P_j}{Q_j} \in R(D_j), \quad \frac{P_j}{Q_j}(D_j) = U, \quad \left(\frac{P_j}{Q_j}(D_j)\right)^{-1}(U) = D_j. \tag{10.9.1}$$

Let $\mathcal{I} = \prod_{j=1}^{q}(P_j - YQ_j)\mathbb{K}(Y, X)$. Then $H(D)$ is isomorphic to $\frac{\mathbb{K}\{Y,X\}}{\mathcal{I}}$.

Proof. Without loss of generality we can obviously assume that $\|x\|_D \le 1$. For each $j = 1, \ldots, q$, we denote by I_j the ideal $(P_j - YQ_j)\mathbb{K}\{Y, X\}$ of $\mathbb{K}\{Y, X\}$. By Theorem 7.7.11, $H(D)$ is isomorphic to the $\mathbb{K}$-algebra $H(D_1) \times \cdots \times H(D_q)$. Each set D_j is an infraconnected set, hence by Theorem 10.9.2, $H(D_j)$ is isomorphic to the Krasner–Tate algebra without divisors of zero $\frac{\mathbb{K}\{Y,X\}}{I_j}$. On the other hand, since $D_k \cap D_l = \emptyset \ \forall k \ne l$, we have $Z(I_k + I_l) = \emptyset$, therefore $I_k + I_l = \mathbb{K}\{Y, X\} \ \forall k \ne l$. Consequently, by Proposition A.1.1, the direct product

$$\frac{\mathbb{K}\{Y,X\}}{(P_1 - YQ_1)\mathbb{K}\{Y,X\}} \times \cdots \times \frac{\mathbb{K}\{Y,X\}}{(P_q - YQ_q)\mathbb{K}\{Y,X\}} \tag{10.9.2}$$

is isomorphic to $\frac{\mathbb{K}\{Y,X\}}{I}$, which completes the proof. $\qquad\square$

Remarks. Given a Krasner–Tate algebra $\frac{\mathbb{K}\{Y,X\}}{(P-YQ)\mathbb{K}\{Y,X\}}$, the form shown in Theorem 10.9.3 is not the only one possible. Indeed, consider the $\mathbb{K}$-affinoid algebra $A = \frac{\mathbb{K}\{Y,Z\}}{(Z^3+4Z^2Y+5ZY^2+2Y^3+Z+1)\mathbb{K}\{Y,Z\}}$. Now, let $X = Z + Y$. Then A becomes $\frac{\mathbb{K}\{Y,X\}}{[(X^3+X+1-Y(-X^2+1)]\mathbb{K}\{Y,X\}}$. Thus, A appears as a Krasner–Tate algebra $H(D)$ where $D = h^{-1}(U)$ with $h = \frac{x^3+x+1}{-x^2+1}$. Now, consider the affinoid algebra $B = \frac{\mathbb{K}\{Y,Z\}}{(Z^2-YZ)\mathbb{K}\{Y,Z\}}$. Apparently, it looks like a Krasner–Tate algebra, except that the polynomials $P = Z^2$ and $Q = Z$ are not relatively prime. Of course, B has divisors of zero. Putting $X = Y - Z$, we obtain $B = \frac{\mathbb{K}\{Y,X\}}{(XY)\mathbb{K}\{Y,X\}}$, so we can check that such a $\mathbb{K}$-algebra has no idempotent different from 0 and 1. Suppose B is a Krasner–Tate algebra $H(D)$. Since B has n non-trivial idempotent, D is infraconnected, hence by Theorem 10.4.2, we know that B is not Noetherian. Consequently, it can't be a Krasner–Tate algebra.

Lemma 10.9.6. *Let $A = \frac{\mathbb{K}\{Y_1,\dots,Y_n\}}{I}$ be a $\mathbb{K}$-affinoid algebra without divisors of zero such that the field of fractions of A is a pure degree one transcendental extension $\mathbb{K}(t)$ of $\mathbb{K}$. Let θ be the canonical surjection from $\mathbb{K}\{Y_1,\dots,Y_n\}$ onto A. For each $j = 1,\dots,n$, let $y_j = \theta(Y_j) = h_j(t)$, with $h_j(t) \in \mathbb{K}(t)$. If t belongs to A, then $sp(t)$ contains no poles of h_j and no points a such that $h'_j(a) = 0\ \forall j = 1,\dots,n$.*

Proof. Suppose that $t \in A$ and let $D = sp(t)$. Suppose that for some $k \in \{1,\dots,n\}$, h_k admits a pole b in D. Let $h_k = \frac{P(t)}{Q(t)}$, with P, Q relatively prime and let $\chi \in \mathcal{X}(A)$ be such that $\chi(t) = b$. Then $\chi(P(t)) = \chi(y_k)\chi(Q(t)) = \chi(y_k)Q(b) = 0$, a contradiction to the hypothesis: P, Q relatively prime. Consequently no poles of the h_j lie in $sp(t)$, whenever $j = 1,\dots,n$. On the other hand let $F(Y_1,\dots,Y_n) \in \mathbb{K}\{Y_1,\dots,Y_n\}$ be such that $t = F(y_1,\dots,y_n)$. Consider the mapping G from $sp(t)$ into $\mathbb{K}^n$ defined as $G(\alpha) = F(h_1(\alpha),\dots,h_n(\alpha))$. Then G is the identical mapping on $sp(t)$. If there exists $\alpha \in sp(t)$ such that $h'_j(\alpha) = 0\ \forall j = 1,\dots,n$, then we have $G'(\alpha) = 0$, a contradiction since G is the identical mapping on $sp(t)$. $\qquad\square$

Theorem 10.9.7. *The $\mathbb{K}$-affinoid algebra $A = \frac{\mathbb{K}\{X,Y\}}{(Y^2-X^3)\mathbb{K}\{X,Y\}}$ is an integral domain but is not integrally closed.*

Proof. Let θ be the canonical surjection from $\mathbb{K}\{X,Y\}$ onto A, let $x = \theta(X)$, $y = \theta(Y)$. Suppose that A is has divisors of zero. Since by Theorem 10.3.7, $\mathbb{K}\{X,Y\}$ is factorial, there exist non-invertible elements f_1, $f_2 \in \mathbb{K}\{X\}[Y]$ such that $f_1 f_2 = Y^2 - X^3$. Then, by Theorem 10.3.3, each f_j is associated in $\mathbb{K}\{X,Y\}$ to a monic polynomial in Y, $P_j(Y)$. Since both P_j are not invertible in $\mathbb{K}\{X,Y\}$, each P_j is of the form $Y + S_j(X)$, with $S_j(X) \in \mathbb{K}\{X\}$. Consequently, we have $Y^2 - X^3 = Y^2 + Y(S_1(X) + S_2(X)) + S_1(X)S_2(X)$, hence by identification, $S_2 = -S_1$, and $S_1^2 = X^3$. But X^3 is not a square in $\mathbb{K}\{X\}$ because it has a zero of order 3 at 0. Consequently we are led to a contradiction proving that $Y^2 - X^3$ is irreducible in $\mathbb{K}\{X,Y\}$, and therefore A is an integral domain.

Suppose now that A is integrally closed. Let $B = \mathbb{K}[x,y]$. The field of fractions of B is clearly the field $\mathbb{K}(u)$ with $u = \frac{y}{t}$. And since u is integral over B, it belongs to A. On the other hand, both x, y are not invertible in A, hence neither is u. But putting $x = g(u)$, $y = h(u)$ we have $g'(u) = h'(u) = 0$, a contradiction with regards to Lemma 10.9.6. $\qquad\square$

Corollary 10.9.8. *The $\mathbb{K}$-affinoid algebra $A = \frac{\mathbb{K}\{X,Y\}}{(Y^2-X^3)\mathbb{K}\{X,Y\}}$ is not a Krasner–Tate algebra.*

Proof. Indeed, a Krasner–Tate algebra without divisors of zero is a principal ideal ring. $\qquad\sqcup$

Proposition 10.9.9 will be useful in further consideration in Section 10.10.

Proposition 10.9.9. *Let $r \in]0,1[\cap|\mathbb{K}|$, let $E = d(0,r)$, $F = d(1,r)$ and let $D = E \cup F$. Let u be the characteristic function of E in $H(D)$, and $A = \frac{H(D)}{ux^2 H(D)}$. Then A is a non-reduced $\mathbb{K}$-affinoid algebra containing a dense Luroth $\mathbb{K}$-algebra.*

Proof. Since D is an affinoid subset of $\mathbb{K}$, $H(D)$ is a $\mathbb{K}$-affinoid algebra, hence so is A. Let θ be the canonical surjection of $H(D)$ onto A. Then, $\theta(ux)$ is a nilpotent element of A, hence A is not reduced. But since $H(D)$ is a Krasner–Tate algebra, it is of the form $\mathbb{K}\{h\}[x]$, with $h \in \mathbb{K}(x), \deg(h) > 0$, and $h(D) = U$, $D = h^{-1}(U)$, so $\mathbb{K}[h,x]$ is a dense $\mathbb{K}$-subalgebra of $H(D)$. Let $\tau = \theta(h)$, $\xi = \theta(x)$. Then $\text{Ker}(\theta) \cap \mathbb{K}\{h\}$ is equal to $\{0\}$, and so is $\text{Ker}(\theta) \cap \mathbb{K}[h,x]$. Consequently, $\mathbb{K}[\tau,\xi]$ is a Luroth $\mathbb{K}$-subalgebra of A isomorphic to $\mathbb{K}[h,x]$, and is dense in A. $\qquad\square$

10.10. Associated idempotents

Theorem 10.10.6, stated in [6], corresponds in ultrametric Banach algebras
to a well known theorem in complex Banach algebras: if the spectrum of
maximal ideals admits a partition in two open closed subsets U and V with
respect to the Gelfand topology, there exist idempotents u and v such that
$\chi(u) = 1 \,\forall \chi \in U$, $\chi(u) = 0 \,\forall \chi \in V$ and $\chi(v) = 0 \,\forall \chi \in U$, $\chi(v) = 1 \,\forall \chi \in V$.

In an ultrametric Banach algebra, it is impossible to have a similar
result because partitions in two open closed subsets for the Gelfand topology
on the spectrum of maximal ideals then makes no sense, due to the total
disconnectedness of the spectrum. B. Guennebaud first had the idea to
consider the set of continuous multiplicative semi-norms of an ultrametric
Banach algebra, denoted by $\mathrm{Mult}(A, \| \, . \, \|)$ instead of the spectrum of
maximal ideals, an idea that later suggested Berkovich theory [3].

Definition: We call *affinoid variety* the multiplicative spectrum
$\mathrm{Mult}(B, \| \, . \, \|)$ of a $\mathbb{L}$-affinoid algebra B.

Let X be an affinoid variety and let F_X be the contravariant functor
which assigns to each affinoid subvariety $\mathrm{Mult}(B, \| \, . \, \|)$ its affinoid algebra
B and to each inclusion $\mathrm{Mult}(B', \| \, . \, \|) \subset \mathrm{Mult}(B, \| \, . \, \|)$ of affinoid subsets
the corresponding restriction homomorphism from B to B'. Then F_X is a
presheaf on X.

We must recall Tate's acyclicity Theorem in the form given by Berkovich
[3] where affinoid varieties are considered in the multiplicative spectrum of
a $\mathbb{L}$-affinoid algebra and its topology is the pointwise convergence topology.

Let $(A, \| \, . \, \|)$ be a $\mathbb{L}$-Banach algebra, let $X = \mathrm{Mult}(A, \| \, . \, \|)$ and let
$\{U_i \; 1 \leq i \leq n\}$ be a finite covering of X by finitely many affinoid varieties,
with $\mathrm{Mult}(A_i, | \, . \, \|_i) = U_i$, $1 \leq i \leq n$. Then $A = \bigcap_{i=1}^{n} A_i$.

That is summarized by Theorem 10.10.1.

Theorem 10.10.1 ([3]). *Let $(A, \| \, . \, \|)$ be a $\mathbb{L}$-Banach algebra, let $X = \mathrm{Mult}(A, \| \, . \, \|)$ and let $F_X = (U_i)_{i \in I}$ be a finite covering by affinoid subvarieties of X. Then the covering is F_X-acyclic.*

Proposition 10.10.2. *Let A be an entire $\mathbb{L}$-affinoid algebra. Then $\mathrm{Mult}(A, \| \, . \, \|)$ is connected.*

Proof. Suppose that $U = \mathrm{Mult}(A, \| \, . \, \|)$ is not connected. It then admits
a partition in two disjoint compact subsets U_1, U_2, hence by Theorem
10.10.1, there exist $\mathbb{L}$-affinoid algebras $(A_1, \| \, . \, \|_1)$ and $(A_2, \| \, . \, \|_2)$ such
that $\mathrm{Mult}(A_1, \| \, . \, \|_1) = U_1$ and $\mathrm{Mult}(A_2, \| \, . \, \|_2) = U_2$.

Then A_1 is the quotient of A through a surjective morphism γ_1 and A_2
is the quotient of A through a surjective morphism γ_2. Given $f \in A$, set

$f_1 = \gamma_1(f)$, $f_2 = \gamma_2(f)$. Then, we have $\|f\| = \sup\{\phi(f) \mid \phi \in U_1 \cup U_2\}$, $\|f_1\|_1 = \sup\{\phi(f_1) \; \phi \in U_1\}$ and $\|f_2\|_2 = \sup\{\phi(f_2) \; \phi \in U_2\}$.

Consider now the $\mathbb{L}$-algebra $A_1 \times A_2$ provided with the norm $\| \, . \, \|^*$ defined by $\|(f_1, f_2)\|^* = \max(\|f_1\|_1, \|f_2\|_2)$. Then we have

$$\|(f_1, f_2)\|^* = \max(\sup\{\phi(f_1), \; \phi \in U_1\}, \; \sup\{\phi(f_2)|, \; \phi \in U_2)$$
$$= \sup\{\phi(f), \; \phi \in U_1 \cup U_2\} = \|f\|$$

which proves that $\|f\| = \|(f_1, f_2)\|^*$. Consequently, A and $A_1 \times A_2$ are isomorphic, a contradiction to the hypothesis "A entire". $\qquad\square$

Lemma 10.10.3. *Let A be a $\mathbb{L}$-affinoid algebra of Jacobson radical $\mathcal{R}$ and let $w \in \mathcal{R}$. The equation $x^2 - x + w = 0$ has a solution in $\mathcal{R}$.*

Proof. Since A is $\mathbb{L}$-affinoid, by Theorem 10.5.5, w is nilpotent, hence we can consider the element

$$u = -\frac{1}{2} \sum_{n=1}^{+\infty} \binom{\frac{1}{2}}{n} (-4w)^n.$$

Now we can check that $(2u - 1)^2 = 1 - 4w$ and then $u^2 - u + w = 0$. $\qquad\square$

Proposition 10.10.4 ([39]). *Let A be a $\mathbb{L}$-affinoid algebra of Jacobson radical $\mathcal{R}$ and let $w \in A$ be such that $w^2 - w \in \mathcal{R}$. There exists an idempotent $u \in A$ such that $w - u \in \mathcal{R}$.*

Proof. We will roughly follow the proof known in complex algebra [8]. Let $r = w^2 - w$. We first notice that $1 + 4r = (2w - 1)^2$. Next, $\frac{r}{1+4r}$ belongs to $\mathcal{R}$ hence by Lemma 10.10.3, there exists $x \in \mathcal{R}$ such that $x^2 - x + \frac{r}{1+4r} = 0$, hence

$$((2w - 1)x)^2 - (2w - 1)^2 x + r = 0.$$

Now set $s = (2w - 1)x$. Then s belongs to $\mathcal{R}$, as x. Then we obtain

$$s^2 - (2w - 1)s + r = 0.$$

Let us now put $u = w - s$ and compute u^2:

$$(w - s)^2 = w^2 - 2ws + s^2 = w + r - 2ws + s^2.$$

But $s^2 = -r + (2w - 1)s$, hence finally:

$$(w + s)^2 = w - r + 2ws + r - (2w - 1)s = w + s.$$

Thus, u is an idempotent such that $u - w \in \mathcal{R}$. $\qquad\square$

Proposition 10.10.5. *Let A be a $\mathbb{K}$-affinoid algebra such that $\mathrm{Mult}(A, \| \cdot \|)$ admits a partition in two compact subsets U, V. There exists an idempotent $u \in A$ such that $\phi(u) = 1 \ \forall \phi \in U$ and $\phi(v) = 0 \ \forall \phi \in V$ and an idempotent v such that $\phi(u) = 0 \ \forall \phi \in U$ and $\phi(v) = 1 \ \forall \phi \in V$.*

Proof. Suppose first that A is reduced. By Corollary 10.5.6, the spectral semi-norm is then a norm. By Corollary 10.3.5, A has finitely many minimal prime ideals: $P_1, \ldots, P_n$. For every $i = 1, \ldots, n$, the algebra $A_i = \frac{A}{P_i}$ is an affinoid $\mathbb{K}$-algebra that has no divisor of zero. Let $\| \cdot \|_i$ be its affinoid norm. By Proposition 10.10.2, $Mult(A_i, \| \cdot \|_i)$ is connected, hence for every $i = 1, \ldots, n$, $\mathrm{Mult}(A_i, \| \cdot \|_i)$ either is homeomorphic to a subset of U or is homeomorphic to a subset of V. Suppose for instance, $\mathrm{Mult}(A_i, \| \cdot \|_i)$ is homeomorphic to a subset of U for every $i = 1, \ldots, r$. Set $P = \bigcap_{i=1}^{r} P_i$, $Q = \bigcap_{i=r+1}^{n} P_i$.

Suppose that P and Q are included in a same maximal ideal $\mathcal{M}$. Then there exists $\psi \in \mathrm{Mult}(A, \| \cdot \|)$ such that $\ker(\psi) = \mathcal{M}$ and there exist $i \in \{1, \ldots, r\}$, $j \in \{r+1, \ldots, n\}$ such that $P_i \subset \mathcal{M}$, $P_j \subset \mathcal{M}$. Therefore, ψ belongs to $U \cap V$, which contradicts $U \cap V = \emptyset$. Consequently, P and Q are not included in a same maximal ideal and hence $P + Q = A$. Therefore, there exist $u \in P$ and $v \in Q$ such that $u + v = 1$ and hence $\phi(u) = 0$ $\forall \phi \in U$, $\phi(1 - u) = 0$ $\forall \phi \in V$. Now, we have $\chi(u) = 0$ $\forall \chi \in \mathcal{X}(A)$ such that $|\chi| \in U$ and $\chi(1 - u) = 0$ $\forall \chi \in \mathcal{X}(A)$ such that $|\chi| \in V$, therefore $\chi(u) = 1$ $\forall \chi \in \mathcal{X}(A)$ such that $|\chi| \in V$. Consequently, $\chi(u) = \chi(u^2)$ $\forall \chi \in \mathcal{X}(A)$, hence $\chi(u - u^2) = 0$ $\forall \chi \in \mathcal{X}(A)$ and hence $\|u - u^2\|_{\mathrm{sp}} = 0$. But since the spectral semi-norm of A is a norm, we have $u = u^2$, hence u is an idempotent such that $\phi(u) = 0$ $\forall \phi \in U$ and $\phi(u) = 1$ $\forall \phi \in V$.

We can easily generalize when A is no longer supposed to be reduced. Let $\mathcal{R}$ be the Jacobson radical of A and let $B = \frac{A}{\mathcal{R}}$. Let θ be the canonical surjection from A onto B. Every $\phi \in \mathrm{Mult}(A, \| \cdot \|)$ is of the form $\varphi \circ \theta$ with $\varphi \in \mathrm{Mult}(B, \| \cdot \|)$. Let $U' = \{\varphi \in \mathrm{Mult}(B, \| \cdot \|)\}$ be such that $\varphi \circ \theta \in U$ and let $V' = \{\varphi \in \mathrm{Mult}(B, \| \cdot \|)\}$ be such that $\varphi \circ \theta \in V$. Then U' and V' are two compact subsets making a partition of $\mathrm{Mult}(B, \| \cdot \|)$. Therefore, B has an idempotent e such that $\varphi(e) = 1$ $\forall \varphi \in U'$ and $\varphi(e) = 0$ $\forall \varphi \in V'$. Let $w \in A$ be such that $\theta(w) = e$. Then we can check that $\phi(w) = 1$ $\forall \phi \in U$ and $\phi(w) = 0$ $\forall \phi \in V$. But by Proposition 10.10.4, there exists an idempotent $u \in A$ such that $u - w \in \mathcal{R}$. Then $\chi(u) = \chi(w)$ $\forall \chi \in \mathcal{X}(A)$ and hence $\phi(u) = \phi(w)$ $\forall \phi \in \mathrm{Mult}(A, \| \cdot \|)$ because, by Theorem 10.5.4, $\mathrm{Mult}_m(A, \| \cdot \|)$ is dense in $\mathrm{Mult}(A, \| \cdot \|)$. Similarly, there exists an idempotent $v \in A$ such that $\phi(v) = 1$ $\forall \phi \in V$ and $\phi(v) = 0$ $\forall \phi \in U$. $\qquad \square$

Proposition 10.10.6. *Let A be a commutative unital $\mathbb{L}$-Banach algebra and assume that $Mult(A, \| \cdot \|)$ admits a partition in two compact subsets X, Y. Suppose that there exist two idempotents u and e such that $|\zeta(u)| = |\zeta(e)| = 1 \; \forall \zeta \in \mathcal{X}(A)$ such that $|\zeta| \in X$ and $\zeta(u) = \zeta(e) = 0 \; \forall \zeta \in \mathcal{X}(A)$ such that $|\zeta| \in Y$. Then $u = e$.*

Proof. Let $\zeta \in \mathcal{X}(A)$ be such that $|\zeta| \in X$. Since $|\zeta(u)| = |\zeta(e)| = 1$, we have $\zeta(u) = \zeta(e) = 1$ hence $\zeta(u - e) = 0$. Now, if $|\zeta| \in Y$, we have $\zeta(u) = \zeta(e) = 0$, hence $\zeta(u - e) = 0$. Consequently, $\zeta(u - e) = 0 \; \forall \zeta \in \mathcal{X}(A)$ and hence $u - e$ belongs to the Jacobson radical of A. Put $e = u + r$. Since $e^2 = e$, we have $(u + r)^2 = u + 2ur + r^2$.

Suppose $r \neq 0$. Then $2u + r = 0$. If $\mathbb{L}$ is of characteristic 2, we then have $r = 0$. Now suppose that $\mathbb{L}$ is not of characteristic 2. Let $\zeta \in \mathcal{X}(A)$ be such that $|\zeta| \in X$. We have $(\zeta(u))^2 = \zeta(u)$ and $|\zeta(u)| = 1$, hence $\zeta(u) = 1$ therefore $2 + \zeta(r) = 0$. But since r lies in the Jacobson radical, $\zeta(r) = 0$, hence $2 = 0$, a contradiction that finishes proving that $r = 0$. $\qquad \square$

Proposition 10.10.7 may be found in [32] with a proof explained in a different way.

Proposition 10.10.7. *Let A be a unital commutative ultrametric $\mathbb{L}$-Banach algebra and assume that $\mathrm{Mult}(A, \| \cdot \|)$ admits a partition in two compact subsets (X, Y). There exists a $\mathbb{L}$-affinoid algebra B included in A, admitting for norm this of A, such that $\mathrm{Mult}(B, \| \cdot \|)$ admits a partition (X', Y') where the canonical mapping Υ from $\mathrm{Mult}(A, \| \cdot \|)$ to $\mathrm{Mult}(B, \| \cdot \|)$ satisfies $\Upsilon(X) \subset X'$, $\Upsilon(Y) \subset Y'$.*

Proof. Since X and Y are compact sets such that $X \cap Y = \emptyset$, we can easily define a finite covering of open sets $(X_i)_{1 \leq i \leq n}$ of X such that $X_j \cap Y = \emptyset \; \forall i = 1, \ldots, n$, where the X_i are in the form $\mathcal{W}(f_i, x_{i,1}, \ldots, x_{i,m_i}, \epsilon_i)$ with $x_{i,j} \in A$.

Let A^* be the finite type $\mathbb{L}$-subalgebra generated by all the $x_{i,j}, 1 \leq j \leq m_i, 1 \leq i \leq n$. Consider the image of $\mathrm{Mult}(A, \| \cdot \|)$ in $\mathrm{Mult}(A^*, \| \cdot \|)$ through the mapping Υ that associates to each $\phi \in \mathrm{Mult}(A, \| \cdot \|)$ its restriction to A^* and let $X^* = \Upsilon(X)$, $Y^* = \Upsilon(Y)$. Since X and Y are compact sets in $\mathrm{Mult}(A, \| \cdot \|)$, so are X^* and Y^* in $\mathrm{Mult}(A^*, \| \cdot \|)$. Moreover, $X \cap Y = \emptyset$ implies $X^* \cap Y^* = \emptyset$.

Let W be a neighborhood of $X^* \cup Y^*$ in $\mathrm{Mult}(A^*, \| \cdot \|)$ which is the union of two disjoint neighborhoods X' of X and Y' of Y. Let $\phi \in \mathrm{Mult}(A^*, \| \cdot \|) \setminus W$. Then ϕ does not belong to $\Upsilon(\mathrm{Mult}(A, \| \cdot \|))$. Consequently, there exists a $\mathbb{L}$-algebra of finite type $A^*(\phi)$ containing A^* such that the image of $\mathrm{Mult}(A^*(\phi), \| \cdot \|)$ in $\mathrm{Mult}(A^*, \| \cdot \|)$ do not contain ϕ.

Now, $\text{Mult}(A^*(\phi), \| \cdot \|)$ is a compact subset of $\text{Mult}(A^*, \| \cdot \|)$. Consequently, there exists a neighborhood of ϕ in $\text{Mult}(A^*, \| \cdot \|) \setminus W$.

In the same way, since $\text{Mult}(A^*, \| \cdot \|) \setminus W$ is compact, there exist $\phi_1, \ldots, \phi_q \in \text{Mult}(A^*, \| \cdot \|) \setminus W$ and neighborhoods $V(\phi_i)$, $1 \le i \le q$ of ϕ_i ($1 \le i \le q$), included in $\text{Mult}(A^*, \| \cdot \|) \setminus W$ making a covering of $\text{Mult}(A^*, \| \cdot \|) \setminus W$.

Let E be the $\mathbb{L}$-algebra of finite type generated by the $A^*(\phi_i)$, ($1 \le i \le q$). Then E is a $\mathbb{L}$-subalgebra of A of finite type which contains A^* and hence is equipped with the $\mathbb{L}$-algebra norm $\| \cdot \|$ of A. Moreover, by construction, $\text{Mult}(E, \| \cdot \|)$ is equal to $W = X' \cup Y'$ and we have $\Upsilon(X) \subset X'$, $\Upsilon(Y) \subset Y'$. The image of $\text{Mult}(E, \| \cdot \|)$ in $\text{Mult}(A^*, \| \cdot \|)$ is then contained in W and E is a $\mathbb{L}$-subalgebra of A of finite type which contains A^*, hence is equipped with the $\mathbb{L}$-algebra norm $\| \cdot \|$ of A.

Let $\{x_1, \ldots, x_N\}$ be a finite subset of the unit ball of E such that $\mathbb{L}[x_1, \ldots, x_N] = E$. Let T be the topologically pure extension $\mathbb{L}\{X_1, \ldots, X_N\}$ and consider the canonical morphism G from $\mathbb{L}[X_1, \ldots, X_N]$ equipped with the Gauss norm, into E, equipped with the norm $\| \cdot \|$ of A, defined as $G(F(X_1, \ldots, X_N)) = F(x_1, \ldots, x_N)$. Since by hypotheses, $\|x_j\| \le 1 \; \forall j = 1, \ldots, N$, then G is a continuous mapping from from $\mathbb{L}[X_1, \ldots, X_N]$ equipped with the Gauss norm, into E, equipped with the norm $\| \cdot \|$ of A. It then has expansion to a continuous morphism $\widehat{G}$ from T into A. Since, by Theorem 10.3.2, all ideals of T are closed, we can consider the closed ideal $\mathcal{I}$ of the elements $F \in T$ such that $\widehat{G}(F) = 0$. Then $\widehat{G}(T)$ is the $\mathbb{L}$-affinoid algebra $B = \frac{T}{\mathcal{I}}$ containing E and included in A. By construction, the $\mathbb{L}$-affinoid norm of B is the restriction of the norm $\| \cdot \|$ of A. Since E is by construction dense in B, we have $\text{Mult}(B, \| \cdot \|) = \text{Mult}(E, \| \cdot \|) = X' \cup Y'$. Consequently, (X', Y') is a partition of $\text{Mult}(B, \| \cdot \|)$ and by construction, we have $\Upsilon(X) \subset X'$, $\Upsilon(Y) \subset Y'$, which ends the proof. $\square$

We can now conclude.

Theorem 10.10.8. *Let A be a unital commutative ultrametric Banach $\mathbb{L}$-algebra such that $\text{Mult}(A, \| \cdot \|)$ admits a partition in two compact subsets (X, Y). There exist a unique idempotent $u \in A$ such that $\phi(u) = 1 \; \forall \phi \in X$ and $\phi(u) = 0 \; \forall \phi \in Y$ and a unique idempotent $v \in A$ such that $\phi(v) = 1 \; \forall \phi \in Y$ and $\phi(v) = 0 \; \forall \phi \in X$.*

Proof. By Proposition 10.10.7, there exists a $\mathbb{L}$-affinoid algebra B and a continuous morphism G from B to A such that $\text{Mult}(B, \| \cdot \|)$ admits a partition in two compact subsets (X', Y') and such that the canonical mapping $\widehat{G}$ from $\text{Mult}(A, \| \cdot \|)$ to $\text{Mult}(B, \| \cdot \|)$ satisfy $\widehat{G}(X) \subset X'$,

$\widehat{G}(Y) \subset Y'$. Now, by Proposition 10.10.5, there exist idempotents u', $v' \in B$ such that $\phi(u') = 1 \ \forall \phi \in X'$ and $\phi(u') = 0 \ \forall \phi \in Y'$. Consequently, putting $u = G(u')$, $v = G(v')$, we have $\phi(u) = 1 \ \forall \phi \in X$, $\phi(u) = 0 \ \forall \phi \in Y$ and $\phi(v) = 0 \ \forall \phi \in X$, $\phi(v) = 1 \ \forall \phi \in Y$. The unicity follows from Proposition 10.10.6. That ends the proof. $\qquad \square$

Corollary 10.10.9. *Let A be a unital commutative ultrametric Banach $\mathbb{L}$-algebra such that $\mathrm{Mult}(A, \| \cdot \|)$ admits a partition in two compact subsets (X, Y). Then A is isomorphic to a direct product of two Banach $\mathbb{L}$-algebras $A_X \times A_Y$ such that $\mathrm{Mult}(A_X, \| \cdot \|) = X$ and $\mathrm{Mult}(A_Y, \| \cdot \|) = Y$. There exists a unique idempotent $u \in A$ such that $\phi(u) = 1 \ \forall \phi \in X$, $\phi(u) = 0 \ \forall \phi \in Y$ and then $A_X = uA$, $A_Y = (1 - u)A$.*

Corollary 10.10.10. *Let A be a unital uniform Banach $\mathbb{K}$-algebra admitting an element x such that $sp(x)$ has an empty-annulus $\Gamma(a, r, s)$. There exist a unique idempotent $e \in A$ such that $\zeta(e) = 1 \ \forall \zeta \in \mathcal{X}(A)$ such that $|\zeta(x - a)| \leq r$ and $\zeta(e) = 0 \ \forall \zeta \in \mathcal{X}(A)$ such that $|\zeta(x - a)| \geq s$ and a unique idempotent $u \in A$ such that $\zeta(u) = 1 \ \forall \zeta \in \mathcal{X}(A)$ such that $|\zeta(x - a)| \geq s$ and $\zeta(u) = 0 \ \forall \zeta \in \mathcal{X}(A)$ such that $|\zeta(x - a)| \leq r$.*

Bibliography

[1] Amice, Y. *Les nombres p-adiques*, P.U.F. (1975).

[2] Amice, Y. and Escassut, A. Sur la non injectivité de la transformation de Fourier relative a $\mathbb{Z}_p$, *C.R.A.S. Paris, A*, 278, pp. 583–585 (1974).

[3] Berkovich, V. Spectral Theory and Analytic Geometry over Non-Archimedean Fields, *AMS Surveys and Monographs*, 33 (1990).

[4] Bosch, S., Guntzer, U., and Remmert, R. *Non-Archimedean Analysis*. Grundlerhen Wissenschaften, Bd. 261, Springer, Berlin-Heidelberg-New York (1984).

[5] Bosch, S. *Lectures on Formal and Rigid Geometry*. Grundlerhen Wissenschaften, Bd. 261, Springer, Berlin-Heidelberg-New York (2014).

[6] Bourbaki, N. Topologie Générale, Chapter 3, *Actualités Scientifiques et Industrielles, Hermann, Paris*.

[7] Bourbaki, N. Algèbre Commutative, *Actualités Scientifiques et Industrielles*, Hermann, Paris.

[8] Bourbaki, N. Theorie Spectrale, Chapters 1 and 2. *Actualités Scientifiques et Industrielles*, Hermann, Paris.

[9] Boussaf, K. and Escassut, A. Absolute values on algebras of analytic elements, *Annales Mathématiques Blaise Pascal*, 2(2) (1995).

[10] Boussaf, K. Shilov Boundary of a Krasner Banach algebra $H(D)$, *Italian Journal of Pure and Applied Mathematics*, 8, 75–82 (2000).

[11] Boussaf, K., Hemdahoui, M., and Mainetti, N. Tree structure on the set of multiplicative semi-norms of Krasner algebras $H(D)$. *Revista Matematica Complutense*, XIII(1), 85–109 (2000).

[12] Chicourrat, M. and Escassut, A. Banach algebras of ultrametric Lipschitzian functions, *Sarajevo Journal of Mathematics*, 14(27), 1–12 (2018).

[13] Chicourrat, M., Diarra, B., and Escassut, A. Finite codimensional maximal ideals in subalgebras of ultrametric continuous or uniformly continuous functions, *Bulletin of the Belgian Mathematical Society*, 26(3), 413–419 (2019).

[14] Chicourrat, M. and Escassut, A. Ultrafilters and ultrametric Banach algebras of Lipschitzian functions, *Advances in Operator Theory*, 5(1), 115–142 (2020)

[15] Chicourrat, M., Diarra, B. and Escassut, A. A survey and new results on Banach algebras of ultrametric continuous function, p-Adic Numbers, *Ultrametric Analysis and Applications*, 12(3), 185–202 (2020).

[16] Diarra, B. Ultraproduits ultramétriques de corps valués, *Annales Scientifiques de l'Université de Clermont II*, Série Math., Fasc. 22, 1–37 (1984).

[17] Escassut, A. Compléments sur le prolongement analytique dans un corps valué non archimédien, complet algébriquement clos, *C.R.A.S.Paris*, A 271, 718–721 (1970).

[18] Escassut, A. Algèbres de Banach ultramétriques et algèbres de Krasner–Tate, *Asterisque*, 10, 1–107 (1973).

[19] Escassut, A. Algèbres d'éléments analytiques en analyse non archimédienne, *Indagationes Mathematicae*, 36, 339–351 (1974).

[20] Escassut, A. T-filtres, ensembles analytiques et transformation de Fourier p-adique, *Annales de Institut Fourier*, 25(2), 45–80 (1975).

[21] Escassut, A. Spectre maximal d'une algèbre de Krasner, *Colloquium Mathematicum*, 38(2), 339–357 (1978).

[22] Escassut, A. The Ultrametric Spectral Theory, *Periodica Mathematica Hungarica*, 11(1), 7–60 (1980).

[23] Escassut, A. Analytic Elements in p-adic Analysis. World Scientific Publishing Inc., Singapore (1995).

[24] Escassut, A. and Mainetti, N Spectral semi-norm of a p-adic Banach algebra, *Bulletin of the Belgian Mathematical Society, Simon Stevin*, 8, 79–61 (1998).

[25] Escassut, A. and Mainetti, N Shilov Boundary for ultrametric algebras, *Bulletin of the Belgian Mathematical Society*, Supplement, 81–89 (2002).

[26] Escassut, A Ultrametric Banach Algebras. World Scientific Publishing Inc., Singapore (2003).

[27] Escassut, A. and Mainetti, N. Multiplicative spectrum of ultrametric Banach algebras of continuous functions, *Topology and Its Applications*, 157, 2505–25015 (2010).

[28] Escassut, A. Value Distribution in p-adic Analysis. World Scientific Publishing Inc., Singapore (2015).

[29] Escassut, A. and Mainetti, N. Spectrum of ultrametric Banach algebras of strictly differentiable functions. *Contemporary Mathematics*, 704, 139–160 (2018).

[30] Garandel, G. Les semi-normes multiplicatives sur les algèbres d'éléments analytiques au sens de Krasner, Indag. Math., 37(4), 327–341 (1975).

[31] Guennebaud, B. Algèbres localement convexes sur les corps, Bull. Sci. Math. 91, 75–96 (1967).

[32] Guennebaud, B. Sur une notion de spectre pour les algèbres normées ultramétriques, Thèse Université de Poitiers (1973).

[33] Haddad, L. Sur quelques points de topologie générale. Théorie des nasses et des tramails. *Annales de la Faculté des Sciences de Clermont*, 44(7), 3–80 (1972).

[34] Krasner, M. Nombre d'extensions d'un degré donn'e d'un corps p-adique. Les tendances géométriques en algèbre et théorie des nombres, Clermont-Ferrand, (1964), pp. 143–169. Centre National de la Recherche Scientifique (1966) (Colloques internationaux du C.N.R.S. Paris, 143).

[35] Krasner, M. *Prolongement analytique uniforme et multiforme dans les corps valués complets. Les tendances géométriques en algèbre et théorie des nombres*, Clermont-Ferrand, (1964) pp. 94–141. Centre National de la Recherche Scientifique (1966) (Colloques internationaux du C.N.R.S. Paris, 143).

[36] Mainetti, N. Spectral properties of p-adic Banach algebras. *Lecture Notes in Pure and Applied Mathematics*, 207, 189–210 (1999).

[37] Morita, Y. On the induced h-structure on an open subset of the rigid analytic space $P^1(k)$. *Mathematische Annalen*, 242, 47–58 (1979).

[38] Motzkin, E. and Robba, Ph. Prolongement analytique en analyse p-adique, Séminaire de theorie des nombres, année 1968–1969, Faculté des Sciences de Bordeaux.

[39] Rickart, Ch. E. *General Theory of Banach Algebras.* Krieger Publishing Company (2002).

[40] Robba, Ph. Fonctions analytiques sur les corps valués ultramétriques complets. Prolongement analytique et algèbres de Banach ultramétriques, *Astérisque*, 10, pp. 109–220 (1973).

[41] Robert, A. *Advanced Calculus for Users.* North-Holland, Amsterdam, New York, Oxford, Tokyo (1989).

[42] Salmon, P. Sur les séries formelles restreintes, *Bulletin de la Société Mathématique de France*, 92, 385–410 (1964).

[43] Sarmant, M.-C. and Escassut A. T-suites idempotentes, Bulletin des Sciences Mathematiques 106, 289–303, (1982).

[44] Tate, J. Rigid analytic spaces, *Inventiones Mathematicae*, 12, 257–289 (1971).

Definitions and Notations

Index